XIII International Research-Technical Conference on the Problems of Designing, Construction and Use of Low Energy Housing (ENERGODOM 2018)

IOP Conference Series: Materials Science and Engineering
Volume 415

Krakow, Poland
11 – 13 September 2018

ISBN: 978-1-5108-7379-7
ISSN: 1757-8981

Printed from e-media with permission by:

Curran Associates, Inc.
57 Morehouse Lane
Red Hook, NY 12571

Some format issues inherent in the e-media version may also appear in this print version.

Printed by Curran Associates, Inc. (2018)

For permission requests, please contact the Institute of Physics
at the address below.

Institute of Physics
Dirac House, Temple Back
Bristol BS1 6BE UK

Phone: 44 1 17 929 7481
Fax: 44 1 17 920 0979

techtracking@iop.org

Additional copies of this publication are available from:

Curran Associates, Inc.
57 Morehouse Lane
Red Hook, NY 12571 USA
Phone: 845-758-0400
Fax: 845-758-2633
Email: curran@proceedings.com
Web: www.proceedings.com

Table of contents

Volume 415

XIII International Research-Technical Conference on the Problems of Designing, Construction and Use of Low Energy Housing

11–13 September 2018, Krakow, Poland

Accepted papers received: 9 August 2018
Published online: 1 November 2018

Preface

Papers

Low Energy Building Design

Building envelope systems and materials

Internal and External Environment

Laboratory and in-situ testing and data acquisition

Lighting and acoustics

Computer aided modeling and design (under auspices of IBPSA)

Building, environment, sustainable development

IOP Conf. Series: Materials Science and Engineering **415** (2018) 011001 doi:10.1088/1757-899X/415/1/011001

PREFACE

The 13-th International Research-Technical Conference on the Problems of Designing, Construction and Use of Low Energy Housing took place on September 11-13-th, 2018, in Kraków, Poland. The conference was organized by the Division of Building and Building Physics of Cracow University of Technology.

ENERGODOM 2018 was intended as an important event to bring together energy conscious scientist and practitioners. Participants from EU Member States and other countries were expected to define the present state of the art that exists in this field, to show not only advantages but also shades of contemporary standards and trends, pointing out the technical feasibility, social acceptance, and economic efficiency of low energy housing.

In 2018, we reemphasized, as we did in 2014, the necessity of extensive retrofitting of existing building stock to the low energy standard. Especially in Central Europe, poorly-insulated, large concrete slab buildings are densely populated, and for the next decades, these buildings will remain an important part of the total building stock. Another important aspect of thermal refurbishment of existing buildings is the improvement of indoor environment quality, including heating conditions as well as protection against overheating.

Prospective implementation of the EU Directive and the urgent necessity to rapidly increase the number of existing (not only newly built) nearly zero-energy buildings demands comprehensive actions and a new awareness of the investors, designers, contractors, and workers. Energodom 2018 conference was intended as a tool helping to create this awareness.

Since the very beginning, this conference stood as a forum for those interested in low energy housing, building physics, and all related fields, allowing us to:

- reveal research findings,
- exchange opinions and gathered experience,
- present new ideas, technologies, and materials,
- introduce local and international contacts and cooperation.

In 2018 55 papers have been accepted by the Scientific Committee to be included in conference proceedings and presented during the conference.

Conference organizers would like to thank:

- all the participants for their decision to write and submit the papers and take part in the conference,
- the active members of the Scientific Committee for paper review,
- the inspiring keynote speakers,
- the sponsors for their support,
- the media sponsors for spreading information about the conference.

ENERGODOM 2018
Organizing Committee

IOP Conf. Series: Materials Science and Engineering **415** (2018) 011002 doi:10.1088/1757-899X/415/1/011002

Division of Building and Building Physics
Cracow University of Technology

13-th International Research-Technical Conference on the Problems of Designing, Construction and Use of Low Energy Housing
September 11-13-th, 2018, Kraków, Poland.

Reviewers

Umberto Berardi, Toronto, Canada
Boris Bielek, Bratislava, Slovakia
Mark Bomberg, Hamilton, Canada
Dorota Bzowska, Warszawa, Poland
Wacław Celadyn, Kraków, Poland
Dorota Chwieduk, Warszawa, Poland
Robert Černý, Prague, Czech Republic
Pavol Ďurica, Žilina, Slovakia
Antonio Frattari, Trento, Italy
Halina Garbalińska, Szczecin, Poland
Dariusz Gawin, Łódź, Poland
Hua Ge, Montreal, Canada
Robert Geryło, Warszawa, Poland
Dariusz Heim, Łódź, Poland
Sabine Hoffmann, Kaiserslautern, Germany
Shuichi Hokoi, Kyoto, Japan
Irena Ickiewicz, Białystok, Poland
Hanna Jędrzejuk, Warszawa, Poland
Dusan Katunsky, Kosice, Slovakia
Katarzyna Klemm, Łódź, Poland
Halina Koczyk, Poznań, Poland
Jan Kosny, Boston, USA
Jarek Kurnitski, Aalto, Finland
Hartwig Künzel, Holzkirchen, Germany
Lech Lichołai, Rzeszów, Poland
Piotr Lis, Częstochowa, Poland
Thomas Lützkendorf, Karlsruhe, Germany
Antoni Malinowski, Lviv, Ukraine
Peter Matiasovsky, Bratislava, Slovakia
Henryk Nowak, Wrocław, Poland
Vojislav Novakovic, Trondheim, Norway
Jacek Nurzyński, Warszawa, Poland
Marco D'Orazio, Ancona, Italy
Vasco Peixoto de Freitas, Porto, Portugal
Michał Piasecki, Warsaw, Poland
Jerzy Z. Piotrowski, Kielce, Poland
Jan Radoń, Kraków, Poland
Anna Sedláková, Kosice, Slovakia
Zuzanna Vranayova, Kosice, Slovakia
David Yarbrough, Cookeville, USA
Daniel Zirkelbach, Holzkirchen, Germany
Tomasz Kisilewicz, Kraków, Poland, CHAIR

IOP Conf. Series: Materials Science and Engineering **415** (2018) 011003 doi:10.1088/1757-899X/415/1/011003

Peer review statement

All papers published in this volume of *IOP Conference Series: Materials Science and Engineering* have been peer reviewed through processes administered by the proceedings Editors. Reviews were conducted by expert referees to the professional and scientific standards expected of a proceedings journal published by IOP Publishing.

IOP Conf. Series: Materials Science and Engineering **415** (2018) 012001 doi:10.1088/1757-899X/415/1/012001

Defining the Polish nearly Zero Energy Building (nZEB) renovation standard

S Firląg, A Chmielewski

The Faculty of Civil Engineering, Warsaw University of Technology, Lecha Kaczyńskiego 16, 00-637 Warszawa, Poland

s.firlag@il.pw.edu.pl

Abstract. The paper describes the process of defining the requirements for renovation of single-family residential buildings to the nZEB standard in Polish conditions. According to the survey results the nZEB renovation standard should include only two indicators: energy need for heating Q_H expressed in kWh/(m²year) and percentage reduction of the primary, non-renewable energy Q_P demand. Process of defining the requirements was divided into two stages: calculation of cost-optimal heat transfer coefficients for renovated elements of building envelope, calculation of cost-optimal renovation standard of the two single-family model houses. The analysis was made for three Polish cities (coldest, medium and warmest) and for 5 different energy prices. As an optimizing criterion the minimum cumulative cost was used, calculated for the 30-year time-scale for different variants of renovation.

1. Introduction

Poles live in homes that are inadequately insulated against heat loss. Heating technology is outdated and the most popular fuel is highly polluting coal, burned in old coal-fired boilers. It is estimated that more than 70% of detached single-family houses in Poland (3.6 million) have no, or inadequate, thermal insulation [1]. Only 1% of all houses in Poland can be considered energy efficient, primarily those that have been built in the last few years [1, 2]. Most of the buildings without thermal insulation had been built before 1989 [3].

In order to support deep renovation of single family-houses, it is crucial to determine the requirements for cost-effective renovation. The method and principles of support should give the investors the incentive to self-incur the costs of carrying out economically viable activities. For example, there should be no support for energy efficiency projects with short payback time, as these can be financed by building owners or by using external capital. It is necessary to introduce a definition for the nZEB renovation standard and define technical requirements for individual energy efficiency measures (e.g. the insulation of external building elements). These must be formulated in a way that in the near future there will be no need for renovation of the currently modernized buildings.

2. Potential indicators of an nZEB renovation standard for Poland

A survey was carried out among 17 experts during a meeting within the Efficient Poland Initiative on potential approaches and indicators that could be used for the nZEB definition of existing single-family houses in Poland. The list of the indicators was created on basis of European nZEB definitions [4]. The results of the survey were the starting point for the definition of nZEB renovation for single family buildings (Table 1.). The results show that 7 of the indicators have scored 10 or more votes –

should be included in the definition. At the same time many indicators have received negative votes. In order to state which of them are most important the number of "No" votes was subtracted for the "Yes" votes. The results are presented in the last column of the table, in order of net score.
It can be seen that the first two indicators scored the highest number of "YES" votes, and among the lowest number of "NO" votes and accordingly can be considered the most important indicators to be used in specifying a nZEB renovation definition.

Table 1. Results of the survey on the potential indicators of nZEB renovation definition.

Should the nZEB definition of renovation include requirements for?	Yes	No	Importance (Yes-No)
Index of non-renewable, primary energy demand Q_P	12	5	7
Index of final (delivered) energy demand Q_F	10	6	4
Index of energy need for heating Q_H	15	2	**13**
CO_2 emission index	10	5	5
Share of renewable energy sources	10	7	3
Thermal transmittance (U-value) of different building elements	12	4	8
Air tightness	7	7	0
Ventilation systems including the efficiencies of heat recovery	9	6	3
Efficiencies of heating and domestic hot water systems	9	6	3
Summertime comfort – risk of overheating	5	9	-4
Energy demand of auxiliary systems	6	8	-2
Energy efficiency of renovation improvement - percentage reduction of the primary, non-renewable energy Q_P demand	13	3	**10**
Index of final (delivered) energy demand Q_F for cooling	1	0	1

According to the findings of the survey, the definition of nZEB renovation should only include two indicators:

- The energy need for heating Q_H, expressed in kWh/(m²year), and
- The percentage of reduction of primary non-renewable energy demand Q_P, (including heating, ventilation, domestic hot water (DHW), cooling and auxiliary systems for the case of residential buildings), determined in relation to the energy demand of building before renovation.

These two indicators would make the definition very flexible (reduction of Q_P can be achieved in many ways, e.g. use of RES, increase of systems energy efficiency, reduction of energy need for DHW, cooling) but at the same time very demanding. The energy need for heating (Q_H) depends among others, on the thermal transmittance of the building elements, the thermal bridges, the air tightness and the ventilation system's type. A low value of the energy need can be achieved in different ways depending on the building's condition. For example, in case it is not possible to insulate the ground floor, other elements of the building's envelope, e.g. external walls, roof, windows or doors, can have a better thermal transmittance. In addition, ventilation with heat recovery can be applied in order to reduce the ventilation heat loss and energy need. The energy need for heating (Q_H) as an indicator gives flexibility and provides the opportunity to choose the best renovation measures.

The indicator referring to the reduction of primary energy demand (Q_P) includes aspects like efficiencies of the heating and DHW system, energy source type and use of Renewable Energy Sources (RES). Depending on the building's conditions, different solutions can be chosen. Additional use of first indicator - the energy need for heating (Q_H), prevents against situation in which only RES (with low primary energy factor PEF) will be implemented in order to reduce the primary, non-renewable energy demand. In Poland value of PEF for biomass is 0,2 and for coal 1,1, so change from coal to biomass will reduce the primary, non-renewable energy demand by about 82% where the energy demand for heating can stay at the same level.

3. NZEB standard based on cost optimal calculations

The process of determining requirements for the nZEB renovation definition was divided in two stages. The cost optimal U-value of the building's envelope is initially calculated and based on this, the energy demand for heating and the reduction of non-renewable primary energy demand are then estimated:

- Stage I – Cost-optimal heat transfer coefficients for renovated elements of the building envelope

The aim of the calculation was to determine the cost optimal U-values for the elements of the building's fabric (e.g. structural material, insulation, windows etc.) for single-family houses to be renovated. The results depend on the type of the building element (external wall, roof, etc.), its initial U-value and the cost of energy. As an optimizing criterion, the minimum cumulative cost (investment + cost of energy losses) was used, calculated over 30-year period. Based on literature review [5, 6, 7] a discount rate of 4.0% was assumed, the energy price increase rate equal to 1.6% above the year-on-year inflation level that was 1.8%. All the values were constant during the analyzed period.

- Stage II – Cost-optimal renovation definition for single-family houses

In the second part of the analysis the cost optimal scenario for renovation of two reference buildings was determined. As an optimizing criterion, the minimum cumulative cost (investment + energy cost of heating, DHW and auxiliary electricity) was used, calculated over 30-year period. 12 scenarios of renovation were defined based on the heat transfer coefficient of the building elements, the ventilation, the central heating and domestic hot water solutions. For each scenario the energy need for heating and ventilation was calculated according the ISO 13790 standard. The reduction of primary non-renewable energy demand for heating, ventilation, domestic hot water and auxiliary systems was also determined following the polish regulations.

4. Building models

Two reference building models were used to determine the requirements for renovation to the nZEB definition (Figure 1):

- a typical two-storey building with a flat roof,
- a one-storey building with an attic.

Figure 1. Reference single family houses: (left) two-storey building with flat roof, (right) one-storey building with an attic [8].

5. Cost-optimal U-values of renovated building envelope

Table 2 shows the calculation results of the discounted cumulative cost depending on the energy cost and U-values for a renovated external wall of U-value 0.82 W/(m^{2}K). Depending on the heating energy cost per 1 GJ the cost-optimal thermal transmittance (characterised with lowest discounted, cumulative cost), is between 0.12 to 0.20 W/(m^{2}K) (Table 2., bold).

IOP Conf. Series: Materials Science and Engineering **415** (2018) 012001 doi:10.1088/1757-899X/415/1/012001

Table 2. The discounted, cumulative cost (k_{Rd}) per m^2 including the cost of energy loss and the unit cost of renovation with tax (k_{VAT}), depending on the renovation variant and the cost of 1 GJ of energy. Insulation of external walls with ETICS using EPS with λ = 0.033 W/(mK).

U-value, insulation cost and energy loss per m^2 of external wall depending on insulation thickness					Heating energy cost in PLN per 1 GJ: 20	30	40	50	60
d	U-value	K	K_{VAT}	Energy loss	Discounted, cumulative cost in PLN per m^2				
cm	W/(m^{2}K)	zł/m^2	zł/m^2	GJ/(m^{2}year)	K_{Rd_20}	K_{Rd_30}	K_{Rd_40}	K_{Rd_50}	K_{Rd_60}
0	0,820	0,00	0,00	0,262	132	199	265	331	397
10	0,231	99,93	107,92	0,074	145	164	182	201	220
12	0,202	103,56	111,84	0,065	145	161	177	194	210
14	0,180	107,19	115,76	0,058	145	159	174	189	203
16	0,163	110,81	119,68	0,052	146	159	172	185	198
18	0,148	114,44	123,60	0,047	147	159	171	183	195
20	0,136	118,07	127,52	0,043	149	160	171	182	193
22	0,125	122,04	131,80	0,040	152	162	172	182	193
24	0,117	125,58	135,63	0,037	154	164	173	183	192
26	0,109	129,99	140,38	0,035	158	167	176	184	193
28	0,102	133,53	144,21	0,033	161	169	177	185	194
30	0,096	137,07	148,03	0,031	164	171	179	187	195

Figure 2 shows the change of discounted cumulative cost depending on the energy cost and U-values for a renovated external wall. The cost optimal U-vale is characterized with lowest discounted, cumulative cost. Increasing trends at the ends of the chart show that either renovation cost or energy loss cost is starting to dominate in cumulative cost.

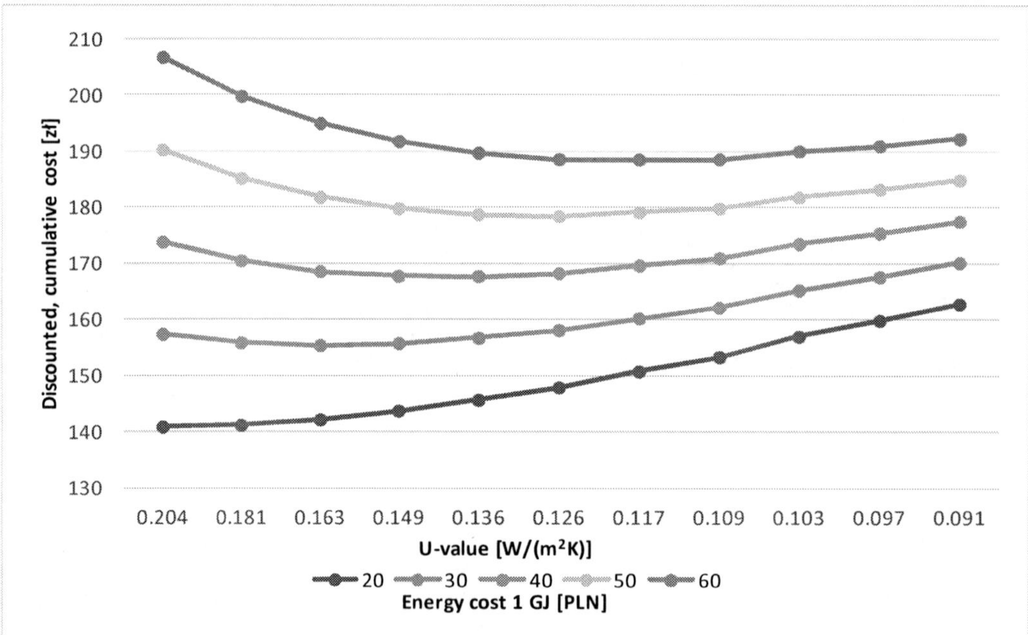

Figure 2. The discounted, cumulative cost (k_{Rd}) per m^2 for renovated external wall depending on the thermal transmittance and the energy cost.

The cost-optimal thermal transmittance for the renovated building elements is presented in Table 3. For each building element the thickness of the additional insulation d, and the cost of renovation is given. The optimal thermal transmittance strongly depends on the energy price. The table shows the values for the lowest and the highest energy price per GJ. The minimum and maximum U-values were used for defining the renovation variants used in the second part of the analysis:

- First variant (W1) – the building envelope was renovated according to the cost-optimal thermal transmittance specified for energy price of 20 PLN per GJ,
- Second variant (W2) – the building envelope was renovated according to the cost-optimal thermal transmittance specified for energy price of 60 PLN per GJ.

Table 3. The cost-optimal renovation variants (depending on the energy price) for the building envelope used in stage II of the analysis.

Building element	W1 - 20 PLN per 1 GJ				W2 - 60 PLN per 1 GJ			
	d	U-value	renovation cost	unit	d	U-value	renovation cost	unit
	cm	W/(m^{2}K)			cm	W/(m^{2}K)		
External wall	12	0.21	112	PLN/m^2	24	0.12	136	PLN/m^2
Floor above an unheated cellar	7	0.25	53	PLN/m^2	10	0.19	76	PLN/m^2
Floor on a ground	14	0.24	129	PLN/m^2	26	0.14	154	PLN/m^2
Flat roof	12	0.22	86	PLN/m^2	26	0.12	111	PLN/m^2
Pitched roof	20	0.18	142	PLN/m^2	35	0.10	164	PLN/m^2
Windows	-	0.9	562	PLN/m^2	-	0.9	562	PLN/m^2
External doors	-	1.3	4088	PLN/door	-	0.9	4347	PLN/door

6. Cost-optimal building renovation standard

On the basis of defining U-values of building elements and solutions referring to the upgrading of the ventilation system and the use of RES, different renovation variants were identified. All of them include upgrading of the existing heating and domestic hot water system together with replacement of the heating source. The symbols used to distinguish the variants are explained below:

- W0: baseline variant, existing reference building before renovation,
- W1: variant I of building's envelope renovation, energy price 20 PLN per 1 GJ,
- W2: variant II of building's envelope renovation, energy price 60 PLN per 1 GJ,
- G: natural ventilation, base case before renovation,
- H: hybrid ventilation, it was assumed that the energy loss through ventilation will be reduced by 20%
- R: balanced ventilation with heat recovery, efficiency of heat recovery: 90%,
- S: solar system used for DHW heating, assumed coverage between 50-60% of DHW energy demand.

The optimal variants of renovation of the two reference buildings are presented in Table 4 (two-storey building) and Table 5 (one-storey building). The choice of the variant depends on a decisive extent on two factors: the price of energy per 1 GJ and the building location (climate).

Table 4. Cost-optimum renovation variants of the two-storey building with flat roof

location	energy price per 1 GJ		
	20 PLN	40 PLN	60 PLN
	cost-optimum renovation variants		
Warszawa	W1/G	W2/G	W2/G/S
Szczecin	W1/G	W2/G	W2/G/S
Suwałki	W1/G	W2/G	W2/R/S

Table 5. Cost-optimum renovation variants of the one-storey building with an attic

location	energy price per 1 GJ		
	20 PLN	40 PLN	60 PLN
	cost-optimum renovation variants		
Warszawa	W1/G	W2/G	W2/H/S
Szczecin	W1/G	W2/G	W2/H/S
Suwałki	W1/G	W2/G	W2/H/S

The calculations show that:

- for both building types at a low energy cost of PLN 20 per GJ, the renovation variant W1/G, was optimal. It is a variant including natural ventilation and the following heat transfer coefficients (U) of external buildings elements: external walls 0.19-0.21 W/m^{2}K, floor on the ground 0.24 W/m^{2}K, flat roof 0.22 W/m^{2}K, pitched roof 0.18 W/m^{2}K, floor above unheated basement 0.25 W/m^{2}K, windows 0.9 W/m^{2}K and external door 1.3 W/m^{2}K.
- At the energy cost of PLN 40 per GJ, the renovation variant W2/G was optimal for both building types. In this case the variant involves natural ventilation and the following heat transfer coefficients (U) of external buildings elements: external walls 0.12 W/m^{2}K, floor on the ground 0.14 W/m^{2}K, flat roof 0.12 W/m^{2}K, pitched roof 0.10 W/m^{2}K, floor above unheated basement 0.19 W/m^{2}K, windows 0.9 W/m^{2}K and external door 0.9 W/m^{2}K.
- In the case of PLN 60 per GJ, for both building types the optimal variant is W2/S, which takes into account solar collector installations supporting the preparation of domestic hot water. Some differences can be seen in regard to ventilation. In the case of the two-storey building with flat roof, the mechanical ventilation with heat recovery is cost-effective in Suwałki (the coldest location), whereas for the second building the use the hybrid ventilation is optimal for all locations.

Table 6 and Table 7 show the energy need for heating (Q_H) (including heating and ventilation) for the cost-optimal variants of renovation of the two reference buildings respectively. It can be seen that the energy need ranges from 30.9 to 91.1 kWh/m2year, corresponding to the optimal variants. The high range of values is a result differences in climate (different locations), energy prices and building types. It corresponds well to the reality in which energy consumption can vary strongly even for the same buildings.

Table 6. Energy need for heating for the two-storey building with flat roof for cost-optimum renovation variants

location	energy price per 1 GJ		
	20 PLN	40 PLN	60 PLN
	Q_H kWh/m^{2}year		
Warszawa	72,9	55,6	55,6
Szczecin	67,6	51,4	51,4
Suwałki	91,1	70,9	30,9

Table 7. Energy need for heating for the one-storey building with an attic for cost-optimum renovation variants

location	energy price per 1 GJ		
	20 PLN	40 PLN	60 PLN
	Q_H kWh/m²year		
Warszawa	66,0	48,5	41,7
Szczecin	60,9	44,5	38,2
Suwałki	83,8	63,6	55,7

Based on calculated energy need for heating primary non-renewable energy demand was calculated for the buildings. At this point renovation of heating and DHW system, auxiliary equipment was included as well as use of solar collectors for DHW. The obtained reduction of the primary energy demand for heating, ventilation, hot water production and the work of auxiliary equipment (this includes energy for fans, pumps, electronics, etc.) ranges from 69% to 86% (Table 8 and Table 9). The higher reduction in primary non-renewable energy can be achieved by using RES and better insulated building envelope. The location does not have a significant effect in this case.

Table 8. Change of primary, non-renewable energy index after renovation to the cost-optimum variant for two-storey building with flat roof

localization	Q_P index for base variant	energy price per 1 GJ		
		20 PLN	40 PLN	60 PLN
		Q_P index for renovation variants (percentage reduction in regard to base variant)		
	kWh/m²year	kWh/m²year	kWh/m²year	kWh/m²year
Warszawa	448	137 (69%)	116 (74%)	96 (79%)
Szczecin	425	130 (69%)	110 (74%)	90 (79%)
Suwałki	527	160 (70%)	135 (74 %)	74 (86%)

Table 9. Change of primary, non-renewable energy index after renovation to the cost-optimum variant for one-storey building with an attic

localization	Q_P index for base variant	energy price per 1 GJ		
		20 PLN	40 PLN	60 PLN
		Q_P index for renovation variants (percentage reduction in regard to base variant)		
	kWh/m²year	kWh/m²year	kWh/m²year	kWh/m²year
Warszawa	574	138 (75%)	117 (79%)	88 (84%)
Szczecin	546	132 (75%)	112 (78%)	84 (84%)
Suwałki	669	160 (75%)	135 (79%)	105 (83%)

7. Conclusions

Based on the above analysis and considering received results the Polish definition of nZEB renovation of single-family house was proposed. Because large range of values, especially in regard to energy need for heating, was obtained, it was assumed that averages will be used for the indicators.

The requirements for nZEB renovation of single-family residential buildings are following:

- The energy need for heating $Q_H \leq 60$ kWh/(m² year),
- The percentage of reduction of the primary, non-renewable energy Q_P demand (including in case of residential buildings heating, ventilation, domestic hot water and auxiliary systems) ≥ 75%.

Defined requirement for energy need for heating can be compared with indicator for low-energy residential buildings in NF40 standard $Q_H \leq 40$ kWh/(m² year) [9] or passive buildings $Q_H \leq 15$ kWh/(m² year) [10]. It is lower, and it should be because reaching low-energy standard in case of existing buildings can be technically complicated (e.g. implementation of mechanical ventilation with heat recovery) and very expensive. Meanwhile, the second requirement, referring to the percentage reduction of the primary, non-renewable energy Q_P demand, is similar as for deep renovation, commonly understood as one that focuses on the building envelope and achieves an energy reduction of 75% or more compared to the situation before the renovation. The absolute values of Q_P are higher than required for new nZEB houses in Poland - primary, non-renewable energy demand ≤ 70 kWh/(m² year) [9].

References

[1] Zaborowski M and Dworakowska 2014 A Energy Efficiency in Poland 2013 Review *Institut of Enviromental* Economics (Cracow) 2014

[2] Węglarz A and Gilewski P 2016 A method of evaluation of polioptimal thermo-modernization schemes of buildings *Procedia Engineering* **vol 153** p 862-865,

[3] Węglarz A and Pierzchalska D 2016 Praktyczna realizacja idei głębokiej termomodernizacji *Materiały Budowlane* **vol 521** (1) p 23-25

[4] Firląg S 2018 Definition of nZEB Renovation Standard *Design Solutions for nZEB Retrofit Buildings* IGI Global p 1-23

[5] Fokaides P A and Papadopoulos A M 2014 Cost-optimal insulation thickness in dry and mesothermal climates: Existing models and their improvement *Energy and Buildings* **vol 68** p 203-21

[6] Zangheri P and Armani R and Pietrobon M and Pagliano L 2018 Identification of cost-optimal and NZEB refurbishment levels for representative climates and building typologies across Europe *Energy Efficiency* **vol 11** p 337–369,

[7] Brambillab A nad Salvalaia G and Imperadoria M and Sesanaa M M 2018 Nearly zero energy building renovation: From energy efficiency to environmental efficiency, a pilot case study *Energy and Buildings* **vol 166** p 271-283

[8] Król P and Firląg S and Węglarz A 2013 Zintegrowana ocena wpływu budynku jednorodzinnego na środowisko *Rynek Instalacyjny* **vol 9** p 20-25

[9] Firląg S 2010 Wpływ rodzaju systemu ogrzewczego na komfort cieplny i zużycie energii w jednorodzinnych budynkach pasywnych *Czasopismo Techniczne* **vol 107/4** p 49-57

[10] Firląg S 2015 How to meet the minimum energy performance requirements of Technical Conditions in year 2021? *Procedia Engineering* **vol 111** p 202-208

IOP Conf. Series: Materials Science and Engineering **415** (2018) 012002 doi:10.1088/1757-899X/415/1/012002

Dormitory thermal retrofitting to nZEB standard

J Kwiatkowski, J Sowa, A Trząski, A Wiszniewski
Faculty of Building Services, Hydro and Environmental Engineering, Warsaw University of Technology, Nowowiejska 20, 00-653 Warszawa, Poland

E-mail: jerzy.kwiatkowski@pw.edu.pl

Abstract. The Recast of the Directive on the Energy Performance of Buildings (the EPB Directive) came into force on 9 June 2010. One of the main requirements of the directive is to set a definition of nearly zero energy building (nZEB). Most of the EU member states have already set a definition of nZEB for new buildings and some of the countries also done it for existing buildings. The measures of thermal retrofitting of a dormitory to nZEB standard are presented in this paper. As there is no definition of nZEB for existing buildings in Poland, the paper presents a considerations of such a standard. Next, a set of thermo-modernization solutions are presented on a basis of existing building of a collective residence. The aim of the paper is to present the barriers that occur in such projects and solutions that can be applied to achieve the set goals. Building chosen for analysis is under historic protection which makes the work even more difficult. It was proved that used solutions were chosen not only to reduce energy demand or increase energy production from renewable energy sources, but also to increase thermal comfort within building.

1. Introduction

The Recast of the Directive on the Energy Performance of Buildings (the EPB Directive) [1] came into force on 9 June 2010. EU member states should until 9 June 2012, publish the relevant laws and administrative regulations necessary to implement its provisions. European regulatory efforts towards increasing energy efficiency of buildings are focusing on building requirements. All new public building after 31.12.2018 must be nearly zero energy buildings. Such requirement can be very hard to implement in public buildings as the energy needs for heating, cooling and lighting to provide internal comfort are typically very high. It is even harder to retrofit existing public buildings to nearly zero energy standard.

In this paper a solutions that can be applied to achieve the set goals are presented. After definition of nZEB standard, thermomodernization measures are presented for a dormitory building. Both energy and comfort analysis are presented, as not only energy decrease but also a comfort increase are required.

2. nZEB definition

Directive 31/2010/EU [1] requires the introduction of nearly zero-energy buildings but does not specify a minimum or maximum harmonized requirements and detailed guidance framework procedure for calculating the energy performance of building. Member states have to define what that means exactly for them. Taking into account the local conditions in the definition does not preclude the adoption of a uniform methodology in all Member States.

The Directive defines a nearly zero-energy building as a high energy performance building and requires the determination of the ratio of primary energy. Very low or almost zero energy demand of the building should be covered from renewable energy sources or renewable sources produced on-site.

In Poland a nZEB standard has been defined by primary energy indicator (EP) and set in Ordinance on technical conditions to be met by buildings and their location [2]. This indicator does not include renewable primary energy for which the primary energy factor is equal to 0. For public buildings, like educational building maximal value of EP has been state as sum of three indicators:

- for heating and domestic hot water – 45 kWh/m^{2}a;
- for cooling – 25*($A_{f,c}$/A_f) kWh/m^{2}a (where $A_{f,c}$ is a cooled area and A_f is a total regulated temperature area);
- for lighting – 25 kWh/m^{2}a for time of light operation <2500 h/year and 50 kWh/m^{2}a for time of light operation ≥2500 h/year.

In total maximum EP may vary from 70 to 120 kWh/m^{2}a.

From the other hand in a Directive 31/2010/EU a requirement of cost optimal parameters is given. Hence in the KODnZEB project [3] a study of the cost optimal EP value for educational building estimation has been done. The calculation has shown that cost optimal EP indicator, that include heating, cooling, hot water, lighting and auxiliary energy, should have value between 40 and 55 kWh/m^{2}a [4].

Despite existing requirements or analysis for the use of KODnZEB project a value of primary energy indicator equal to 20 kWh/m^{2}a has been set as nZEB standard for two chosen buildings. This target value include heating cooling, hot water, lighting and auxiliary energy. Such a stringent requirement was defined as a goal was to prove that educational building can be retrofitted with very high energy requirement reduction.

3. Case building description

The analysed building is Muszelka dormitory located in Warsaw, Poland. The building volume is 9 562 m^3 and usable area of 3 366 m^2. The building is a part of a dormitory complex of Warsaw University of Technology. The building offers 151 living places for students.

The building has 4 over-ground and one underground storey. There are technical rooms in the basement (warehouses, workshops, heating ducts). On each floor living rooms, a common kitchen and toilet are located. In addition on the ground floor administrative rooms are situated. Student rooms are designed in segments of 2-3 rooms in segment, and in each segment a bathroom with shower is placed. Communication in the building is provided by corridors and 2 staircases.

The shape of the building is similar to a rectangle with dimension 60m x 13m, the longer sides have the shape of arches, consistent with the curvature of the axis of Mochnacki Street. From the east, the building adjoins "Akademik" dormitory, and from the west "Bratniak" dormitory. The main entrance is on the south side. Figures 1 and 2 show current look of building facades.

The building was built in 1956 in the construction of a reinforced concrete frame filled with ceramic brick walls. Over the last storey of the building, the ventilated roof is designed.

Figure 1. Photo of the south façade: west part (on the left) and east part (on the right)

IOP Conf. Series: Materials Science and Engineering **415** (2018) 012002 doi:10.1088/1757-899X/415/1/012002

Figure 2. Photo of the north façade: west part (on the left) and east part (on the right)

In 1989, the roof was renovated and a insolation of ceiling over last floor was done. Then in 1996, all windows were replaced with airtight windows with an estimated U-value of 2.0 W/m^{2}K. The external walls are made of bricks without an insulating layer, and the estimated heat transfer coefficient ranges from 0.90 to 2.02 W/m^{2}K.

In the building only gravity ventilation system consisting of exhaust ventilation ducts is designed. The air inlet takes place through the gaps in the window joinery, and the outflow through gravity ventilation ducts extended above the roof of the building. The existing central heating installation is a water, two-pipe pump system. Heating installation is equipped with steel panel radiators. In 1993, thermostatic valves were installed. Currently, some of these valves are not working properly, some have been dismantled and replaced with ordinary shut-off valves, while others have been vandalized. The radiators are corroded.

In different types of rooms of the analysed building, different types of lighting fixtures are used. The largest variation in the types of lighting fixtures can be noticed in the living segments. In addition, students independently exchange light sources in luminaires in the segments, adapting them to their needs and preferences. Lighting of communication zones in the building, common kitchens and bathrooms is unified, and properly maintained by the administration of the facility.

The installed capacity of the lighting system in Muszelka dormitory (without individual table lamps in living rooms) is 27.152 kW, while nominal capacity for whole building is equal to 11.22.W/m^2.

The building is under constant operation for whole year, during students year as dormitory and during summer as hostel.

The climate in Warsaw is moderate with lowest temperature of -12.3°C and highest of 33.2°C. The mean yearly temperature is equal to 8.2°C and global horizontal irradiation about 980 kWh/m^2.

As part of the analysis of thermal retrofitting, simulations of energy consumption for heating in the building of the Muszelka dormitory were carried out. The results of numerical calculations were compared with measures values. The comparison showed that the difference is only 15.0%. Therefore, it can be concluded that the constructed dynamic model is correct and can be used for further energy analyses. Table 2 presents the energy performance of the building in the existing state.

Table 1. The energy performance of the building in the existing state.

Primary energy	Unit	Value
Heating	kWh/m^{2}year	140.8
Domestic hot water	kWh/m^{2}year	69.1
Lighting	kWh/m^{2}year	85.3
Auxiliary energy	kWh/m^{2}year	3.3
Photovoltaic	kWh/m^{2}year	0
Sum	kWh/m^{2}year	298.5

The total primary energy indicator for a building in the existing state is equal 298.5 kWh/m^{2}year and 255.0 kWh/m^{2}year if students' rooms are considered as independent living units. In the calculation a primary energy factors of 0,68 for district heating and 3,0 for electricity were used.

4. Modernizations

The following paragraphs present modernizations proposed for the Muszelka building introduced to meet the nZEB requirements set for the KODnZEB project. Proposed modifications include the thermal envelope of the building, technical systems, lighting as well as systems that influence behaviour of the occupants and the use of renewable energy sources.

4.1. Thermal envelope

The reduction of heat loss due to heat transfer and infiltration several modernisations were proposed:

- insulation of external walls,
- insulation of the ceiling between the top floor of the building and the unheated attic,
- replacement of windows.

4.1.1. External walls. It was proposed to insulate external walls with aerogel panels having the thermal conduction value of λ=0,015 W/mK and the thickness of 4 cm. This modernization should result in a reduction of the heat transfer coefficient of walls from U=1,09 W/m^{2}K down to U=0,279 W/m^{2}K. The type of the insulating material was dictated by the restrictions set by the monument conservator that allowed only a 4 cm increase of the external walls thickness. Due to the limited thickness of the insulating material the best possible material, in regards to the thermal conductivity coefficient, was chosen.

4.1.2. The top floor ceiling. To insulate the ceiling below the unheated attic and the roof over the staircase a mineral wool with a thermal conduction coefficient λ=0,035 W/mK and a thickness of 16 cm was proposed. This modernization should result in a reduction of the heat transfer coefficient of the ceiling from U=2,02 W/m^{2}K down to U=0,197W/m^{2}K. This would be enough to meet the technical requirement for the year 2014. The analysis indicated that an increase of the insulation thickness up to 22 cm in order to meet the requirements set for the year 2021 would not result in a noticeable heat demand reduction.

4.1.3. Windows. It was proposed to replace the existing windows with a new set with a better heat transfer value. This would result in an improvement of the heat transfer coefficient of windows from U=1,87 W/m^{2}K to U=0,70 W/m^{2}K. This would be enough to meet the technical requirement for the year 2021, moreover it would greatly reduce the heating demand. Simultaneously to the thermal insulation modernization works there would be conducted works leading to the improvement of the air tightness of the building. It is estimated that this modernization should allow to decrease the air tightness coefficient from n_{50}=3,0 1/h down to n_{50}=1,5 1/h.

4.2. Ventilation

The concept of ventilation in modernized dormitory foresees several different types of systems:

- mechanical exhaust ventilation with air to water heat recovery from living spaces;
- balanced mechanical ventilation with air to air heat recovery in other zones (corridors and staircase, shared kitchens, shared bathrooms, service rooms);
- natural ventilation of unused rooms in the basement of the building.

4.2.1. Ventilation of living spaces. All living spaces are ventilated by decentralized mechanical exhaust ventilation system. Outdoor air will be supplied to the rooms via air vents that will be installed in window frames. In some cases small reduction of glazing will be required. Than used air will pass through small entrance corridor and will be exhausted from a shower recess. Exhaust fans installed in these spaces will work with several speeds according to demand. Collective vertical ducts will transport air the attics. Ducts serving rooms of the same purpose will be connected there and then before release air to the atmosphere heat will be recovered to the hot water tank.

Two concepts of ventilation rate control will be used. For single rooms the main assumptions are as follows:

- nominal supply ventilation rate 20 m^3/h person (occasionally increased in case of using shower);
- nominal exhaust ventilation rate from shower recess 50 m^3/h (higher fan speed), but ventilation rate can be reduced to 20 m^3/h (lower fan speed);
- operation of the fan in a higher speed is activated by a light switch in the recess (with an additional delay of 15 minutes after switching off), when the light is switched off, the fan operates in a lower speed.

For units when one shower recess is dedicated for 1 double and 2 single rooms the main assumptions are as follows:

- nominal supply ventilation rate 20 m^3/h person;
- exhaust fan can remove 80 m^3/h (the highest fan speed), 50 m^3/h (middle fan speed) or 20 m^3/h; (the lowest fan speed);
- fan operation is controlled by a contactor system (central power switches for power sockets in the rooms); assumed fan operation scheme: 0 active contactors – ventilation rate 20 m^3/h, 1 active contactor - ventilation rate 50 m^3/h, 2 or 3 active contactors - ventilation rate 80 m^3/h.

These schemes of control allow significant savings in energy demand for ventilation (both heat and electricity).

4.2.2. Ventilation of other rooms. All spaces in the dormitory with open access (common toilets and kitchens on the floors as well as corridors and staircases) will have balanced mechanical ventilation system with heat recovery of minimum efficiency 85%. The air flows have been designed to prevent the dispersal of undesirable odors. The assumed ventilation rates are:

- collective toilet rooms - 50 m^3/h per 1 toilet bowl and 25 m^3/h per 1 urinal;
- collective kitchen rooms - 70 m^3/h for a gas stove;
- corridors and stairwells - 0.5 air changes per hour.

The modernization of ventilation in these rooms will solve existing problem of unbalanced air flows will lead to significant savings of heat demand for the preparation of ventilation air.

4.3. Space heating system

The heating system modernization includes replacement of the radiators and installation of regulating valves under the vertical piping and for the new radiators. The rest of the system, like piping, would remain unchanged.

4.4. Domestic hot water system

The domestic hot water system modernization includes installation of a water meter in each residential unit. This should allow occupants to monitor the use of water and therefore should result in hot water use savings. It is estimated that water meters installation should result in a 20% reduction of the hot water consumption. Additionally the old hot water tank is going to be replaced. The rest of the system, like piping, would remain unchanged.

4.5. The lighting system

The concept of the room lighting system in the Muszelka building includes a use of several different lighting fixtures, accordingly to the purpose of the rooms. The modernisation includes replacement of the existing lighting fixtures and light sources as well as installation of a lighting control system, regulated both by a time schedule and the presence of the occupants. Basing on the capacity of the installed lighting in the rooms, the total and nominal capacity of the lighting system on each floor and for the whole building was calculated.

The average nominal capacity of the lighting system in Muszelka building after the renovation is 3,34 W/m^2. Simultaneously it is proposed to install an automatic lighting control system that would allow further reduction of the energy consumption, it would include:

- residential rooms – no additional control system;
- technical rooms – no additional control system;

- corridors and staircases – installation of movement sensors (standby lighting reaching full capacity at a presence of movement)
- shared kitchens – installation of presence sensors (lighting active only if the occupants are present);
- shared bathrooms – installation of presence sensors (lighting active only if the occupants are present).

4.6. Heat source

The planned renovation does not include a replacement of the existing heat source. The heat would be still delivered by a heating unit supplied by the heat from the district heating network. The heating unit is planned to be equipped with a heat meter for the space heating system.

In case of the domestic hot water system, it is planned to install a heat pump system that would use the heat from the exhaust ventilation air from the residential rooms to pre-heat the water. Due to the varying ventilation air flow, the pump was selected for the average airflow value of 2 084 m^3/h. The remaining heat demand for the hot water preparation will be delivered from by the existing district heating unit.

4.7. Renewable energy sources

It is planned to install a PV system on the roof of the Muszelka building, that would produce energy for the internal needs when possible and for the rest of the buildings in the complex or sold to the external energy grid in case of over generation. The capacity of the system was selected taking into account the limitations set by the monument conservator in regard to the visibility of the system from the street level near the building. The figure 3 presents the results of the visibility analysis of the roof system.

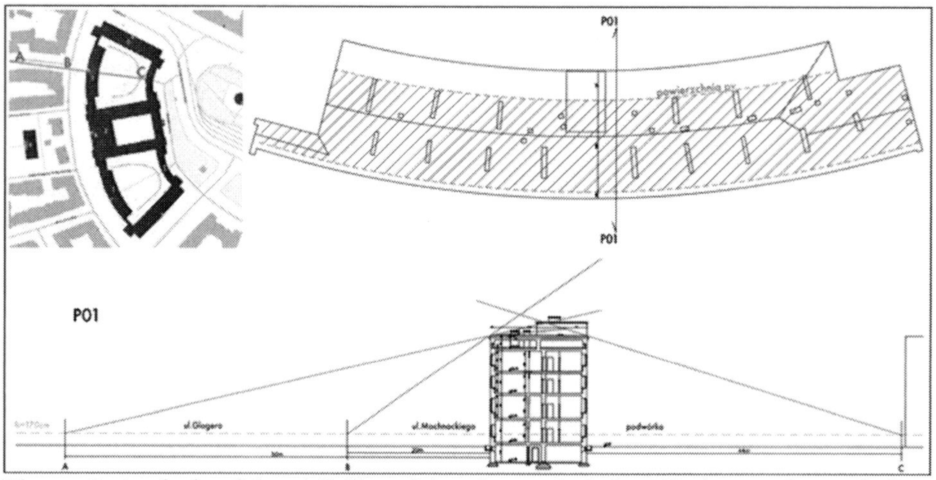

Figure 3. Analysis of the visibility of the photovoltaic installation on the roof of the building

The photovoltaic system with a nominal capacity of 79,35 kW_p and an area of the PV panels of 374,8 m^2 would produce annually 69 953 kWh of electricity. The production is limited due to the available area of the roof and the possible slope of the panels dictated by restrictions regarding visibility.

5. Results of energy calculation

As part of the modernization of the Muszelka dormitory, which is under historical protection, a measures interfering in the building envelope (insulation of partitions and replacement of windows), technical systems (exhaust ventilation with variable air stream, mechanical ventilation with heat recovery, thermostatic valves, water meters or energy-saving sources and lighting fixtures), alternative and renewable heat sources (heat pump on exhaust air or installation of solar panels on the roof) has been proposed. Each of the applied solutions does not go beyond the limitations resulting from conservation protection restrictions.

For the proposed modernizations, simulations of dynamic energy demand of the building were carried out and the results of calculations are presented in table 2.

Table 2. The energy performance of the building after modernization

Primary energy	Unit	Value
Heating	kWh/m^{2}year	19.7
Domestic hot water	kWh/m^{2}year	26.0
Lighting	kWh/m^{2}year	26.5
Auxiliary energy	kWh/m^{2}year	4.2
Photovoltaic	kWh/m^{2}year	-58.6
Sum	kWh/m^{2}year	17.9

The total primary energy indicator for a building after modernization would be equal 17.9 kWh/m^{2}year. It can be noticed that the total primary energy demand indicator is almost in line with the concept of nZEB definition for the purposes of the KODnZEB project, and as a result of modernization, its value has been reduced by more than 16 times.

6. Comfort analysis

The required temperature in the building during heating season was always maintained but due to lack of cooling system during late spring or summer an overheating could occur. Thus in addition to energy analysis a verification of summer comfort has been done.

For the assessment of the thermal comfort the number of hours with indoor temperature exceeding 24°C for I class, 25°C for II class, 26°C for III class has been calculated for six students rooms. Those values has been establish on a basis of PN-EN 15251 [5] adaptive comfort. For the purpose of this analysis two rooms on the ground floor, two on the 1st floor and 2 on the last floor has been chosen. On each storey one of the room was with the windows oriented to the south and other with windows oriented to the north. In the table 3 and 4 the results of comfort analysis is presented.

Table 3. Number of hours with indoor temperature exceeding required value for given comfort class – existing state

Comfort class	I	II	III
Ground floor – north oriented room	0	0	0
Ground floor – south oriented room	0	0	0
1st floor – north oriented room	6	0	0
1st floor – south oriented room	49	10	0
Last floor – north oriented room	60	19	0
Last floor – south oriented room	158	52	12

Table 4. Number of hours with indoor temperature exceeding required value for given comfort class – after modernization

Comfort class	I	II	III
Ground floor – north oriented room	0	0	0
Ground floor – south oriented room	0	0	0
1st floor – north oriented room	1	0	0
1st floor – south oriented room	9	0	0
Last floor – north oriented room	78	0	0
Last floor – south oriented room	222	4	0

It can be noticed that the thermal comfort on in the rooms on the ground floor is always obtained and temperature do not exceed 24°C. For the 1st floor the number of hours with temperature that exceeds required level for given comfort class has been decreased for the state after modernization. There is no hours with temperature over 25°C and the number of hours with temperature over 24°C is decreased by six times. In the case of last floor the total number of hours with temperature over assumed limit has not change but there was a shift form II and III class to I class limit. On the basis of presented analysis it can be concluded that due to applied modernization a thermal comfort has been increased and overheating phenomena has been limited.

7. Conclusions

In the paper a measures of modernization of existing building of a collective residence where presented. The aim was to obtain a nZEB standard after modernization. As the definition of nZEB standard in Poland is not very demanding the new, more sharp definition has been used. Next the applied measures has been presented.

It was shown that all the modernization allows to decrease the primary energy demand indicator by over 16 times, from 298.5 up to 17.9 kWh/m2year. The highest decrease in primary energy demand was for heating and lighting. Also a huge impact on final results has a photovoltaic system, that coves almost 20% of primary energy demand. It was also showed that due to building modernization thermal comfort of students room has been increased. The overheating risk has been limited almost to zero.

It was presented that due to integrated energy design, common work of architects and engineers the nZEB standard with increased thermal comfort can be achieved by modernization of building under historic protection.

Acknowledgment

This paper was performed with the financial support of EEA Financial Mechanism 2009-2014 – Operational Program PL04 "Saving energy and promoting renewable energy source" within the framework of the" project KODnZEB: "Thermo-modernization of two chosen Public Buildings according to nZEB Standards".

Through the funds EEA and Norwegian funds, Iceland, Liechtenstein and Norway contribute to reducing social and economic inequality and strengthening bilateral relations with beneficiary countries in Europe. These three countries closely cooperate with the EC on the basis of the Agreement on the European Economic Area (EEA Agreement).

References

[1] Directive 2010/31/EU of the European Parliament and of the Council of 19 May 2010 on the energy performance of buildings (OJ L 153, 18.06.2010).
[2] Ordinance of the Minister of Transport, Construction and Maritime Economy of 5 July 2013 amending the regulation on technical conditions to be met by buildings and their location (Dz.U. 2013 pos. 926)
[3] KODnZEB project 2016 Thermo-modernization of two chosen Public Buildings according to nZEB Standards, Poland,
[4] Kwiatkowski J and Rucinska J 2016 *6th International Symposium on Energy Challenges & Mechanics (ISECM) – towards a big picture*, 14-18.08.2016, Inverness, Scotlad
[5] Standard EN 15251:2012. Indoor environmental input parameters for design and assessment of energy performance of buildings- addressing indoor air quality, thermal environment, lighting and acoustics. CEN, Brussels.

A probabilistic tool for evaluating the effectiveness of financial measures to support the energy improvements of existing buildings

M D'Orazio[1], E Di Giuseppe[1], R Esposti[2], S Coderoni[2] E Baldoni[2]

[1] DICEA Dept. Università Politecnica delle Marche, via Brecce Bianche 12, 60131 Ancona, Italy
[2] DISE Dept. Università Politecnica delle Marche, Piazza Martelli12, 60121 Ancona, Italy

Abstract

LCC procedures are commonly used in building retrofit interventions for the evaluation of the costs and benefits of different solutions. LCC methods are also commonly applied to energy retrofit interventions according to UNI-EN 15459. However, these procedures can give significantly different results in relation to the methods used to define the economic and technical parameters used in the calculation. Several authors have highlighted how the uncertainties related to the quantification of the costs of initial investment and maintenance, the economic scenarios (inflation and interest rates), the costs of energy and labor and the "service life" of the components strongly influence the results and the assumption of different data sets can result in significantly different results. Several authors pointed out that it is necessary, prior to these procedures, to perform sensitivity analyzes, useful for understanding if and how each of these parameters can influence the results, which leads to the need to use probabilistic methods. The study reported here analyses the influence that different assumptions about the service life of building components have on the economic benefits brought by interventions of energy requalification. Specifically, through a specially developed probabilistic procedure, the influence on the global costs of different assumptions about the probability distributions and the specific parameters characterizing each of them related to the service life of components useful to realize the thermal insulation was evaluated. the envelope of existing buildings. In particular, four different probability distributions (normal, uniform, left triangular, right triangular), different values of the median have been taken into consideration for the same distribution amplitude and the global costs and the return time values have been evaluated. investment considering different alternative economic scenarios for inflation and interest rate.

1. Introduction

Buildings are responsible for the largest share of European final energy consumption (about 40%) and they represent the greatest potential to save energy, as 75% of buildings standing in the EU were built during periods with no, or minimal, energy-related building codes. However, aggregated projections of EU-27 for 2020[1] show that, even if the additional measures currently planned by Member States are adopted and fully implemented, greenhouse gas emissions will reach a level approximately 2% higher than in 2005, and only 6% below their 1990 level. This is a significantly higher level than the unilateral commitment of a 20% reduction, compared to 1990 levels, decided by the European Council in December 2008. Without a public intervention and the strong commitment of all actors, these ambitious objectives cannot be reached. EU commission has increasingly favored economic or market-based instruments ("MBI"), such as indirect taxation and targeted subsidies, for such policy purposes.

Direct fiscal incentives are, or have been, used in a number of EU Member States as tool to improve energy reduction, specially of existing buildings. Effectiveness of direct fiscal incentives have been analyzed by several authors[1–3], showing the necessity to improve costs and benefits tools to evaluate in a systematic way impacts and efforts for each policy instrument.

Life Cycle Costing (LCC) calculations, based on EN 15459:2007[4], have been implemented in Europe at national level in compliance with the Directive 2010/31/EU. However, LCC procedures based on EN 15459 have some limitations, mainly related to a deterministic approach in the cost items selection and the quantification and forecast of the macroeconomic variables[5]. It is necessary instead taking into account the uncertainties[6–14] associated with the initial and periodic costs related to the energy efficiency measures under future economic scenarios and to assess in a probabilistic way the benefits deriving from the adoption of direct tax incentives policies.

The present paper proposes a LCC probabilistic methodology and a related software tool able to evaluate cost and benefits of direct fiscal measures on building energy efficiency measures (EEM), considering alternative macro-economic scenarios. The methodology is based on an uncertainty analysis via Monte Carlo (MC) approach. After a preliminary description of the methodological approach, the application is illustrated through a case study, comparing the impact of specific direct fiscal measures under different macro-economic scenarios.

2. Methodological approach and calculation tool

The LCC tool is based on the Global Cost method described in the European Standard EN 15459[4] with some modifications to take into account, in probabilistic terms, the initial and periodical costs (investments, replacements, energy costs) and possible alternative evolutions of the macro-economic scenario (how the scenario affects the inflation rate, the discount rate and the price escalation rate). The Global Costs $C_g(cp)$ at the end of the calculation period (cp), referred to the starting year, are calculated through the following equation:

$$Cg_{cp} = \sum_{j=1}^{N}\left\{ CI_j + \sum_{t=1}^{CP} \left[\left(CM_{j,t} * R_{disc} * R_t \right) + \left(CE_{j,t} * R_{disc} * R_t \right) \right] + CR_{j,t_j} - Val_{j,cp} \right\} \qquad (1)$$

Where: CI are the initial investment costs of the EEM j; $CM_{(i)}$ are the annual maintenance costs of the EEM j; CEj is the annual energy cost due to the EEM j; $R_{disc(i)}$ is the discount rate; $R_{t(i)}$ is the prices escalation rate; $CR_{(i)}$ are the replacement costs with a frequency depending on the service life $SL_{(j)}$ of the EEM concerned; Val_F is the residual value of the EEM at the end of the calculation period.

The discount rate $R_{disc(i)}$ depends on the inflation rate $\pi_{(i)}$ and interest rate $i_{(i)}$. Since a dynamic calculation according to EN 15459[4] is taken into account, we consider annual variations of the discount rate as well as annual variations of the rate of development of prices for any of the costs considered in the annual costs (i.e. energy costs, periodic or replacement costs, maintenance costs).

The Global Cost assessment of the energy efficiency measures proposed in this work includes uncertainty propagation via Monte Carlo (MC) methods. While a deterministic LCC analysis approach requires input variables that are fixed in their "deterministic" value, in Monte Carlo approach variables are modelled using a Probability Density Function (PDF) and the quantification of the uncertainty of the outputs is a result of possible variance of the input parameters. MC method consists in randomly selecting input variables (or using sampling schemes) and inserting them into the output-equation a proper number of times, depending on the envisaged accuracy level, to predict the corresponding output distributions.

In order to introduce a MC based approach to Life Cycle Costing, we quantified the PDFs of the model's input parameters to lead the Monte Carlo simulation. Model input included in the assessment are listed in **Table 1**.

The calculation of the Global Costs is performed under three alternative macro-economic scenarios: Regular Growth (RG), Intense Growth (IG) and Deflation (DE). For each scenario, the economic variables necessary to carry out the LCC calculation have been forecasted over the calculation period of the renovation project. These variables are the nominal interest rate (i), the inflation rate (π) and the rate of growth of the GDP expressed in real terms (g). From historical data, multivariate statistical distributions for these scenarios are estimated using the VAR methodology. Consistently with the simulation-based LCC calculation procedure, a MC approach have been used to generate an array of alternative prediction of the variables for each scenario. Thus, projections are generated as draws from appropriate probability distributions. The MC forecast approach and the multi-scenario approach have been used to address the uncertainty inherent in the behaviour of economic variables.

Table 1. LCC input parameters included in the probabilistic assessment

LCC Stage		LCC Parameter description
Financial data	Duration of the calculation	Duration of the calculation [years]
	Financial rates	Inflation rate [%] Market interest rate [%]
	Rate of development of prices	Rate of development of prices [%]
System characteristics	Component Investment cost	Design option investment cost [ ]
	Periodic costs for replacements	Design option Service Life [years] Design option replacement costs [ ]
	Running Costs	Design option annual Maintenance cost [ ]
Energy Costs	Energy consumption	Energy need [kWh/y] Building overall efficiency for heating [-]
	Energy Costs	Energy source national tariff [/kWh/y]

While the RG scenario represents the dynamics of an economic system with a balanced growth path, i.e., with moderate growth in inflation and in GDP and moderate nominal interest rate, the IG and DE scenarios account for more extreme cases in which relationships between economic variables deviate remarkably from the RG case.

As expected, the RG scenario presents a more balanced evolution of the economic variables with respect to the other two scenarios. This scenario is characterized by a real GDP growth around the 2.5% (sd 1.6%) and an inflation rate around the 2.2% (sd 0.9%). Interest rate, in nominal terms, is around the 5% (sd 1.5%).

The IG scenario is characterized by a more robust growth with an annual average rate of real GDP growth over the 3.3% (sd 1.2%). Inflation rate (mean 2.6%; sd 0.6%) and nominal interest rate (mean 6%; sd 0.7%) are both higher than that in the RG scenario.

The DE scenario is instead characterized by a near-zero annual inflation rate. This is the result of many periods of negative annual inflation rates. In such scenario, interest rate and the growth of the economy are expected to be low. Consistently, the average nominal interest rate is around the 1.9% (sd 1.5%) and the average annual real GDP growth is around 1.3% (1.6%).

As a consequence of the procedure, the resulting Global Costs in the probabilistic analysis is to be evaluated based on its probability distribution. Since the quality of the outcome (the PDF of Global Cost) is dependent on the number of simulations carried out and the sampling scheme used, in this work, we use Sobol's sequences as quasi-random sampling technique, in order to generate samples as uniformly as possible.

Data fitting for the input uncertainty characterization, sample generation, uncertainty propagation, and sensitivity analysis are performed through a developed web tool based on the data analysis software "R", a free software environment for statistical computing and related Shiny web libraries.

3. Application to a case study: comparison of direct fiscal measures

3.1 Case study description

In order to show the potential of the methodology and tool, we evaluate the effectiveness of the financial measures in use in Italy to support the energy improvement of single family houses built during the first part of last century. A large part of the Italian building stock is characterized by the presence of single-family houses of beginning XX Century Figure 1. They are typically two floor masonry buildings of about 60-90 m^2/floor, with uninsulated brick masonries of different thickness - about 20-35 cm (U value = 1.5-2.3 W/m2K) - wooden slabs and roofs (U=1.2-1.5 W/m2K) with clay tiles, single glazing windows with timber frames (U=5-6 W/m2K), combined heating and DHW with traditional gas non- condensing boiler. The typical whole energy consumption could range from 110 to 170 kWh/m^2 depending on the Italian climate zone, on the geometry and on specific constructive features[15]. Actually in Italy about 7 millions of buildings with these characteristics are in use, with at least 200-250 millions of m^2 of uninsulated external wall.

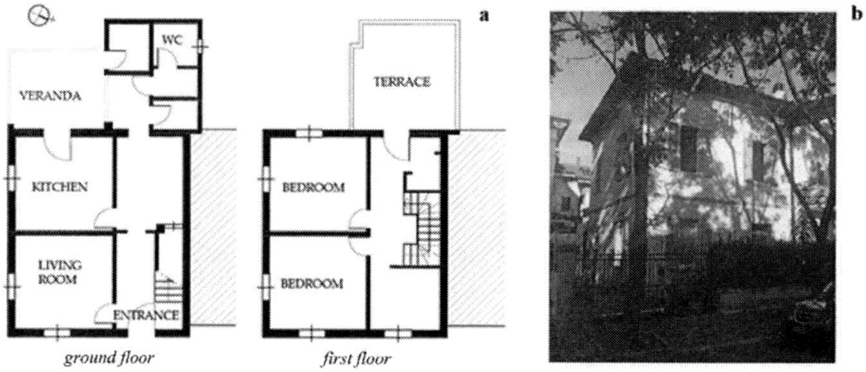

Figure 1. A typical Italian single-family house of beginning XX Century. (a) Plans; (b) View of the north facade.

Energy consumption of this building typology is largely due to the poor performance of the exterior wall. Thus, a typical energy retrofit measure is the internal insulation of the exterior wall, in order to preserve the architectonic features of the façade, considering that these buildings are often locally listed-buildings. This energy measure may benefit of a specific contribute from the Italian government in terms of tax deduction. The percentage of the deduction varied over the time from 36% to 65% and is actually equal to 50% of the initial cost.

To evaluate the effectiveness, from a user perspective, of the financial measures adopted by the Italian Government in a life cycle perspective, we defined a set of data able to describe the energy performance of this "building class" after and before an internal insulation intervention (10 cm EPS board).

The LCC assessment requires input data on operational energy use before and after the renovation measure in order to take into account the "use phase" (energy costs) and determine the costs savings. The energy cost (*CE*) is the annual cost for the delivered energy, including national taxes. It is obtained multiplying the annual energy consumption by the tariff ($EnT_{(e)}$) for the energy carrier considered (*e*). The delivered energy depends on the energy need (Q_h) and the building global efficiency for heating (ETAh). As the LCC is performed considering only the envelope intervention, the operational energy use is considered due to the only heat transmission losses through the wall. Heat losses through the wall were calculated using a HDD (annual heating degree-days) method. HDD data were extracted from Eurostat database for Italy for years from 2000 to 2016, considering their variability during time and space. For the original wall thermal resistance, normal distributions are implemented in the software tool, based on the variable thickness range. Thermal resistance of the insulation system was considered as deterministic. PDFs of initial and maintenance costs and environmental impact of the proposed intervention were estimated using Italian market data. Table 2 reports the LCC data inputs distributions.

Table 2. Class data and energy source data. Par1 and Par2 characterise the distributions: they are respectively the mean and standard deviation for the normal distribution and the minimum and maximum values for the uniform distribution.

		Distribution type	Par1	Par2	Unit
	Qh pre	Normal	100.63	6.57	kWh/m2
	Qh post	Normal	14.12	2.35	kWh/m2
	CI	Normal	40.42	5.59	/m2
	CM	Normal	0.88	0.12	/m2
	SL	Normal	30.00	2.98	Years
Gas	EnT	Uniform	0.065	0.085	
	ETAh	Uniform	0.6	1	-
Ele	EnT	Uniform	0.1584	0.2143	
	ETAh	Uniform	2.5	4	-
Oil	EnT	Uniform	0.115	0.1354	
	ETAh	Uniform	0.4	0.8	-

Three different heating equipment and related energy sources (gas, oil, electricity) were selected to evaluate the influence of the overall efficiency of alternative equipment (gas boilers, oil furnaces, electric heat pumps). Three different economic scenarios were considered, including alternative trends of inflation rates, discount rates, prices escalation rates. Finally, three Tax deduction measures were considered (36-50-65%).

3.2 Impact of direct fiscal measures on global cost

Figure 2 and 3 show the box-plots of the expected net global cost (subtracted by the tax deduction) for each wall sqm, respectively during 30 and 10 years. Each graph reports the results obtained in the different economic scenarios (RE, IG, DE) and with the alternative energy sources, identified by colors. These graphs allow to identify the uncertainty ranges and the median values of the economic indicators for the insulation systems considered. As shown, these results are associated with considerable uncertainty, if we consider that the outcome is included within the ranges of the blue box plots only with a 50% probability. What in general emerges in all economic scenarios and time periods is that the wall renovation coupled to the electricity energy scenario is the one able to guarantee minor Global Costs, followed by the natural gas scenario.

Comparing the global costs median values in a "regular growth" economic scenario and gas energy scenario (cases 1 to 4), results show that, from the user perspective, as expected, the increase of the tax deduction level has a positive impact on the global costs. For these cases, 30-years net global costs range, as mean value, from about 122 /m^2 (no tax deduction) to 112 /m^2 (36%) to 108 /m^2 (50% tax deduction) to 104 /m^2 (65% tax deduction). The standard deviation for all these cases is about 13 /m^2.

IOP Conf. Series: Materials Science and Engineering **415** (2018) 012003 doi:10.1088/1757-899X/415/1/012003

The different macro-economic scenarios affect the results, by increasing the Global Cost on average by 12% (deflation scenario) or decreasing it by 5% (intense-growth scenario).

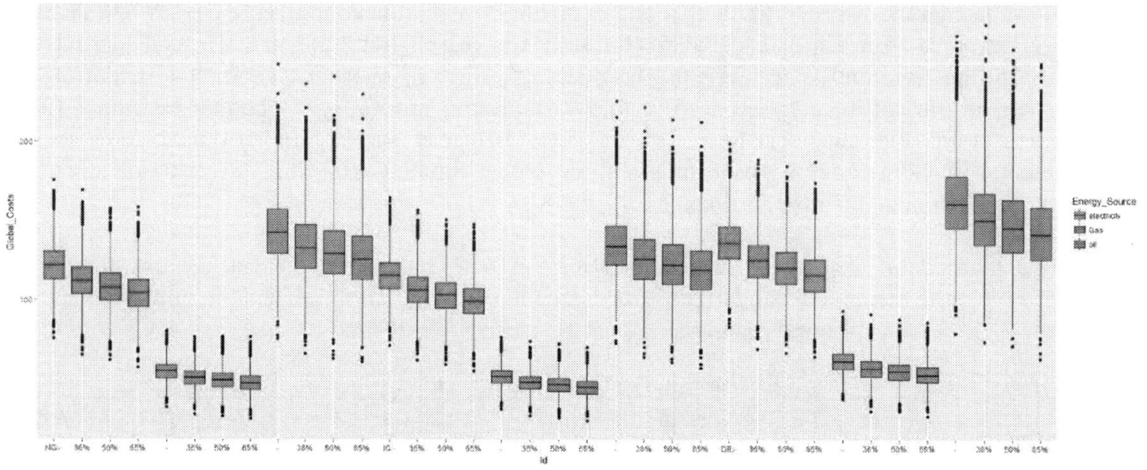

Figure 2. Boxplot of cases studies characterized by a calculation period of 30 years

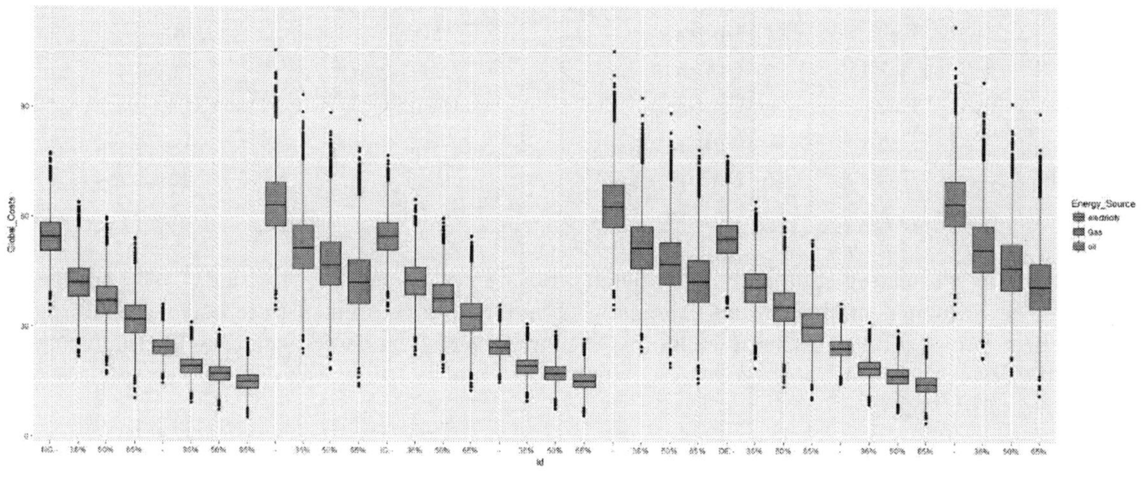

Figure 3. Boxplot of cases studies characterized by a calculation period of 10 years

IOP Conf. Series: Materials Science and Engineering **415** (2018) 012003 doi:10.1088/1757-899X/415/1/012003

The PDFs of the cases with and without deductions (from 1 to 4) are reported in **Figure 4**. They are almost overlapped, showing that tax deduction are always positive for the user, but if results variances due to data input uncertainties are taken into account, deduction policies may produce few differences on the results.

Different energy sources, and the related tariffs and energy efficiency, have the effect of increasing the difference between the cases with and without deduction fees. **Figure 5** shows the CDFs of the cases characterized by 65% tax fee deduction with different energy sources. In case of electricity as energy source, deduction allows to cover more than 100% of the investment cost in all economic scenarios. The median values of global costs range for these cases from about 104 /m^2 (gas) to 47 /m^2 (electricity) to 127 /m^2 (oil). This depends on the low electricity tariffs in Italy and the high equipment efficiency set as data input.

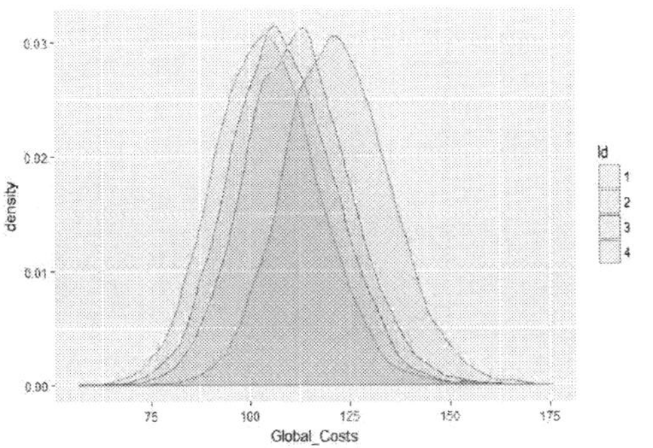

Figure 4. Global Costs PDFs for cases 1 (no deduction), 2 (36%), 3 (50%), 4 (65%).

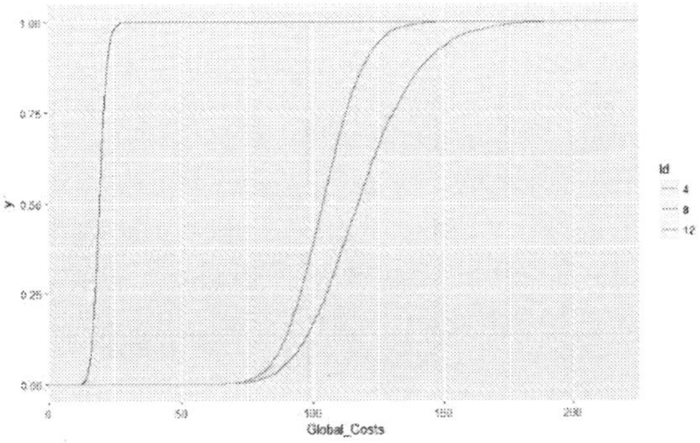

Figure 5. Global Costs CDFs for cases 4 (gas), 8 (electricity), 12 (oil) - "normal growth" scenario

4. Conclusions

This work presents the development and the application of a probabilistic LCC method to a case study of historic building retrofitting through internal insulation. Through a Monte-Carlo based LCC methodology for uncertainty analysis, this paper particularly focuses on the impact of fiscal measures. We compared under different economic scenario the economic impact of tax deduction actions actually in use in Italy, but potentially valuable in other UE countries. Results demonstrate the great potential of probabilistic assessment methodologies in providing more realistic information about results uncertainties. Applying the developed methodology, we obtained more realistic information about the opportunity to introduce very high tax deduction fees as in Italy. Results show that high deduction fees have a little impact on the increase of the probability to spend less money in the LCC, in a user perspective and that is better to maintain for longer period lower deduction fees, in a policy maker perspective. The developed methodology comprised the characterization of several UE potential economic scenarios (regular growth, inflation, deflation, stagflation), hence it is applicable to any UE country to evaluate alternative fiscal measures.

Acknowledgements

This project has received funding from the European Union's Horizon 2020 research and innovation programme under grant agreement No 637268.

IOP Conf. Series: Materials Science and Engineering **415** (2018) 012003 doi:10.1088/1757-899X/415/1/012003

References

[1] Katri Kosonen, G. N. *Taxation papers. The role of fiscal instrumentis in environmental policy.* (2009).

[2] Keay, M. & Hammes, K. *Fiscal policy for decarbonisation of energy in Europe.* (2017).

[3] Lindkvist, C., Karlsson, A., Sørnes, K. & Wyckmans, A. Barriers and Challenges in nZEB Projects in Sweden and Norway. *Energy Procedia* **58,** 199–206 (2014).

[4] *EN 15459:2007 Energy performance of buildings - Economic evaluation procedure for energy systems in buildings.*

[5] Ilg, P., Scope, C., Muench, S. & Guenther, E. Uncertainty in life cycle costing for long-range infrastructure. Part I: leveling the playing field to address uncertainties. *Int. J. Life Cycle Assess.* 1–16 (2016). doi:10.1007/s11367-016-1154-1

[6] Rahman, S. & Vanier, D. J. Life cycle cost analysis as a decision support tool for managing municipal infrastructure. in *CIB 2004 Triennial Congress* 1–12 (2004).

[7] Pittenger, D., Gransberg, D., Zaman, M. & Riemer, C. Stochastic Life-Cycle Cost Analysis for Pavement Preservation Treatments. *Transp. Res. Rec. J. Transp. Res. Board* **2292,** 45–51 (2012).

[8] Wang, N., Chang, Y.-C. & El-Sheikh, A. Monte Carlo simulation approach to life cycle cost management. *Struct. Infrastruct. Eng.* **8,** 739–746 (2012).

[9] Das, P., Van Gelder, L., Janssen, H. & Roels, S. Designing uncertain optimization schemes for the economic assessment of stock energy-efficiency measures. *J. Build. Perform. Simul.* **1493,** 1–14 (2015).

[10] Goh, Y. M., Newnes, L. B., Mileham, A. R., McMahon, C. A. & Saravi, M. E. Uncertainty in through-life costing-review and perspectives. *IEEE Trans. Eng. Manag.* **57,** 689–701 (2010).

[11] Menassa, C. C. Evaluating sustainable retrofits in existing buildings under uncertainty. **43,** 3576–3583 (2011).

[12] Sesana, M. M. & Salvalai, G. Overview on life cycle methodologies and economic feasibility for nZEBs. *Build. Environ.* **67,** 211–216 (2013).

[13] Gluch, P. & Baumann, H. The life cycle costing (LCC) approach: a conceptual discussion of its usefulness for environmental decision-making. *Build. Environ.* **39,** 571–580 (2004).

[14] Burhenne, S., Tsvetkova, O., Jacob, D., Henze, G. P. & Wagner, A. Uncertainty quantification for combined building performance and cost-benefit analyses. *Build. Environ.* **62,** 143–154 (2013).

[15] TABULA Project. Typology Approach for Building Stock Energy Assessment. 16 (2012).

IOP Conf. Series: Materials Science and Engineering **415** (2018) 012004 doi:10.1088/1757-899X/415/1/012004

NZEBs built of elements based on styrofoam re-granulate

M Wesołowska[1], **P Szczepaniak**[1], **J Gajewski**[2]

[1] UTP University of Science and Technology in Bydgoszcz, Department of Building Engineering and Building Physics, Al. prof. S.Kaliskiego 7, 85-796 Bydgoszcz, Poland

[2] UTP University of Science and Technology in Bydgoszcz, Department of Building Construction, Al. prof. S.Kaliskiego 7, 85-796 Bydgoszcz, Poland

Maria.Wesolowska@ utp.edu.pl

Abstract. In the article there was introduced an idea of NZEB constructed from large size wall elements mounted with dry connection and horizontal joints filled with polyurethane mortar. The proposed element was 50 cm thick with set of layers made of styro-concrete with EPS re-granulate and EPS-032 separation layers. The result was a solution with high thermal isolation and traditional division for construction and isolation zones. Additional elements of the system are layered lintels and intermediate moldings with increased pressing strength. The material solutions of the system elements are the effect of combined construction and thermal analyses.

1. Introduction

Expanded polystyrene is currently the most often used material in thermal insulation of construction. During modernization of thermal insulation, repairs, and demolition work a lot of this material is amounted. Ono of possibilities for its use is to recycle it as ecostyrene[5].In this form it can be applied as filling for construction, insulation for construction partitions, elastic bedding for roads and sport arenas, and as insulation for thermal exchangers and swimming-pools.Ecostyrene is also used as water-protective insulation[3](styrozolcreated during solution of polystyrene with solvent, and next enriched with plasticizers and fillers for water-resistant features). There is also re-granulate on the market, sized from about 2 to 10 mm[6] used as bulk material or concrete filler[4]. The resulted concrete is weaker[1], but with enhanced thermal properties.It opens possibility of using this material for production of elements for NZEB(nearly zero-energy building).According to the Directive 2010/31/EU[2] 'nearly zero-energy building' means a building that has a very high energy performance (...). The nearly zero or very low amount of energy required, should be covered to a very significant extent by energy from renewable sources, including energy from renewable sources produced on-site or nearby.

2. Solution description

The goal of this project was creation of an alternative system of low-budged passive building using production and execution work wastes of expanded polystyrene (EPS). Multi-layer wall elements were designed with vertical tongue-and-groove joints made from two materials:

- 50 mm thick boards from styro-concrete of900density classwith 30% addition of organic filler of the EPS re-granulate, from crushed wastes,

- 50 and 200mm thick boards from the EPS 70-031 graphite styrofoam.

The layers are bound with polyurethane mono-component polyurethane glue and additionally screwed with steel tie-rods of 6 mm in diameter, equipped with polyamide collar nuts(Figure 1). The tie-rods ensure the proper pressure during the glue settingand additional reinforcement of the element during construction. After building-in the protruding part of a tie-rod is cut out and masked with a styrofoam plug.

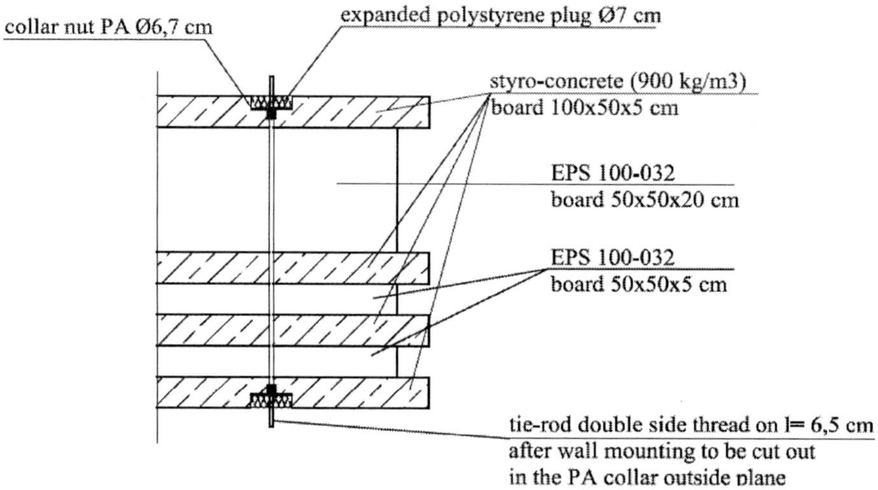

Figure 1. Tie-rod solution in a multi-layer wall element.

The set includes: base wall elements sized 105x50x50cm (Figure2), supplement half elements (longitudinal 105x25x50 cm,and transverse 55x50x50cm) and quarter base elements reinforced with ferroconcrete plate under the lintel support.In the elements there was preserved the traditional division of functions:

- construction layer 25cm thick,
- thermal insulation layer 20cm thick,
- styro-concrete of 5cm as a stabile base for thin-layered plate with a steel reinforcement net.

Figure 2. Ekoblok wall system elements.

The fixed measurements allow to erect walls with required translation of vertical contacts by using a half-element and without necessity of cutting:

- window sill level: 2 basic elements (2x50=100cm),

- lintel level: 4 basic elements + quarter-element,
- ring beam level: 6 basic elements or 5 basic elements + quarter-element.

The proposed layer layout of a large-block element will allow to support the ceiling and building ring beam with necessary insulation minimizing influence of the thermal bridge in this node. Open joints created by vertical hole edges will be filled with styrofoam inserts on building site which minimizes heat loss in area of jambs. The system is prepared for using modular hole widths: 90 cm, 120 cm, 220 cm, and 270 cm. This solution allows using monolithic and prefabricated door lintels (Figure 3).Basic solution of building setting predicts linear foundation with both sides of perimetric insulation.

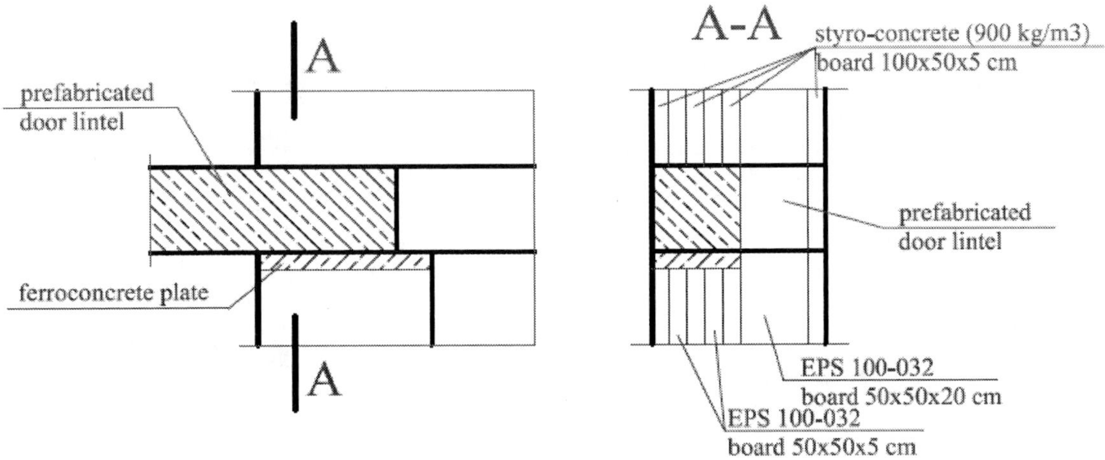

Figure 3. Ekoblok system node of door lintel support.

3. Construction designs analysis

The system is designated for erecting of passive buildings. The basic requirements for this type of buildings are listed in Table 1. The heat permeation coefficient of a wall element is U=0.100W/(m^2·K). The division of a construction layer in the element ensures continuity of thermal insulation in most nodes of walls. However, the construction solution used (steel tie-rods, joint, tying of wall with layer elements in the corner, door lintel support) interfere with the specific thermo-insulating layer.

Table 1. Minimal energetic requirements for buildings

Energy standard	**NZEB**	**Passive[7]**
	thermal transmittance U [W/m^2·K]	
wall (T_i>16 °C)	0.20	0.10
roof	0.15	0.10
slab-on-ground	0.30	0.12
	linear thermal transmittance Ψ [W/m·K]	
all thermal bridges	no requirements	recommended value: 0.01 maximum value: 0.05
	energy factors	
primary energy factor [kWh/(m^2·rok)]	70	70
heating energy factor [kWh/(m^2·rok)]	-	15
T_i - inertia temperature		

IOP Conf. Series: Materials Science and Engineering **415** (2018) 012004 doi:10.1088/1757-899X/415/1/012004

3.1. Analysis of steel tie-rod influence

In connection with reaching the assumed limit value of a heat transfer coefficient for the element it was necessary to propose a tie-rod design which minimized its influence during the wall functioning. In order to achieve this goal there was assumed a piercing the wall element with a steel connecting rod in half of the wall element volume (Figure 4). The spot heat transfer coefficient $\chi$was calculated with the formula:

$$\chi = L_{3D} - \sum_{i=1}^{N_i} U_i \cdot A_i - \sum_{j=1}^{N_j} \Psi_j \cdot l_j \qquad (1)$$

where:

L_{3D}–heat coupling coefficient from calculation of 3-D component, separating the two discussed environments,

U_j–heat transfer coefficient of 1-D j component, separating the two discussed environments,

l_j–length along which the value of U_j is applied.

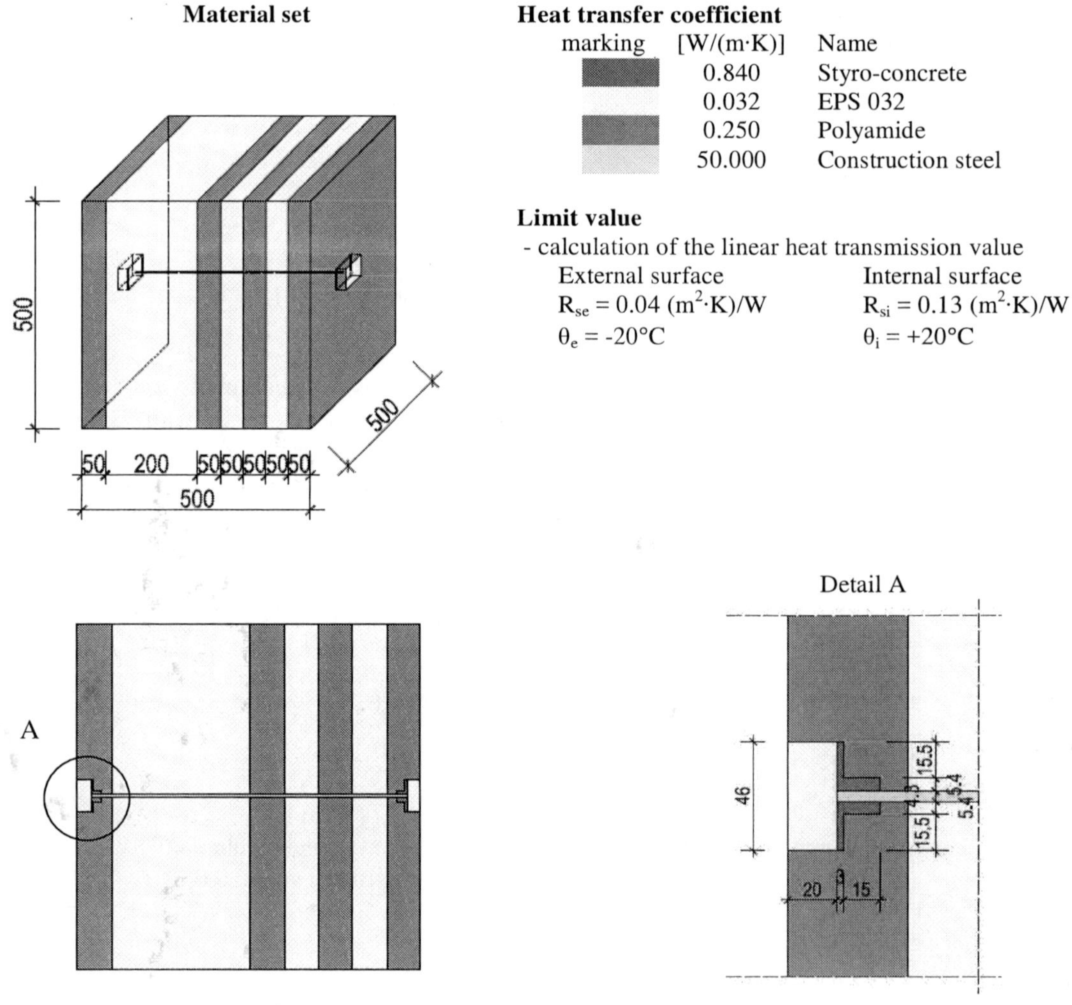

Figure 4. Calculation modelfor definition of mechanic connector influence.

In the discussed case there are no values of linear heat conductivity coefficient Ψ, so the value of $\chi$is:

$$\chi = L_{3D} - \sum_{i=1}^{N_i} U_i \cdot A_i = 0.028 - 0.100 \cdot 0.25 = 0.002\ W/K \tag{2}$$

The coefficient $U_c$for the basic element with tie-rods amounts to 0.104 W/(m^2·K).

3.2. Selection of material set for joints

The joint passing through all layers of wall element is in contact with internal and external environment which interfere the continuity of the isolated thermal insulation layer.On the horizontal joint of wall elements there is created a linear bridge transferring into spatial bridge in the corner. In this place the geometric bridge and mutual penetration of horizontal and vertical joints overlay (Figure 6).

In order to limit the influence of the spatial thermal bridge, thin joints were proposed (3 mm). Two material solutions were discussed:

- Mineral mortar for thin joints (λ = 0.900 W/(m·K)),
- Polyurethaneglue (λ = 0.036 W/(m·K)).

In a 2D analysis of joining wall elements with a supporting joint, with assumed thickness of 3 mm, heat conductivity of mortar has secondary significance. The joint does not influence the heat transfer coefficient value U for the wall. The influence of mortar is visible in the2Dand 3D node model of the corner (Figure 5, Table 2).

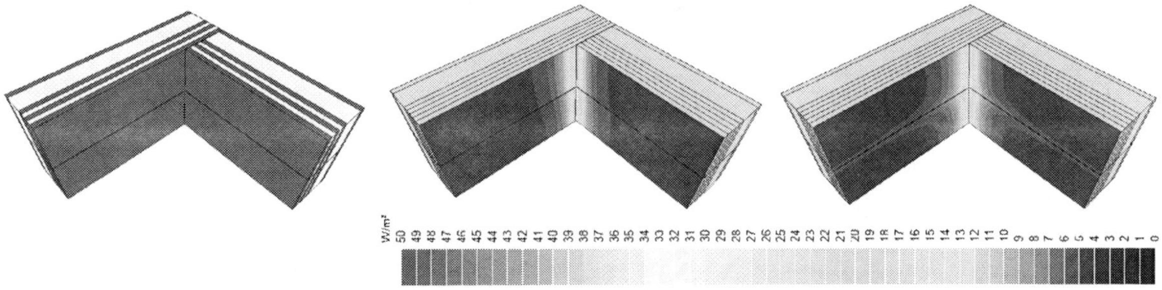

Figure 5. Mortar influence on heat flux distribution in the joint.

Table 2. Evaluation of heat quality for connecting node of external walls

	Mineral mortar for thin joints	Polyurethane glue
$\Psi_{vertical}$ [W/(m·K)]	0.056	0.050
$\Psi_{horizontal}$ [W/(m·K)]	0.014	0.000
χ [W/K]	0.040	0.028

In both material solutions of the mortar, the obtained value of the linear heat transfer coefficient for the analyzed node is below the acceptable value (Table 2). However, the node geometry and the path of materials in the node do not allow to reach the recommended value. The picture of the stream field in a3D model (Figure 5) indicates that it is more proper to use polyurethane mortar both in the horizontal and vertical joints.

3.3. Solution of door lintel supporting on wall

The places of door lintel support on walls are zones especially threatened with damage. The essential issue is to define the minimum depth of support ensuring safe transfer of load from the door lintel to walls. To achieve this goal there was performed a detailed material strength analysis. A FEM three-dimensional model of the door lintel was prepared using volume elements with a linear shape function.

IOP Conf. Series: Materials Science and Engineering **415** (2018) 012004 doi:10.1088/1757-899X/415/1/012004

a)

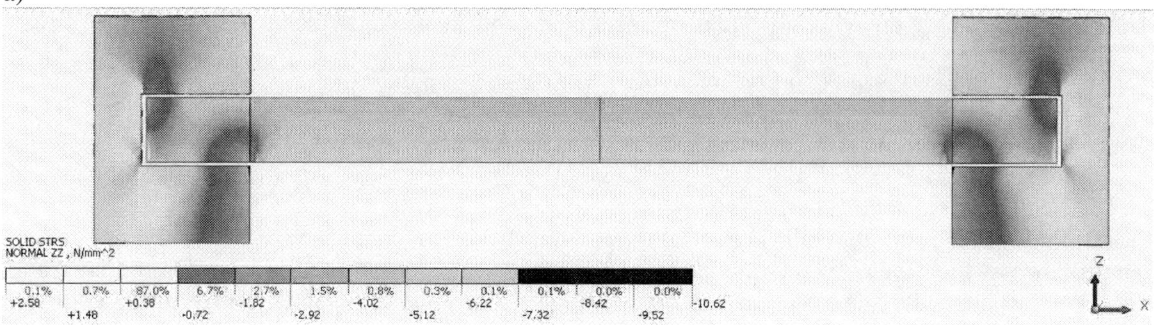

b)

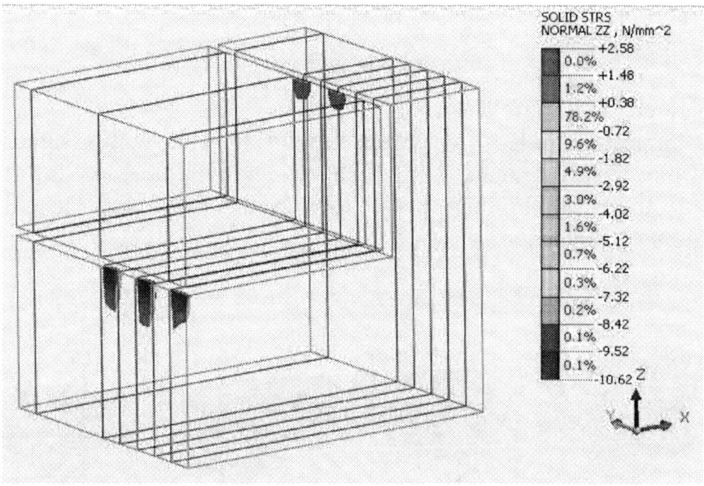

c)

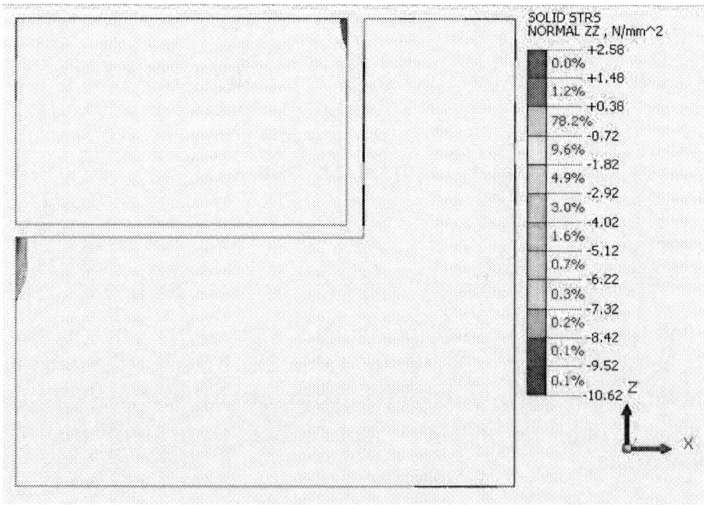

Figure 6. Results of material strength calculations for a 270cm door lintel, with 40 cm support a) entire lintel (strain$\sigma_{ZZ,max}$= -10.62 MPa), b) axonometric projection (σ_{ZZ} from -7 to -10.62 MPa), c) side view (σ_{ZZ} od -7 do -10.62 MPa).

In analyses there was assumed a linear and resilient material model for styrofoam and concrete. In thethe FEM model MES the following values of Young's modulus were assumed as: 30GPafor concrete, 15 GPafor styro-concrete, 1.5MPa for styrofoam.
Based on material strength analysis the minimum depths of door lintel support were defined:

- 40 cm for the case of 270 cm holedoor lintel,
- 30 cm for the case of 220 cm holedoor lintel,
- 20 cm for the case of 120 cm holedoor lintel.

It was also defined that in the support edge zone there is a strain concentration which can lead to a creation of scratches in styro-concrete boards. Thus the half-element was modified by insertion of a horizontal ferroconcrete plate in construction zone.
The created material set introduces a heat flux concentration in a hole corner area (Figure 7). This concentration results in lowering of temperature on the surface, but it is not threatened by mildew development.

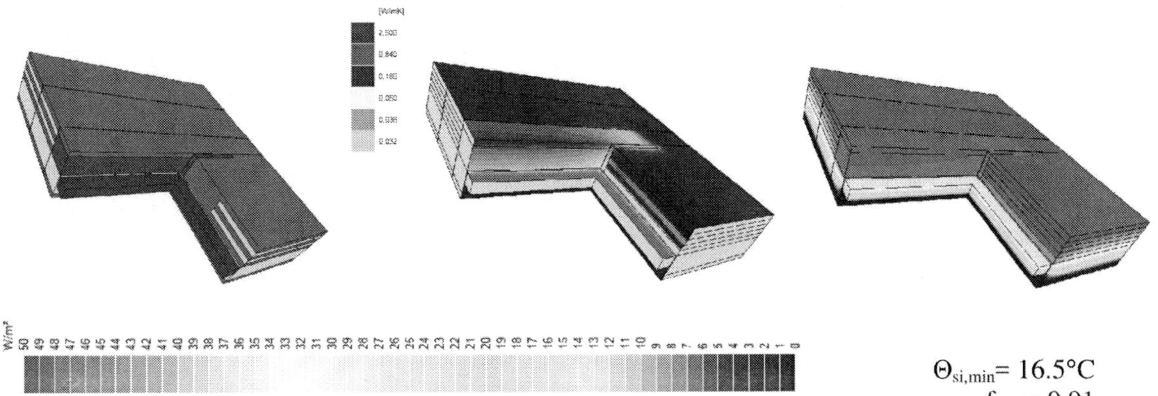

$\Theta_{si,min}$= 16.5°C
f_{Rsi} = 0.91

Figure 7. Distribution of heat flux in the door lintel construction node

4. Summary

The heat calculations performed for the element indicate that the thermal bridges generated by spot anchoring do not introduce correction whichresults in exceeding the limit values of heat transfer coefficient U. The translation of boards forms a tongue and groove joint which allows dry connection and air tightness of the wall.The thin-layered joint does not affect the heat transfer coefficient U for the wall. The material solutions of the mortar used for connection of wall elements are decided by spatial bridges. The bearing zone distinguished in the element has a layer layout. The introduction of necessary construction solutions,though they are closing within its borders has its consequences in heat transport.This is hardly visible in a typical 2D analysis. The flux disturbances are noticeable in a spatial temperature field.

The sizes of elements and erection technology gives possibility for execution of a passive building in shorter time than for traditional solutions.It also utilizes industrial wastes.However, because of its construction and shape it has not equivalents in standards. Its implementing into building practice is conditioned by detailed regulations. In this area the binding regulation is the Regulation of the European Parliament and of the Council (EU) No 305/2011 of 9March 2011 laying down harmonized conditions for the marketing of construction productsand repealing the Council Directive 89/106/EEC.

References

[1] AmiantiM.,et al Recycling of EPS: A new methodology for production of concrete impregnated

with polystyrene(CIP) Cement & Concrete Composities 2008, 30, pp 23-28

[2] *Directive 2010/31/EU of the European Parliament and of the Council of 19 May 2010 on the energy performance of buildings (recast)*, Official Journal of the European Union, 2010.

[3] Geryło, R.,*Styrofoam recycling – needs and possibilities for application (Recykling styropianu – potrzeby i możliwości zastosowania)*,InżynierBudownictwa, No 3, 2013, pp. 42-45.

[4] Haghi, A.K, et al., *Applications of expanded polystyrene (EPS) beads and Polyamide-66 in civil engineering* Part1: Lightweight polymeric concrete. Composite Interfaces No 13, 2006, pp. 441-450.

[5] Juraszek, J.Grzywa, A.Walusiak, H.*The concept of deformation tests for new generation prefabricatebuildingsperformed with FBG fiberglasssystems (Koncepcja badań odkształceń budynku z prefabrykatów nowej generacji za pomocą systemów światłowodowych FBG)*, Materiały Budowlane No 5, 2017, pp. 33-34.

[6] Kaya A. et al 2016 *Properties of concrete containing waste expanded polystyrene and natural resin.* Construction and building materials, 105, pp 572-578

[7] Priority Program of the NFOŚIGW (National Fund for Environment Efficiency and Water Management), *Energy Efficiency Programme*, available online:http://nfosigw.gov.pl/en/priority-programmes, 2015.

Acknowledgments

The research was done on financial support by Smart Growth Operational Programme 2014-2020, Implementation of innovation by SMEs. Project no. POIR.02.03.02-04-0017/17 for PPU Ekobud, sp.zo.o. Grudziądz.

The choice of the optimal level of the primary energy index for the collective residential buildings

M Basińska[1,a] **and H Koczyk**[1]

[1] Faculty of Civil and Environmental Engineering, Institute of Environmental Engineering, Poznan University of Technology, 4 Berdychowo St., Poznan 60-965, Poland
[a] Corresponding Authors: malgorzata.basinska@put.poznan.pl

Abstract. This paper presents an optimal method of choosing a primary energy ratio, depending on the solutions regarding the building's technical equipment. A global cost method was used in the analysis, as that method allows for determining the requirements in relation to the type of the building and the way of energy supply. Analysis regards five –storey building, where 4 floors were repetitive. The building represents a standard of the collective residential buildings. 4 chosen scenarios were analysed: basic one (thermal protection requirements following the requirement valid in Poland in 2008 – **SEB**), presently valid (including a modernisation helping to fulfil the actual building rules) – **SEB***, low energy building – **LEB** and passive building standard – **VLEB**.

1. Introduction

In accordance with Article 5 of valid Polish Building Law [1] the parties of the building process are required to undertake the activities necessary to minimize a building's negative impact on the environment. The mentioned above activities shall include designing process, building process and exploitation of the buildings. It refers to such stages as: building material production, structure erection, building's exploitation, it's demolition and utilisation or after-demolition material recovery.

In order to evaluate the efficiency of the applied in the building solutions one can use an analysis of building's life cycle costs related to the calculation of the future cash flows. Recently, the most popular method of evaluation of energy performance of the building is LCC - global cost method – C_G. "Optimum level in relation to the costs" means such an energy performance efficiency level, which provides the lowest costs during the estimated life cycle. The lowest costs are established, taking under consideration the investment costs related to the energy, maintenance and exploitation costs including the energy and savings costs. This is why an analysis optimal in terms of the costs is a powerful tool promoting the energy related modernisation of the buildings [2-8]. It provides a possibility of choosing the optimal solution for a low value of the global cost. Kurnitski and others have prepared a number of the papers regarding the mentioned above subject, presenting the analyses of residential and office buildings [9-11]. The analyses of the residential buildings are focused mainly on establishing the optimum value of primary energy ratio during a typical way of usage (one-family or multi-family buildings). The collective residential buildings are the special cases of the residential buildings. A share of energy necessary for DHW preparation increases significantly in such type of objects. It results mainly from smaller surface of the apartments and lower surface ratio per one inhabitant. In this paper the optimal value of primary energy ratio was established for a collective residential building based on the global cost method, depends on the applied solutions for the technical equipment. An influence of the primary assumptions on the gained value of primary energy ratio was also analysed.

2. Description of method

In the analyses conducted for the collective residential building and its technical equipment an algorithm of global costs calculation was used, according to the standard regarding the economic evaluation of the building's energy systems [12]. The method was used to establish an optimal level of primary energy and the analysis of primary parameters impact on the searched optimum.

2.1. Establishing usable, final and primary energy

An usable energy was established by means of the Polish requirements resulting from building's energy performance certificate law [13]. The monthly balance sheet method is taking into account the heat losses resulting from transfer or ventilation as well as the solar and indoor heat gains.

2.2. Establishing global cost

The global costs of the building and its elements are calculated as a sum of initial investment costs and discounted annual exploitation costs in the calculation period [12]. The analysis of global costs was conducted using the below presented formula:

$$C_G(\tau) = C_{in,inv} + \sum_{j=1}^{j_x}\left[\sum_{i=1}^{\tau}\left(C_{a,i}(j) \cdot R_d(i)\right) - V_{f,\tau}(j)\right] \quad (1)$$

where: $C_G(\tau)$, global cost in relations to initial year, PLN; $C_{in,\ inv}$, investment costs, PLN; τ, calculation period, a; $C_{a,i}(j)$, component's annual costs in i-year, PLN; $R_d(i)$, discount rate (for i-year), -; $V_{f,\ \tau}(j)$, residual value, PLN.

3. Sensitivity analysis

This paper presents a sensitivity analysis for establishing the energy optimum by means of global cost method, including the changeability of the initial parameters:

- installation solutions – 4 heat sources were analysed: gas central boiler (GCB), district heating (DH), ground heat pump (GHP), pellet boiler (PB),
- money value – annual inflation rate, WIBOR (VM),
- increase of investment costs (IC),
- electricity price development (EPD).

3.1. Description of the analysed variants – building's technical equipment

3.1.1. *Quality of building's envelope.* Table 1 presents the analysed variants of changeability of heat transfer coefficients for building's envelope, participation of thermal bridges and air tightness of the building responding with the process of achieving the requirements regarding the low energy building. SEB* variant is approximately 15% better than SEB variant.

Table 1. Heat transfer coefficient of the external partitions U ($W \cdot m^{-2} \cdot K^{-1}$), thermal bridges (%) and building's tightness depending on energy standard (h^{-1}).

Variant	Wall	Ground floor	Flat roof	Door	Window	Thermal bridges	Air tightness
SEB	0.30	0.45	0.25	2.60	1.80	15	3.0
SEB*	0.25	0.35	0.20	1.80	1.30	15	3.0
LEB	0.22	0.30	0.16	1.80	1.30	10	3.0
VLEB	0.15	0.20	0.12	1.30	0.90	5	1.0

3.1.2. Quality of ventilation system. As a part of the conducted analyses the following variants of ventilation systems were foreseen [14]:

- V1 natural, the designed ventilation air stream was established at the level of 5050 $m^3 \cdot h^{-1}$
- V2 hybrid, for the calculations of hybrid ventilation it was assumed to decrease the air stream by 59% in relations to the required heat stream, appointed due to the hygiene reasons
- V3 mechanical, exhaust, air flow control was installed according to a load resulting from usage profile, decrease of air stream to 50% was assumed in relations to the required heat stream (following the hygiene reasons)
- V4 mechanical, supply and exhaust in relations to the load, heat recovery 50%,
- V5 mechanical, supply and exhaust in relations to the load, heat recovery 70%,
- V6 mechanical, supply and exhaust in relations to the load, heat recovery 80%.

3.1.3. Quality of heat source. for all analysed variants it was assumed to use 4 systems of energy supply to the building based on:

- boiler room with gas condensing boiler (GCB variant),
- district heating supply system (DH variant),
- boiler room with a ground source heat pump (GHP variant),
- boiler room with pellet boiler (PB variant).

3.1.4. Analysed variants. Table 2 presents the analysed variants depending on the quality of the building's envelope and ventilation system.

Table 2. Summary of analysed technical variants.

TV_1	**TV_2**	**TV_3**	**TV_4**	**TV_5**	**TV_6**	**TV_7**
SEB	SEB*	LEB	LEB	VLEB	VLEB	VLEB
V1	V2	V3	V4	V4	V5	V6

3.2. Description of analysed economic variants

3.2.1. Value of money – (VM). The changeability of discount rate and annual inflation rate according to table 3 were assumed in the analyses. A reference point in the analysis is the values described as „*base". The value of money for base parameters was assumed according to the economic data valid on December 2015 – from 3 months period.

Table 3. Variants of the interest rates (inflation or WIBOR), (%).

	VM_1	VM_1,5	VM_base	VM_2	VM_2,5	VM_3
annual inflation rate -R_i	1.0	1.5	-0,7	1.8	2.5	2.6
WIBOR	4.0	4.3	1.75	4.1	4.6	4.4
discount rate –R_R	3.0	2.75	2.47	2.25	2.00	1.75
VAT	23	23	23	23	23	23

In the analyses of the impact of a changeability of interest rates level on the value of global cost the coefficient of exploitation prices increase and change of energy prices was assumed to be 0%.

3.2.2. Increase of investment costs (IC) in time. In the conducted analysis the initial costs of the investment (year zero) were defined according to the status for I quarter of 2012 as the average costs incurred for building's erection and handing over for use, according to the material outlays catalogue SekoCenbud. The individual unit investment costs were grouped into following categories:

- IC_1 –construction elements influencing the energy consumption – sun blinds,
- IC_2 –investment costs of the installed devices related to energy consumption – chimney, heat source, central heating system, DHW preparation or ventilation,
- IC_3 –investment building costs having an impact on the energy consumption – windows, thermal insulation,

- IC_4 –investment building costs – body of the building, water and sewage systems, gas system and electricity.

Depending on the technical variant the investment costs related to the building and heat system are the subjects of modification.

Table 4. Unit building costs for base variant IC_base, PLN·m^{-2}.

		TV_1	TV_2	TV_3	TV_4	TV_5	TV_6	TV_7
Total investment costs divided into parts regarding building, systems and energy source								
GCB	BUILDING	1884	1894	1909	1875	1901	1901	1901
	SYSTEMS	239	255	225	561	561	559	558
DH	BUILDING	1883	1893	1907	1873	1899	1899	1899
	SYSTEMS	239	257	227	566	565	564	562
GHP	BUILDING	1883	1893	1907	1873	1899	1899	1899
	SYSTEMS	272	287	251	587	588	587	585
PB	BUILDING	1884	1894	1909	1875	1901	1901	1901
	SYSTEMS	224	243	214	554	556	554	549

The sensibility analysis includes the coefficients of investment costs increase in time according to table 5, considering in that way the behaviour of trading parties.

Table 5. Variants of growth rates of investment costs over time.

IC_1	IC_base	IC_2	IC_3
90% base	100%	110% base	120% base

The analyses of impact of investment costs changeability on the value of global cost do assume the basic financial parameters but do not assume the electricity price development.

3.2.3. Energy price development (EPD). In Polish market economy it is difficult to estimate the average annual rates of energy and it's carriers prices increase, so some scenario of the changes was assumed, according to table 6.

Table 6. Variants of rising energy costs over time.

	EPD_base	EPD_1	EPD_2	EPD_3
gas	0.0 %	0.5 %	1.8%	3.3%
electricity	0.0 %	0.5 %	1.2%	2.5%
district heating	0.0 %	0.5 %	0.8%	1.3%
pellets	0.0 %	0.5 %	1.1%	2.4%

In the analyses of the impact of energy costs changeability on the value of global costs the coefficient of exploitation costs increase at level 0% was assumed.

3.3. Multi-variants analysis

Considering the changeability of financial end economic coefficients according to tables (tab. 3, tab. 5, tab. 6) the analysis was taking into account the relations between the investment costs (IC) and electricity price development (EPD) for two money values in time:

- base parameters (VM_base),
- variant with the highest inflation value (VM_3) suits the lowest discount rate.

4. Description of analysed building

4.1. Building and architectural characteristics

The analysis was conducted for a five-storey building with 4 repetitive floors, representing the class of the collective residential buildings. The building is rectangular with the dimensions: 53.9 m and 11.9 m, the height of one level is 2.5 m and the building shape coefficient A/V_e=0.33 m^{-1}. In the heated part of building $A_f = 2483\ m^2$ with the regulated temperature there is 77 double rooms including toilets. The surface of a typical room (with toilet) - 17.9 m^2. The assumed load of the building is at level of 80%, the unit consumption of hot water is 25 dm^3 per person and per day.

4.2. *Energy characteristics*

4.2.1. Usable energy. Usable energy for heating and ventilation $Q_{H,nd}$ was determined assuming the following:

- the internal exploitation temperature +20^{0}C,
- unit heat capacity at level of 370000 $J{\cdot}K^{-1}{\cdot}m^{-2}$
- indoor heat gains based on the building's usage profile at the level of 3.5 $W{\cdot}m^{-2}$,
- localisation in Poznan,
- sun heat gains not including shading of the building.

Table 7 presents the change of coefficient of usable energy depends on the given variant. Following the improvement of building characteristics a share of usable energy for DHW increases in the balance.

Table 7. The value of coefficient of heat losses, transfer and ventilation related heat loses as well as usable energy for a purpose of heating, ventilation and DHW preparation depending on given variant.

	H_{tr}/A_f $W{\cdot}m^{-2}{\cdot}K^{-1}$	H_{ve}/A_f $W{\cdot}m^{-2}{\cdot}K^{-1}$	$Q_{H,nd}/A_f$ $kWh{\cdot}m^{-2}{\cdot}a^{-1}$	$Q_{W,nd}/A_f$ $kWh{\cdot}m^{-2}{\cdot}a^{-1}$	t_{sg} h
TV_1	0.59	0.79	91	24	5875
TV_2	0.46	0.44	50	24	5196
TV_3	0.40	0.42	42	24	4992
TV_4	0.40	0.19	23	24	4318
TV_5	0.26	0.19	13	24	3898
TV_6	0.26	0.13	9	24	3602
TV_7	0.26	0.10	7	24	3367

4.2.2. Final energy. The coefficient of final energy was determined assuming the changeability of efficiency coefficients. Determining the final energy for the purpose of auxiliary energy, the time of auxiliary devices work was taken into consideration and, based on the catalogue data, the unit powers of technical devices were defined.

4.2.3. Primary energy. The primary energy was defined depending on the technical variant and the type of applied heat source. At the figures 1a to 1d the proportions between the particular elements of primary energy were presented – for the analysed building, the changing variants of energy performance characteristics and used heat sources. An improvement of building's envelope and supplied air flow control result in decrease of primary energy ration by 61% in case of gas boiler: from 151.6 to 59.2 $kWh{\cdot}m^2{\cdot}a^{-1}$, in case of district heating system by 58%: from 115.5 to 48.7 $kWh{\cdot}m^2{\cdot}a^{-1}$, in case of ground heat pump by 56% from 120.4 to 53.3 $kWh{\cdot}m^2{\cdot}a^{-1}$. In case of pellet boiler the balance sheet includes the share of electricity with an input coefficient equal, in Polish conditions, $w_{el}=3$ at level of 66.5% of total primary energy for variant TV_7. The improvement of energy characteristics up to variant TV_7 means simply a decrease of the primary energy from PE = 37.3 to 28.1 $kWh{\cdot}m^2{\cdot}a^{-1}$, which means a change at the level of 24.6% in relation to PE for TV_1 variant.

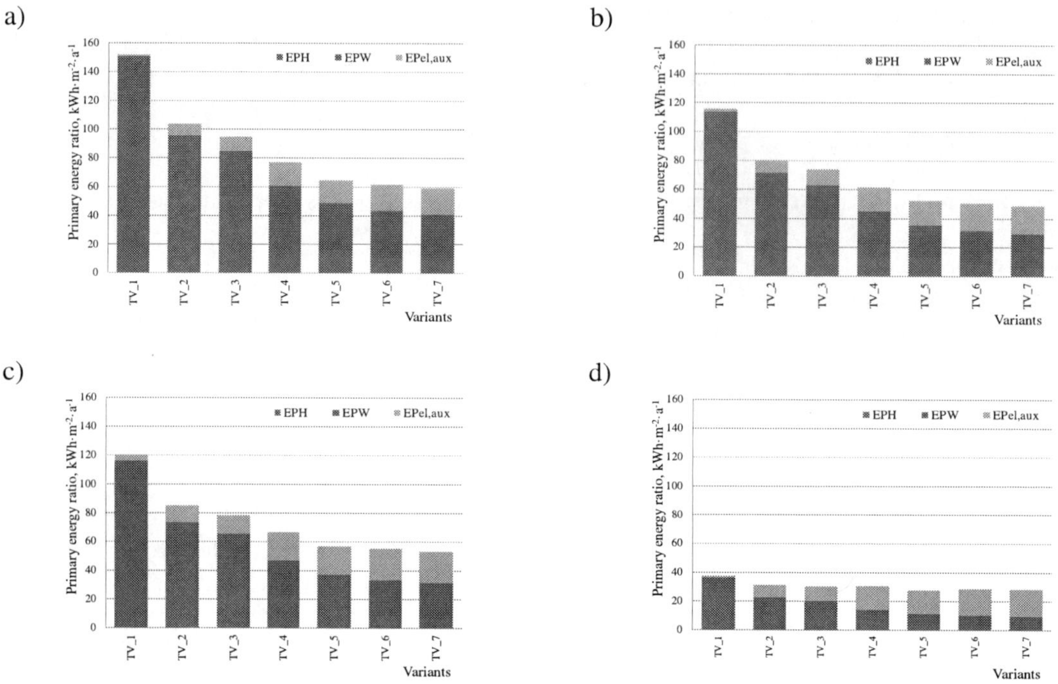

Figure 1. Changeability of primary energy in relations to the technical equipment of the building, dividing into the separate elements a) GCB – gas condensing boiler, b) DH – district heating, c) GHP – ground heat pump, d) PB – pellet boiler

5. Choice of EP_{opt} for basic economy coefficients

PE_{opt} (fig. 2), was defined, using the basic macro-economy parameters, applying the global costs method, based on 7 items of changeability of primary energy ration for the collective residential building and the analysed heat source referring the designated values to the changeability of global cost appointed for gas condensing boiler.

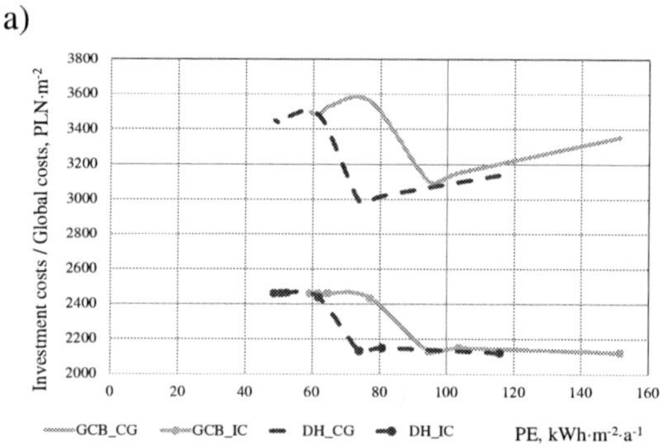

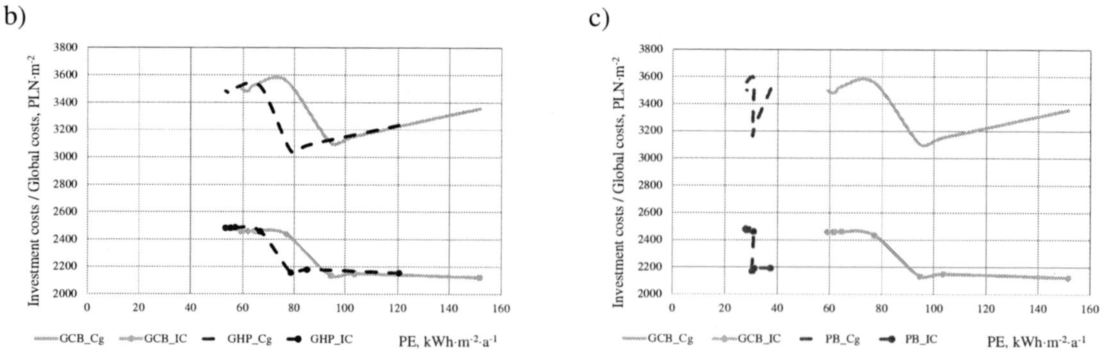

Figure 2. Variability of global costs for the building and the heat sources a) DH in relations to GCB, b) GHP in relations to GCB, c) PB in relations to GCB

IOP Conf. Series: Materials Science and Engineering **415** (2018) 012005 doi:10.1088/1757-899X/415/1/012005

Point PE_{opt}, optimal in terms of global costs, changes depending on the heat source used in the building. The lowest value it achieves in case of pellet boiler, the highest - for the gas boiler, the solution mostly used in Polish new boiler rooms. In case of district heating system, in comparison to the gas boiler, PE_{opt} shifts by 21 $kWh \cdot m^{-2} \cdot a^{-1}$, while the global cost for the optimum values decreases up to 115 $PLN \cdot m^{-2}$.

For the ground heat pump in relation to the gas boiler PE_{opt} changes by 16 $kWh \cdot m^{-2} \cdot a^{-1}$, when the global cost decreases by 50 $PLN \cdot m^{-2}$. For the pellet boiler PE_{opt} point is strongly indicated and equals 65.5 $kWh \cdot m^{-2} \cdot a^{-1}$, when the global cost increases by 120 $PLN \cdot m^{-2}$.

A difference of the global costs, in case of a different heat sources in the building, results, among the other, from the costs of heat source and costs of energy – their variability for the analysed variants were presented by means of a line with the tags. In the same time for each of the heat sources the value of global cost changes.

6. Sensitivity analysis

6.1. Sensitivity analysis of global costs results in relation to value of money (VM) in time

The detailed analysis of relationship between the global costs and character of financial parameters changes in reference to variant TV_1 and VM – base was presented on the below figures, respectively to the analysed heat sources. The analysis indicates an insignificant sensitivity to the change of financial coefficients.

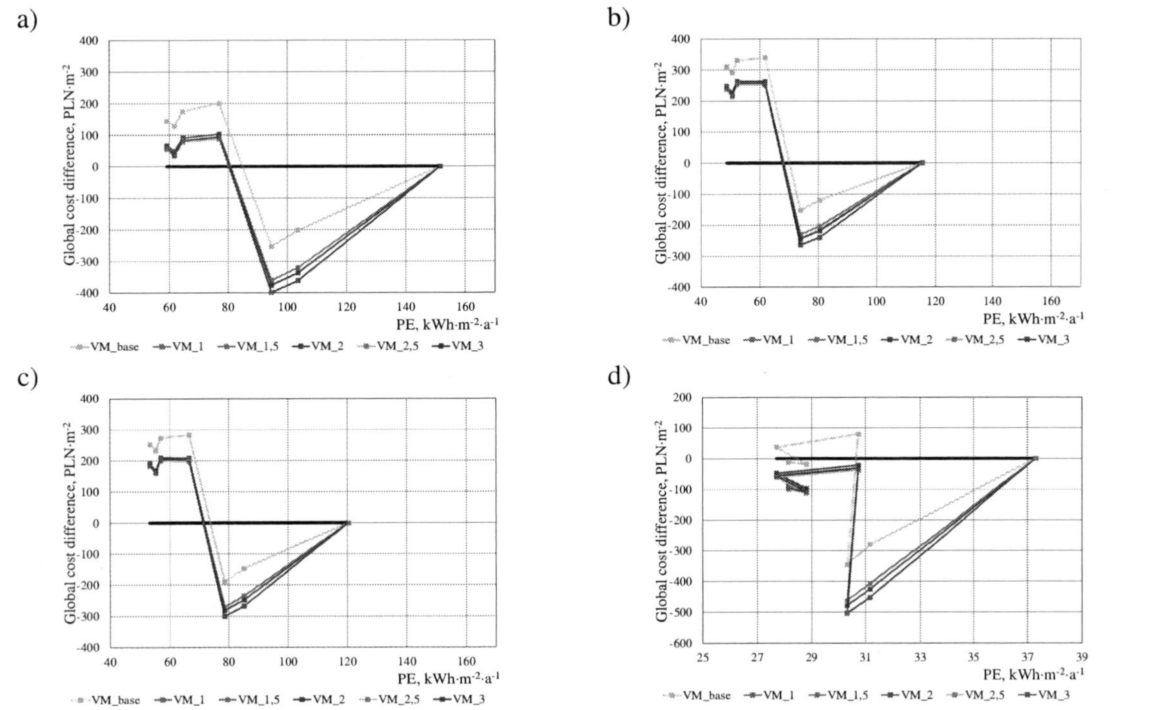

Figure 3. Absolute change of global costs in relations to the basic parameters (VM_base) and building's technical variant TV_1 a) GCB - gas condensing boiler, b) DH- district heating, c) GHP – ground heat pump d) PB – pellet boiler

Analysing the variability of global costs increase including the improvement of energy characteristics of the building up to a base variant the primary cost for the collective residential building decreases to variant 3, and then it rises. An application of better air handling unit causes the increase of the global costs. The improvement of technical equipment of the building impairs the impact of value of money in time.

6.2. Sensitivity analysis of global cost results in relation to the change of the initial investment costs (IC) in time

The variability of value of money in time translates into foreseen increase or decrease of the goods and services prices. The increase of the investment costs (IC) up to 120% of the initial value results in the increase of the global costs by 18% for each analysed variant.

The difference of the global costs increases when the insulation of the building is improved. Below the detailed analysis of the relationship between the global costs and price increase character is presented (fig. 4)

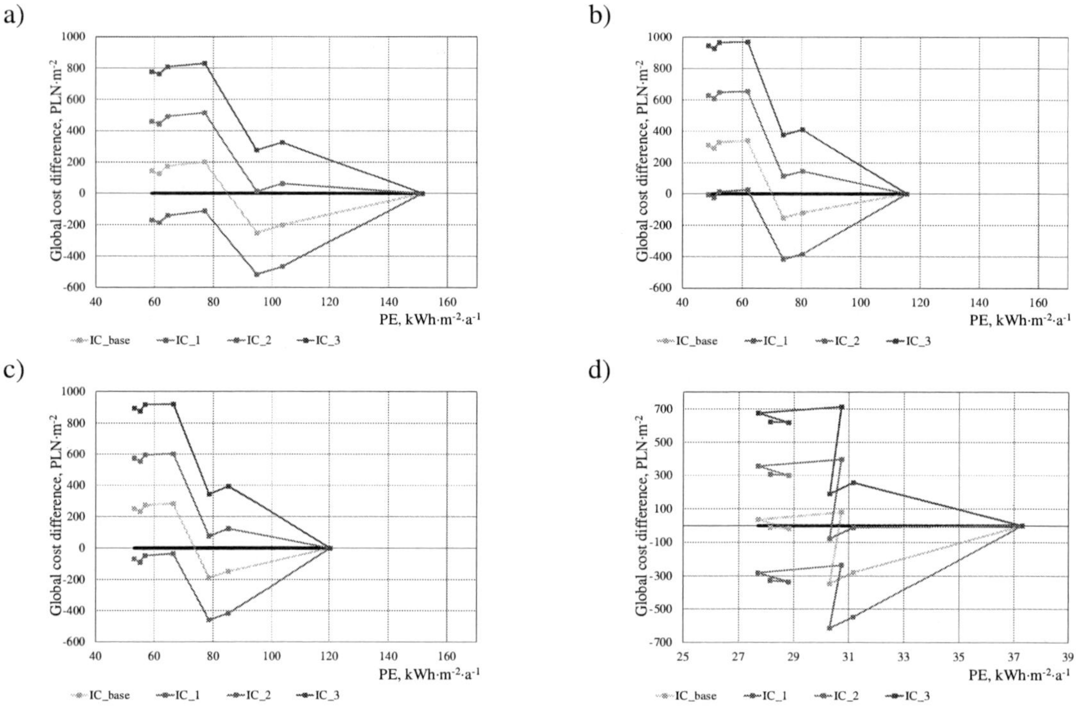

Figure 4. Absolute change of the global costs in relations to the basic parameters (IC_base) and building's technical variant TV_1, a) GCB – gas condensing boiler, b) DH – district heating, c) GHP – ground heat pump, d) PB – pellet boiler

6.3. Sensitivity analysis of the global cost results in relation to electricity performance development (EPD) in time

When the energy prices rise the variable increase of global cost is noticeable, along with the improvement of the building's technical equipment in relation to the analysed heat source.

Moving from EPD_base to EPD_3 variant for each of analysed source shows the movement into lower value of primary energy with the simultaneous increase of the global costs.

The highest increase of the global costs can be recognised for a gas condensing boiler , while the lowest for the building with pellet boiler.

The detailed analysis of the relationship between the global costs and the character of prices increase is presented in the below figures (5a to 5d).

The decrease of relative global costs change results from the improvement of the building technical equipment and the decrease of final energy supplied to the building. A consideration of an increase of the energy costs (from variant EPD_base to EPD_3 variant) in the change of the global costs depends on the building's energy performance characteristics and for a gas condensing boiler it is 14% (TV_1 variant) and 5% for TV_7 variant.

IOP Conf. Series: Materials Science and Engineering **415** (2018) 012005 doi:10.1088/1757-899X/415/1/012005

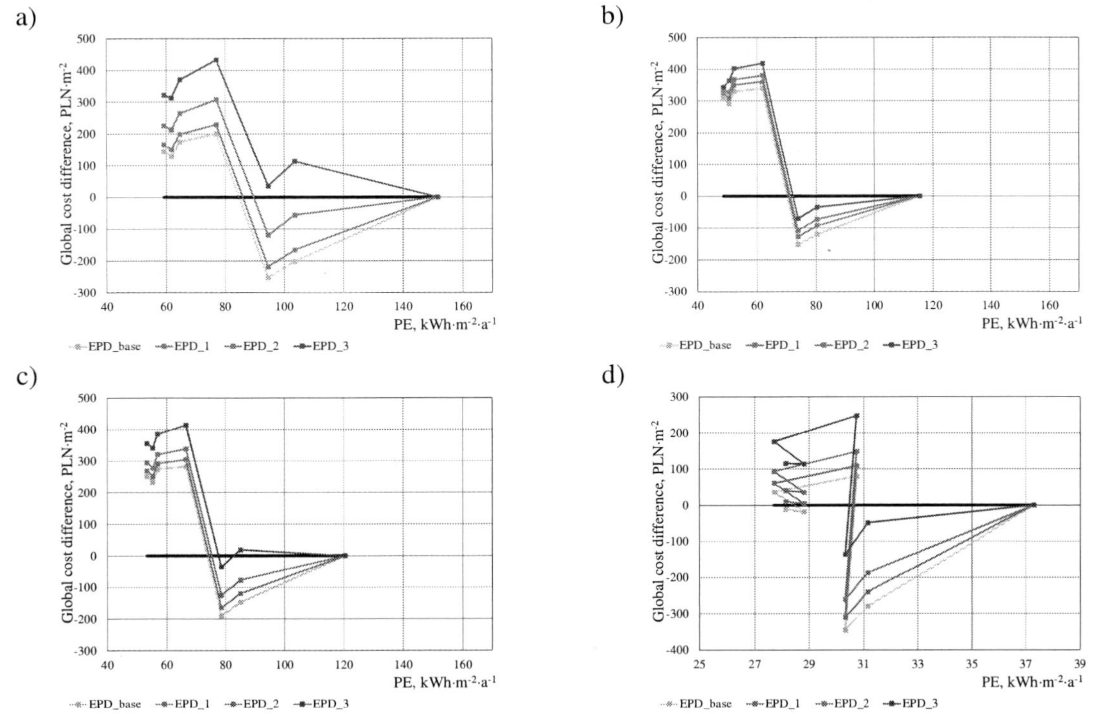

Figure 5. Absolute change of the global costs in relations to the basic parameters (EDP_base) and building's technical variant TV_1, a) GCB – gas condensing boiler, b) DH – district heating, c) GHP – ground heat pump, d) PB – pellet boiler

7. Summary

Appointing the optimal value of PE_{opt}, in terms of the global costs, the initial assumption shall be analysed. A change of macro-economic parameters (discount rate, WIBOR) from the low inflation (VM_base) to so called high (VM_3) results in a shifting into lower value of primary energy with simultaneous increase of the global costs. More important is an increase or decrease of goods and services prices and energy costs. The general regularities are similar to these observed in the analyses for the residential buildings. However the detailed values, zero places or slope of the curves of variability of the costs are different. The differences result, in the particular way, from the different initial assumptions regarding the value of the money and costs of energy.

References

[1] Ordinance of Minister of Infrastructure dated 12th of April 2002 reg. Technical conditions of the buildings and its locations, Journal of laws 2015 item 1422 and its subsequent amendments, (in polish)

[2] Ascione F, Bianco N, De Stasio C, Mauro GM and Vanoli GP 2016 Multi-stage and multi-objective optimization for energy retrofitting a developed hospital reference building: A new approach to assess cost-optimality *Appl Energ* **174** 37–68

[3] Basińska M 2017 The use of multi-criteria optimization to choose solutions for energy-efficient buildings *Bulletin of the Polish Academy of Sciences Technical Sciences* **65(6)** 812-826

[4] Basińska M, Koczyk H and Kosmowski A 2015 Assessment of thermo modernization of the global cost method *Enrgy Proced* **78** 2040-2045

[5] Basińska M, Koczyk H and Szczechowiak E 2015 Sensitivity analysis of determining the optimum energy for residential buildings in Polish conditions *Energ Buildings* **107**, 307-318

[6] Hamdy M, Hasan A and Siren K 2013 A multi-stage optimization method for cost-optimal and nearly-zero-energy building solutions in line with the EPBD-Recast 2010 *Energ Buildings* **56** 189–203
[7] Mauro GM, Hamdy M, Vanoli GP, Bianco N and Hensen JLM 2015 A new methodology for investigating the cost-optimality of energy retrofitting a building category *Energ Buildings* **107** 456–478
[8] Pisello AL, Goretti M and Cotana F 2012 A method for assessing buildings' energy efficiency by dynamic simulation and experimental activity *Appl Energ* **97** 419–429
[9] Kurnitski J, Saari A, Kalamees T, Vuolle M, Niemela J and Tark T 2013 Cost optimal and nearly zero energy performance requirements for buildings in Estonia *Estonian Journal of Engineering* **19(3)** 183–202
[10] Kurnitski J, Saari A, Kalamees T, Vuolle M, Niemela J and Tark T 2011 Cost optimal and nearly zero (nZEB) energy performance calculations for residential buildings with REHVA definition for nZEB national implementation *Energ Buildings* **43(11)** 3279–3288
[11] Pikas E, Thalfeldt M and Kurnitski J 2014 Cost optimal and nearly zero energy building solutions for office buildings *Energ Buildings* **74** 30–42
[12] PN-EN 15459 Energy performance of buildings–Economic evaluation procedure for energy systems in buildings. Warsaw: Polish Standards Institution
[13] Ordinance of Minister of Infrastructure and Development dated 27th of February 2015 reg. Methodology of building or its part energy performance indication and energy performance certificates, Journal of Laws 2015 item 376 and its subsequent amendments, (in polish)
[14] Szczechowiak E 2010-2012 Strategic research project Research task no. 2. Development of optimal energy sources of typical structural-material and installation solutions *NCBiR* SP/B/2/76638/10, (in polish)

IOP Conf. Series: Materials Science and Engineering **415** (2018) 012006 doi:10.1088/1757-899X/415/1/012006

Energy monitoring of low-energy houses in northern Poland Part one: *Characteristics of buildings and their energy demand*

W Sarosiek[1], **B Sadowska**[1], **R Stachniewicz**[1] **and A Święcicki**[1]

[1] Faculty of Civil and Environmental Engineering, Bialystok University of Technology, 45E Wiejska Street, 15-351 Bialystok, Poland
e-mail: wsarosiek@op.pl

Abstract. Slowing development of low energy buildings in Poland and other EU is the lack of recommendations, compliance with which would guarantee the achievement of specified energy level. An attempt of a comprehensive approach to such a standard was made in 2013 by a program and the subsidy system for passive residential buildings. Two energy requirements were defined (NF15 and NF40) with appropriate specifications for building assemblies and installations. The maximum of the energy consumed during the heating season was specified as well as limits for other energy use. This publication is the first of two articles on the assessment of selected characteristics of several buildings with a reduced energy demand, which were built in north-eastern and central Poland. It is an introduction in which the assessed buildings were characterized both in terms of their designed technical and functional parameters as well as real energy supplies related to their heating and ventilation.

1. Introduction

The development of construction with low energy demand is a fact in Poland and other EU countries. The observed increase in interest in energy-efficient buildings results in a regular increase in the number of buildings with good energy performance [1, 2, 3]. However, because the share of building maintenance needs in global energy consumption is estimated at approx. 40%, the dynamics of launching buildings with reduced energy demand in the market should still be considered too slow.

A difficulty in popularizing low energy buildings in Poland was the lack of recommendations, compliance with which would guarantee the achievement of specified energy level. The first attempt of a comprehensive approach to such a standard was made in 2013 by a program to improve the energy efficiency of new residential buildings - priority program of the NFEPWM [4, 5].

The subsidy system was intended for in this program. Practically it was an award for achieving a certain level of energy quality. Two energy requirements were defined (NF15 and NF40) with appropriate specifications for building assemblies and installations. The numbers in energy requirements (15 and 40) specify the maximum value of usable energy demand for heating during the heating season divided by the heated area of the building [kWh/m^2·yr]. There are two groups of detailed requirements:

- regarding the building envelope - affecting the index of seasonal energy demand for heating EU,
- regarding the efficiency of central heating, mechanical ventilation and hot water installations - affecting the final energy index EK.

Several low-energy buildings in Poland, are described in this article. Among them were both buildings meeting the standards NF15 and NF40 as well as energy-efficient buildings (hereinafter referred to as LE) built in 1999-2001, more than a decade before the creation of these standards. As part of environmental observations conducted in these buildings, the actual values of selected parameters shaping the energy performance were measured or determined based on measurements. From the point of view of the energy quality of buildings, only the energy used for heating is relevant. The energy used to heat domestic hot water may cause a blurring of the result of the energy quality assessment of the building. Therefore, this article does not deal with the energy used hot water preparation.

2. Characteristics of the analyzed buildings

Low-energy buildings (LE) located in Bialystok are identical in terms of their construction [6]. Both are two-story, without a basement and with an unheated garage thermally insulated. They only have a different heating system. The first of LE buildings (figure 1) has a traditional heating system with panel radiators supplied from a gas boiler, while the second one uses accumulation electric heaters with dynamic discharge.

Parameters of LE houses:

- usable heated area - 177 m^2,
- garage area - 21 m^2 (unheated),
- building volume - 732 m^3.

Solutions and materials used in external partitions in low-energy buildings are:

- three-layer walls of 47 cm thickness (made of 18 cm calcium-silicate brick, 18 cm mineral wool and 11 cm calcium-silicate brick),
- floor on the ground (concrete floor slab 10 cm + polystyrene 10 cm + concrete floor 6 cm + parquet / terracotta),
- roof in a wooden structure (with a layer of 2 cm polystyrene + glass wool 18 cm),
- double glazed, tight windows with gas inside the space between the panes,
- tight entrance door, insulated.

Figure 1. View of the front façade (NW direction) of the low-energy building LE1 (the LE2 building in addition to the color details is identical) - and the air intake to the ground heat exchanger

LE houses are equipped with mechanical ventilation system with heat recovery. The external air is supplied to the recuperator via a ground heat exchanger in which the ventilation air entering the interior of heated zone is preheated. In the summer season, after exchanging the contribution in the recuperator to the "summer" ground heat exchanger is used to cool the air blown into the building. The effect is obviously far from the standard air conditioning, but it allows to obtain a reduced internal temperature during the 24 hours during the highest outdoor temperatures.

The building of NF40 class - NF40 (1) (figure 2) located in Warsaw is a semi-detached building, without a basement, three-story, with an area of 269,4 m^2 (with a garage in the block of the building) and a volume of 759.8 m^3. A foundation slab was made in this building with polystyrene of 20 cm thickness. The walls were made of 18 cm calcium-silicate blocks and polystyrene 20 cm (15 cm in the garage). The building has a gable roof, wooden with a layer of thermal insulation in the form of mineral wool 30-40 cm thick. The windows are three glazed, tight, with a noble gas. Doors are insulated. The source of heat in the building is a dual-purpose, condensing gas boiler. In most heated rooms, a floor system was used, and only in bathrooms on the second floor, wall heating.

Figure 2. View of the front façade (S direction) of the building - NF40 (1)

Buildings NF40 (2), NF40 (3) and NF40 (4) located in Bialystok and around are two-story, without basements. Their usable areas are: 177 m^2, 238,7 m^2 and 180,4 m^2. These buildings have wooden roofs insulated with mineral wool 30-40 cm thick.

Building NF40 (2) has an unheated garage with an area of 32,4 m^2. The walls are made of hollow bricks (thickness 30 cm) and 20 cm of mineral wool. Under the foundation slab 25 cm thick polystyrene was used. The walls of the NF40 (3) building were made of hollow bricks (25 cm thick) and polystyrene (20 cm thick). In the floor on the ground polystyrene 10 cm was used. Building NF40 (3) has a garage with an area 35 m2. In the NF40 building (4), the walls were made of an aerated concrete (40 cm thick) and polystyrene (22 cm thick) as well as 15 cm of polystyrene is used in the floor on the ground.

In the building of NF15 class (with the areas of 215 m^2 and 140,5 m^2), thicker thermal insulation layer was used, which resulted in lower heat transfer coefficients (table 1). In the NF15 (1) building (figure 3), the walls are made of cellular concrete (24 cm thick) and polystyrene (32 cm thick). In the double-pitched roof with a wooden structure used 50 cm mineral wool. Floor on the ground is insulated by polystyrene of 40 cm thick. In the NF15 (2) building, the walls were made of hollow bricks (25 cm thick) and polystyrene (40 cm thick). Flat roof was insulated with mineral wool (50 cm thick), and the floor on the ground with thermal insulation layer of 40 cm.

Figure 3. View of the building (S and E direction) of NF15 class - NF15 (1)

The basic thermal characteristics of the baffles of the analyzed buildings are shown in Table 1.

Table 1. Thermal characteristics of selected partitions baffles of the analyzed low-energy buildings [1]

level of thermal protection requirements	thermal transmittance coefficient U (W/m²·K)				
	wall	roof, sloping roof	floor on the ground	window	door
technical requirements WT1997	**≤ 0,30**	**≤ 0.30**	**≥ 1.50** [1)]	**≤ 2.00**	**≤ 2.60**
low-energy (LE1)	0.195	0.176	0.184	1.90	2.00
low-energy (LE2)	0.195	0.176	0.184	1.90	2.00
requirements of the NF40 standard	**≤ 0.20**	**≤ 0.15**	**≤ 0.25**	**≤ 1.10**	**≤ 1.30**
NF40 (1)	0.14	0.11	0.10	1.00	1.30
NF40 (2)	0.103	0.110/0.140	0.092	0.73 – 0.81 basement: 0.84 / 0.91	0.87
NF40 (3)	0.10	0,138/ 0.118	0.151/0.145	0.74 – 0.90	1.20/ 1.30
NF40 (4)	0.118	0.109/0.11	0.124/0.125	0.75/0.80	0.97/1.1
requirements of the NF 15 standard	**≤ 0.20**	**≤ 0.15**	**≤ 0.25**	**≤ 1.10**	**≤ 1.30**
NF15 (1)	0.081/0.082 /0.089	0.089/ 0.086	0.090	0.63 – 0.70	0.65
NF15 (2)	0.072	0.071	0.071	0.55 – 0.65	0.57

[1)] Required resistance of floor and ground layers.

Estimates of investment costs of buildings built in various energy standards are discussed in articles [7, 8, 9]. The cost of a low-energy house (LE1) was about 5% higher than the cost of an object designed in accordance with the requirements of Polish technical conditions. In the case of a building in the NF40 standard, this difference can be about 13%, and in the case of the NF15 building even 30%.

3. Description of the measurements and results

In the context of the thematic scope of this publication, the research carried out is aimed at determining the demand indicators for final energy as well as determining the effectiveness of using a ground heat exchanger and recuperator in mechanical ventilation system. In some of the buildings referred to at the beginning of paper, a constant registration of consumption of energy carriers was carried out. It allowed to determine the real energy consumption. The effectiveness of a mechanical ventilation system with a ground heat exchanger and a recuperator was assessed by analyzing the results of measurement of air temperature of external environment, behind the ground heat exchanger, behind the recuperator and the supply air temperature to the heated zone of building LE1.

day / month	1	2	3	4	5	6	7	8	9	10	11	12	13	14	15	16	17	18	19	20	21	22	23	24	25	26	27	28	29	30	31
September	1	2	2	2	1	2	1	2	1	1	2	2	1	2	2	2	2	2	2	2	3	1	2	2	2	2	2	3	1	2	
October	2	2	2	3	2	2	2	2	2	2	2	3	2	2	2	2	2	2	3	2	3	2	2	2	3	2	2	2	3	2	2
November	3	3	2	2	2	2	3	2	2	2	2	2	4	11	10	10	9	7	7	12	4	1	4	2	6	9	3	7	4	6	
December	9	5	9	8	6	9	9	6	7	7	9	10	6	7	7	9	7	9	7	9	8	3	8	8	11	10	10	8	6	7	9
January	8	9	10	12	12	11	13	12	13	13	10	10	9	10	9	8	9	8	10	9	11	11	13	10	12	14	11	11	12	12	10
February	9	10	8	8	8	6	7	7	9	10	9	10	13	11	9	10	9	8	8	11	10	11	8	9	9	8	8	10	10		
March	9	8	10	9	9	9	9	8	8	8	6	7	8	8	8	6	6	5	5	5	5	4	5	3	4	6	7	6	6	7	6
April	6	6	6	5	7	5	5	4	6	5	5	5	4	6	3	4	4	5	3	2	3	3	3	3	3	3	2	4	2	3	
May	3	4	3	2	3	2	3	3	3	3	3	2	3	2	2	2	1	2	1	2	2	2	1	2	1	2	2	1	1	2	1

Figure 4. Diagram of gas consumption in the LE1 building in one of the heating seasons

In one of the low-energy buildings (LE1) described in this article, long-term measurements of gas consumption were carried out, as well as measurements allowing to evaluate the effectiveness of the ground exchanger used for preheating the ventilation air. The measured gas consumption in this building in 2015-2017 amounted to 2,855 m^3, or 29,341 kWh. In individual years it was as follows: in 2015: 808,3 m^3 (8,308 kWh), in 2016: 1028.3 m^3 (10,657 kWh), and in 2017: 1,018.1 m^3 (10,463 kWh). Table 2 shows the daily gas consumption in LE1 building in an exemplary heating season. On the basis of the measurement of instantaneous gas consumption for heating and hot water, it was possible to determine the actual length of the heating season and it was possible to divide energy for heating the building and for preparing hot usable water. The analysis of the results recorded by the gas consumption recorder clearly shows the shortening of the heating season in relation to the standard season. The yellow field in the diagram (figure 4) indicates the actual heating season, while the extension of the yellow area by blue fields is the standard heating season in the IV climate zone, appropriate for the location of the LE1 building. White fields show gas consumption from May to September, except for the standard heating season.

In the building of the NF40 class (NF 40 (1)) average annual gas consumption for heating and ventilation amounted to 937.3 m^3, which gives 10 003 kWh. The measured gas consumption in the heating season for heating and ventilation in the NF15 building (1) amounted to 314.4 m^3, which gives 3,187.5 kWh. The results of these measurements allowed to estimate the actual demand for usable energy for heating and ventilation, as shown in Table 2.

The usable energy for heating and ventilation of other buildings described in Table 1, determined on the basis of calculations, is [1]:

- LE1: 36.7 kWh/(m^2·yr),
- NF40 (2): 29.2 kWh/(m^2·yr),
- NF40 (3): 17.6 kWh/(m^2·yr),
- NF40 (4): 18.9 kWh/(m^2·yr),
- NF14 (2): 10.4 kWh/(m^2·yr).

Table 2. Index of seasonal energy demand for heating of monitored low-energy buildings

standard of building	EU_H (kWh/(m^2·yr))		
	calculated value	based on measurements	requirements
low-energy (LE1)	39,8	41,7	-
NF 40	19,9	37,13	40
NF 15	14,6	14,83	15

To be able to determine the efficiency of the ground heat exchanger in the LE building (1), the temperature of external air, behind the ground heat exchanger, behind the recuperator and the internal was measured (Figure 5 and Figure 6).

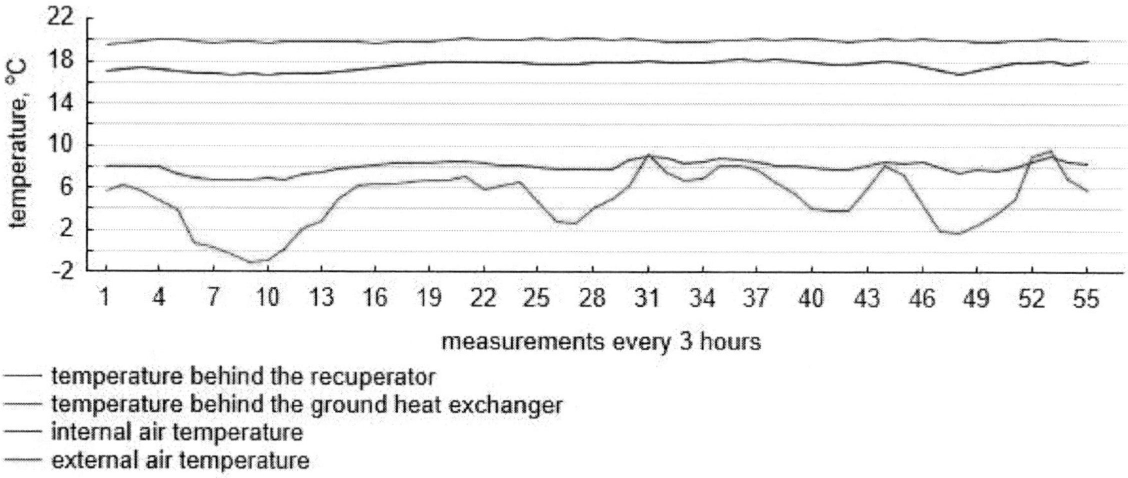

Figure 5. Temperature measurement results in the winter season - measurements every 3 hours.

There is a noticeable effect of the ground heat exchanger on the temperature of the air supplied to the recuperator, as well as on the "smoothing of the temperature line". In the winter season, there is also a significant, because by approx. 8°C, "boost" temperature line behind the recuperator. In total, both tested elements (ground heat exchanger and recuperator) cause (in the case of the winter period visible in Figure 5) supplying ventilation air to the flat with a temperature of approx. 15-18°C instead of the outside temperature, that is approx. -1 to + 9°C.

The effect of using a ground heat exchanger and recuperator in the ventilation system is also observed in the summer period (Figure 6), however, the difference is that the work of the ground heat exchanger causes the inflow of air to the building at a temperature lower than the outside temperature (on average, considering the night and day). A significant decrease of the temperature of the ventilation air supplied to the building takes place during the day, while during the night the lower external temperature causes that the cooling effect may not occur (figure 5).

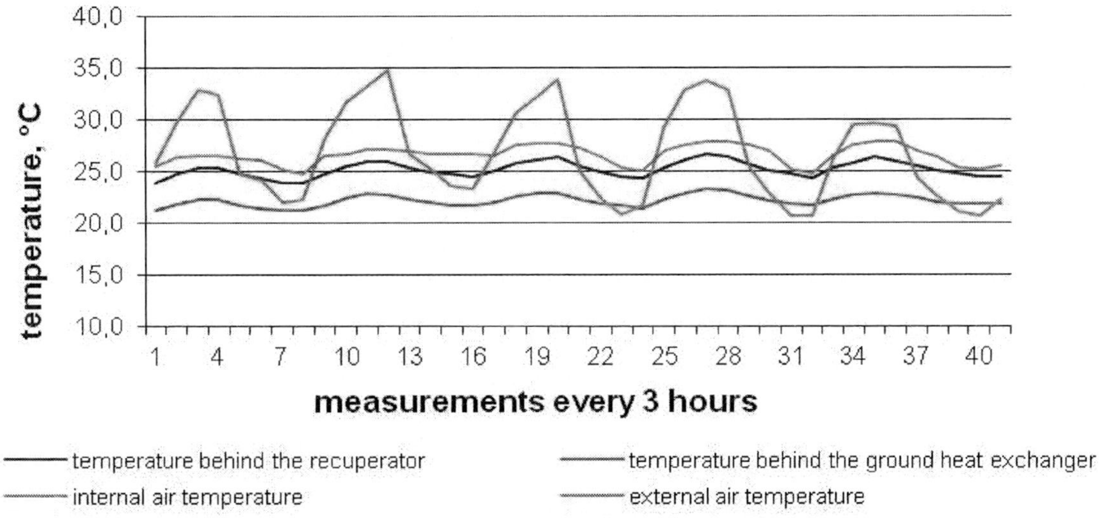

Figure 6. Temperature measurement results in the summer season - measurements every 3 hours

In summer, the temperature measured behind the recuperator with a summer insert is higher than just after the ground heat exchanger by only approx. 4°C, which is an unavoidable, unfavorable effect of the work of fans that force the air circulation. The same effect occurs in winter, resulting in increased efficiency of the recuperator. Generally, the described ventilation system results in a slight (0.2°C) in the assessed summer period (figure 5), a decrease in the average internal temperature of the air relative to the outside, but (which is definitely more important) its stabilization at +/- 1.6°C with respect to fluctuations in outdoor temperature +/-0.1°C.

4. Summary

The analysis of energy consumption monitoring carried out in low-energy buildings showed the possibility of obtaining energy indicators characterizing the assumed standard, previously calculated in advance (energy performance and energy performance certificate). It also turned out that the mechanical ventilation system with a ground heat exchanger and a recuperator is relevant in this respect. The effect of their use in the LE (1) building is not only to reduce energy consumption, but also to stabilize the internal temperature, which is important from the point of use and thermal comfort. In the summer period the ground heat exchanger also assumes the function of a quasi-air conditioner. In the period analyzed in this paper, its application in LE (1) used without significant increase in investment cost (no air conditioning system and related costs) decreased the internal temperature in relation to the external air temperature by 7.7°C. This was especially noticeable in periods of high outside temperatures and high insolation.

On the example of a low energy building LE (1) in which detailed monitoring was carried out and observation from almost twinning as to the construction and operating parameters of the low-energy building LE (2), it can be concluded that the conscious and rational use of a low energy building allows obtaining energy performance indicators at the standard level NF40, despite the fact that these buildings (erected in the years 2000-2002) do not meet the detailed requirements of this standard. This is important because the investment costs of the low-energy building were only slightly higher (by approx. 5%) than the cost of building erected in accordance with the requirements of TR [10].

The cost of constructing a building in the NF 40 or NF 15 standard is already much higher (by 13 and 30% respectively) and, in the absence of subsidies from the state, it is often the reason not to invest in improving the energy standard.

The research was carried out as part of the work no. S/WBiIŚ/3/2016 and financed from funds for the education of the Ministry of Science and Higher Education.

References:

[1] Dubas W 2006 *Przegląd budowlany* Warsaw 5 19-24
[2] Low Energy Buildings Database (http:/www.lowenergybuildings.org.uk/)
[3] Passive House Database of the Passive House Institute, the Passivhaus Dienstleistung GmbH, the IG Passivhaus Deutschland and the iPHA (International Passive House Association) and Affiliates (http://www.passivhausprojekte.de/index.php?lang=en)
[4] Guidelines for the priority program "Efficient use of energy. Subsidies for loans for the construction of energy-efficient houses ", the National Fund for Environmental Protection and Water Management 2015
[5] Energy-efficient houses. Good practice guide. National Energy Conservation Agency 2012
[6] Technical documentations, including energy performance certificates for the analyzed buildings
[7] Sadowska B 2010 Operational model of design of energy-saving residential buildings, Rozprawa doktorska, Bialystok Technical University
[8] Sarosiek W, Sadowska B 2012 *Materiały budowlane* 1 54-56
[9] Żurawski J 2013 *Izolacje* Program dopłat do domów energooszczędnych 7/8 18-22
[10] Regulation on technical specifications of buildings and their locations, Regulation of the Minister of Internal Affairs and Administration of 30 September 1997, Journal of Laws No. 132, item 878 of 1997 as amended

Energy monitoring of low-energy houses in northern Poland Part two: *Verification of the thermal quality of cooling baffles*

R Stachniewicz[1], A Święcicki[1], B Sadowska[1] and W Sarosiek[1]

[1] Faculty of Civil and Environmental Engineering, Bialystok University of Technology, 45E Wiejska Street, 15-351 Bialystok, Poland
e-mail: r.stachniewicz@pb.edu.pl

Abstract. This article is the second of a two-part cycle describing the assessment of some characteristics of selected buildings with a reduced energy demand that were built in north-eastern and central Poland. Two low-energy buildings (built in 1999-2001) and several buildings of NF15 and NF40 class were analyzed. Projected demand indicators for useful energy of these buildings range from 10.4 and 14.6 kWh/(m^2·yr) for NF40 to 30,2 kWh/(m^2·yr) for NF15. One the most important factors affecting the efficiency of low-energy buildings are: thermal quality of the housing and its tightness. This paper presents the results of the tests carried out to verify design assumptions, in particular thermal insulation of partitions, weight of thermal bridges and air tightness and its impact on the energy demand for heating and ventilation.

1. Introduction

In the first part of the case study on several low-energy buildings in north-eastern and central Poland, the continuation of which is this paper, the analyzes made to verify the planned energy performance of some of the buildings under consideration are described. On the example of the low energy building (LE1) the assessment of the effect of using a ground heat exchanger and a recuperator in the mechanical ventilation system for the energy consumption for ventilation was also presented. The important role of these elements in reducing the thermal energy consumption necessary to heating the ventilation air in winter and the positive effect of the ground heat exchanger on the thermal comfort of occupants in the summer period has been demonstrated.

The assessment of the actual energy performance of the considered buildings [1] showed the achievement of the projected indicators required by the adopted target energy standard. This would not be possible without providing adequate thermal quality of the partitions and good air tightness of the building's envelope. These are factors that must not be underestimated at the design stage. The practice that the authors met with shows, however, that the implementation stage is equally important, because even the best theoretical assumptions can be squandered by their improper implementation.

2. Description of the tests carried out and their results

In this publication, the assessment of thermal characteristics of the elements of building envelope was taken. The thermal quality of external partitions of selected buildings was verified by infrared thermography and analysis of recorded thermograms. The assessment of the actual values of thermal transmittance coefficients U of the external walls was made on the basis of measurements of heat flux as well as internal and external temperature. The values of exemplary, linear thermal bridges and their compliance with the requirements of building standards have also been analyzed [2].

For the purpose of verification of the "thermal quality" of the envelopes of analyzed buildings, elevation thermograms were made in the winter period (figures 1, 2, 3 and 4). The evaluations were

IOP Conf. Series: Materials Science and Engineering **415** (2018) 012007 doi:10.1088/1757-899X/415/1/012007

carried out with the cloudy sky and at external air temperatures close to 0 °C. The measurement time was chosen so that the conditions specified in the standard [4] were described in [5][6][7][8].

Figure 1. Thermogram of entrance elevation (N) and the detail of the ground floor window (elevation N); low-energy building LE1.

Figure 2. Thermogram of entrance (NE) and side elevation (NW) of the low-energy building LE2.

Figure 3. Thermogram and photo of the building in the NF40 standard - front elevation (S).

IOP Conf. Series: Materials Science and Engineering **415** (2018) 012007 doi:10.1088/1757-899X/415/1/012007

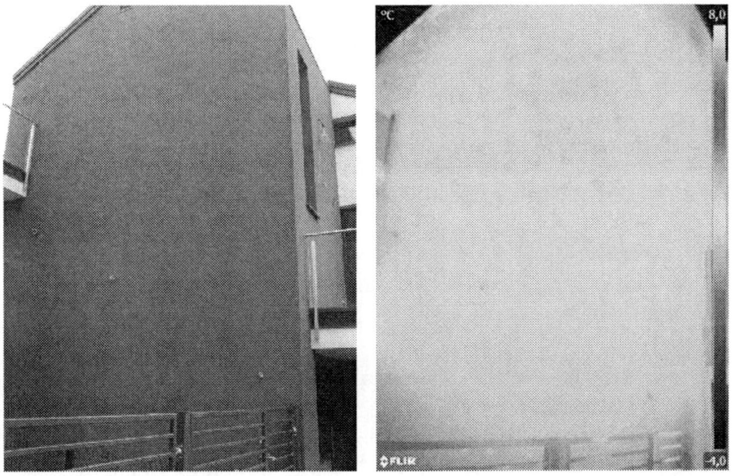

Figure 4. Thermogram and photo of the building in the NF40 standard - side elevation (W).

In buildings NF40 (1), in addition to the registration of gas consumption and the performed thermographic assessment, control measurements were made to determine the actual values of thermal transmittance coefficient of the external partitions. They were determined based on the obtained values of thermal flux and temperatures of internal and external air. N and W orientation barriers were selected for the tests to compensate for the effect of insolation. During the measurements, heat meters, thermocouples and Ahlborn recorder were used. The measurements were carried out in accordance with the recommendations given in the ISO 9869 [3].

The heat setting points were selected using a thermal imaging camera, which allowed to eliminate the influence of thermal bridges, gaps and wiring of the electrical system that could affect the correctness of the heat transfer coefficient U (figures 5 and 7) [9]. The registration was carried out for a period of 13 days at five-minute intervals. A total of 3,744 readings were obtained. On this basis, the thermal transmittance coefficient was determined using the mean rise (Figure 6 and 8) method. Thanks to this, the influence of momentary fluctuations of thermal flux and temperature difference was reduced [8].

Figure 5. The location of the heat meter on the tested western wall of the NF40 building (1) - a thermogram and photo.

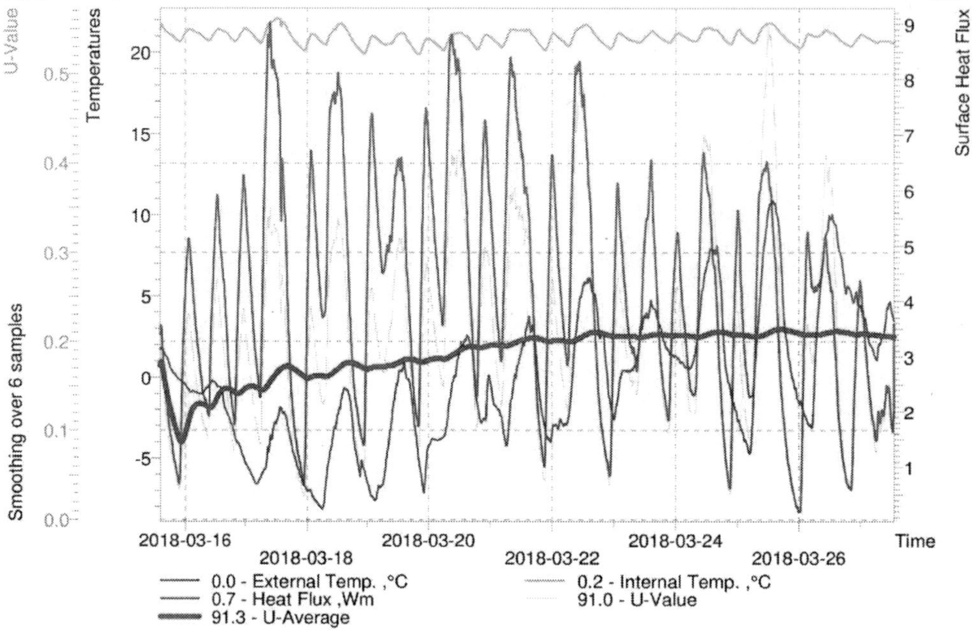

Figure 6. Graph of the thermal transmittance coefficient for the western wall in the NF40 (1) building.

Thermal transmittance coefficients U determined by the HFM method in a building erected in the NF40 standard amounted to 0.20 $W/(m^2 \cdot K)$ and 0.42 $W/(m^2 \cdot K)$. Values obtained from measurements of heat fluxes and temperature of baffles differ from the designed values of 0.14 $W/(m^2 \cdot K)$. The value of thermal transmittance coefficient U for the first measuring point is worse than assumed but still consistent with the maximum value for the NF40 standard. In the case of the second measuring point assessed (pier between window openings), it was found to be less than allowed by the above-mentioned standard. It is the effect of such and not the other position of the measuring point. Above 100% increase in the thermal transmittance coefficient U is related to the overlap in the tested cross-section of the influence of thermal bridges from both windows - the lower thickness of the door frame insulation [1].

Figure 7. Heat meter on the tested northern wall of the LE (1) building (pier between window openings) - thermogram and photo.

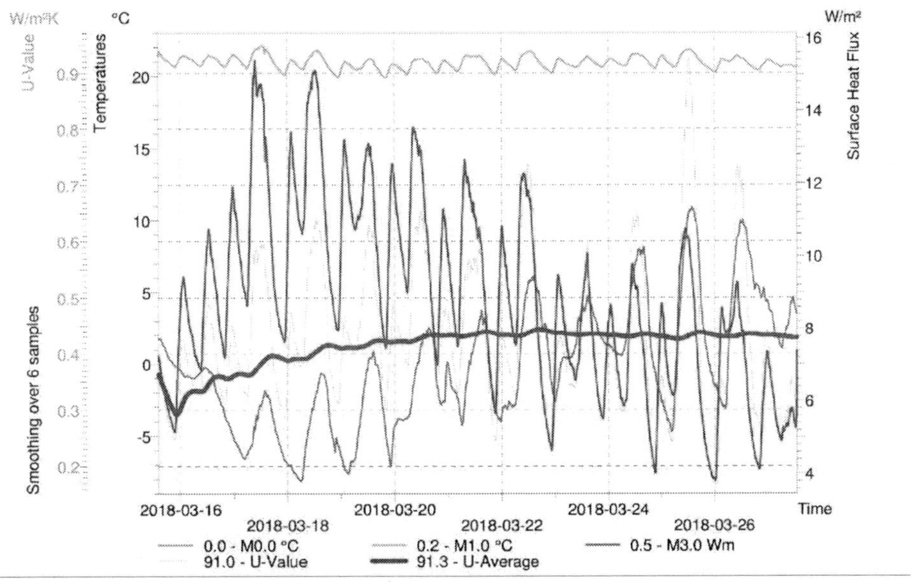

Figure 8. Determination of the thermal transmittance coefficient for a wall with a pier between window openings in building NF40 (1).

Thermal imaging of the internal temperature distribution in the vertical profile of the north and west wall corner was also carried out. The recording of thermograms was carried out with a thermography camera for a period of 13 days. From the whole registration period, 5 days were selected with the most stable outside temperature - day / night differences around ± 5°C, overcast throughout the registration period [3]. Thermograms were analyzed taking into account internal conditions such as: air temperature, reflected temperature, relative air humidity, emissivity of the tested surfaces. The results of the distribution of temperatures recorded for five days are shown below (figure 9 - right).

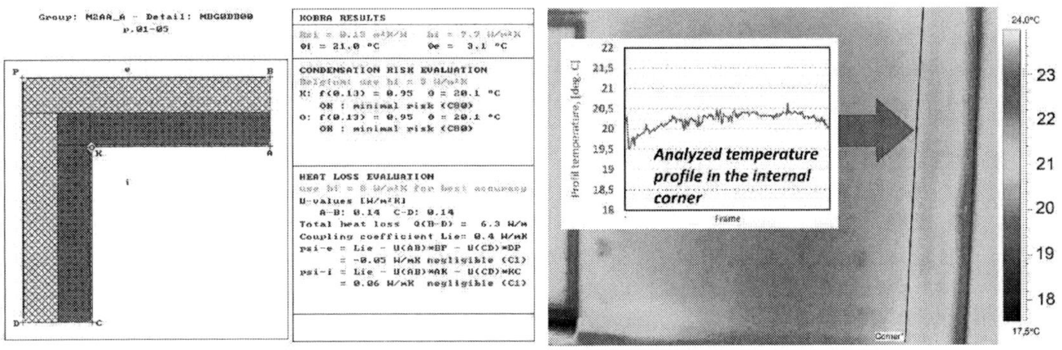

Figure 9. Calculation of thermal bridge in the corner of the tested room in the NF40 building (1) and the average temperature value from the vertical profile in the examined corner (period of 5 days - 1440 thermograms).

On the example of external corner of the NF40 (1) building, the calculated theoretical values of heat transfer coefficients in the places of occurrence of the linear bridge were verified. This was done by comparing the internal temperature of the bridge surface determined by the KOBRA program with the value obtained from thermography measurements.

The analysis took into account the average internal temperature of 5 days of 20.95 °C and external 3.1 °C, i.e. conditions during the thermography tests. The temperature value on the inner surface of the external corner calculated with the Cobra program is 20.1 °C (figure 9 - left), while the average value obtained from thermographic measurements 20.2 °C. This suggests the convergence of computational values of the thermal transmittance coefficients of linear thermal bridges with real values.

In the case of the buildings in the NF40 (1) and NF15 (1) standards, the calculated values of linear thermal transmittance coefficients of analyzed thermal bridges (table 1) do not exceed the recommendations of the NF standard [2]. For comparison, there are given the values of linear heat transfer coefficients of the thermal bridges considered in the thermal insulation variant U, baffles in which they occur, at the level of current thermal protection requirements (TR) [10]. According to the minimum obligatory technical requirements for a single-family building in NF40 standard, the limit value of linear thermal transmittance coefficients is 0.10 W/(m·K) (0.30 W/(m·K) for balcony slabs), and in the NF15 standard 0.05 W/(m·K) for all cases of linear thermal bridges [2]. With regard to buildings in the standard of thermal protection requirements, there are no such recommendations.

Table 1. Calculated values of linear thermal transmittance coefficients of exemplary thermal bridges for selected buildings.

standard of the building	external corner of the wall connection				window jamb			
	the assumed energy standard		in the thermal protection standard (TR)[1)]		the assumed energy standard		in the thermal protection standard (TR)[1)]	
	Ψ_i	Ψ_e	Ψ_i	Ψ_e	Ψ_i	Ψ_e	Ψ_i	Ψ_e
LE (1)	0.08	-0.11	0.11	-0.14	0.01	0.05	0.01	0.05
NF40 (1)	0.06	-0.05	0.09	-0.06	0.05	0.10	0.04	0.08
NF15 (1)	0.04	-0.05	0.06	-0.10	0.02	0.05	0.02	0.04

[1)] Current technical requirements for buildings [10].

According to the analysis plan, the influence of the level of airtightness of monitored buildings on the amount of usable energy demand for heating and ventilation was examined. The air leakage tests show that the building in the LE1 standard is characterized by the indicator $n_{50} = 1.20\ h^{-1}$, in the NF40 (1) standard $n_{50} = 1.00\ h^{-1}$, and NF15 (1) $n_{50} = 0.44\ h^{-1}$ [2]. In both NF buildings, therefore, the formal criterion defined by the guidelines was met, which is $n_{50max} = 1.00\ h^{-1}$ for NF40 and $n_{50max} = 0.60\ h^{-1}$ for NF15. In the case of the LE1 building, reference can only be made to the recommendations in the applicable technical conditions and $n_{50} \leq 1.50\ h^{-1}$ for mechanical ventilation. As you can see, the LE1 building meets these recommendations. Assuming the results of the leak tests carried out in the above-mentioned buildings, computational indices of demand for usable energy were obtained:

- LE1: 40.2 kWh/(m^2·yr),
- NF40 (1): 30.2 kWh/(m^2·yr),
- NF40 (3): 14.6 kWh/(m^2·yr).

Airtightness is an important factor in shaping the energy balance of a building [11]. The graph (figure 10) shows the impact of the airtightness level on the usable energy demand of the buildings LE1, NF40 (1) and NF15 (1). The change in the level of the n_{50} parameter from 0.6 h^{-1} to 1.0 h^{-1} and 1.5 h^{-1} corresponds to the increase in the calculation of the usable energy demand by 3.7 and 8.2% respectively.

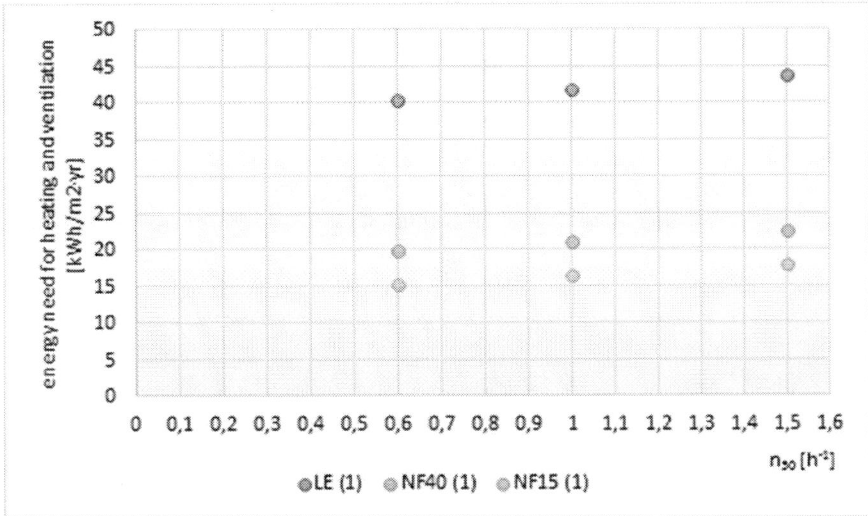

Figure 10. Energy demand for heating and ventilation at various levels of air tightness of buildings.

3. Summary

The measurements and the calculations made on their basis indicate generally good thermal performance of the analyzed buildings allowing for obtaining detailed parameters assumed at the design stage, i.e. thermal transmittance coefficients U and low values of linear thermal bridges Ψ, air tightness and thermal homogeneity of external baffles.

The energy indicators determined on the basis of multi-seasonal energy measurements were within the limits of the values forecasted for different standards of low-energy buildings. Undoubtedly, this is to ensure high airtightness of the assessed buildings at a much lower level than the recommendations of technical conditions ($n_{50} \leq 1.5$ 1 h^{-1}) and meet the requirement of minimizing the effect of linear thermal bridges. The refinement of sensitive details of construction nodes and joinery fixation means that the values of linear heat transfer coefficients do not exceed the required 0.05 W/(m · K) for buildings NF15 and 0.10 W/(m · K) for the NF40 standard. In view of the standard of thermal protection requirements, this criterion is currently not specified. The correctness of the detail solution for linear thermal nodes is confirmed by the performed thermography inspection.

The research was carried out as part of the work no. S/WBiIŚ/3/2016 and financed from funds for the education of the Ministry of Science and Higher Education.

References:

[1] Technical documentations, including energy performance certificates for the analyzed buildings
[2] Guidelines for the priority program "Efficient use of energy. Subsidies for loans for the construction of energy-efficient houses " 2015 The National Fund for Environmental Protection and Water Management
[3] International Organization for Standardization (ISO). Thermal Insulation—Building Elements—In-situ Measurement of Thermal Resistance and Thermal Transmittance; ISO 9869-1 2014, ISO: Geneva, Switzerland
[4] PN-EN 13187 2001 Właściwości cieplne budynków -- Jakościowa detekcja wad cieplnych w obudowie budynku -- Metoda podczerwieni, *PKN*
[5] Wróbel A 2010 Termografia w pomiarach inwentaryzacyjnych obiektów budowlanych, *Rozprawy i monografie AGH*, **no. 209**, Kraków
[6] H. Nowak 2012 Zastosowanie badań termowizyjnych w budownictwie, Oficyna Wydawnicza Politechniki Wrocławskiej
[7] Walker N Nowicki A N 2012 Infrared Thermography Handbook, *The British Institute of Non-Destructive Testing*, **vol. 1&2**
[8] Kisilewicz T, Wróbel A 2009 Diagnostyka termowizyjna przegród w zmiennych warunkach brzegowych *Fizyka budowli w teorii i praktyce* **t. IV**, 75-78

[9] Evangelisti L, Guattari C, Gori P and De Lieto Vollaro R 2015 In Situ Thermal Transmittance Measurements for Investigating Differences between Wall Models and Actual Building Performance, *Sustainability*, **no. 7**, 10388-10398

[10] Regulation on technical specifications of buildings and their locations, Regulation of the Minister of Transport, Construction and Maritime Economy of 5 July 2013, Journal of Laws item 926 of 2013 as amended.

[11] Domy energooszczędne. Podręcznik dobrych praktyk 2012 *Krajowa Agencja Poszanowania Energii S.A.*

Solar energy impact on space heating and cooling needs in moderate climate

D A Chwieduk

Institute of Heat Engineering, Faculty of Power and Aeronautical Engineering, Warsaw University of Technology, 00665 Warsaw, Nowowiejska 21/25, PL

Abstract. The paper describes impact of solar energy on a standard new building of a low energy consumption for space heating, and shows a problem of overheating of such a building in summer, that occurs quite often nowadays in moderate climate. The paper presents how important is to analyse in details the solar energy availability for a building envelope, especially for glazing of different orientation and inclination. Impact of solar energy on the energy balance of a building is analysed. Results of numerical simulations of dynamics of a building and its surrounding are presented. The focus is put on monthly changes of the space heating and cooling energy needs for every month of the averaged year – with regard to the solar energy influence depending on the different orientation of the rooms of a building. It turns out that cooling demand previously very small, now can become very significant and can dominate the total energy demand of new buildings, especially for some location of rooms and apartments. The results of simulation studies presented in the paper indicate necessity of changes in existing regulations on determination of energy performance of buildings, especially for certification of thermal energy performance of residential buildings.

1. Introduction

In high latitude countries for many years the design and construction of buildings have been focused on the heating season only, mainly on reduction of the heat losses through building envelopes. In the past the heat losses used to be very high. Solar energy gained by the interior of a building was lost very quickly. However, nowadays situation is different. New buildings are constructed according to new codes and regulations [1]. They are very well insulated and heat losses are reduced a lot, not only through external walls but through windows, as well. Solar energy impact on thermal energy balance becomes more evident, especially in some rooms locations and with big windows. Cooling demand can become the dominant part of the energy balance in high latitudes. Therefore knowledge of solar energy availability and its impact on the energy balance of a building should be considered carefully to design buildings to be low energy throughout the year.

The existing regulations for residential buildings require determination of final and primary energy demand indices only for heating energy needed for the space heating and hot water [2]. However, it is necessary to determine the energy consumption of residential buildings throughout the whole year. It means that the energy consumption for space cooling (air-conditioning) should be also determined and indices of cooling energy should be introduced as soon as possible. The paper presents the problem of summer overheating of rooms and apartments in specific location in modern low energy buildings in Poland.

2. Solar energy in a building in high latitudes

Reduction of the energy intensity of a building starts with decreasing its energy demand. In the case of space heating and cooling demand not only the use of sufficiently good insulation [3] and high thermal capacity of the building is required, but also a well thought through architectural and civil engineering concept. Any building can utilise energy available in the environment, mainly solar energy, when the building and its surrounding are well designed. Every building is under the influence of solar radiation. It is both a solar collector and a store of gained heat [4-5].

It is necessary to be aware of the climatic conditions of the place where the building is located. To determine the space heating and cooling load it is necessary to know the ambient temperatures, solar irradiation, wind speed etc. In Poland the average annual global solar irradiation is in the range 950 - 1150 kWh/m^2. There are big differences between summer and winter months, e.g. in Warsaw in June the monthly the solar irradiation is about 150 kWh/m^2 and in December about 11 kWh/m^2. It is evident that in winter solar energy availability is rather poor, but in summer is really good. However, it turns out that even so small a solar irradiation level in winter can have positive impact on the energy balance of a building and can lead to a reduction of space heating needs [6]. In summer the solar energy impact can be large if nothing is done to protect a building against too much solar radiation.

To analyse solar energy availability for different surfaces of a building it is necessary to determine solar irradiation of surfaces with different azimuth (orientation) and inclination angles. To determine solar energy availability the averaged representative hourly solar radiation data on horizontal surface located in Warsaw has been used as input data for calculations of hourly irradiation of inclined surfaces [7]. Hemispherical solar irradiations have been calculated using the HDKR anisotropic sky model (Hay - Davies - Klucher – Reindl [8]). A mathematical model of the dynamics of a building has been developed [4]. The model takes into account all energy fluxes flowing into and out of the building. During a day time solar radiation flows into a building. Variations of ambient air temperature, direct and diffuse solar radiation have been considered as hourly values averaged for a representative day of every month of the averaged year. The energy balance of air in a room or in a building takes into account the following energy fluxes coming out or into the room/building, or existing in a building:

- heat transferred through windows,
- solar energy transmitted through windows,
- heat transferred through walls,
- heat transferred through a roof,
- heat transferred through a floor on the ground or basement,
- heating/cooling needed for ventilation (supplied directly by a ventilation system or by a floor heating system or radiators to fresh air),
- internal heat gains,
- heating/cooling energy supplied by space heating/cooling system.

A bespoke MATLAB model has been developed for numerical simulations of a building dynamics. The simulations have been used to calculate all components of the energy balance mentioned above. Some results of simulation studies are presented in this paper to show the evident impact of solar radiation on space heating/cooling demands.

Monthly heating and cooling energy loads resulting from the energy losses and gains of rooms of different locations in the building are presented in Figures 1-4, which refers to attic rooms (inclined roof windows of large size: 2 x 2m^2). These figures include the following components of the energy balance of air in the rooms: energy flows through external walls, energy flows through windows and energy needed for ventilation, for selected rooms with particular window sizes, orientations and inclinations. The calculations of heating/cooling needed for ventilation take into account heat recuperation (recuperative heat exchangers are used). It is evident that energy flow through a window, including solar energy entering the room directly, has a major influence on the energy balance of the room, and as a result on the heating/cooling demand. Location of a room strongly affects the contribution of energy transferred through a window to the energy balance of the room. It is characteristic that the large windows under consideration (with all four main orientations and two

IOP Conf. Series: Materials Science and Engineering **415** (2018) 012008 doi:10.1088/1757-899X/415/1/012008

inclinations) cause the energy gains, resulted from the solar radiation, to be higher than the heat losses during the entire year. What is more, the distribution curves of heating/cooling loads are very close to the curves representing distribution of energy transferred through windows. The influence of ventilation loads (heat recuperation is assumed to be applied) and then of the heat transferred through opaque walls on the heating/cooling demand is not significant.

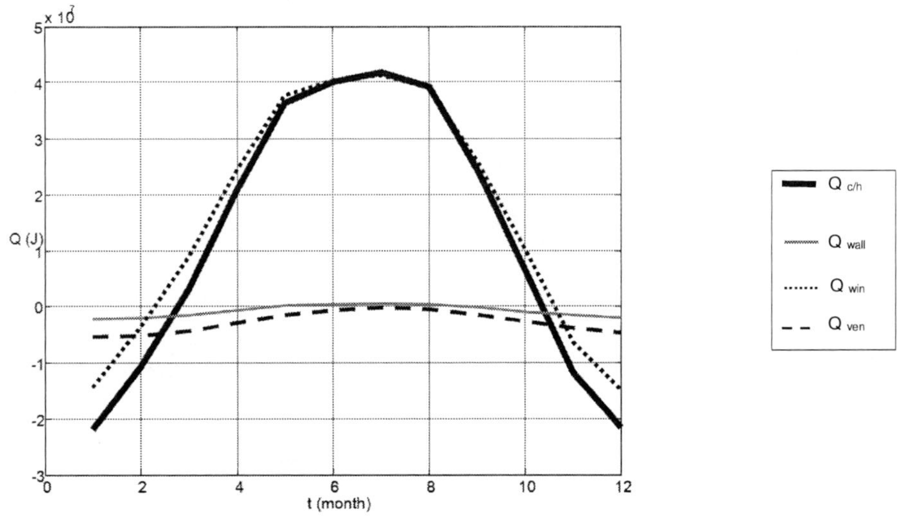

Figure 1. Distribution of monthly heating/cooling demand for the south facing attic room.

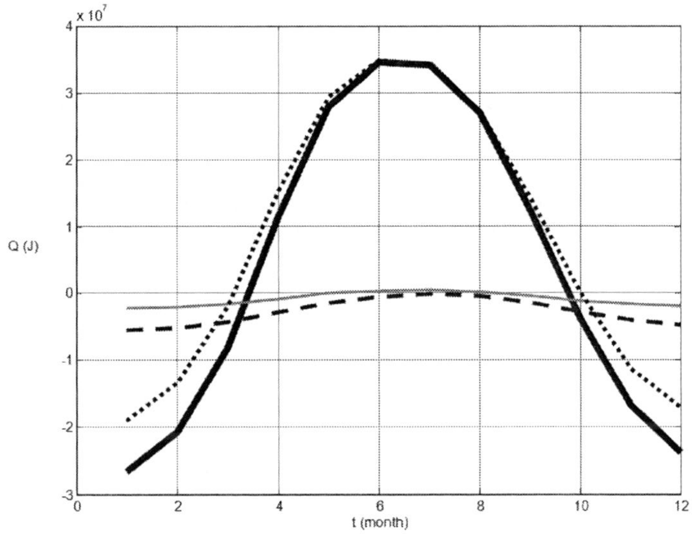

Figure 2. Distribution of monthly heating/cooling demand for the east facing attic room.

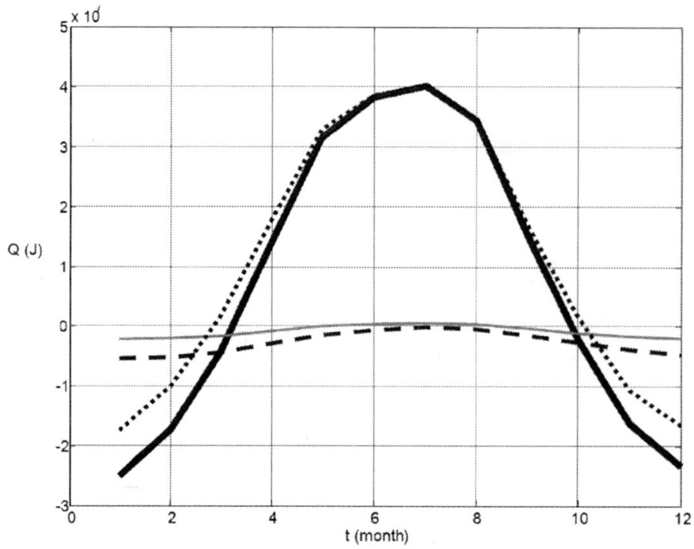

Figure 3. Distribution of monthly heating/cooling demand for the west facing attic room.

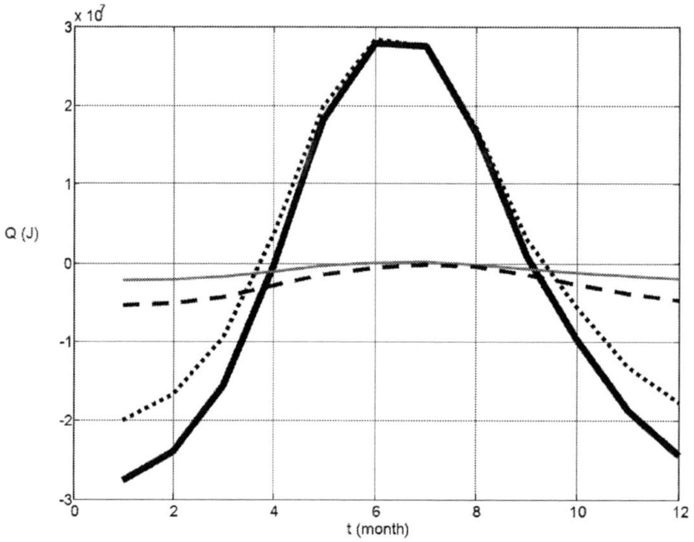

Figure 4. Distribution of monthly heating/cooling demand for the north facing attic room.

Thanks to the use of large windows the duration of the heating season is reduced to 4 months for the south oriented room and to 5 months for the west and east oriented rooms, and to 7 months for the north oriented rooms. (For standard windows size the heating season usually lasts 7 months). A shorter heating season can mean a longer cooling (air conditioning) season. In addition, the energy needs for cooling in warm months for the south and west oriented rooms are higher than the heating demands (for south orientation the cooling demand is twice the heating demand), if a constant room temperature is to be maintained for the whole year.

For south and west oriented rooms the energy needs for cooling, during 8 months from March till the end of October, are three times larger than heating demand in the rest of the year. The cooling demand is especially high from May to August. Generally, attic rooms require more energy for heating in winter (because of the radiative heat exchange between roof windows and the sky in the night) than standard rooms with vertical windows and more cooling in summer (due to the strong influence of solar radiation, which is larger on inclined surfaces than that on vertical surfaces). However, in spring and autumn the energy required to maintain thermal comfort is at a low level.

Figure 5 presents the annual heating and cooling demand of apartments with attic or regular rooms. The figure also includes the energy losses or gains through windows. It is evident that depending on location of apartments the impact of energy transfer through the window (mainly solar energy) on the heating and cooling load can be very different.

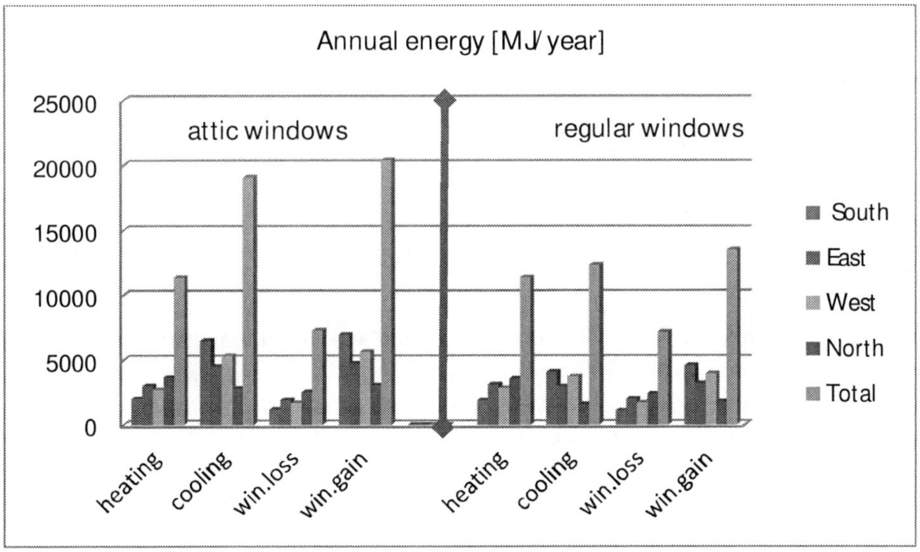

Figure 5. Annual heating and cooling demand, and energy losses and gains through windows for attic rooms - left side and regular rooms – right side.

Analysing Fig. 5 it can be seen that in case of the regular rooms the same pattern, as for the attic rooms, of the impact of the orientation of rooms and windows on the heating and cooling demand can be seen. However, the quantity of energy needed for heating and cooling demand is much lower, and the cooling demand is only a little higher than the heating demand. Energy gains through windows are nearly two times larger than heat losses. Heating demands for attic rooms and regular rooms are very similar (for regular rooms the demand is slightly higher), but there is a large difference in cooling demands, which for attic rooms are more than two times larger. This results from too much solar energy gains through the roof windows in summer (the size and structure of opaque walls of regular rooms and rooms of attics are the same).

3. Conclusions

The simulation results show that overheating in summer due to the high solar irradiation can be a real problem for the design of some buildings even in high latitude countries. It is noticeable that in new buildings, especially in attic apartments the cooling demand can be at the same level or even higher than heating demand. Overheating can be a problem of the south and west rooms in summer. However, in winter, solar gains are needed to reduce space heating demand. Therefore, it is necessary to pay special attention to windows. It would be good to apply windows which can change their optical parameters with time, high solar transmissivity in winter and low in summer. External shading elements should be used at the south and west side of buildings. Such shading elements, e.g. overhangs, light shelves, should be an obstacle for direct solar radiation in summer. In winter external shading elements should not cause any shading, at all. In winter solar energy availability is very important and its impact on a building helps to reduce space heating demand in a natural very simple way.

In new buildings very well insulated and of high thermal capacity the energy flow through windows, mainly solar energy, strongly affects the energy balance of such rooms. South, west and east attic rooms require much more cooling energy than heating energy. The annual energy demand of attic rooms is much higher than the demand of regular rooms of the same orientation.

The results of simulation studies presented in the paper indicate necessity of changes in existing regulations on determination of energy performance of buildings, especially for certification of thermal energy performance of residential buildings, at the national level, as well as European. It is necessary to develop the energy efficiency certification procedure for residential buildings, which considers annual operation of energy systems and requires determination of final and primary energy indices for cooling demand. Limitation on cooling energy demand should be put into force for residential buildings in every geographical location (latitude).

4. References

[1] Directive 2010/31/EU of the European Parliament and of the Council of 19 May 2010 on the energy performance of building.

[2] 7] Rozporządzenie Ministra Infrastruktury w sprawie warunków technicznych, jakim powinny odpowiadać budynki i ich usytuowanie Warszawa, Dziennik Ustaw dnia 13 sierpnia 2013 r. z późniejszymi zmianami. Poz. 926. *National Regulation on technical conditions for buildings and their location with amendments* (in Polish) 2017.

[3] Jelle B P 2011 Traditional, state-of-the-art and future thermal building insulation materials and solutions – Properties, requirements and possibilities *Energy & Build.* 43 10 pp. 2549-2563.

[4] Anderson B 1975 *Solar Energy: Fundamentals in Building Design*, Total Environmental Action, Inc., Harrisville New Hampshire,

[5] Kolokotsa D Santamouris A Karlessi 2014 *Passive Solar Architecture*, in Sayigh A Ed. *Comprehensive Renewable Energy,* Elsevier, Vol. 3, pp.637 – 665.

[6] Chwieduk D 2014 *Solar Energy in Buildings. Thermal Energy Balance for Efficient Heating and Cooling* Elsevier Academic Press.

[7] Chwieduk D Bogdańska B 2004 Some recommendations for inclinations and orientations of building elements under solar radiation in Polish conditions *Renew. Energy* 29, pp. 1569 –1581.

[8] Reindl D T Beckman W A Duffie J A 1990 Evaluation of Hourly Tilted Surface Radiation Models *Sol.Energy* 45 1 pp. 9-17.

Retrofitting of building envelopes: Evaluation of effectiveness using weather-affected material parameters

V Kočí[1], J Maděra[1] and R Černý[1]

[1]Department of Materials Engineering and Chemistry, Faculty of Civil Engineering, Czech Technical University in Prague, Thákurova 7/2077, 166 29 Prague 6, Czech Republic

vaclav.koci@fsv.cvut.cz

Abstract. Thermal performance of a brick wall after retrofitting is investigated by means of computational modelling of coupled heat and moisture transport. In the first step, temperature and moisture distribution over a reference year (Prague) is estimated, being subsequently used for the calculation of heat transmission through the envelope. Then, the annual energy balance is expressed. The simulations are carried out using both reference material parameters and weather-affected data of the brick. The comparison of results indicates that weathering may have a positive effect on the thermal performance of brick walls as it can contribute to a decrease of annual energy balance by ~ 2%. It means that this change can be considered on the safe side, without a risk of overestimation of thermal insulating capabilities of brick walls after retrofitting. The thermal assessment of several retrofitting alternatives reveals that hydrophobic mineral wool applied on the exterior side presents the most favourable option..

1. Introduction

Retrofitting of old building envelopes represents nowadays the only way to comply with standards which have been making more and more stringent demands on thermal protection of buildings [1, 2]. While the thickness or composition of new envelopes can be easily adapted, the possibilities in case of old buildings are limited. However, the thermal retrofitting of existing building envelopes should be in the owners' or investors' interest as it contributes to substantial financial savings [3].

A thermal insulation system is usually selected based on several criteria. The economic point of view or availability of materials on the market might play nevertheless the most decisive role. Unfortunately, this approach does not consider several very important aspects such as environmental conditions, the interaction between newly connected materials or proper hygric function of the system [4]. These aspects, that can negatively affect the service life, may be investigated only by sophisticated techniques, such as semi-scale experiments or computational modelling. The latter, in particular, has come through a large development during recent decades which made it a very popular tool for the assessment of building materials or constructions.

Even if computational modelling is able to provide very fast results of long-term performance, the accuracy of results strongly depends on the quality of input data. Among all the input parameters, material characteristics represent the most decisive factor in that respect. Despite well-developed methods that are being used, the computational results might be still distorted as most of researchers use the data in its reference form [5, 6]. It means that environmental effects on the material properties are neglected although the parameters are more likely changing in time.

In this paper, brick masonry is considered to be a subject of retrofitting under different weather conditions. Several thermal insulation systems, both exterior and interior, are applied and their effectiveness is investigated from the point of view of energy savings using computational modelling. Despite other researchers who use only reference material data, the computations are carried out also with brick parameters affected by the environment. For that purpose, the samples were exposed to natural weathering for three years prior to the experimental investigation of their parameters.

The obtained results could serve for a recommendation of the most suitable thermal insulation system for the wall retrofitting. Additionally, comparing the results obtained using reference and weather-affected data, this approach brings a better view on thermal performance of building envelopes after exposure to environmental effects.

2. Materials and methods

2.1. Materials

Solid clay brick was selected as a representative material of old building envelopes requiring thermal retrofitting according to current thermal standards. The experimental characterization of the brick was performed in reference state and then after three years of natural weathering between 2012 and 2015. The summary of selected weather parameters in the 2012-2015 period is given in Figure 1.

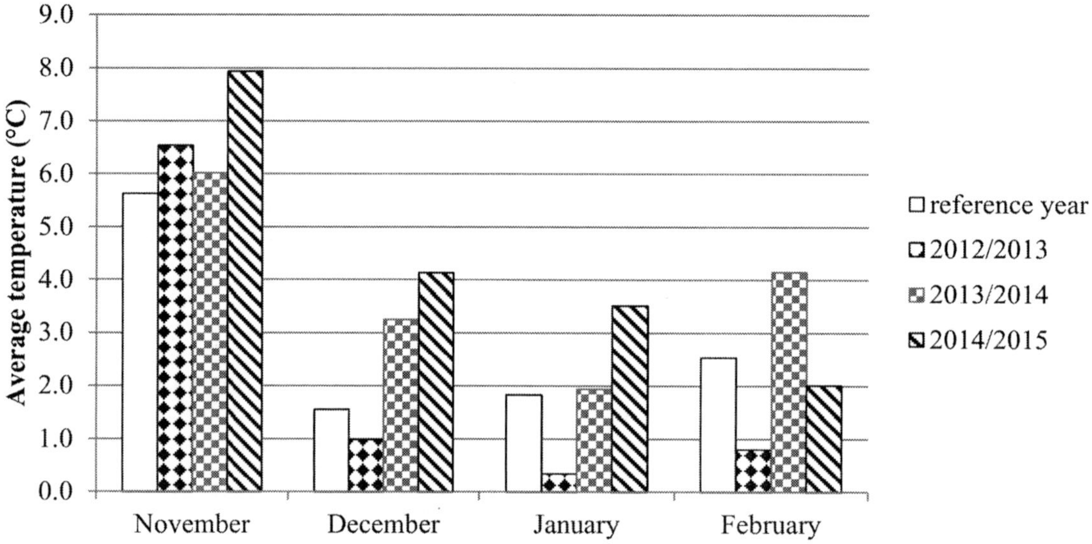

Figure 1. Average winter temperatures in Prague (2012-2015).

The material characteristics of the brick [7] are summarized in Table 1 together with other components of the thermal insulation systems: two types of mineral wool (hydrophobic and hydrophilic) [8], M2 connecting layer [9] and plasters with pozzolanic admixtures [10]. The meaning of used symbols is as follows: ρ – bulk density, ψ – total porosity, μ - water vapor diffusion resistance factor (measured using dry cup and wet cup methods), κ_{app} – apparent moisture diffusivity, c – specific heat capacity (in dry and water saturated state), λ – thermal conductivity (in dry and water saturated state).

Table 1. Material characteristics of solid brick.

	Brick		Mineral wool		M2	Lime-cement
	reference	3y-weathering	hydrophobic	hydrophilic	connecting layer	plaster
ρ (kg·m^{-3})	1831	1860	100	70	1430	1690
ψ (%)	27.9	30.9	96.6	96.9	42.6	43.9
$\mu_{\text{dry cup}}$ (-)	22.1	11.4	2.6	2.4	6.6	17.8
$\mu_{\text{wet cup}}$ (-)	8.8	8.3	1.3	1.2	6.1	12.2
κ_{app} (m^2·s^{-1})	$1.08\cdot10^{-6}$	$7.23\cdot10^{-7}$	$6.10\cdot10^{-11}$	$8.50\cdot10^{-6}$	$3.92\cdot10^{-7}$	$2.49\cdot10^{-10}$
c_{dry} (J·kg^{-1}·K^{-1})	825	813	790	810	1020	1050
c_{sat} (J·kg^{-1}·K^{-1})	1254	1287				
λ_{dry} (W·m^{-1}·K^{-1})	0.590	0.541	0.036	0.037	0.481	0.610
λ_{sat} (W·m^{-1}·K^{-1})	1.735	1.896	0.701	0.790	2.022	1.692
w_{hyg} (m^3·m^{-3})	0.00360	0.00896	0.00021	0.00026	0.021	0.01

2.2. Computational modelling

The main objective of the computational modeling is determination of moisture and temperature field distribution in time across the studied envelopes when exposed to dynamic boundary (climatic) conditions. This data subsequently serves as a background for the assessment of effectiveness of retrofitting measures. Besides the effectiveness of particular insulation types and materials, expressed as a rate of energy reduced, the influence of weathering of load bearing material can be also assessed by comparing the results obtained using reference and weather affected data.

The computations were performed using the computer code HM Transport [11] which works on the basis of the general finite element package SIFEL [12]. A 500 mm thick wall made of solid brick was analyzed, being provided with exterior or interior thermal insulation system. The system consisted of hydrophobic/hydrophilic mineral wool (100 mm) that was connected using M2 connecting layer (10 mm). However, the hydrophilic mineral wool was assumed only on the interior side. Being a representative of capillary active thermal insulations, it is supposed to be used primarily in the interior. In all cases, the wall was finished from both sides by lime-cement plaster (10 mm). The scheme of analyzed detail is depicted in Figure 2. The material parameters were described in Section 2.1.

The studied wall was exposed from the exterior side to dynamic climatic conditions of Prague, Czech Republic and to standard conditions (21 °C, 55 % RH) from the interior side [13]. The climatic data were applied in the form of the Test Reference Year which contains long-term average hourly values of temperature, relative humidity, rainfalls, wind direction and orientation, and several kinds of sun radiation. A diffusion-type mathematical model was used [11] for the expression of balance equations of heat and moisture. The simulation period was selected to be 5 years.

The assessment of thermal performance was presented in the form of quantification of the annual amount of energy transported through 1 m^2 of the particular envelopes. It was calculated as

$$Q = \int_{1.Jan}^{31.Dec} q(t)\,dt, \tag{1}$$

where

$$q(t) = \lambda(w,t)\frac{\Delta T(t)}{\Delta x}. \tag{2}$$

In Eq. (2), $\lambda(w,t)$ is the moisture dependent thermal conductivity as a function of time (W·m^{-1}·K^{-1}), $\Delta T(t)$ the temperature difference (K) between two nodes defining the boundary element and Δx is their distance (m). The hourly values of $\lambda(w)$ and T were obtained as the results of computational modeling.

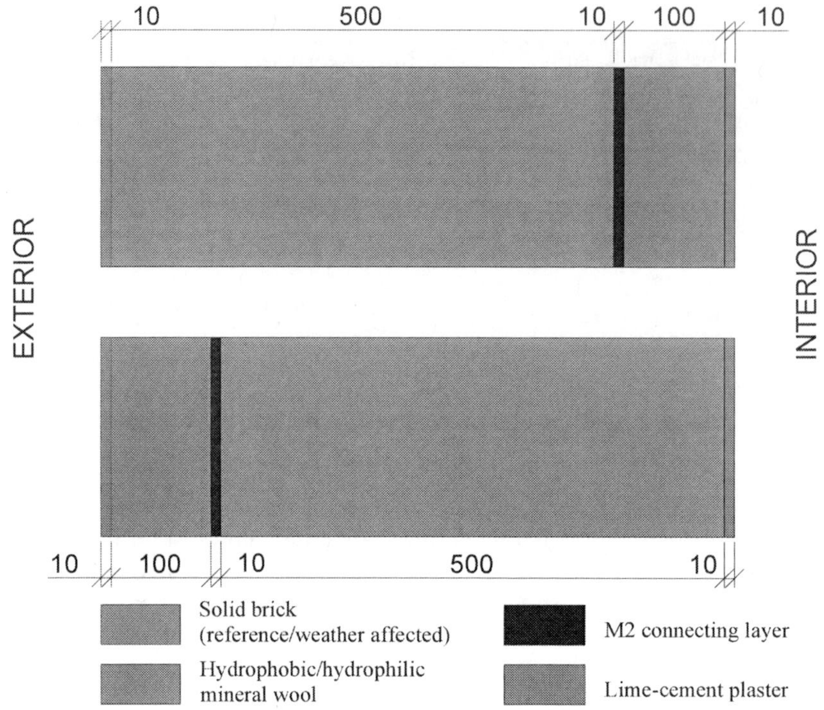

Figure 2. Scheme of the analysed detail.

3. Results and discussion

According to the weather data presented in Figure 1 it is apparent that the winters between 2012 and 2015 were relatively mild, mostly warmer than the reference weather year. The weather-induced changes of brick properties were therefore not as high as it might be expected. However, some changes have been detected, as it is shown in Table 1. Thermal conductivity of the brick in dry state decreased by ~8.3 %, accounting for 0.541 $W \cdot m^{-1} \cdot K^{-1}$ after weathering. On the other hand, the water saturated value increased by 9.3 % to 1.896 $W \cdot m^{-1} \cdot K^{-1}$. Since the thermal conductivity is the most important parameter from the point of view of energy balance calculations (see Eqs. (1) and (2)), it is very important to pay attention to its changes. The weathering of the brick led to an increase of its porosity, thus more air was contained in the dry state, resulting in better thermal insulating capabilities. However, more voluminous porous space was able to accommodate more water, which explained higher values of thermal conductivity in the water saturated state. A detailed analysis of moisture behaviour is therefore essential to understand the thermal performance of the wall.

In all the investigated variants of building envelopes the brick masonry was protected either by thermal insulation or by plaster from the exterior side. Since both of these layers have very low moisture diffusivity, an effective protection against excessive moisture income due to rainfalls is ensured. The average relative humidity of air contained in bricks in reference state was 64.7 % when insulated from the interior side and 61.9 % when insulated from the exterior side. Assuming the weather-affected bricks, these values were 65.5 % and 61.8 %, respectively. Recalculating relative humidity in the brick's pore space to volumetric moisture content, one obtains very low values of ~0.11 – 0.18 $\%m^3 \cdot m^{-3}$. Within this range, thermal conductivity of weather-affected brick is lesser by ~7.7 % which predisposes it to reach better results from the point of view of the annual energy balance. The summary of annual energy balance of particular variants of building envelopes is given in Table 2, the comparison of heat flux of the worst (No 3) and best (No 4) variants is depicted in Figure 3.

Table 2. Energy balance summary.

Variation No.	Brick	Mineral wool	Placement	Annual energy balance (kWh·m^{-2})
1	reference	hydrophobic	exterior	25.24
2	reference	hydrophobic	interior	25.26
3	reference	hydrophilic	interior	25.71
4	weather-affected	hydrophobic	exterior	24.80
5	weather-affected	hydrophobic	interior	24.82
6	weather-affected	hydrophilic	interior	25.25

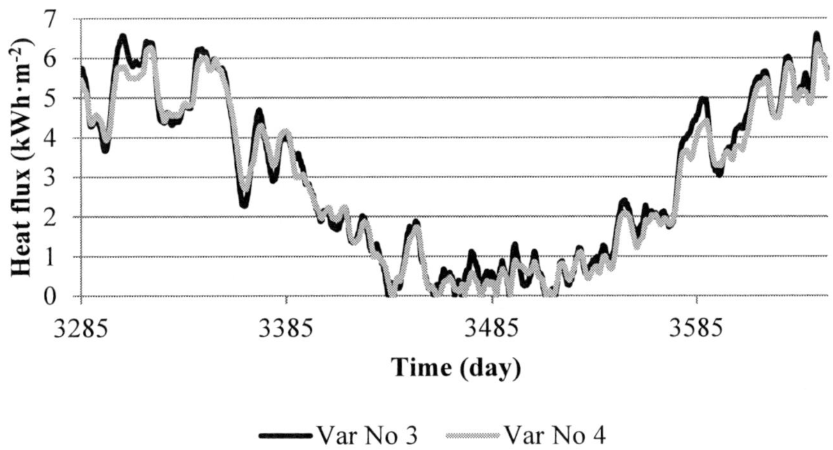

Figure 3. Comparison of annual heat flux: variants No 3 and 4.

Based on the summary of annual energy balance presented in Table 2, hydrophobic mineral wool showed the best thermal performance when applied to the brick masonry, which made it the most suitable material among the investigated thermal insulations. The results presented in Table 2 also proved that weather-affected bricks reached better thermal performance as their energy balances were lower than those of the reference brick. The difference after weathering was, however, very small, accounting only for ~ 2 %. Comparing the heat flux of variations No 3 and 4, one could notice slightly less heat transmission through envelope No 4, which resulted in the annual energy balance lower by 0.91 kWh·m^{-2}.

Generally it can be concluded, that weathering of bricks before their retrofitting did not have any significant influence on their thermal performance. The material changes that occurred after weathering were very small, almost negligible. Moreover, they were on the safe side as they contributed to heat flux decrease. Such a finding could be considered as positive because weathering of masonry did not lead to overestimating the thermal insulating capabilities of brick walls after retrofitting. One could question the mild winters in 2012-2015 and thus a minor influence on brick's properties changes. However, our previous research revealed, that even mild winter periods may have significant effects on hygrothermal performance of bricks [7]. The minimal differences of annual energy balance could be therefore ascribed to the presence of thermal insulation after retrofitting which minimized the differences between particular envelopes.

4. Conclusions

Thermal performance of brick envelope after retrofitting was investigated using computational modelling of coupled heat and moisture transport in this paper. Based on the hourly values of heat flux, calculated from the heat and moisture distribution, annual energy balance of one square meter of the envelope was expressed. The results indicated that the best thermal performance by means of the minimal heat flux was achieved when hydrophobic mineral wool had been applied on the exterior side of the brick wall. In this case, the annual energy balance reached 24.80 $kWh \cdot m^{-2}$.

In order to assess the influence of weathering on thermal performance, the calculations were carried out using reference and weather-affected brick properties. However, the differences between the performance of reference and weather-affected brick were very small, accounting for ~2 % only. Moreover, this difference was on the safe side. This finding can be considered as positive as it means that weathering of masonry does not lead to overestimating of thermal insulating capabilities of brick walls after retrofitting. It can be explained by the presence of thermal insulation that minimizes the differences in thermal performance. Additionally, the mild winter periods between 2012 and 2015 did not contribute to significant changes of material properties of the bricks, so one can conclude that the weathering had negligible effect on the bricks weathering.

Acknowledgement
This research has been supported by the Czech Science Foundation, under project No P105/12/G059.

References

[1] Iwaro J and Mwasha A 2010 A review of building energy regulation and policy for energy conservation in developing countries *Energy Policy* **38** 7744-55
[2] Adamczyk J and Dylewski R 2017 Changes in heat transfer coefficients in Poland and their impact on energy demand - an environmental and economic assessment *Renewable & Sustainable Energy Reviews* **78** 530-8
[3] Rakhshan K and Friess W A 2017 Effectiveness and viability of residential building energy retrofits in Dubai *Journal of Building Engineering* **13** 116-26
[4] Madera J, Koci V, Koci J, Vyborny J and Cerny R 2010 *Advanced Computational Methods and Experiments in Heat Transfer Xi,* ed B Sunden, *et al.* (Southampton: Wit Press) pp 291-301
[5] Kunzel H M and Zirkelbach D 2013 Advances in hygrothermal building component simulation: modelling moisture sources likely to occur due to rainwater leakage *Journal of Building Performance Simulation* **6** 346-53
[6] Skerget L, Tadeu A and Brebbia C A 2018 Transient simulation of coupled heat and moisture flow through a multi-layer porous solid exposed to solar heat flux *International Journal of Heat and Mass Transfer* **117** 273-9
[7] Kočí V, Čáchová M, Koňáková D, Vejmelková E, Jerman M, Keppert M, Maděra J and Černý R 2018 Heat and Moisture Transport and Storage Parameters of Bricks Affected by the Environment *International Journal of Thermophysics* **in press**
[8] Jerman M and Cerny R 2012 Effect of moisture content on heat and moisture transport and storage properties of thermal insulation materials *Energy and Buildings* **53** 39-46
[9] Kočí V, Jerman M, Fiala L and Černý R 2017 Effect of hygric and thermal properties of connecting layers on the performance of interior thermal insulation systems *AIP Conference Proceedings* **1906** 140005
[10] Vejmelkova E, Maca P, Konvalinka P and Cerny R 2011 *Advances in Innovative Materials and Applications,* ed M Soueidan, *et al.* pp 372-5
[11] Kočí V, Kočí J, Maděra J, Pavlík Z, Gu X, Zhang W and Černý R Thermal and hygric assessment of an inside-insulated brick wall: 2D critical experiment and computational analysis *Journal of Building Physics* **0** 1744259117750495
[12] Kruis J, Koudelka T and Krejci T 2010 Efficient computer implementation of coupled hydro-thermo-mechanical analysis *Mathematics and Computers in Simulation* **80** 1578-88
[13] ČSN 2011 73 0540-2 Thermal Protection of Buildings - Part 2: Requirements. (Prague: Czech Office for Standards, Metrology and Testing)

IOP Conf. Series: Materials Science and Engineering **415** (2018) 012010 doi:10.1088/1757-899X/415/1/012010

Design of envelopes for timber buildings in terms of sustainable development in the low-energy construction

A Iringová[1*]

[1]Department of Building Constructions and Urban Planning, Faculty of Civil Engineering, University of Žilina, Univerzitná 1, 010 26 Žilina, Slovakia.

[*] E-mail: agnes.iringova@fstav.uniza.sk

Abstract. The paper focuses on theoretical analysis of physical determinants affecting construction of envelopes for wood-based structures in the intentions of sustainable development under the current and expected legislation in Slovakia after 2020. It deals with the optimization of model building envelopes in terms of heat loss according to the requirements of the Europe 2020 Strategy and consequently calculates the values of the fire resistance considering Eurocode 5. It assesses the application and fire safety of analysed envelopes in construction of buildings with a combustible construction system, i.e. wood-based buildings intended for housing and accommodation, in accordance with the Slovak legislation. The paper also concentrates on the optimisation of built-in thermo-insulating and facing materials of wooden load-bearing members in the model building envelopes in terms of the environmental burden throughout their life cycle and the impact on the resulting fire resistance. It assesses the possibilities of using building components made from recycled materials in the construction of timber buildings as a full-valued replacement of standard building materials from non-renewable sources.

1. Introduction

At the beginning of this millennium, the world's communities began to deal with the issue of sustainable energy consumption due to the pollution of macroclimate by production-based emissions. Possibilities for the global use of renewable energy sources are being sought. Builders now prefer low-emission manufacturing processes and products to maintain the sustainable development and the balance in all areas of human life including global responsibility of each individual for the current and future human coexistence with the ecosystem. The energy consumption and production-based emissions should be significantly reduced in the construction of buildings. The environmental burden associated with building´s construction comprises the three stages - preparation and construction, operation, and disposal at the end of its lifetime. The choice of materials and compositions strongly affects the production of emissions in all the stages.

2. Slovak legislation in sustainable construction

Since 1992, when the UN Conference on Environment & Development was held in Rio de Janeiro, the main objective of the international community has been to achieve sustainable development. The document from this summit outlines the principles for sustainable land development in terms of social, environmental, and economic growth. Twenty seven principles of sustainable development result from Agenda 21, the declaration on environment and development adopted by more than 178 Governments at this conference. Chapter 7 contains the basic concept for sustainable construction. It specifies main problems, determines the resulting tasks for the individual participants of the construction process, and defines the criteria of sustainable construction considering their significance during all phases of the building's life cycle. It deals with the following issues: dilution, product quality, building´s environmental impact, resource consumption - energy, water, soil, materials, impact of building construction on sustainable urban development, ecological burdens, social, cultural, and economic issues. This document is a follow-up to the Europe 2020 Strategy which is one of the main economic reforms of the EU.

One of the main objectives is to construct all new buildings as nearly-zero-energy ones as of 31 December 2020. As for the buildings in state ownership, this requirement has been in effect since 2018. The obligation of all EU members is to reduce greenhouse gas emissions by min. 20% by 2020 and to increase the energy efficiency of buildings including the use of energy from renewable sources by 20%. [1] Specification of the requirements for eco-friendly products is defined in the Regulation of the European Parliament and European Council No. 305/2011 [2] replacing the European Council Directive No. 89/106 /EEC on construction products. The basic requirements for the buildings defined therein are supplemented by the new seventh requirement - the sustainable use of natural resources. This requirement is supported by a set of European Standards incorporated into Slovak legislation. [2] The standards for designing sustainable buildings are as follows: STN EN 15643-Part 1-4, STN EN 15942 (73 0911) Sustainable construction, Environmental product declarations, Communication Formats in the Business Environment (1 April 2012); STN EN 15804 (73 0912) Sustainable construction, Environmental product declarations, Basic rules for Building Products (1 August 2012); TNI CEN/TR 15941 (73 0910) Sustainable Construction, Environmental product declarations, Methodology for the selection and use of generic data (1 August 2010).

3. Analysis of determinants affecting the construction of lightweight wood-based envelopes in the intentions of Europe 2020 Strategy

3.1. Thermal protection

The first determinant affecting the construction of a lightweight wood-based envelope is the **thermo-insulating efficiency** and the subsequent energy performance of a building under the given climatic conditions in both winter and summer. Currently, the legislative requirements for thermal insulation of building envelopes are specified in STN 735040-2:2012, Z1-2016 so that the building would meet the condition of the global indicator - primary energy in kWh/(m^2.a) within class A1of Energy Efficiency Index (EEI) Classification. The limited values depend on the building´s category, i.e. its functional use. From 2020 onwards, all the designed buildings, including those for housing, will have to meet national requirements for energy consumption in the category of nearly-zero-energy buildings within the global indicator - primary energy in kWh /(m^2.a) within class A0 of EEI. The increasing demands for reduction of heat loss through the building envelopes means the increasing demands for thermo-insulating efficiency of the thermal insulation related to its thickness and thermal conductivity coefficient. Apart from the thermo-insulating effectivity, the flammability class, lifetime, and recyclability are also important. [2]

3.2. Fire safety

The second determinant fundamentally affecting the construction of a lightweight wood-based envelope, both in terms of the composition and use, is the **fire safety**.

Wood-based construction systems are in terms of fire safety classified as combustible structural units. The fire height of the buildings with the timber supporting system according to legislation valid in Slovakia is limited to maximal 12 m, except the buildings intended for housing and accommodation (typically 4-storeyed). This goes provided that the fire load calculated in the most critical fire section of such buildings is up to 20 kg/m^2. If the fire load is higher than 60 kg/m^2, the maximal allowable fire height is 9 m. The requirement for maximal fire resistance of the load-bearing members is REI 90.

According to the currently valid standards, residential buildings and houses may contain wood-based load-bearing systems up to the 3rd storey (see Table 1). The fire resistance requirement for the first level of fire safety is REI 30.

Considering the above limitations, the non-combustible construction system, e.g. the concrete skeleton, is combined with the envelope made of timber sandwich panels. The use of this system is limited by the requirement for non-combustible fire sections to be situated between the fire areas in both vertical and horizontal direction. This system may be used in buildings with the fire height of 12 m where the fire sections are not required. It does not go for the residential buildings where the fire sections are always required.

The resulting fire resistance of wooden load-bearing structures depends on the covering and filling materials and is a criterion for the possibility of their application in a building.

Table 1. Fire safety degrees for residential buildings and buildings from B group and the required fire resistance of the load-bearing building envelope with the combustible construction unit.

Construction unit	Maximal number of above ground storeys	Required fire safety	Minimal fire safety degree of fire sections	
			above ground storey + 1st below ground storey	2nd and further below ground storey
Combustible	1	REI 30	I	inadmissible
	2	REI 45	II	
	3	REI 60	III	

3.3. Environmental burden

The third determinant important for choosing materials and construction of a lightweight wood-based envelope is the **environmental burden** at the time of its construction as well as disposal. The environmental burden is defined by the built-in energy consumption and the emissions produced throughout its life cycle.

It is determined by the embodied energy consumption and emissions produced throughout the life cycle. The aspect of embodied energy consumption is monitored - the energy consumed throughout the life cycle of a product given in MJ/m^2 PEI and the amount of emissions produced in the production, use, and disposal of a product: CO_2ekv - GWP (global warming potential), SO_2ekv - AP (acidification potential), PO_4ekv - EP (eutrophication potential), R-11ekv - ODP (ozone depletion potential), C_2H_4 - POCP (ground-level ozone formation potential).

In terms of emission reduction, it is optimal to use slab covering and recycled thermo-insulating materials for the load-bearing structures in the construction of lightweight wood-based envelopes.

4. Composition optimization of wooden sandwich structures in terms of thermal and fire protection in the context of sustainable efficiency in the model solution

4.1. Composition optimization of wooden sandwich structures in terms of thermal protection for the construction after 2020

Nowadays, wood-framed construction systems are the most commonly used in the construction of wood-based structures. In the low-storey structures, the frame construction acts as both a supporting system and a load-bearing structure of the envelope. The load-bearing structure of the envelope is made of vertical wooden posts with an air gap filled with thermal insulation.

The choice of thermal insulation for sandwich timber structures is conditioned by its thermo-insulating efficiency, combustibility class, durability, and recyclability. The building envelope and timber load-bearing framework must be designed so that the water condensation would not occur on joints of timber elements. As for hygienic demands, the layers of thermal insulation and its physical parameters must be designed to keep the minimal surface temperature of the interior wall above the critical surface temperature for mould growing, depending on the boundary environmental conditions and the type of heating. The type and thickness of thermal insulation is essential for the fire resistance of a timber panel. Its melting temperature should be above 1000 °C and a minimal mass weight should be above 30 kg/ m^3. Thermal insulation based on mineral-wool or glass-fibre meets these physical parameters. [3]

Each of the compositions analysed is designed considering the approximately required standard heat transfer coefficient U ≤ 0.15 W/m^{2}K which will be in force for buildings in Slovakia after 2020. All the model structures have the same supporting timber construction differing in the thermal insulation as well as in the interior and exterior surface finishing. Detailed compositions of model solutions are given in Table 2.

Table 2. Construction solutions of envelope model compositions with physical parameters of layers.

	Type of a layer (from inside to outside)	Thickness [mm]	ρ [kg/m^3]	λ W/mK	R m^{2}K/W	ΔOI3 Pkt/m²
Wall 1	Gypsum plasterboard	15	900	0.25	0.06	4
Glass	90% Glass wool MW	60	18	0.038	1.57	6
wool	10% Timber – frame a´= 625 mm	(60/40)	475	-	-	0
	Polyethylene (PE) vapour brake	1	-	0.5	0	5
	OSB boards	18	650	0.13	0.14	4
	80% Glass wool MW	140	18	0.038	3.68	8.4
	20% Timber – frame a´= 625 mm	(140/60)	475	-	-	-1
	OSB boards	18	650	0.13	0.14	4
	Mineral adhesive	5	-	1.0	0.01	3
	Mineral thermo-insulating panel	80	112	0.044	1.8	9.5
	Silicate plaster with synthetic resin additive, reinforced	5	-	0.8	0.01	9
Wall 2	Gypsum plasterboard	15	900	0.25	0.06	4
Wood	Wood fibre WF-W Fibreboard soft	60	50	0.042	1.43	2.4
fibre	10% Timber – frame a´= 625 mm	(60/40)	475	-	-	0
	Polyethylene (PE) vapour brake	1	-	0.5	0	5
	OSB boards	18	650	0.13	0.14	4
	Wood fibre WF-W Fibreboard soft	140	50	0.042	3.33	5.6
	20% Timber – frame a´= 625 mm	(140/60)	475	-	-	-1
	OSB boards	18	650	0.13	0.14	4
	Wood fibre WF-W (130 kg/m³)	80	130	0.046	1.74	7
	Silicate plaster (without synthetic resin additive)	10	-	0.8	0.01	3
Wall 3	Gypsum plasterboard	15	900	0.25	0.06	4
Cellulose	90% Glass wool MW	60	18	0.038	1.57	6
	10% Timber – frame a´= 625 mm	(60/40)	475	-	-	0
	Polyethylene (PE) vapour brake	1	-	0.5	0	5
	OSB boards	18	650	0.13	0.14	4
	89% Cellulose vertical cavity wall insulation	140	54	0.041	3.41	6
	(11%) Timber	(140/60)	-	-	-	-1
	OSB boards	18	650	0.13	0.14	4
	Mineral thermo-insulating panel	80	112	0.044	1.8	9.5
	Silicate plaster (without synthetic resin additive)	10	-	0.8	0.01	3
	Clay plaster	15	900	0.81	0.02	1

Wall 4 Sheep wool	Wood wool panel WW, magnesite-bonded	60	550	0.14	0.43	6
	Polyethylene (PE) vapour brake	1	-	0.5	0	5
	OSB boards	18	650	0.13	0.14	4
	90% Isolena Schafwolle, Optimal,	140	18	0.043	3.25	1
	20% Timber – frame a´= 625 mm	(140/60)	475	-	-	-1
	OSB boards	18	650	0.13	0.14	4
	Wood fibre WF-W (130 kg/m³)	120	130	0.046	2.6	13.9
	Silicate plaster (without synthetic resin additive)	6	-	0.8	0.01	3
Wall 5 Hemp fibre	Gypsum plasterboard	15	900	0.25	0.06	4
	90% Isolena Schafwolle, Optimal,	60	18	0.043	1.4	1
	10% Timber – frame a´= 625 mm	(60/40)	475	-	-	0
	Polyethylene (PE) vapour brake	1	-	0.5	0	5
	OSB boards	18	650	0.13	0.14	4
	80% Hemp fibre insulation	140	41	0.045	3.11	8.4
	20% Timber – frame a´= 625 mm	(140/60)	475	-	-	-1
	OSB boards	18	650	0.13	0.14	4
	Wood fibre WF-W (130 kg/m³)	80	130	0.046	1.74	9.3
	Silicate plaster (without synthetic resin additive)	10	-	0.8	0.01	3

4.2. Composition optimization of wooden sandwich structures in terms of fire protection

The fire resistance of the designed perimeter walls is directly dependent on the materials used. All the compositions consider the fire resistance of the interior covering up to timber supporting structure.

The required time for the fire resistance and the criteria for limit states of timber load-bearing elements of external walls are directly related to the static load. In the case of frameworks, the timber vertical elements are part of the supporting walls and are linked up by the horizontal ceiling or roof supporting elements. These structures, including all bracing and anchor elements, must meet the REI criterion.

The composite supporting elements of the fire-separating structures, as well as all the elements related to the supporting ones in terms of load, are required to meet the criteria for the fire resistance time. The interaction of all the elements provides the fire resistance of the whole structure. In the case of timber supporting elements, the value of fire resistance can be determined by calculation methods depending on the static load using STN EN 1995-1-1 (Eurocode 5). The fire resistance of timber or wood-based load-bearing and construction elements depends on the burning and charring speed.

As the analysed compositions are model solutions and the static stress of the supporting wall elements is unknown, the assessed fire resistance comprises layers up to the timber supporting wall structure. The chart in Figure 1 compares the individual construction solutions in terms of fire resistance and the extent of their use in multi-storey structures according to the requirements given in Table 1.

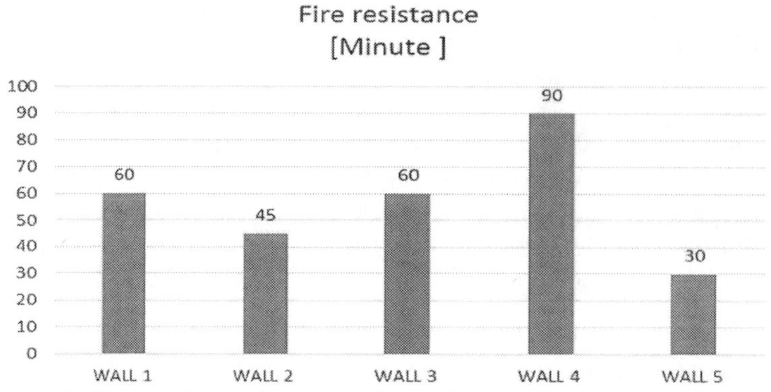

Figure 1. Comparison of construction solutions in terms of fire resistance.

5. Composition optimization of wooden sandwich structures in terms of environmental burden

The environmental burden is determined by the built-in energy consumption and emission production throughout the life cycle. The generally accepted equivalent for environmental impact assessment of building materials is the environmental index $OI3_{KON}$; it is given in Table 2 for individual envelope layers. It refers to $1m^2$ of a structure and takes into account the third-weights of eco-indexes drawn on $1m^2$ of a structure. www.baubook.at

The chart in Figure 2 compares the individual construction solutions in terms of the environmental burden of a member and it is expressed by the environmental index $OI3_{KON}$. Mainly recyclable materials with a high recycling rate were used in the proposed solutions.

Generally accepted equivalent for assessing the impact of building materials and structures on the environment is the value of CO_2ekv emission production along with the assessment of primary energy tied, and, from the regional point of view, the value of SO_2ekv emission production. OI3 calculates these three important categories of environmental protection for each squared meter of a building element on a scale 0 to 100.

Buildings are rated better if their ecological impact measured by OI3 is lower. Parametric data of the environmental burden for the individual layers were taken from the website www.baubook.at. Compositions are classified as A+ and A in terms of construction sustainability. [1]

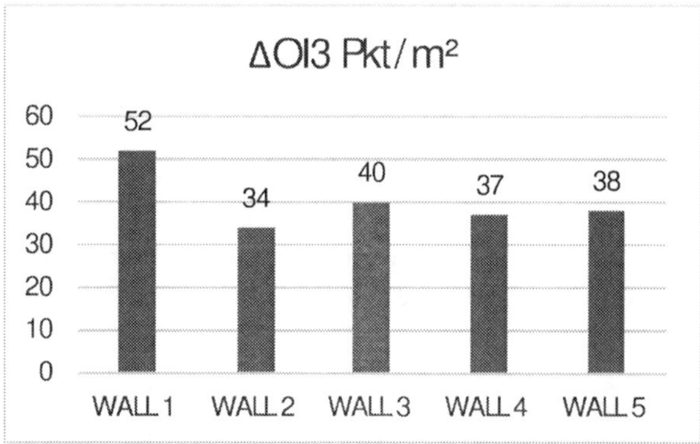

Figure 2. Comparison of individual construction outer solutions in terms of construction sustainability

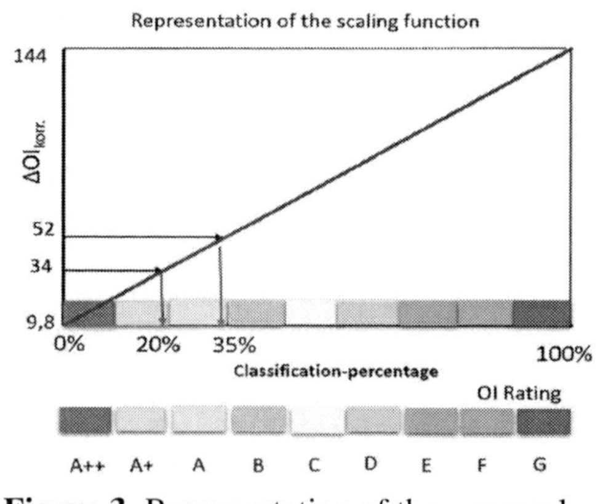

Figure 3. Representation of the - example. [1]

6. Possibilities of using building elements from recycled materials in the construction of wood-based houses

Environmental impact of buildings in the contemporary building standard is considerable. Optimization of building materials and products is important because the environmental burden in the building sector should be the lowest. The model composition solutions are based on using thermal insulation and covering from renewable resources such as wood, cellulose, hemp, sheep wool etc.
They are excellent thermal insulators and accumulators with low eco-index values. It is possible to use boards made from recycled municipal waste, e.g. TetraPack carton packages, as interior coverings. The production technology consists in grinding and pressing it under the high pressure and temperature into the shape of boards. The important fact is that no binding agents or chemical fillers are added to the base mixture, i.e. there is no environmental burden, and waste does not pollute the environment. The Slovak market currently offers TETRA K boards of size 2.700 x 1.200 mm and thickness up to 20 mm. The boards are made from a crushed polyethylene (PE) and have a paper cover layer. Their utility and physical parameters are comparable to the commonly used material such as plasterboards or cement-wood boards. They are re-recyclable after the lifetime period.

7. Conclusion

The increasing thickness of a thermal insulation, in terms of energy performance optimization, causes increasing fire resistance of the building envelopes where the non-combustible thermal insulation with the melting temperature of 1000 °C is used. The low-storey panel buildings, as well as kit homes with framework systems, are provided with the adequate fire-resistant load-bearing members of a building envelope, if thermal insulation is chosen optimally.

As for sustainability, it is important to reduce amount of waste from a building construction to an environmentally acceptable level, which means to use renewable and recycled materials.

Acknowledgements

The paper presents the results of the project VEGA 1/0945/16 - Theory and creation of energy effective and environmental friendly envelope constructions of wooden buildings.

References

[1] Denz P, Gmeiner H, Rohrer F, Sutter Ch 2017 *Oekoindex Bauteilbewertung, Leitfaden zur ökologischen Bewertung von Bauteilkonstruktionen mittels Oekoindex, Leitfaden V1.4* (Dornbirn: Energieinstitut Vorarlberg)
[2] Ďurica P 2012 *Defects in buildings - diagnostics and remedies* (Žilina: Žilinská univerzita)
[3] Makovická-Osvaldová L, Gašpercová S 2015 The evaluation of flammability properties regarding testing methods *Civil and Env. Eng.* **11(2)** 142-146
[4] Iringová, A.: Lightweight building envelopes in prefabricated buildings in terms of fire resistance. In: *Matec Web of Conferences : Vol. 117 (2017) : RSP 2017 - XXVI R-S-P Seminar 2017 Theoretical Foundation of Civil Engineering : Warsaw, Poland, August 21-25, 2017* (online).
[5] STN EN 1995-1-2, Eurocode 5, Design of wooden structures - Part 1 - 2, General rules – Design of structures for fire effect. SUTN, Bratislava, 2008.
[6] Lokaj, A. et al.: Wooden houses and structures, Volume I. and II. (in Czech). CERM, Brno, 2010.
[7] Vilcekova S., Burdova E.: Multicriteria analysis of building assessment regarding energy performance using a life cvycle approach / 2014
In: International Journal of Energy and Environmental Engineering. Vol. 5, no. 2 (2014), p. 1-9. ISSN 2251-6832
www.baubook.at

IOP Conf. Series: Materials Science and Engineering **415** (2018) 012011 doi:10.1088/1757-899X/415/1/012011

Effect of two different types of building envelope insulation on future heating demand of a family house using simulation software

P Turcsanyi [1], **A Sedlakova** [2]

[1] Faculty of Civil Engineering, Institute of Architectural Engineering, Vysokoskolska 4, 042 00 Kosice, Slovakia
[2] Faculty of Civil Engineering, Institute of Architectural Engineering, Department of Building Physics, Vysokoskolska 4, 042 00 Kosice, Slovakia

peter.turcsanyi@tuke.sk

Abstract. Computer programs allowing determining accuracy of building design in advance, or its parts, from the thermal engineering point of view are on the rise. Finding the most suitable project design for optimizing future energy performance of building significantly contributes to implementation of European Directive on Energy performance of buildings 2010/31/EU in Slovakia. Using simulation software, DesignBulder in this case, became a very useful tool on the road to energy effective design. In this paper, we have placed a virtual family house (which will be built in Kosice, Slovakia) to the simulation software DesignBuilder and were finding out the most suitable design in case of future heating demand. Family house was designed taking in account architectural, environmental and constructional requirements of today´s directives focusing on energy performance and energy efficiency. Results of this project are displayed in numbers as well as in graphic figures. Our goal was to find out the difference between two different types of envelope wall insulation.

1. Introduction

The result of architectural and construction solution affect many parameters. Those parameters either can or cannot be affected by designer; nevertheless, first raw design usually needs to be optimized according to importance of energy efficiency of building. Finding optimal solution means knowing correlation between key parameters which are affecting energy performance of buildings (i.e. heat demand, cooling demand, thermal comfort, etc.). After simulation on energy performance of building, designers are able to optimize parameters to find the best ratio between initial cost and long term energy save. Currently, there are more than 400 applications that can be applied to analyzing building energy and thermal simulation [1].

Building thermal simulation tools predict the thermal performance of a given building and the thermal comfort of its occupants. In general, they support the understanding of how a given building operates according to certain criteria and enable comparisons of different design alternatives [2]. Evaluation of thermal comfort involves assessment of at least six factors: human activity levels, thermal resistance of clothing, air temperature, mean radiant temperature, air velocity and vapour pressure in ambient air [3]. In this paper, we are comparing two different types of external wall insulation and their effect on heat demand for heating. First is commonly used wall thermal insulating material EPS polystyrene, part of ETICS system. On the other hand, as the age and research go further every day and new materials are being introduced into everyday construction life, insulation based on phenol insulation board is being assessed in this paper.

2. Theory

House that is being evaluated is located in Košice-Krasna, Slovak Republic. City of Košice lies at an altitude of 206 meters above sea level and covers an area of 242.77 square kilometers. It is located in eastern Slovakia, about 20 kilometers from the Hungarian borders, 80 kilometers from the Ukrainian borders, and 90 kilometers from the Polish borders (figure 1). It is about 400 kilometers east of Slovakia's capital Bratislava. Košice city is situated on the Hornád River in the Košice Basin, at the easternmost reaches of the Slovak Ore Mountains. More precisely it is a subdivision of the Čierna hora Mountains in the northwest and Volovské vrchy Mountains in the southwest. The basin is met on the east by the Slanské vrchy Mountains [4]

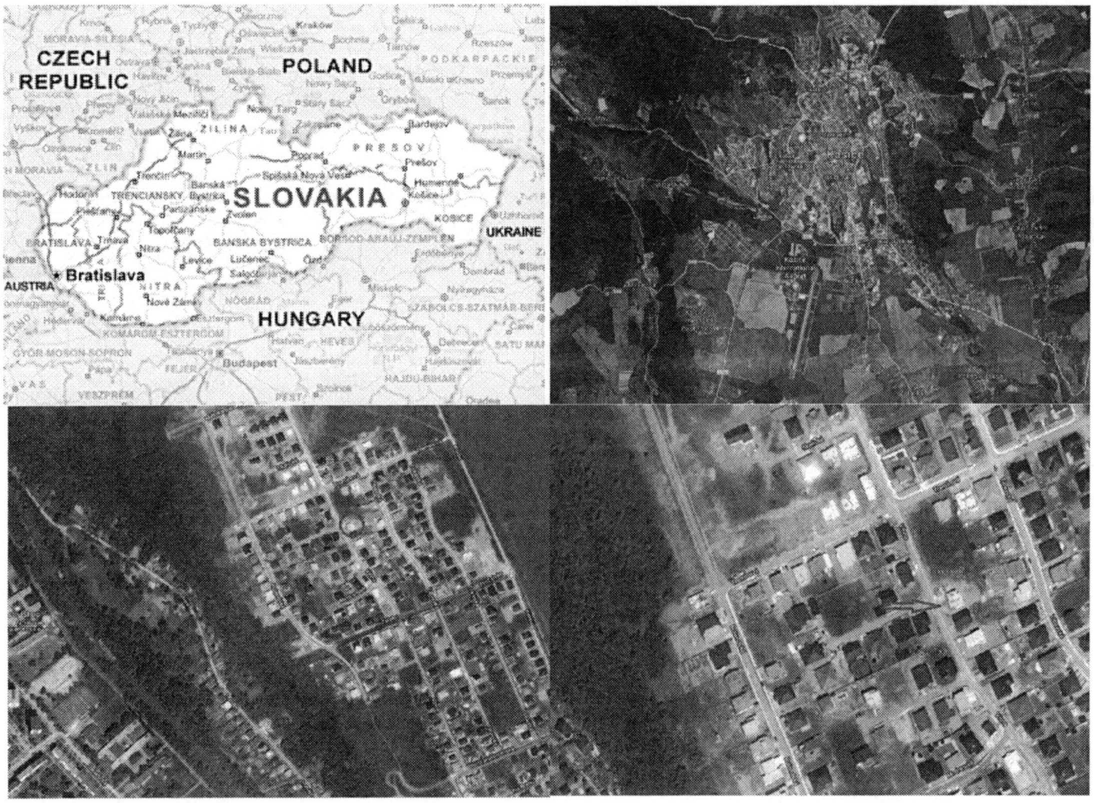

Figure 1. Position of Kosice city and detailed location of evaluated family house

Two-storey family house represents a typical type of residential buildings built in this area. The floor layouts were designed for a family of two parents and three kids and the concept of a whole house was made in according to the latest energy standards. Building is based on a concrete footing foundation. The family house envelope was designed using materials with high thermal capacity. As for thermal insulation - thick layer of polystyrene (ETICS system)is being used, to minimize the heat loses during the heating period as well as to prevent the heat transfer during the summer period. Roof structure creates unconditioned spade between insulated ceiling and top of the roof. Transparent parts of building envelope were designed as wood-aluminum, triple glazed windows Internorm with outside louvers for a regulation of solar gains through the windows.

2.1. Phenolic insulation

The main use of phenolic foam in construction is as thermal insulation, to improve is thermal efficiency. The thermal efficiency of the existing building stock is of prime concern, if carbon emissions targets are to be met. For those homes with solid walls, external wall insulation can be an effective method of improving their thermal efficiency. Phenolic foam is starting to be popular material choice for external wall insulation systems owing to its low thermal conductivity and good fire performance [5]. Phenolic foam is made from three main components: phenolic resin, a blowing agent and an acid catalyst; a number of additives can also be utilised to develop specific properties within the foam. There are patents and other information available at [6] that describe the chemicals used to manufacture phenolic foam.

3. Materials and methods

The external wall insulated with polystyrene represents the most common insulation type of ETICS insulating system. On contrary, phenolic insulation materials are used quite rarely – but they are on the rise. The main reason is its price, which is extensively higher than polystyrene (or any other classical insulation material). However, phenolic insulation boards allow us to design energy efficient building while keeping the external wall construction as subtle as possible due to its low coefficient of thermal conductivity of λ=0,020 W/m·K. With its thickness, phenolic insulation board of thickness 6 cm equals to 12 cm of polystyrene.

Table 1 Physical and thermal parameters of building

Shape factor (A/V ratio)	0.78 1/m
Volume of building space	726.18 m^3
Total heat transfer surface	569.48 m^2
Total floor area	222.09 m^2
Wall U-Value	0.148 W/(m^2·K)
Insulated ceiling U-Value	0.123 W/(m^2·K)
Ground floor R-Value	4.38 (m^2·K)/W
Wood-aluminum U-Value Internorm HF 350	0.60 – 0.95W/(m^2·K)

3.1. Technical parameters of an envelope wall type A – polystyrene insulation

Envelope wall type A is a classic type of the external envelope wall insulated with EPS polystyrene. Polystyrene insulation is the most common insulating material used in Slovak Republic. Polystyrene insulation is being put on masonry or any other wall structure, using adhesive mortar and anchors.

Table 2. Thermo-physical properties of an envelope wall type A

	d [m]	λ [W/m·K]	c [J/kg·K]	ρ [kg/m^3]	m [kg/m^2]
Plaster	0.020	0.570	1000.0	1300.0	10.0
Porotherm 38 Ti Profi	0.380	0.134	1000.0	750.0	29.0
EPS insulation	0.120	0.038	1050.0	18.0	45.0

Adhesive mortar	0.005	0.800	920.0	1300.0	18.0
Silicon render	0.003	0.700	920.0	1700.0	37.0

3.2. Technical parameters of an envelope wall type B – phenolic insulation

Envelope wall type B is a type of the external envelope wall insulated with phenolic boards, in this case Kingspan KOOLTHERM K5. The core of Kingspan Kooltherm K5 External Wall Board is a fibre–free performance rigid thermoset phenolic insulant manufactured with a blowing agent that has zero Ozone Depletion Potential (ODP) and low Global Warming Potential (GWP). The core of Kingspan Kooltherm K5 External Wall Board has a 90% closed cell structure. Phenolic board insulation is being put on masonry or any other wall structure, using adhesive mortar and anchors.

Table 3 Thermo-physical properties of an envelope wall type B

	d [m]	λ [W/m·K]	c [J/kg·K]	ρ [kg/m^3]	m [kg/m^2]
Plaster	0.020	0.570	1000.0	1300.0	10.0
Porotherm 38 Ti Profi	0.380	0.134	1000.0	750.0	29.0
Phenolic insulation	0.060	0.020	800.0	35.0	300.0
Adhesive mortar	0.005	0.800	920.0	1300.0	18.0
Silicon render	0.003	0.700	920.0	1700.0	37.0

4. Simulation

Using simulation software DesignBuilder, the assessed family house was simulated. In the first simulation, polystyrene was used as wall insulation. Second simulation shows the alternative to EPS insulation - phenolic insulation board.

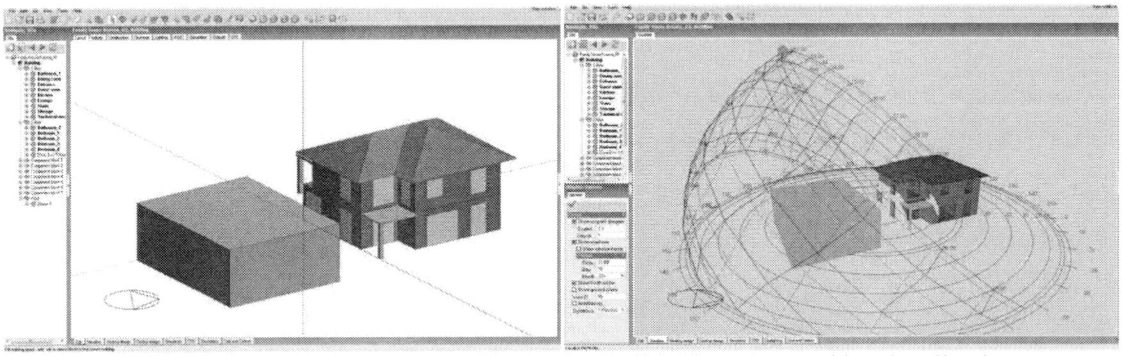

Figure 2. Family house created in DesignBuilder interface and its visualization

Weather data of the reference year for Kosice were used in this simulation. They provide inputs of dry bulb air temperature, relative humidity, solar radiation, wind speed, wind direction etc., for achieving the most accurate and realistic simulation.

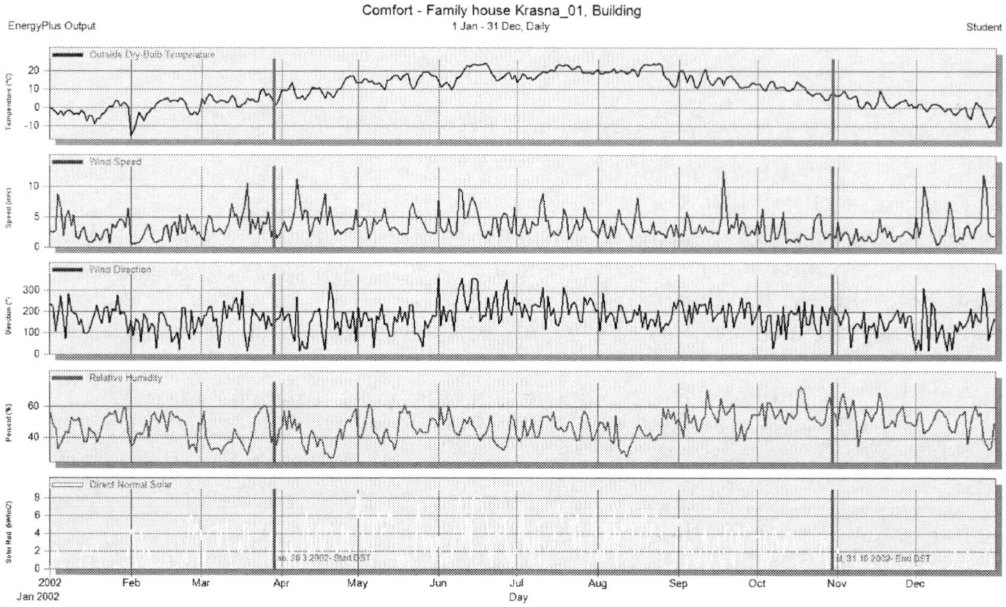

Figure 3. Weather data for the reference year in Košice, Slovakia. From the top: Outside dry-bulb temperature, wind speed, wind direction, relative humidity, solar intensity

5. Results

Building simulations were set up with the DesignBuilder v4 software, in which building performance data were generated by the simulation engine EnergyPlus.

5.1. Envelope wall type A – polystyrene insulation

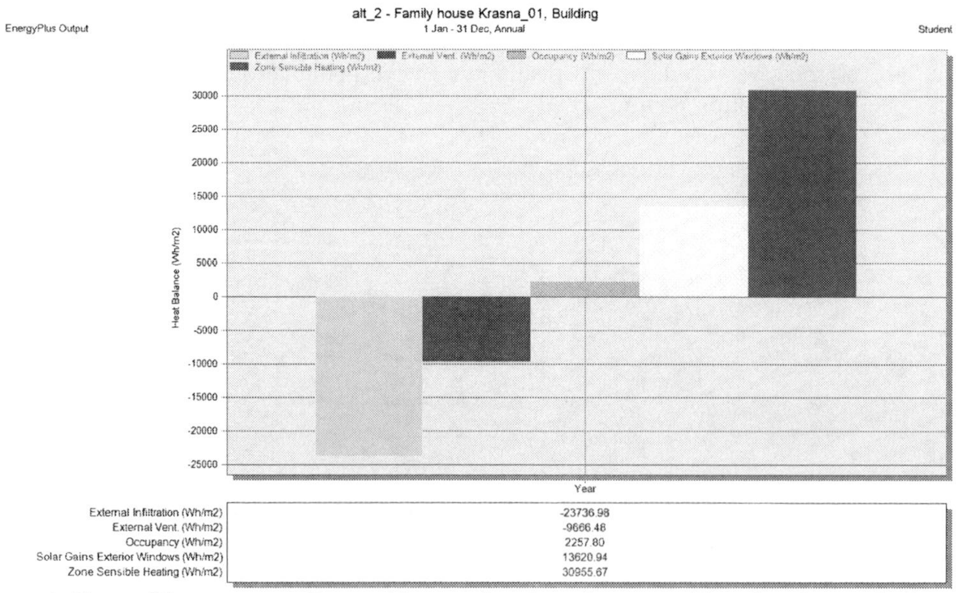

Figure 4. EnergyPlus outputs of annual heat gains, heat losses and heating demand of type A

Figure 4 shows annual heating demand of a family house, with envelope wall insulated by EPS polystyrene alternative, on square meter of floor area is 30,96 kWh/m^2·a. Solar gains through the exterior windows are 13,62 kWh/m^2·a and a heat gains from the occupancy 2,2 kWh/m^2·a. On the other hand heat losses via external infiltration and ventilation are little higher than 32 kWh/m^2·a.

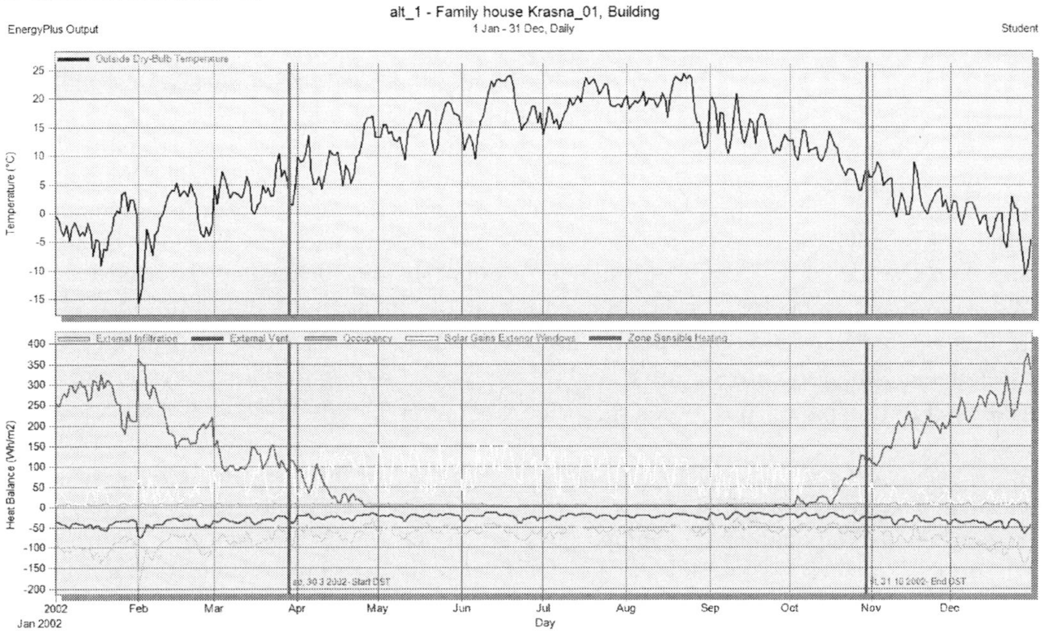

Figure 5. Daily data of heating demand, solar gains through exterior windows, heat gains from occupancy, losses from infiltration and ventilation

As seen on figure 5, daily extremes on heating demand peak on 1st of February, with over 0,35 kWh/m^2. This, of course is a result of the outside dry-bulb temperature drop.

5.2. Envelope wall type B – phenolic insulation

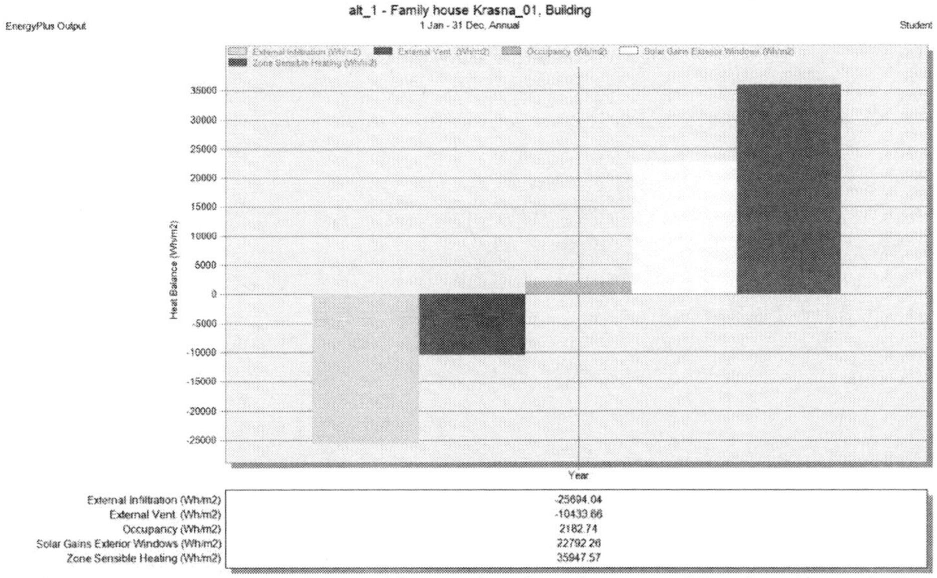

Figure 6. EnergyPlus outputs of annual heat gains, heat losses and heating demand of type B

Figure 6 shows annual heating demand of a family house, with envelope wall insulated by phenolic insulation board alternative, on square meter of floor area is 35,9 kWh/m^2·a. Solar gains through the exterior windows are 13,62 kWh/m^2.a and a heat gains from the occupancy 2,2 kWh/m^2·a. On the other hand heat losses via external infiltration and ventilation are little higher than 32 kWh/m^2·a.

6. Conclusion

Nowadays, not only in Slovakia but worldwide, building sector is heading towards emissions-free goal. Every day, more and more new and efficient materials are being implemented into construction process. One of these materials with future is phenolic insulation. Classical insulation used in ETICS systems such as polystyrene, mineral wool, etc., with thickness over 20cm, are being slowly replaced by thinner and subtle insulation materials – such as phenolic boards. As seen from results, envelope wall insulated by polystyrene with thickness of 12cm, can be replaced by phenolic insulation board with the thickness equal to half of polystyrene – 6cm. Heating demand is affected by 5 kWh/m^2 annually. Given the price for 1 kWh, total amount of finances saving equals to 35 € a year, which is not considered as a significant improvement in saving costs on heating. One of the biggest cons of phenolic insulations is its high price. Compared to polystyrene insulation material used in this paper, it doubles the price with over 17 €/m^2.

With more energy efficient materials with very low coefficient of thermal conductivity, such as phenol based insulation systems, it is possible reaching goals as net-zero energy buildings or even energy plus buildings while making building envelope structure more subtle and modern. With the help of simulation tools, such as DesignBuilder, it is feasible to find the most efficient option for building structures, taking into account costs on future energy demand as well as costs on construction itself.

Acknowledgments
This work was supported by the Grant Agency of the Slovak Republic No. 1/0307/16.

References

[1] Energy Efficiency and Renewable Energy (EERE). Building Energy Software Tools Directory, U.S. Department of Energy. Available online: http://apps1.eere.energy.gov/buildings/tools_directory/(accessed on 18 January 2013).

[2] Maile, T.; Fischer, M.; Bazjanac, V. *Building Energy Performance Simulation Tools—A Life-Cycle and Interoperable Perspective*; Stanford University: Stanford, CA, USA, 2007.

[3] Fanger, P.O. *Thermal Comfort, Analysis and Applications in Environmental Engineering*; McGraw-Hill Book Company: New York, USA, 1970.

[4] *Available: https://en.wikipedia.org/wiki/Ko%C5%A1ice.*

[5] Densley Tingley, Hathway, Davison, Allwood. *The environmental impact of phenolic foam insulation boards,* Proceedings of the Institution of Civil Engineers, Volume 170 Issue 2, April, 2017, pp. 91-103, 2015.

[6] Kingspan Holdings Limited (2006) A Phenolic Foam. International Patent Application No. PCT/IE2006/000096, 8 September.

Strength and hygric properties of ETICS applied at various temperature, using adhesives with different cellulose ethers

P Konca[1]

[1]Lodz University of Technology Poland

Abstract. The ETICS comprises a prefabricated insulation product bonded to the wall or a combination of adhesive and mechanical fixings. The insulation is faced with a rendering consisted of one or more layers (site applied), one of which contains a reinforcement. Adhesives used for bonding the insulation to the wall and base coat include cellulose ethers. These chemical compounds are used in building materials as thickeners, binders, film formers, and water-retention agents. Cellulose ethers influence on the physico-mechanical characteristics of a mortar. Temperature application and then conditioning of adhesives strongly affect the water retention. This paper presents a study on the impact of different temperatures on the setting and hardening adhesives with three types of cellulose ethers (HEC, HPMC and HEMC). The tests included water retention in freshly – mixed mortars and bond strength between the adhesive and substrate in mortars. Water transport in freshly-mixed mortars containing cellulose ethers were examined with classical and new methods Time Domain Reflectometry. During the laying and curing time the ambient temperature was hold on to 5±2 °C (reduced), 23±2°C (standard), 30±2°C(increased) in the relative humidity 50±5% or 80±5%. Adhesives were cured for 4 weeks. The pull-off test was performed on the following samples: without supplementary conditioning (dry condition) and after immersion of the adhesive in water for 2 days and 2 h drying at laboratory conditions. The tests were performed on the substrates: a smooth concrete slabs and the expand polystyrene insulation boards. The hygric properties were measured in the sample thickness 12 mm and 9 mm during the first 24 hours.

1. Introduction

Mortars in ETICS systems contain the two most important admixtures: redispersible polymer powder and one of three types of cellulose ethers. The redispersible polymer powder primarily improves adhesion to the substrate. Cellulose ethers increase viscosity and water retention in thin layers of mortar. The evaporation of water in the materials depends on the temperature and the relative air humidity. The paper presents the influence of temperature on water retention during bonding of cement mortar.

2. Materials

A typical adhesive formulation was used to investigate the effect of three various types of cellulose ethers (HEC, HPMC and HEMC) on the physico-mechanical characteristics of a mortar. The following commonly applied materials were used in the investigations over adhesives for ETICS. Basic adhesive composition shown in table 1 did not contain cellulose ethers.

Table 1. Mix design for basic adhesive "0".

Component	Content (*% mass of dry mixture*)
Portland cement CEM I 42.5 R	20.0
Quartz sand with a grain size of 0.1–0.5 mm.	65.8
Calcium carbonate fine powder with a grain size below 0.1 mm.	12.2
Redispersible powder consisting of copolymer ethylene and vinyl acetate (VAC/Et)	2.0
Cellulose ether	-

The properties of the cellulose ethers depend on both the molecular weight of the polymer and on the degree of etherification. The mechanical properties of the ethyl cellulose are reported to be influenced more by the former, whereas other physical properties such as solubility and water absorption are influenced more by the latter [1]. This information is not usually provided as it is considered to be confidential by manufacturers. It has been demonstrated that the average molecular weight of the polymer is linked to its solution viscosity, when it is dissolved in water at a set concentration and reference temperature [2].

Three sets of cellulose ether were selected: hydroxyethyl cellulose (HEC), methyl hydroxyethyl cellulose (MHEC), hydroxypropyl methyl cellulose (HPMC). Data of the cellulose ether (viscosity - in solutions 2%, method Brookfield RV 20 rpm) and content in adhesive are summarized in table 2.

Table 2. Composition of adhesives with content of cellulose ethers and water.

Name	Component name (-)	Viscosity (*mPa·s*)	Content of (*% mass of dry mixture*)	Water/powder ratio (*% mass of dry mixture*)
Adh 0	-	-	0.0	16
Adh 1	HEC	20 000	0.2	20
Adh 2	MHEC	30 000	0.2	18
Adh 3	HPMC	7 000	0.2	18

3. Methods

3.1. Test condition and mixing of the adhesive

Standard conditions were maintained at (23 ± 2) °C and (50 ± 5) % relative humidity and a circulation of air in the testing area of less than 0,2 m/s. After application mortars were cured for 4 weeks at temperature 5±2°C (reduced) in the relative humidity RH 80±10%, 23±2°C (standard) in RH 50±5%, 30±2°C (increased) in 50±5%. At the same temperature retention capacity and adhesion tests were performed. The samples were stored in three different climate chambers (Uni-Mors, Weiss Technik C1000, ITEPIB Radom).

The amount of water required for preparing the adhesive was stated in proportion to weight (e.g. the amount of liquid, in kilograms and the amount of powder, in kilograms). A minimum quantity of 2 kg of the adhesive was prepared in a mixer of the type described in 4.4 of EN 196-1 [3], using the slow speed settings, (140±5) r/min mix for 30 s; scrape down the paddle and pan within 1 min and next (62±5) r/min planetary movement for 1 min. The adhesive was left at rest for 5 minutes before a final 15 s homogenization.

3.2. Consistency

Water/powder ratio was established experimentally at level 120 – 140 mm by flow tables for mortar and building lime [4]. The flow was measured in two directions.

3.3. Water retention

The fresh mortar with accepted level of consistence was subjected to a suction treatment under defined conditions using filter paper as a substrate. The water retention of the mortar is the mass of water retained in the mortar after the suction treatment and is expressed as a percentage by mass of its original water content [5].

The adhesive was poured into ring previously placed on a filter paper and non-fibrous textile. The assembly was weighted and covered with a plastic plate, next left to stand for 5 minutes. Water retention capacity WRV (in %) was calculated according to Eq. (1)..

$$WRV = 100 - \frac{W_3}{W_2} \times 100 \qquad (1)$$

where: W_2 - the mass of water in the mortar,
W_3 - the mass of water absorbed by filter paper as a substrate.

3.4. TDR method for the measurement of water content

The Time Domain Reflectometry (TDR) method for the measurement of soil water content becomes more and more popular because of its simplicity of operation, accuracy and non-destructiveness, as compared to other methods [6,7]. This measuring technique takes advantage of three physical phenomena characteristic to the materials, i.e.:

- in the frequency range of 1 GHz the complex dielectric constant of the material is reduced to its real value and the electromagnetic wave propagation velocity, v,
- a dielectric constant of the material liquid phase has much higher value than other phases, i.e., about 80 against 2÷4 for the solid and 1 for the gas phase;
- the relation between the material moisture and its dielectric constant is highly correlated [8,9].

TDR measurements were made using LP/ms Laboratory Operated Meter (LOM) probes and TDR/MUX Integrated Measuring Module produced by E-Test company under the license of the Institute of Agrophysics, Polish Academy of Sciences in Lublin. The measuring system was calibrated with water and ethocel as calibration media, according to the manufacturer instructions [10]. All samples were 9 mm thin and surface that allowed complete immersion needles length of 52 mm. The results were then correlated with the initial water content and after test with the adhesive dried to constant weight.

3.5. Bond strength test between adhesive and substrate

The tests were performed on the following substrates:

- A substrate consisting of a smooth concrete slab at least 40 mm thick. The water/cement ratio was contained in the range of 0.45 to 0.48. The tensile strength of the slab was at least 1.5 N/mm². The moisture content of the slab prior to the test was a maximum of 3 % of the total weight.
- A substrate consisting of expanded polystyrene (EPS) foam sheet at density 19,5 kg/m^3.

The adhesive was spread on the substrate to a thickness of 3 mm and with an area of 19,6 cm². After allowing the adhesive to cure at three temperatures and corresponding humidity for at least 28 days, metal plates of appropriate size were bonded to it using a suitable adhesive. The pull-off test was performed at a tensioning speed of 1 to 10 mm/minute on the following samples:

- without supplementary conditioning (dry condition),
- after immersion of the adhesive in water for 2 days and 2 h drying at (23±2)°C and (50±5)% RH,

The individual and mean values were recorded and the results expressed in N/mm² [11].

4. Results

4.1. Water retention

Right amount of water is important for the proper setting of the cement in the adhesive. Water retention is essential especially if thin layers of mortar do not contain redispersible powder. The increase of water retention is interpreted as an increase of viscosity of the water phase in modified mortar. The standard EN instruct to the use of 12 mm thick discs. Result for adhesives are presented in figure 1.

Figure 1. Water retention capacity WRV (in %) at different temperatures and humidities.

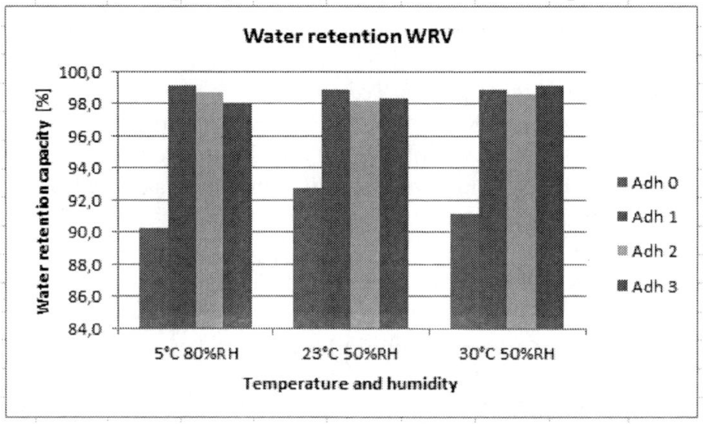

Before the test materials were stored in various appropriate conditions. Mortar without cellulose ethers kept the water at about 91% and the other materials at about 98%. At this thickness, in such a short time, the effect of temperature was small.

4.2. TDR method for the measurement of water content

The Time Domain Reflectometry was alternative method to weight method of water mass loss. One cannot use the precision balance in continuous measurement by vibration and low temperature in the climate chamber. The results of water content changes in the first 24 hours are shown in figure 2.

Figure 2. Changes in the content of water in the first 24 h at various temperatures and humidities.at different temperature and humidity.

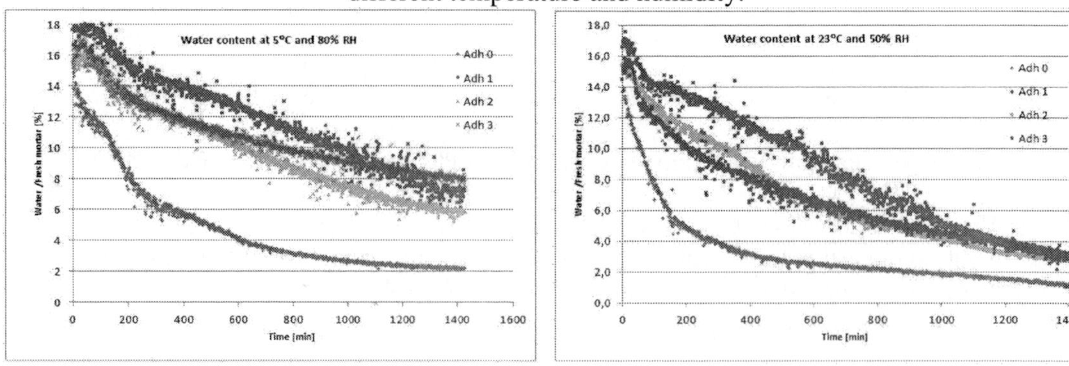

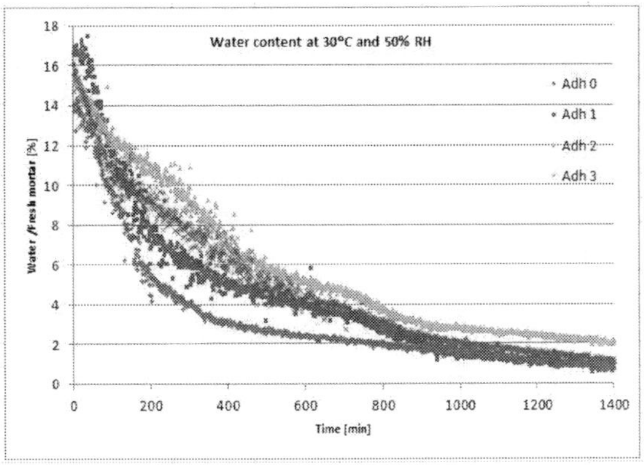

The initial water content in the mortar as wall as temperature and humidity had an impact on the obtained results. High humilities and low temperatures improved water retention in mortar. High temperature and low humidity increased water loss, but positive effects of the cellulose ethers were observed during the first six hours. The author observed effect of surface drying in adhesive with cellulose ethers. These admixtures sucked water from the surface to the interior of the mortar. Such a phenomenon was not observed in the mortar Adh 0, for which the surface remained wet for a long time.

4.3. Bond strength test between adhesive and substrate

Before applying the adhesive plate provided in appropriate chambers for 24 hours to obtain the desired temperature, one had to maintain a constant temperature and humidity; 5°C at 80±5% RH and 23°C or 30°C at 50±5% RH. After 28 days in dry condition and after immersion of the adhesive in water for 2 days and 2 h drying, samples were executed pull-off test (figure 3).

Figure 3. Bond strength adhesive to: (a) smooth concrete slab, (b) expanded polystyrene (EPS) foam sheet after immersion in water.

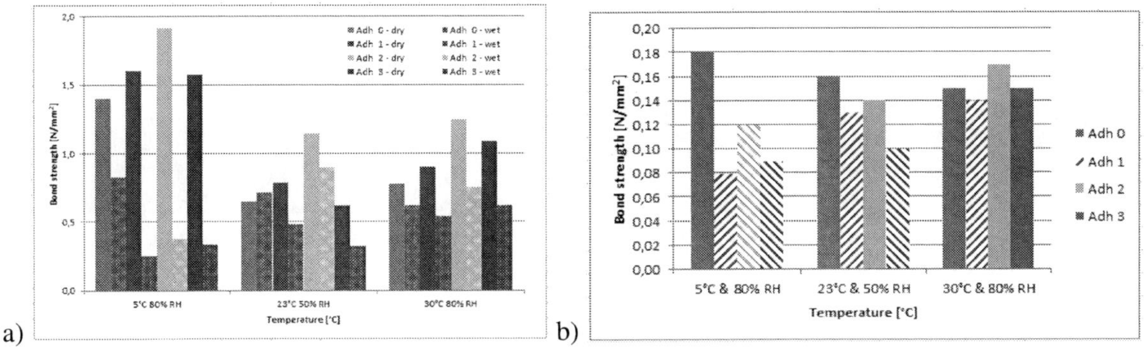

Three types of failure were observed: Cohesive Failure within the Substrate (CF-S), Cohesive Failure within the Adhesive (CF-A), Adhesive Failure between adhesive and Substrate (AF-S). In the case of expanded polystyrene (EPS) in dry condition all damages were of type CF-S. The obtained results were contained in the range of 0,15 to 0,18 N/mm^2 and it was EPS tensile strength. In the case of the concrete slabs all failure types were CF-A. Hydration mortar after water immersion weakened adhesive but strength level is higher than EPS tensile strength. The situation was different when we had a wet mortar on EPS. Type of destruction depended on how much moisture was kept in the sample at the time of the test. Adhesive without cellulose ethers dried faster (method provides for two hours in a laboratory after removal from the water). In this case all destruction were of types of CF-S (solid fill in figure 3b). Lower temperature of storage caused destruction type AF-S (texture fill in figure 3b) for adhesives with cellulose ethers.

IOP Conf. Series: Materials Science and Engineering **415** (2018) 012012 doi:10.1088/1757-899X/415/1/012012

5. Conclusions

Based on the results presented herein the following conclusions may be drawn:

- cellulose ethers change the plastic properties and consistency of the adhesive.
- temperature and humidity do not affect significantly on short time water retention WRV. Large differences are observed in mortars with or without cellulose ethers.
- TDR method allows the observation of changes in the amount of water in the long term. Character of change is strongly dependent on humidity and temperature in the first 24 hours. The hygral properties are especially important in thin layers of mortar. The pure cement mortar (without any chemical admixtures) quickly evaporates the water, lack of it later in the hydration process.

6. References

[1] Lorand E J 1938 Cellulose ethers: variations of physical properties with composition. *Industrial & Engineering Chemistry* **30** (5) 527-530.

[2] Petit J-Y, Wirquin E 2013 Evaluation of various cellulose ethers performance in ceramic tile adhesive mortars. *International Journal of Adhesion and Adhesives* **40** 202-209

[3] EN 196-1. Methods of testing cement. Determination of strength

[4] EN 1015-3. Methods of test for mortar for masonry. Determination of consistence of fresh mortar (by flow table)

[5] EN 459-2. Building lime. Test methods

[6] Malicki M A 1993 The influence of physical soil properties on the electrode soil parameters in the aspect of the measurement of soil moisture and salinity (in Polish). Habilitation Thesis,. *Acta Agrophysica, Institute of Agrophysics,* Polish Academy of Sciences

[7] Malicki M A, Plagge R, Roth C H 1996 Reduction of soil matrix effect on TDR dielectric moisture determination by accounting for bulk density or porosity. *European J. Soil Sci.* **47** 357-366

[8] Malicki M A, Skierucha W M 1989 A manually controlled TDR soil moisture meter operating with 300 ps rise-time needle pulse. *Irrigation Sci.* **10** 153-163

[9] Topp G C, Davis J L, Annan A P 1980 Electromagnetic determination of soil water content: measurements in coaxial transmission lines. *Water Resour. Res.* **16** 574-582

[10] Skierucha W M, Wilczek A, Alokhina O 2008 Calibration of a TDR probe for low soil water content measurements. *Sensors and Actuators A: Physical.* **147** 544–552

[11] ETAG 004 Guideline For European Technical Approwal of External Thermal Insulation Composite Systems With Rendering , European Organisation for Technical Approvals, Edition 2011. 42-43.

IOP Conf. Series: Materials Science and Engineering **415** (2018) 012013 doi:10.1088/1757-899X/415/1/012013

Hygrothermal assessment of internally insulated brick wall based on numerical simulation

D Kaczorek

Building Research Institute, Department of Thermal Physics, Acoustics and Environment, Warsaw, Poland

Email: d.kaczorek@itb.pl

Abstract. Massive walls exposed to external climate are characterized by low levels of thermal insulation. Unfortunately in many cases, due to the historic character of the building, the only possible way to improve their thermal insulation is the use of internal insulation. However, such solutions are associated with the risk of frost damage, mould growth or interstitial condensation. Calculations regarding the occurrence of these risks require complex calculation methods. Therefore, in the present work, the evaluation of the hygrothermal behavior of the internally insulated solid brick wall was performed based on the numerical simulations using the program WUFI Pro. Various material solutions of internal insulations, based on both synthetic and natural materials, were compared. The focus was on a particularly sensitive interface between the insulation and the original wall

1. Introduction

The improvement of thermal insulation for external partitions is an important and effective means to achieve energy savings in relation to existing buildings [10]. The process of building's retrofit is complex and require integrating different aspects, which should be taken into account, i.a: the interaction between the building envelope and its internal environment to balance energy costs, feasibility, ecology, aesthetics and user comfort [1]. In the case of heritage buildings, and all other where there is no possibility of interference in the appearance of external façade, the only possible solution becomes the use of interior insulation. However, the improper interior insulation thermal modernization of the walls can lead to undesirable thermal bridges, interstitial condensation, and frost damage in outer layer. Therefore, all issues related to the presence of moisture in a partition should be considered at the design stage.

Among the interior insulation systems, there are both vapour-tight and vapour permeable solutions [5]. The former type includes the diffusion-tight, synthetic foams, which impede the vapour diffusion from the indoor space. In the case of this type of insulation it is necessary to lay it very carefully, ensuring a perfect tightness, and avoiding an uncontrolled flow of vapour through the partition. Such solutions limit the drying potential, leading to increased moisture content in the underlying wall [7]. On the other turn, the vapour-permeable solutions allow the exchange of water vapour between the masonry and the indoor environment. However, it may lead to interstitial condensation, especially during the heating season, when the warm vapour is penetrating from the indoor to the cold masonry through the insulation layer [8]. Therefore, in order to avoid the excessive moisture accumulation in the masonry, as an alternative solution, the capillary-active systems are promoted, allowing moisture to dry out towards the indoor space [2, 6]. However, the question is, what their performance will be when they are exposed to uncontrolled indoor climate conditions.

The hygrothermal simulation software WUFI Pro was used to evaluate the moisture performance of traditional masonry wall with internally added thermal insulation with both synthetic and natural materials such as: cellulose, aerogel, phenolic foam, calcium silicate. As an indicator for the assessment, the relative humidity level at the interface between the insulation and masonry was used. The presented analyses related only to a typical one-dimensional cross-section, and did not take into account structural details such as corners, joints of floor beams, etc., simplifying the masonry wall to the homogenous layer. According to Vereecken and Roels [9], such approach is allowed for assessing the hygric behaviour of masonry wall, provided that the analyses are made in relation to real climatic data.

2. Materials and methods

As a case study, a traditionally constructed masonry brick wall, insulated internally was used. In its original form without any insulation, the wall was referred as a base wall. The base wall, named "historical" has an overall thickness of 380 mm and internal lime plaster. For testing the wall's hygrothermal performance the WUFI Pro 5.3 [12] software tool, based on numerical simulation in accordance with EN 15026:2007 [3] was used. The analyses were done for four different insulation product, characterized in the table 1.

Table 1. Comparison of the insulation products assessed in the case study.

Product name	Cellulose fibres, sprayed	Aerogel blankets	Phenolic foam boards	Calcium silicate boards
Description	sprayed cellulose fibres, finished with plaster-board	aerogel-coated polyester mesh, bonded to plasterboard	boards made from phenolic resin, foil and plaster-board	calcium silicate boards, finished with lime plaster
Natural or synthetic	natural (plant fibres)	synthetic (aerogel & polyester)	synthetic (phenolic resin)	natural (sand &lime)
λ value ranges [W/(m·K)]	0.035-0.046	0.013-0.014	0.020-0.025	0.045-0.065
Vapour resistance	air	air	air	air
Absorption behaviour	highly permeable	permeable	impermeable	hygroscopic
Capillary behaviour	hygroscopic	hydrophobic	not hygroscopic	hygroscopic
Capillary behaviour	capillary active	capillary inactive	capillary inactive	highly capillary
AVCL[a] incorporated in product	no	no	yes	no
Fixing method	bonded to masonry (no separate adhesive)	mechanically fixed to battens	mechanically fixed to battens	bonded to masonry with adhesive
Air gap at insulation	no	yes	yes	no

[a] AVCL-air and vapour control layer.

Depending on the insulation product, the materials include insulation, plasters, plasterboards, adhesives, vapour barrier and air cavities. The material properties used in the case study are listed in table 2 and 3, respectively with data provenance.

Table 2. Properties of insulation and brick

Material properties	Cellulose fibres, sprayed	Aerogel blankets	Phenolic foam boards	Calcium Silicate boards	Historical brick	Vienna brick
Bulk density [kg/m^3]	55 ^{W}	146 ^{W}	43 ^{N}	222 ^{P}	1800 ^{W}	1560 ^{W}
Porosity [m^3/m^3]	0,93 ^{P}	0,92 ^{W}	0,95 ^{W}	0,92 ^{P}	0,31 ^{W}	0,38 ^{W}
Specific heat capacity [J/(kg·K)]	2544 ^{P}	1000 ^{W}	1400 ^{N}	1303 ^{P}	850 ^{W}	850 ^{W}
Thermal conductivity [W/(m·K)]	0.036 ^{P}	0.014 ^{W}	0.023 ^{N}	0.057 ^{P}	0,6 ^{W}	0,6 ^{W}
Vapour diffusion resistance factor [-]	2,0 ^{P}	4,7 ^{W}	50 ^{N}	5,4 ^{P}	15 ^{W}	15 ^{W}
Free water saturation (100%·RH) [kg/m^3]	494 ^{W}	213 ^{W}	0 ^{W}	815 ^{P}	230 ^{W}	368 ^{W}
Liquid transport coefficient [m^2/s]	2.3x10^{-7} ^{W}	1.3x10^{-11} ^{W}	0 ^{W}	4.9x10^{-6} ^{P}	5.0x10^{-4} ^{W}	1,5x10^{-5} ^{W}

^{W} – WUFI data
^{N} – EN ISO10456:2009 [4]
^{P} – manufacturer's data

Table 3. Properties of materials used in the case study.

Material properties	PE membrane	Intello membrane	PCV foil	Lime plaster	Air cavity	Adhesive	Gypsum plaster-board
Bulk density [kg/m^3]	130 ^{W}	115 ^{W}	130 ^{W}	1600 ^{W}	1,3 ^{W}	1410 ^{P}	700 ^{N}
Porosity [m^3/m^3]	0.001 ^{W}	0.086 ^{W}	0.001	0.3 ^{W}	0.999 ^{W}	0.468 ^{P}	0.65 ^{W}
Specific heat capacity [J/(kg·K)]	2200 ^{W}	2500 ^{W}	2300 ^{W}	850^W	1000 ^{W}	1059 ^{P}	1000 ^{N}
Thermal conductivity [W/(m·K)]	1.65 ^{W}	2.4 ^{W}	2.3 ^{W}	0.7 ^{W}	0.155 ^{W}	0.06 ^{P}	0.21 ^{N}
Vapour diffusion resistance factor [-]	87 000 ^{W}	26 000 ^{W}	20 000 ^{W}	7 ^{W}	0,51 ^{W}	22,9 ^{P}	8,3 ^{N}
Free water saturation (100% RH) [kg/m^3]	0 ^{W}	85 ^{W}	0 ^{W}	250 ^{W}	0 ^{W}	280 ^{P}	400 ^{W}
Liquid transport coefficient [m^2/s]	0 ^{W}	0 ^{W}	0 ^{W}	1.5 x 10^{-7} ^{W}	0 ^{W}	7x 10^{-10} ^{P}	4.5 x 10^{-7} ^{W}

^{W} – WUFI data
^{N} – EN ISO10456:2009 [4]
^{P} – manufacturer's data

The chosen insulation products were compared for different target U-values: U=0.23 W/(m^2·K), U=0.2 W/(m^2·K), regardless of typical thicknesses with which they are supplied to the market. Installing an air and vapour control layer (AVCL) between plasterboard and insulation is often considered as a good practice, therefore assessment undertaken with and without AVCL was made.

Boundary conditions were chosen as follows: The investigated wall was assumed to face North-West direction. The building was modelled as a high building, and the wall was vertical. Driving rain water absorption was set to 70%. Weather data, including hourly values for Warsaw, according to WUFI, were used. The indoor moisture and temperature conditions was taken as normal according to

EN 15026:2007 [8]. The initial temperature in the existing wall was constant and equal to 20°C. The initial moisture content profile was the final profile in the base wall after equilibrium was reached. The simulation started on the first day of October so as to take into account the whole heating season and was carried out for the period of ten years, assuming that it is the adequate length for solid wall to reach equilibrium.

As high relative humidity levels can lead both to mould growth and moisture related damage, the relative humidity at the interface between insulation and masonry wall were investigated. To assess the conditions for the mould growth and frost risk, according to WTA guideline [11], the critical level RH>95% for fully bonding systems and RH>80% for the other solutions was adopted.

3. Results

3.1. Influence of internal insulation product

Figure 1 shows the changes of relative humidity over time, at the condensation critical location for all analyzed cases.

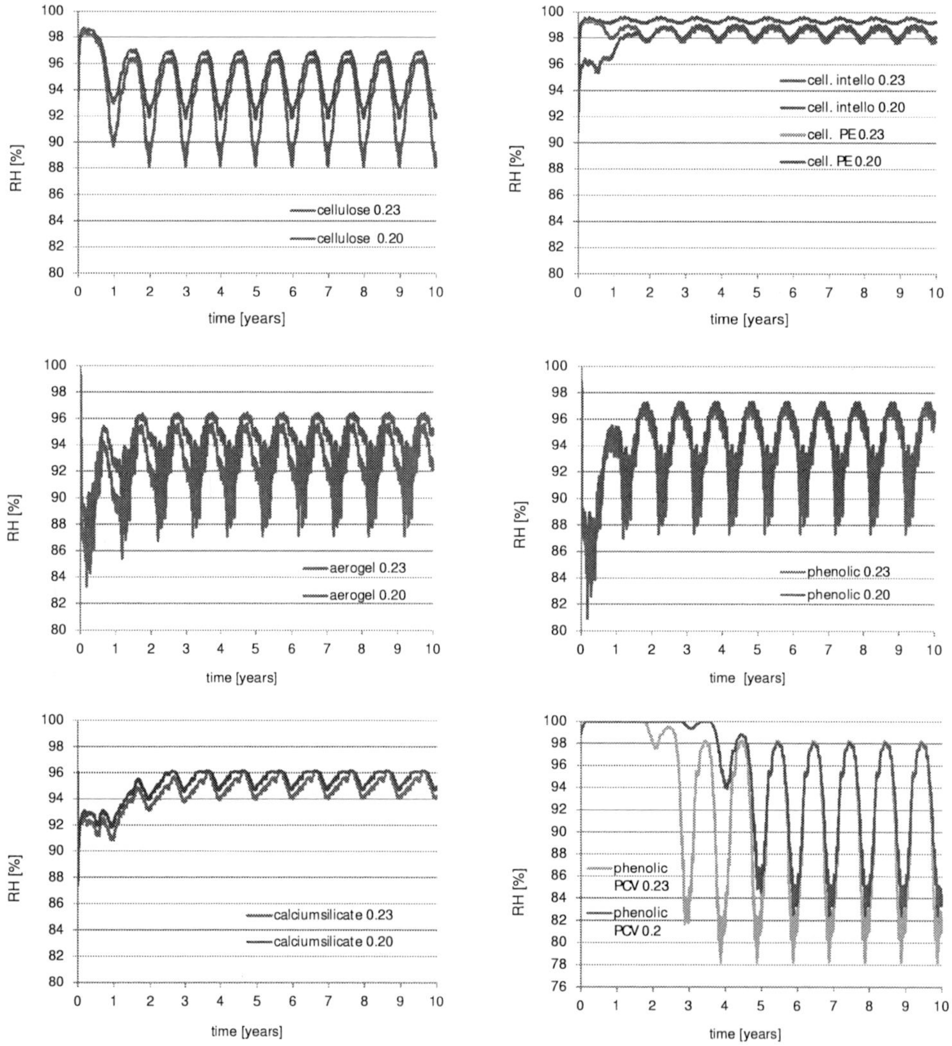

Figure 1. Relative humidity at the interface between insulation systems and base wall (the numbers after the names of materials refer to the target of U-values).

The application of internal insulating layer caused in all cases an increase in the relative humidity at the insulation-masonry interface, due to the lowering the temperature in this cross-section. This increase does not always mean damage. The key to assessing a given solution is the analysis of RH level obtained and the consequences of using this solution.

For the cellulose fibre without AVCL figure 1 shows rapid increase of relative humidity in the initial period, than the humidity fluctuate annually between 92-97% for U=0.20 W/(m^2·K) and 88-96% for U=0.23 W/(m^2·K). This initial surge is associated with the immediate temperature drop within the insulated wall. Because the cellulose fibre insulation is highly permeable, this initial moisture content decreases by the water vapour diffusion to the room. The humidity at the insulation masonry interface varies seasonally but there is no an increase in the fallowing years and the moisture is not accumulating at the critical location. The adhered nature of the cellulose fibre insulation without AVCL ensures that the probability of air gaps and air paths is negligible so this solution according to WTA key criterion (95% RH threshold) could be acceptable.

The graphs in figure 1, for cellulose fibre with AVCL shows that the use of AVCL changes the performance considerably. After using the Intello membrane, the relative humidity increased to 97.5% and applying PE foil up to 99.2%. With regard to 95% threshold, the retrofits with AVCL exceed this threshold for the whole analyzed period, so this solutions have the potential to cause moisture damage. We can see that the variable character of diffusion for Intello membrane with masonry wall is less advantageous than in lightweight construction because the reverse diffusion effect is much weaker.

Figure 1 for aerogel blanket retrofits, shows that the ~~oscillation~~fluctuation in relative humidity is very similar to that of cellulose fibre without AVCL, both material have the similar vapour permeable. But the 80% threshold is exceeded for both options.

For phenolic foam option, figure 1 shows that the relative humidity increases over the first two years and then reaches equilibrium, with fluctuations caused by the seasonal changes in temperature. In the case of AVCL membrane, the situation is different. In the first years of operation, the high level of relative humidity is maintained due to the increase of moisture in the air gap caused by the high diffusion resistance, additionally increased by the use of PVC foil, limiting the diffusion of water vapour through the insulation back into the room. It would seem that this solution will cause an increase in moisture content over time. However, after the period of increasing, the relative humidity reaches a state of dynamic equilibrium.

For calcium silicate boards, as Figure 1 shows, the relative humidity reaches an equilibrium quickly with annual fluctuation between 94 to 96%. Although the 95% threshold is exceeded only temporarily and the maximum relative humidity is similar to the case of cellulose fibre retrofits, the amplitude of these fluctuation is smaller. The ability of calcium silicate boards to easily transport moisture through capillary action increases the drying potential of the wall and therefore greatly reduces peaks in moisture at the critical location, but the average relative humidity level is higher than in case of cellulose fibre.

3.2. Influence of base wall material

It is often extremely difficult to assign the right input data to the material that was actually used in the solution being analyzed. In order to determine what effect the external layer has on the moisture conditions in the wall, the analyses were also performed for wall solutions with special type of brick named "Vienna". "Historical" brick has the higher capillary transport coefficient, i.e. it absorbs water faster, but at the same relative humidity stores less water than "Vienna" brick. In the case of exposure of both bricks to external moisture, e.g. to heavy rain, "historical" brick will be much more susceptible to damage than "Vienna" brick. Table 2 presents the basic thermal and moisture properties of these two materials. The results of the simulations are presented in Figure 2. As previously, the graphs show the relative humidity at the insulation-masonry interface but for "Vienna" brick.

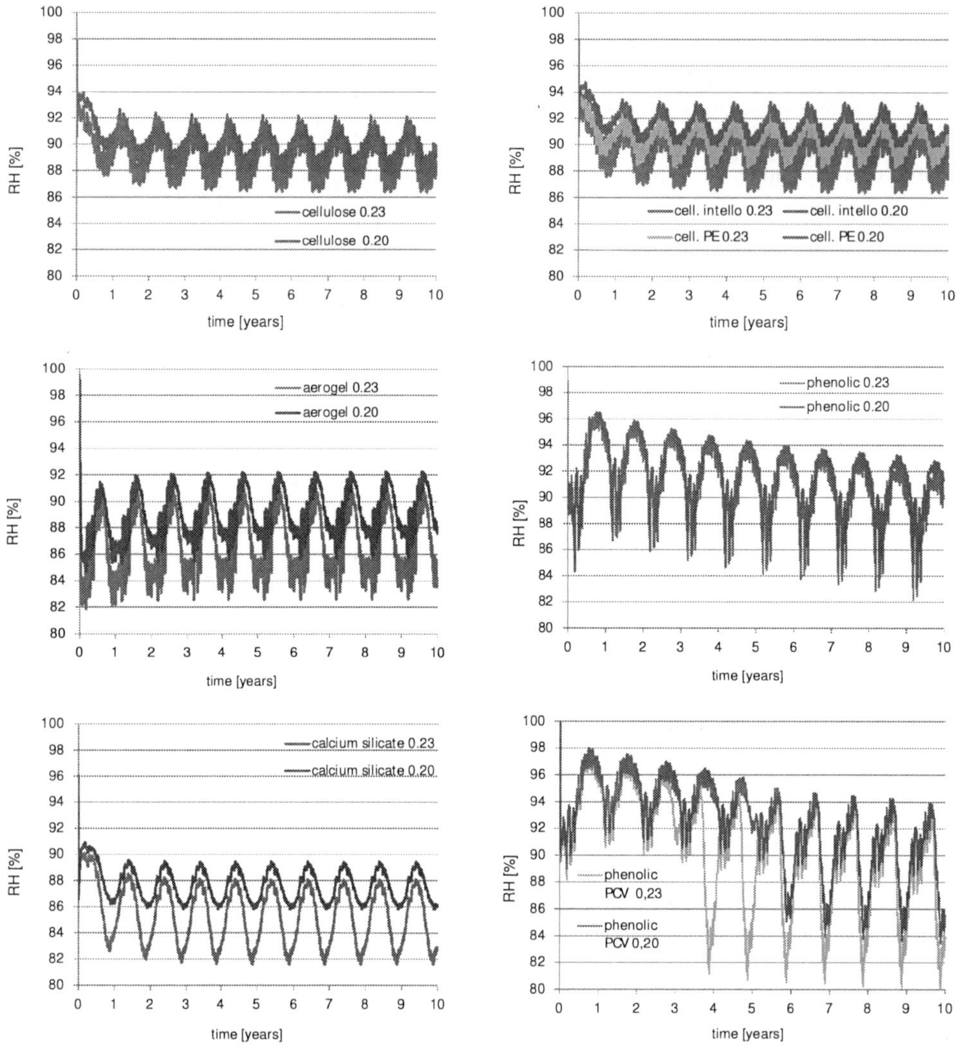

Figure 2. Relative humidity at the interface between insulation systems and base wall, with external surface from "Vienna" brick (the numbers after the names of materials refer to the target of U-values).

We can see that applying the external layer of wall made from the materials with reduced rain water absorption results in lower relative humidity at the insulation-masonry interface. Comparing the corresponding options (figure 1 with figure 2), the biggest differences in relative humidity occurs in the less water vapour permeable retrofit option – phenolic foam board. With "historic" brick, relative humidity exceeds 100% in the initial years and in subsequent years, this value stabilizes but is subject to high seasonal fluctuations. On the other hand, in the case of "Vienna" brick, the initial level of relative humidity is lower than 100% and in the subsequent years shows a downward trend. In addition, the amplitudes of seasonal variations in the case of "Vienna" bricks are also smaller.

For cellulose fibre retrofit without AVCL in turn, the type of the material forming external layer does not seem to have such an important effect on the humidity conditions inside the wall. The range of relative humidity for "historic" brick is 97-88%, while for "Vienna" brick amounts to 92-86%, so these are not significant differences. However, when using AVCL, the differences become more visible i.e. from 99-98% for "historical" bricks to 93-87% for "Vienna" brick. It indicates that the solution based on cellulose fibre insulation with AVCL, combined with the more absorbent external layer can lead to significant water accumulation at the insulation-masonry interface.

4. Summary and conclusion

In this paper, the hygrothermal influence of different interior materials was investigated by a numerical simulation. It was shown that applying internal insulation has a high impact on the hygrothermal behavior of the wall, leading to a higher moisture level in the masonry wall and a higher risk of frost and mould damage.

The case study has demonstrated that the moisture absorption characteristic of external wall surface determines the importance of the different transport mechanisms. When moisture transport in internally insulated brick wall occurs, AVCLs appear to have a poorer performance than vapour–open or capillary active insulation retrofits. Reducing the absorption characteristic of wall outer surface can get more freedom to choose the insulation product.

Hence, the other important parameter which has a high influence on the magnitude of these hygrothermal changes is the brick type. It was shown that the moisture storage function is insufficient to estimate the hygrothermal impact of interior insulation, more detailed material properties like liquid conductivity coefficient should be known.

References

[1] Kaklauskas A, Zavadskas E K, and Raslanas S 2005 *Energy Build.* **37** 361-372
[2] Pavlik Z and Cerny R 2010 *Appl. Therm. Eng.* **29** 1941-46
[3] EN 15026:2008 Hygrothermal Performance Of Building Components And Building Elements. Assessment of Moisture Transfer By Numerical Simulation
[4] EN ISO 10456:2009 Building materials and products. Hygrothermal properties. Tabulated design values and procedures for determining declared and design thermal values
[5] Scheffler G A 2011 *Proc. 5th Int. Conf. on Autoclaved Aerated Concrete* 15-16 September (Bydgoszcz, Poland) pp 323-335
[6] Scheffler G and Grunewald J 2003 *in Research in Building Physics* eds Carmeliet, Hens, Vermeir and Lisse (A.A. Balkema Publishers) pp 463-469
[7] Tariku F, Simpson Y and Iffa E 2015 *Build. Environ.* **92** 368-379
[8] Vereecken E and Roels S 2015 *Mater. Struct.* **48** 3009-21
[9] Vereecken E and Roels S 2013 *Constr. Build. Mater.* **41** 697-707
[10] Weiss J, Dunkelberg E and Vogelpohl T 2012 Improving policy instruments to better tap into homeowner refurbishment potential: lessons learned from a case study in Germany, *Energy Policy* **44** 406-415
[11] WTA 2009 Innendämmung nach WTA I Merkblatt 6-4 Ausgabe 05.2009/D Fraunhofer IRB Verslag
[12] WUFI PRO 5.3 software (Wärme und Feuchte instationär – Transient Heat and Moisture), The Fraunhofer Istitute for Building Physics IBP.

Thermal simulation of building performance with different loadbearing materials

J Strzałkowski,[1] Halina Garbalińska [1]

[1] Department of Building Physics and Building Materials, Faculty of Civil Engineering and Architecture, West Pomeranian University of Technology Szczecin, al. Piastów 50, 70-311 Szczecin, Poland

jstrzalkowski@zut.edu.pl

Abstract. In recent years, overheating of buildings during the summer has become a serious problem. This is the result of a significant increase in the number of cooling degree days over the last thirty years in Europe. As a consequence, increasing energy is consumed by buildings to keep the temperature at a relatively low level in the summer. This prompted us to examine the thermal simulation performance of a typical 50m^2 flat with different loadbearing materials used for the external walls. Calculations were made for lightweight cement composites aerated with air-entraining admixture or with the addition of aerogel particles. The results were compared with those obtained for walls based on conventional materials. Simulations were carried out using the Wufi Plus software. The components had a variable width of thermal insulation so that thermal transmittance of all the tested walls was constant. This assumption made it possible to evaluate parameters related to the thermal accumulation of the tested walls. These results demonstrate that the use of lightweight cement-based materials not only improves the thermal insulation of the whole building, but also significantly increases thermal mass of the walls. As a consequence, these type of wall component improve the microclimate, i.e. by lowering the internal temperature of buildings during the summer.

1. Introduction

Currently, specialists look for material and technology solutions that would contribute to radical reductions of energy consumption of buildings. An objective evaluation of the obtained results requires the application of more accurate methods of assessment of energy consumption and, consequently, more complex models of calculation.

To base solely on data concerning the thermal conductivity of the constituent materials, and, consequently, on the thermal transmittance U of separate component is, therefore, rendered insufficient.

Currently, testing the material and construction solutions in non-stationary conditions of heat flux which correspond to the real climate data is considered more justified.

Such analyses, on the one hand, increase the accuracy of the results, but also allow to reflect the characteristics of new material-related solutions (for example, PCM) whose properties are not properly recognised by the stationary models. Moreover, these analyses allow for comprehensive use of the data on heat accumulation and its effect on short-term and long-term heat losses and gains in buildings.

In the papers [1] and [2] the authors review present thermal energy storage active and passive systems. The main aim of the papers is to review and identify thermal storage building integrated systems and to classify them depending on the location of the thermal storage system. The most commonly used systems has been described, such as: buildings core activation (walls and floors) systems, thermal mass ventilation systems, external solar façades, TES solar collectors, photovoltaics and water tanks systems. In relation to passive systems the authors reviewed different types of high thermal mass materials and latent heat storage materials.

Another review which describes thermal energy storage systems is presented in the paper [3]. The prime intention of this paper is to review the potential research studies pertaining to a variety of latent heat energy storage (LHES) and cool thermal energy storage (CTES) systems solely dedicated for building heating, cooling and air conditioning (A/C) applications. Technical revelations regarding the integration and performance evaluation of heat storage materials in building fabric elements have been gleaned from numerous research contributions and presented.

Different types of TES systems has also been shown in [4]. The goal of this paper is to give a comprehensive review of a wide variety of TES technologies, with a clear focus on the combination of storage technology and building type. The authors emphasize that it is important to remember that there are specific demands and conditions that have to be fulfilled for each type of a building, and this is something that has to be evaluated separately for every new construction or renovation project. It is therefore important to analyze storage aspect early in the process of design.

Another paper [5] reviews the aspect of thermal dynamics which in most cases is omitted. Many of the current numerical models for building energy systems assume empty rooms and do not account entirely for the internal thermal inertia of objects like furniture. This review article points out that such assumption is not valid for dynamic calculations. The furnishing elements and other internal content can have a significant impact on the indoor thermal dynamics and on the occupants' comfort.

The next paper [6] focuses on the energy storage materials. The manuscript reviews recent advances in the development of thermal energy storage materials for building applications oriented towards zero energy buildings. Two main types of materials have been evaluated: latent energy storage and thermochemical energy storage materials. The authors suggest that applications of the first group of materials in free-cooling ventilation systems, solar energy storage solutions for short and long-term storage periods, and demand-side management strategies towards the road to zero energy buildings are highlighted as promising, leading to a reduction of energy consumption of more than 30%. On the other hand, for the second group there is currently no available material for thermochemical energy storage that satisfies all the requirements for building operations.

Another issue related to use of thermal mass is its impact on thermal comfort [7]. In the paper four methods of mass activation are discussed: surface activation, forced-air activation, hydronic activation and electrical activation. The performance and control for such systems with peak shaving and shifting as an objective is discussed and barriers to its application and development is also addressed.

In the review paper [8] the authors describe different approaches to quantify thermal inertia using specific properties and parameters of materials and whole components. This paper also presents a literature review focussing on the reported impacts of building thermal inertia on thermal comfort and energy use for space heating and cooling.

The most commonly used material in buildings for thermal mass is concrete. The authors of [9] reviewed factors affecting the heat storage capacity of concrete. In addition, common measurement methods of cement-based materials' thermal conductivity, thermal diffusivity and specific heat capacity are presented in paper. Various studies presented in the paper reveal that temperature, humidity, aggregate type, cementitious material type as well as phase change material (PCM) used influence the thermal properties of concrete [10] to be used as thermal mass.

Numerous studies are also conducted that aim at determining the behaviour of buildings in non-stationary conditions.

In the paper [11], the object of assessment was an office building for which the analysed parameter was the time constant dependent of the specific heat of the thermal mass of the building, specific heat of internal air and air-exchange rate of night ventilation. It was shown that with the increasing time constant, the energy demand for cooling purposes decreases significantly.

The paper [12] presents a comparative analysis of buildings with external walls made of normal concrete with the insulation variably located on the outside or the inside. All the wall components were characterised by a fixed value of total thermal resistance. In the transitional months (spring/autumn), the values of the dynamic thermal resistance of walls were clearly different from the values of resistance determined using stationary calculations. The obtained results proved that high thermal mass can reduce heat flux in the months in which the outdoor temperature is very variable and fluctuates around the internal temperature.

Experiments with various kinds of external walls were also conducted for locations with warm climate, such as Australia [13]. The study tested solutions for components with variable thermal mass, such as: light wooden cavity wall with facade brick exterior finish, ceramic brick wall with fibre cement facade, hollow brick wall and a lightweight wooden cavity wall. In the warm climate of Australia characterised by large daily temperature fluctuations, the lowest energy consumption was observed for variants with massive inner bearing layer of ceramic brick. Moreover, the window area had a very large impact on the results. The greater the window area, the greater the differences in the behaviour of light and heavy walls.

The paper [14] assessed the influence of location of thermal insulation in the concrete wall on the total energy consumption of the studied building. Variants of outer concrete layer walls were characterised by a significantly higher thermal admittance. Thanks to this fact, these variants could more effectively reduce the fluctuations of room temperature.

The reports [15], [16] analysed a one-story building consisting of five rooms/zones. Different variants of exterior walls with variable heat capacity were tested. The assessment concerned the impact of the building heat capacity on the number of hours with exceeded levels of adaptive comfort (80% and 90%). The differences between the massive and the low heat capacity objects was significant, and the results are definitely in favour of the massive walls.

The study [17] compares the dynamic properties of three types of external walls and their influence on the internal temperature of the building. A comparison was made of variants of walls made of ultra-lightweight 70 cm thick concrete and two walls of ordinary concrete with a thickness of 20 cm and a 19 cm thick insulation layer on the inner and the outer side. While the homogeneous ultra-lightweight concrete wall demonstrated very good damping properties (high damping ratio and heat transmittance), it had a mediocre internal heat capacity.

Variants of PCM material exterior walls are also the subject of research, for instance in [18], [19]. According to the authors, an additional PCM layer in the appropriate climatic conditions allows for a reduction of heat loss in winter and a decrease of heat gains during the summer season.

The work [20] analysed the influence of additives to reduce gypsum plaster heat conductivity. An improvement of thermal properties of inside plasterwork may potentially influence the indoor microclimate as well as the energy efficiency of the entire building.

Studies are also being conducted over the influence of various additives and admixtures which greatly affect the thermal properties of the whole composite. Papers [21], [22] examine the effect of addition of recycled glass cullet aggregate. The resulting composite is characterised by a significantly higher degree of thermal insulation compared to concrete materials based on natural quartz aggregate. Solutions are also checked for using slag from sewage sludge as a replacement for natural aggregates [23].

The additives and admixtures used in cement composites also largely affect other properties, i.e. water vapour transport, capillarity and frost resistance. The evaluation of these parameters is discussed in detail in i.a. [24] and [25]. How these processes occur is of great importance given the effect of composite moisture on their thermal conductivity, which issue was the subject of research presented in the publications [26], [27]. In this context, great importance is attached to the processes of sorption and desorption of moisture in building materials and determination of isotherms linking the extent of material humidity with the thermal and moisture conditions in the environment, for example [28], [29].

The authors decided to evaluate the effects of different wall construction materials on the energy demand for cooling purposes. A comparison was made of different solutions for traditionally finished external walls - using normal concrete, sand-lime bricks, autoclaved aerated concrete, extruded ceramic bricks and lightweight composite walls - based on artificial aggregates further modified by an air-entraining admixture [30] or aerogel particles [31].

2. Materials and methods

The analysis was conducted for a flat with an area of 50 m^2. Simulations were performed using the Wufi Plus thermal calculation module. The analysed object was modelled as repeatable part of a multifamily residential building located in Kołobrzeg. The simulation was performed for the full calendar year. The object model is presented in figure 1.

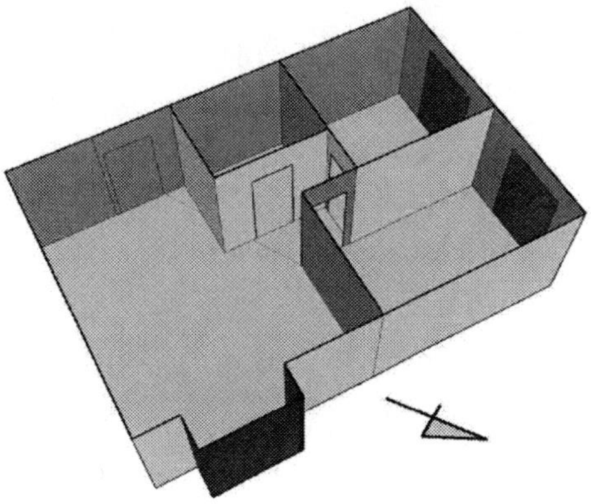

Figure 1. The model of the analysed part of the building – flat.

The internal load-bearing walls and floors above and below the apartment were assumed to be adiabatic boundaries due to the proximity of flats with similar internal temperature. All the inner walls of the flat were defined as partition walls of thickness of 12 cm with plaster gypsum finish. Exterior walls were modelled as double-layer components with polystyrene foam insulation. For the all variants the external walls thermal transmittance U was assumed equal to 0.15 (W/m^2·K). To maintain the constant value of U, the thickness of the insulation with thermal conductivity λ = 0.036 W/(m·K) was modified accordingly. This made it possible to assess the dynamic properties of different solutions which, however, had the same value of the coefficient U.

The windows in the model with the total area of 16.28 m^2 were modelled as triple glazed windows with thermal transmittance U = 0.80 W/(m^2·K). The model did not take into account any sunshade or solar control systems.

The simulations were performed in two variants: with the cooling system and without it. The assumed minimum internal temperature was equal to 20°C and the maximum temperature, in the variant with cooling, was 21°C.

The analysis included a mechanical ventilation providing air exchange at 1.5 1/h rate periodically increased for the summer period (from 16.05 to 15.09) to the level of 2.0 1/h. In addition, separate calculations with the use of night ventilation from 01.06 to 31.08 were made. Ventilation heat recovery with efficiency of 80% was included in the period from 16.10 to 15.05.

It was further assumed that the heating system would be disabled in the period from 16.05 to 15.09.

A total of eight types of load-bearing wall materials were tested. These included four traditional materials: autoclaved aerated concrete (AAC), silicate (S), normal concrete (NC) and extruded ceramic brick (CB). In addition, calculations were made for four types of lightweight composites based on expanded clay aggregate (EC) and sintered fly ash aggregate (FA), modified by air-entraining admixture (A) or aerogel particles (Aero). Data on the density and thermal properties of individual materials are given in table 1.

Thermal parameters of the first four lightweight concretes have been gathered during earlier experimental research [30], [31]. Data for the remaining materials have been adopted on the basis of general standards.

The thermal properties of the lightweight concretes have been determined during a wide range research project. The main aim of the project was to analyse thermal and dynamic properties of composites modified by different additions and admixtures. For the selected four of them, it was decided to perform an additional simulation analysis in comparison to traditional materials

Table 1. The properties of the tested materials.

Material	Symbol	ρ (kg/m^3)	λ (W/m·K)	c_p (J/kg·K)	c_v (10^6·J/(m^3·K))
Sintered fly ash based concrete with air-entraining admixture	**FA/A**	1164.1	0.273	1239.80	1.443
Sintered fly ash based concrete with aerogel particles	**FA/Aero**	1358.5	0.431	1144.60	1.555
Expanded clay based concrete with air-entraining admixture	**EC/A**	880.8	0.269	1684.00	1.483
Expanded clay based concrete with aerogel particles	**EC/Aero**	1032.5	0.380	1560.90	1.612
Autoclaved aerated concrete	**AAC**	500.0	0.170	850.00	0.425
Silicate	**S**	1973.0	0.619	800.00	1.578
Normal concrete	**NC**	2234.0	1.650	797.67	1.782
Extruded ceramic brick	**CB**	600.0	0.120	850.00	0.510

3. Results and discussion

Figure 2 shows the results regarding energy demand for cooling purposes obtained with different variants of external walls of the studied apartment. The difference between the variants with and without night ventilation is clearly seen. The simulations assumed a five-hour period of intensive ventilation at the level of 4 air exchanges per hour. The use of night ventilation leads to a decrease in cooling demand from 25% in the variant with autoclaved aerated concrete by up to 31% in the case of normal concrete.

There are also clear differences in the demand for energy for cooling purposes between the individual variants. In the comparison, normal concrete wall and silicate wall demonstrated the best results. Similar results were also obtained in the case of the use of lightweight concrete composites.

Furthermore, the simulations examined the differences in terms of energy consumption for heating purposes. In this case, the differences between the variants are less than 1%, and the average value of heat consumption amounted to approx. 1500 kWh/year (30 kWh/(m^2·year)), when ventilation heat recovery was applied.

The results point to the justifiability of using massive components only in the case when the outdoor temperature fluctuates around the internal temperature. In the situation when the outdoor temperature is significantly higher or lower than the room temperature, additional heat capacity generally does not affect the energy efficiency of the building.

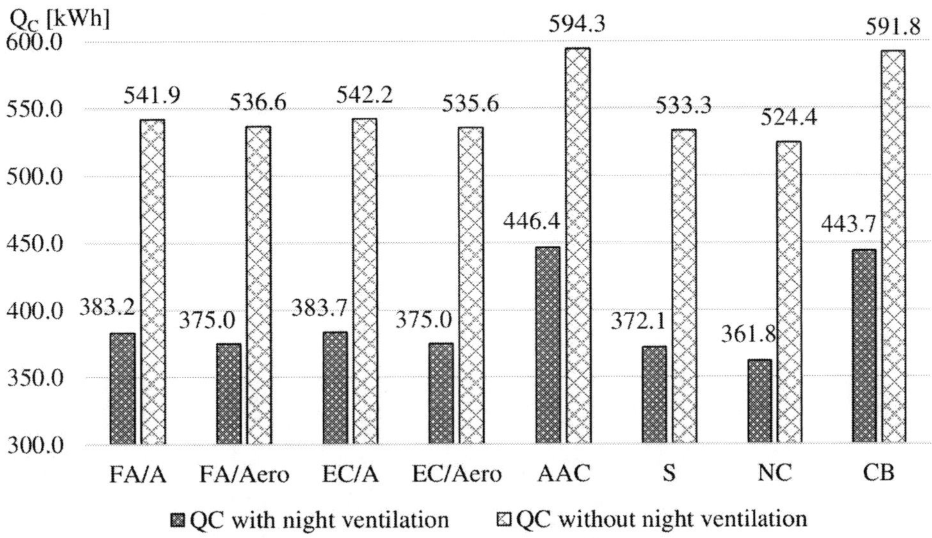

Figure 2. Cooling demands for each variant.

Figure 3 shows the required theoretical thicknesses of the thermal insulation necessary for the thermal transmittance U to be 0.15 W/(m^2·K) in each case. Also, the percentage difference of the energy demand for cooling purposes between the normal concrete variant and other solutions was presented.

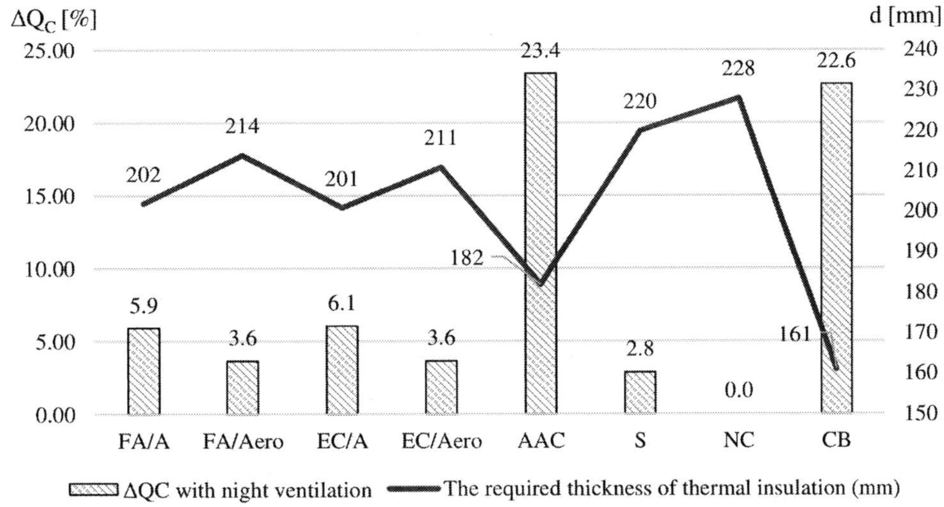

Figure 3. The relative differences in cooling energy demand compared to the normal concrete variant and the required thicknesses of thermal insulation.

The greatest differences were exhibited in the case of autoclaved aerated concrete and extruded brick ceramic variants (23.4% and 22.6% respectively). The differences between normal concrete and lightweight concretes were significantly lower.

However, it should be noted that the normal concrete variant would require the use of the thickest theoretical layer of insulation (228 mm), and the autoclaved aerated concrete variant would need insulation thinner by 46 mm compared to normal concrete.

Figure 4 shows the maximum internal temperatures obtained in the simulations without the cooling system. Similarly to previous results, night ventilation led to the lowering of maximum temperature by about 1 degree. Also, significantly large differences between the individual variants can be observed. The lowest temperatures were obtained for the massive normal concrete walls.

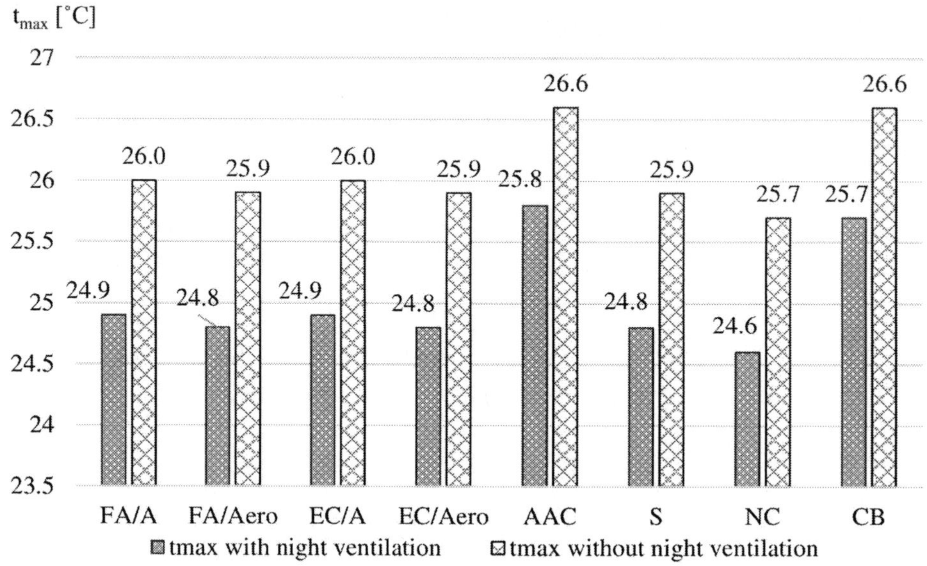

Figure 4. Maximum internal temperatures for each variants.

Figure 5 shows the overall amount of hours with temperatures exceeding 23°C and 24°C. The differences between the variants of lightweight materials with low thermal capacity (AAC, CB) and other systems are very high. In particular, in case of temperature exceeding 24°C, between the best (NC) and the worst variant (AAC) the difference is more than 230%.

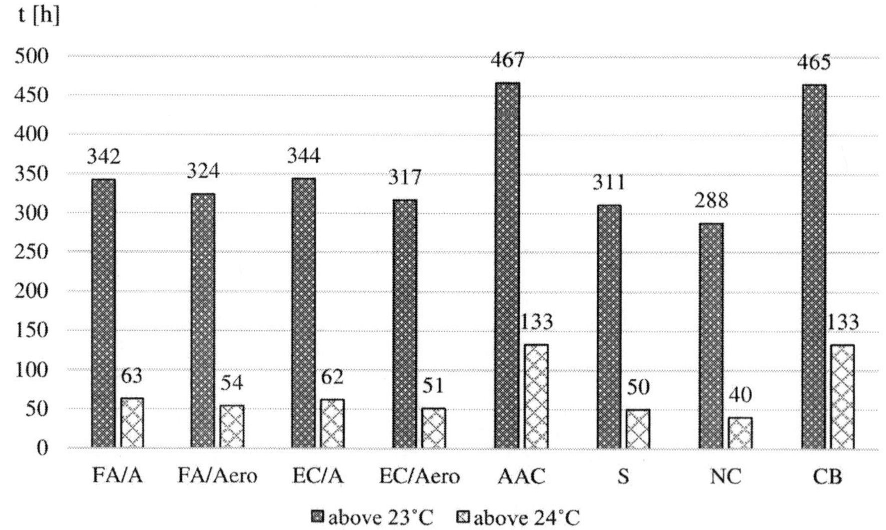

Figure 5. The number of hours during the year with exceeded temperature over 23°C and 24°C.

4. Conclusions

The article presents the results of simulations concerning an object with different wall variants. The following main conclusions arise from the results:

- the use of such materials as normal concrete, silicates or lightweight concrete composites results in a reduction in energy demand for cooling purposes in the given climate,
- the type of load-bearing material has virtually no effect on the energy demand for heating purposes in the given climate,
- the materials used for constructing walls with large heat capacity cause a decrease in the maximum internal temperature during the summer period, helping to reduce the load on the air conditioning system,
- the high heat capacity materials reduce room overheating and minimise more efficiently the amount of hours with exceeded thermal comfort temperature.

Acknowledgement(s)

The research of lightweight cement composites was realized as a part of research project no 2014/13/N/ST8/00091 financed by the National Science Centre, Poland.

References

[1] Navarro L, de Gracia A, Colclough S, Browne M, McCormack S J, Griffiths P and Cabeza L F 2016 Thermal energy storage in building integrated thermal systems: A review. Part 1. active storage systems *Renewable Energy* **88** 526–47

[2] Navarro L, de Gracia A, Niall D, Castell A, Browne M, McCormack S J, Griffiths P and Cabeza L F 2016 Thermal energy storage in building integrated thermal systems: A review. Part 2. Integration as passive system *Renewable Energy* **85** 1334–56

[3] Parameshwaran R, Kalaiselvam S, Harikrishnan S and Elayaperumal A 2012 Sustainable thermal energy storage technologies for buildings: A review *Renewable and Sustainable Energy Reviews* **16** 2394–433

[4] Heier J, Bales C and Martin V 2015 Combining thermal energy storage with buildings – a review *Renewable and Sustainable Energy Reviews* **42** 1305–25

[5] Johra H and Heiselberg P 2017 Influence of internal thermal mass on the indoor thermal dynamics and integration of phase change materials in furniture for building energy storage: A review *Renewable and Sustainable Energy Reviews* **69** 19–32

[6] Lizana J, Chacartegui R, Barrios-Padura A and Valverde J M 2017 Advances in thermal energy storage materials and their applications towards zero energy buildings: A critical review *Applied Energy* **203** 219–39

[7] Olsthoorn D, Haghighat F, Moreau A and Lacroix G 2017 Abilities and limitations of thermal mass activation for thermal comfort, peak shifting and shaving: A review *Building and Environment* **118** 113–27

[8] Verbeke S and Audenaert A 2018 Thermal inertia in buildings: A review of impacts across climate and building use *Renewable and Sustainable Energy Reviews* **82** 2300–18

[9] Shafigh P, Asadi I and Mahyuddin N B 2018 Concrete as a thermal mass material for building applications - A review *Journal of Building Engineering* **19** 14–25

[10] Zhang D, Li Z, Zhou J and Wu K 2004 Development of thermal energy storage concrete *Cement and Concrete Research* **34** 927–34

[11] Yang L and Li Y 2008 Cooling load reduction by using thermal mass and night ventilation *Energy and Buildings* **40** 2052–8

[12] Al-Sanea S A, Zedan M F and Al-Hussain S N 2012 Effect of thermal mass on performance of insulated building walls and the concept of energy savings potential *Applied Energy* **89** 430–42

[13] Gregory K, Moghtaderi B, Sugo H and Page A 2008 Effect of thermal mass on the thermal performance of various Australian residential constructions systems *Energy and Buildings* **40** 459–65
[14] Kossecka E and Kosny J 2002 Influence of insulation configuration on heating and cooling loads in a continuously used building *Energy and Buildings* **34** 321–31
[15] Kisilewicz T 2015 Passive Control of Indoor Climate Conditions in Low Energy Buildings *Energy Procedia* **78** 49–54
[16] Kisilewicz T 2015 Performance of Building Materials and Whole Enclosures in Non-Stationary Thermal Conditions *Procedia Engineering* **108** 445–52
[17] Roberz F, Loonen R C G M, Hoes P and Hensen J L M 2017 Ultra-lightweight concrete: Energy and comfort performance evaluation in relation to buildings with low and high thermal mass *Energy and Buildings* **138** 432–42
[18] Heim D and Wieprzkowicz A 2017 Attenuation of Temperature Fluctuations on an External Surface of the Wall by a Phase Change Material-Activated Layer *Applied Sciences* **8** 11
[19] Wieprzkowicz A and Heim D 2016 Energy performance of dynamic thermal insulation built in the experimental façade system *Management of Env Quality* **27** 681–94
[20] Heim Dariusz, Mrowiec Andrzej, Pralat Karol and Mucha Maria 2018 Influence of Tylose MH1000 Content on Gypsum Thermal Conductivity *Journal of Materials in Civil Engineering* **30** 04018002
[21] Chung S-Y, Abd Elrahman M, Sikora P, Rucinska T, Horszczaruk E and Stephan D 2017 Evaluation of the Effects of Crushed and Expanded Waste Glass Aggregates on the Material Properties of Lightweight Concrete Using Image-Based Approaches *Materials (Basel)* **10**
[22] Sikora P, Horszczaruk E and Rucinska T 2015 The Effect of Nanosilica and Titanium Dioxide on the Mechanical and Self-Cleaning Properties of Waste-Glass Cement Mortar *Procedia Engineering* **108** 146–53
[23] Głowacka A, Rucińska T and Kiper J 2017 The slag original from the process of sewage sludge incineration selected properties characteristic *E3S Web Conf.* **22** 00054
[24] Wygocka A and Garbalinska H 2013 *Assessment of scale of the microstructure impact on capillary transport in cement-based composites with polypropylene fibers* ed A Mahdavi and B Martens (Vienna: Okk-Editions)
[25] Garbalinska H and Wygocka A 2014 Microstructure modification of cement mortars: Effect on capillarity and frost-resistance *Constr. Build. Mater.* **51** 258–66
[26] Stolarska A and Garbalińska H 2017 Assessment of suitability of some chosen functions for describing of sorption isotherms in building materials *Heat Mass Transfer* **53** 1603–17
[27] Siwińska A and Garbalińska H 2011 Thermal conductivity coefficient of cement-based mortars as air relative humidity function *Heat Mass Transfer* **47** 1077–87
[28] Garbalińska H, Bochenek M, Malorny W and von Werder J 2017 Comparative analysis of the dynamic vapor sorption (DVS) technique and the traditional method for sorption isotherms determination — Exemplified at autoclaved aerated concrete samples of four density classes *Cement and Concrete Research* **91** 97–105
[29] Garbalińska H and Bochenek M 2017 Experimentelle Ermittlung der Desorptionsisothermen von Porenbeton unterschiedlicher Dichte mit der Standardmethode SSS und mit der DVS-Technik *Bauphysik* **39** 191–6
[30] Strzałkowski J and Garbalińska H 2017 Porosimetric, Thermal and Strength Tests of Aerated and Nonaerated Concretes *IOP Conf. Ser.: Mater. Sci. Eng.* **245** 032017
[31] Strzałkowski J and Garbalińska H 2016 Thermal and strength properties of lightweight concretes with the addition of aerogel particles *Advances in Cement Research* **28** 567–75

Evaluation of thermal performance of window lintel construction detail

J Fořt[1], Z Pavlík[1], M Jerman[1] and R Černý[1]

[1]Department of Materials Engineering and Chemistry, Faculty of Civil Engineering, Czech technical University in Prague Thákurova 7, 166 29, Prague, Czech Republic

Corresponding email adress: jan.fort@fsv.cvut.cz

Abstract. Ceramic bricks belong to the oldest and most used building material since the ancient ages. Notwithstanding, this favorable building material passed through many improvements and currently used ceramic brick is substantially improved compared to bricks from ancient production. The contemporary demands are paid especially on the mechanical resistance, high vapor permeability, outstanding thermal insulation properties, thermal accumulation function and noise insulation. On this account, the hygrothermal properties of construction detail composed by a contemporary lightweight hollow brick together with window lintel is investigated within this paper. Compared to laboratory methods aimed at the measurement of small samples, the semi-scale analysis represents a valuable approach for building materials testing. For this purpose, a system composed of two climatic chambers separated by the connecting tunnel with desired building construction detail is used. The brick block and window lintel is placed in a chamber connection tunnel, thermally insulated by mineral wool to ensure one dimensional transport and studied in sense of loading by real climatic data from Czech Republic. This valuable analysis allows long-term monitoring of the wall cross-section temperature and relative humidity profiles of particular elements in real scale. The obtained results represent valuable information for the practical application of the studied brick block in building practice. Such comparison allows optimization of building envelopes in order to improve their energy efficiency.

1. Introduction

Increased attention of this period is paid on the thermal stability of building envelopes. The optimization and improvements focused on decrease of the thermal conductivity is one of the most investigated topic by building designers and engineers. The ceramic burnt bricks are still the most often used building material across developed countries and their history can be traced back to antiquity. Despite of development of many modified or competing product based on cementitious material, wood construction, sandwich element made from recycled materials, the popularity of application of ceramic bricks is still high. This fact can be assigned to their favorable properties based on very good mechanical resistance, thermal insulation performance, thermal accumulation parameters, resistance to outdoor climatic fluctuations, high water permeability and no health risk [1]. However, building envelopes are not consisted only by bearing elements such as bricks or building blocks from various materials but also by specific construction elements with different characteristics, technical purpose and properties [2]. In this sense, it can be mentioned all atypical construction detail such as niche, door, windows and related lintels. The displacement of these elements pose significant risk for occurrence of thermal bridges and consequently associated phenomena [3].

The advanced design of individual building elements is necessary for reduction of energy consumption associated to the building operations without deterioration of indoor thermal comfort.

Based on the sustainability principle, which paid attention in building sector especially on construction of low and passive houses, the thermal performance of the whole building envelope still requires research. The latter mentioned building standards are sensitive on the site, building orientation, ventilation, applied insulation materials, thermal storage capacity of used materials and design of windows or other similar elements. The evaluation of energy efficiency of used elements is essential for construction of modern building envelopes with desired functional properties. Moreover, these parameters are important for maintenance of interior comfort and is related to living condition for residents [4].

The assessment of the thermal conductivity or thermal resistance can be perceived as one of the most often used parameter for building materials classification. In this respect, the thermal properties and thermal insulation plays an important role in current building industry [5]. The thermal performance of the building envelopes is closely linked to the moisture content contained in pores of building materials. Namely, the water presence in the pore space is responsible for deterioration of thermal performance due to significantly higher thermal conductivity of water compared to thermal conductivity of air. Therefore, the definition of thermal performance in relation to the humidity level is very important [6].

In summary, the proper understanding of correlation between outdoor conditions, thermal performance of building materials, quality of the indoor environment together with energy consumption associated with building operations is essential for compliance of sustainable development principles within the building design.

The determination of the thermal conductivity of materials is usually carried out within laboratory conditions by transient or steady-state methods. The main advantage of the transient method compared to the steady-state is its shorter measurement time. However, the accuracy and reproducibility of the measurement is questionable. On the other hand, the steady-state method is used as a reference method. The thermal conductivity is determined in this case by Fourier law equation. The arrangement is based on the measurement of heat fluxes induced by the temperature difference between heated and cooled plate of the measurement device. The weak point of this method is consisted of measurement of small samples and thus the transferability of the obtained data is limited, especially when the heterogeneity of particular building elements is taken into account.

In the light of these assumptions, this paper is focused on determination and comparison of thermal performance of window lintel together with hollow brick block. Since these materials are utilized for construction of building, the determination of their thermophysical properties represent important parameter for their application. Here, the studied construction detail is placed in the climatic chamber system and real scale measurement is carried out.

2. Studied materials

The window lintel fabricated by Heluz CZ, was chosen for this study. This element was together with hollow brick block subjected to experimental assessment of hygric and thermal properties during the winter period determined by the semi scale experiment.

The basic material properties of the hollow brick (Table 1) were determined by the measurement of the bulk and matrix density. Bulk density was accessed gravimetrically, matrix density by means of water vacuum saturation method. Consequently, the total open porosity was calculated. The moisture dependent thermal conductivity is presented in Table 2. Provided results were measured by commercial device ISOMET 2114 (Applied Precision) working on dynamic measurement principle.

Table 1. Basic physical properties of particular materials

Bulk density (kg/m^3)	Matrix density (kg/m^3)	Total open porosity (%)
1 389	2 830	50.9

Table 2. Thermal conductivity of the brick body

Moisture content (m^3/m^3)	Thermal conductivity (W/mK)
0.0	0.3
0.11	0.44
0.13	0.57
0.29	0.75
0.36	0.88
0.46	1.03
0.49	1.06

3. Semi-scale experiment

The light-weighted hollow ceramic brick as well as window lintel are produced by the Heluz Brick industry (Czech Republic). This type of elements is usually used for the thermal insulation masonry with a width of 500 mm. The window lintel is composed by four elements. Three of them are prestressed concrete reinforcements with about 100 mm thickness completed with polystyrene thermal insulation with thickness about 200 mm. The detail of construction detail placed in the connection tunnel of climatic chamber system is depicted in Figure 1.

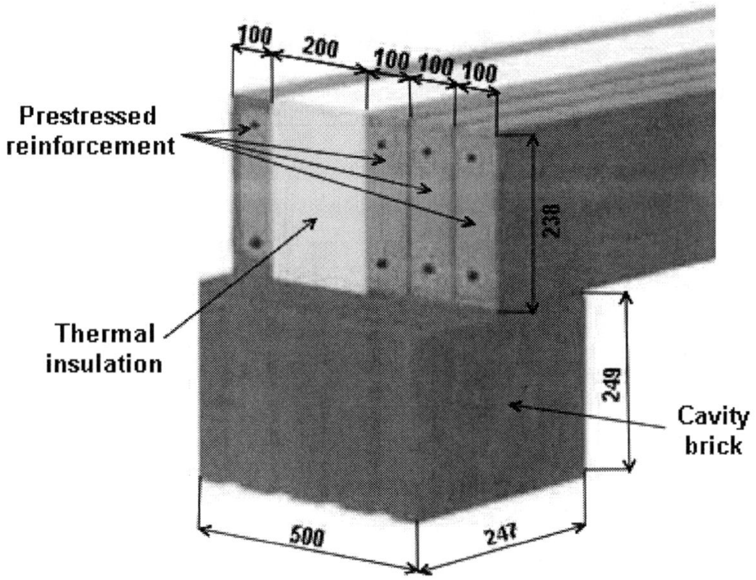

Figure 1. Description of studied construction detail

The semi-scale experiment was realized in a climatic chamber system assembled in laboratory of Department of Materials Engineering and Chemistry [7]. System is composed of two climatic chambers equipped by connecting tunnel for placing the studied construction (Figure 2). Maximum volume of the tested specimen, which is possible place into the connecting tunnel, is about 0.35 m^3 with surface exposed to the simulated outdoor conditions of dimension 0.7 x 0.9 m. The construction of the individual chambers is based on common commercial solutions for controlling temperature and relative humidity conditions but the solution of connections between the chambers and the tunnel and of the organization of additional admission holes for parallel measurements are designed in our laboratory.

Within this study the investigated window lintel construction detail of real dimensions was placed into the tunnel, equipped by the combined temperature and relative humidity sensors and thermally insulated from sides in order to achieve dominant heat and moisture transport through the studied elements (Figure 3). For realization of the semi-scale experiment simulating outdoor climatic conditions, the temperature and relative humidity in the first climatic chamber used for simulation of indoor climate was maintained at the constant temperature at 21± 0.3 °C and relative humidity at 30 ± 2 %RH. The second chamber was used for simulation of difference outdoor conditions based on the reference year (TRY) [8] by meaning of hourly changes. TRY use averaged real weather data for specific locations from several decades from the past. Here, the winter period was chosen due its importance and demanding conditions for testing of thermal performance of building envelope. Finally, the climatic chamber system was closed, and required temperature and relative humidity values were set. On the exterior side, there were simulated typical climatic conditions for Prague, whereas the climatic loading started with data corresponding to 1st November and the experiment was stopped on date corresponding to 31st February. Hence, the whole winter period, which is the most critical part of the year from the point of view of heat losses, was simulated.

The data logged by measuring devices were evaluated and critical morning data (corresponding with 8 am) were chosen for comparison.

Continuous monitoring of the temperature and relative humidity distribution along the hollow brick thickness was carried out by combined sensors from Ahlborn, Germany. Sensors have measuring range from 5% to 98% RH with a 2% uncertainty. The temperature measurement is done using NTC thermistors with an uncertainty of 0.1 °C. The temperature profiles and relative humidity profiles were logged in 10-minute interval.

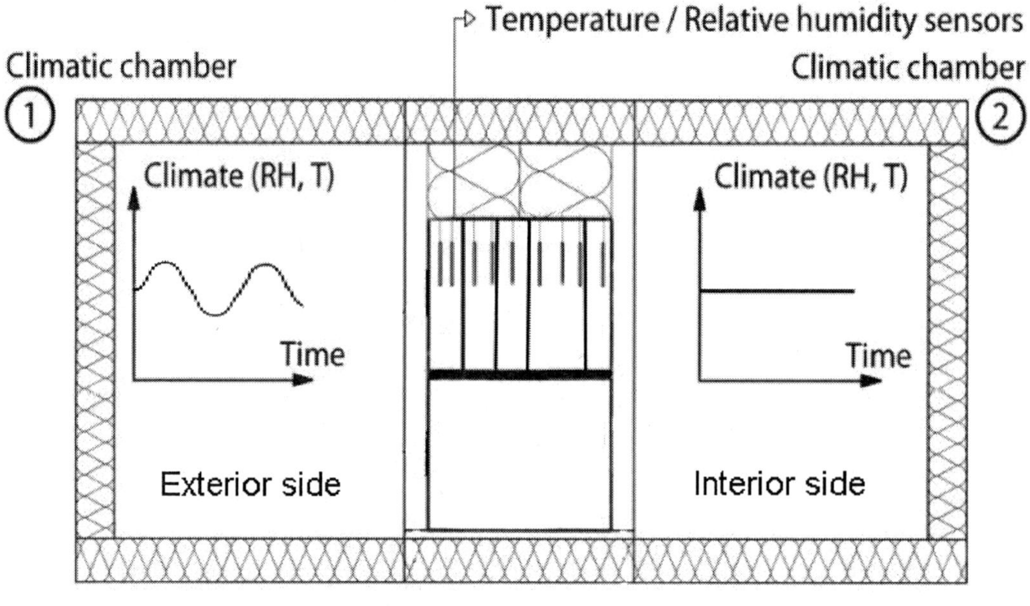

Figure 2. Schema of the experimental setup in cross-section

IOP Conf. Series: Materials Science and Engineering **415** (2018) 012015 doi:10.1088/1757-899X/415/1/012015

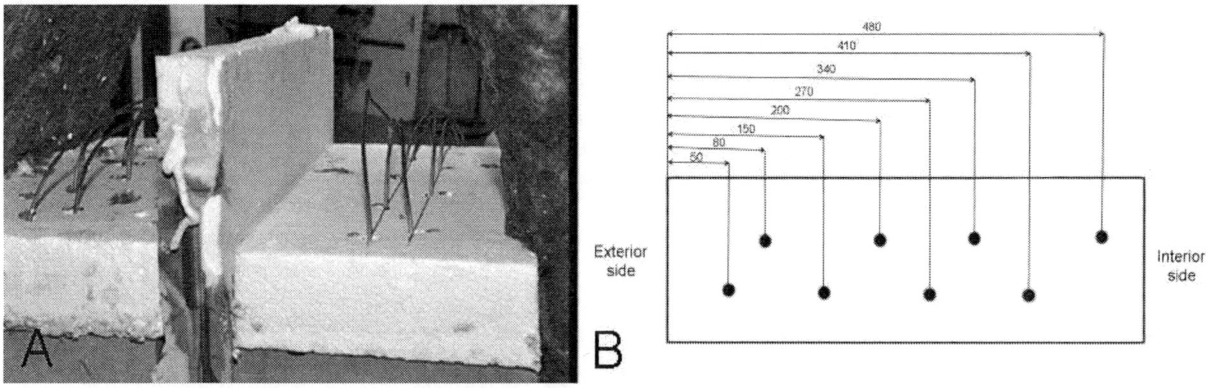

Figure 3. Sample insulation (A) and sensor positioning (B)

4. Results and discussion

The temperature and relative humidity wall cross-section profiles of window lintel and hollow brick block are plotted in Figures 4 and 5. Looking at the temperature profiles one can observe high sensitivity of the researched construction detail on the exterior temperature variations. On the other hand, only slight changes can be distinguished between the window lintel and the hollow brick. While the thermal conductivity of the concrete part has worse thermal performance, the applied insulation layer with 200 mm of polystyrene insulation improved insulation properties of the window lintel. From this point of view, the overall thermal response of lintel is slightly better compared to hollow brick despite the poor thermal insulation properties of prestressed concrete reinforcement which composed main part of the element. The thermal insulation function of both brick block and window lintel can be considered good in general, what is promising for practical application of the developed material in building practice. At 150 mm distance from the interior surface, there were measured temperatures of the brick about 18 °C, what is very good for the conditioning of the interior climate. Basically, there were only small differences between interior temperatures and the temperatures of the brick at the distance of 150 mm from the brick surface. This observation is even more distinct for window lintel with the polystyrene insulation layer.

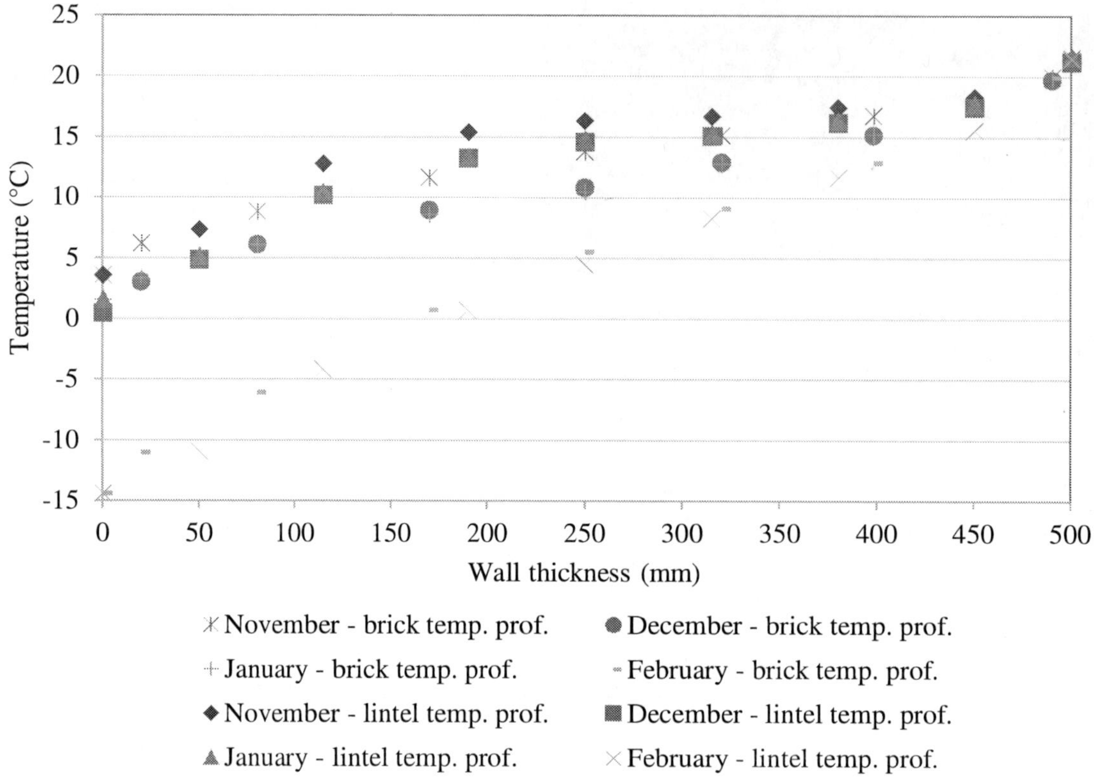

Figure 4. Measured temperature profiles in wall cross-section

Relative humidity profiles reflected the simulated relative humidity in the exterior chamber. Typically, the high values of relative humidity were measured for climatic loading corresponding to period from November to February. Here, the relative humidity values were < 60% up to 400 mm distance from the interior surface of the window lintel. Contrary to this finding, the relative humidity profiles obtained for the hollow brick were significantly different. Here, gradual decrease of the relative humidity was achieved and even 100 mm from the interior surface was relative humidity higher than 80 %. Based on the obtained data, the moisture transport was pronouncedly affected by the polystyrene insulation layer. The very low water vapor permeability of polystyrene, the relative humidity dropped from about 95% to about 60 %. This fact can be correlated with the worsening of the thermal performance of the hollow brick. The increase of the water vapor in bricks cavities induced shift in the thermal conductivity thus, insulation performance of the hollow brick was decreased. Moreover, different material response to the varying exterior conditions could be perceived as the potential risk especially on the contact surfaces of the particular elements.

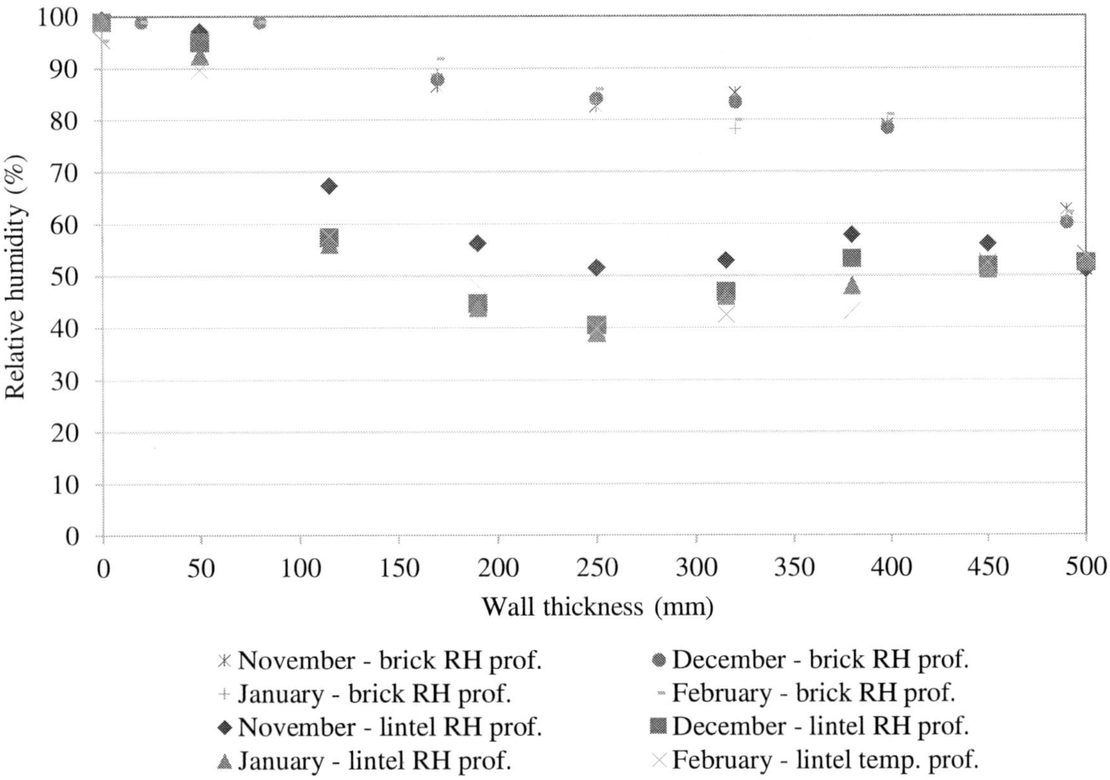

Figure 5. Measured relative humidity profiles in wall cross-section

5. Conclusions

The utilization of semi-scale analysis for experimental assessment of hygrothermal performance of the large construction element was found to be effective way for such type of analysis. The relatively low cost of the experiment in a combination with the suitable accuracy of the temperature and relative humidity profiles during long-term continuous measurement makes good prerequisites for an extensive use of the technique.

Applied climatic load has made it possible to evaluate the functionality of the tested construction detail during the winter period, which can be perceived as the most demanding and critical season especially from the thermal losses and related energy demands. The obtained results revealed a good hygrothermal performance of the window lintel. Concurrently, the significant influence of applied insulation layer in the window lintel to water vapor transport properties was revealed.

Acknowledgement

Authors gratefully acknowledge the financial support of the Czech Science Foundation in the frame of Projects for promotion of excellence in basic research, under project No P105/12/G059.

6. References

[1] Pavlik Z, Fort J, Pavlikova M, Pokorny J, Trnik A and Cerny R 2016 *Energ. Buildings* **126** 113
[2] Al-Sanea S A and Zedan M F 2011 *Appl. Energ.* **88** 3113
[3] Vasile C, Lorente S and Perrin B 1998 *Energ. Buildings* **28** 229
[4] Zhang N, Cao B, Wang Z J, Zhu Y X and Lin B R 2017 *Build. Environ.* **117** 208
[5] Mlakar J and Strancar J 2013 *Build. Environ.* **60** 185
[6] Grey C N B, Schmieder-Gaite T, Jiang S Y, Nascimento C and Poortinga W 2017 *Indoor Built. Environ.* **26** 902
[7] Pavlik Z, Jerman M, Fort J and Cerny R 2015 *Int. J. Thermophys.* **36** 557
[8] Koci J, Madera J, Jerman M and Cerny R 2015 *Energ. Buildings* **99** 61

Selected aspects of design for internal partitions and construction joints in low energy consumption buildings – a case study

K. Pawłowski[1]
[1] Faculty of Civil and Environmental Engineering and Architecture, UTP University of Science and Technology Al. S. Kaliskiego 7, 85-796 Bydgoszcz, Poland

Abstract. Sustainable building is construction approach which is friendly to both humans and the environment, saving natural resources and counteracting pollution. The most essential design and execution features in this kind of construction approach include: the use of renewable energy sources to preserve natural resources; lowering the energy consumption of technological processes; decreasing the amount of material waste (by changing or modernising the building material technologies used); limiting of energy consumption of existing buildings and those being designed; modern installations in buildings; computer aid in sustainable building design.
This paper presents principles for the design of external partitions and their joints using innovative thermal and insulation materials regarding their thermal and humidity requirements. Integral design elements in this range include certified computer programs which use numerical calculations allowing for parameters of external and internal air. The introduction of new thermal and insulation materials has a crucial impact on processes inside partitions and building joints. That is why the choice of a material layers system should be based on individual calculations considering 'the specifics of the situation in which it will be applied. Based on the performed calculations and analyses, certain recommendation regarding design and execution were formulated in this paper for design of low energy buildings.

1. Introduction

The national plan for low energy building development [1] contains a definition recommended for use in practice: ***a 'low-energy building'*** is a building which fulfils the requirements for energy saving and isolation contained in the technical regulations mentioned in article 7, paragraph 1, point 1 of the Regulation – Construction law [2], especially Chapter X and Appendix 2 for Regulation [3] from 31 Dec 2020 (in the case of buildings owned and occupied by public authorities – from 1 Jan 2019).
On 13 Aug 2013 a regulation by Ministry of Transportation, Building and Marine Economy was published in Government Monitor on 5 July 2013, changing the regulation on technical requirements for buildings and their location [3]. In this legal act, lower maximum values for heat transfer coefficient $\mathbf{U_{c(max)}}$ [W/(m^2·K)] were defined concerning external partitions of buildings and the lower limit values of annual demand for non-renewable primary energy $\mathbf{PE_{(max)}}$ [kWh/(m^2·year)] for periods **2014-2016**, **2017-2020** and after **31.12.2020**. Requirements concerning the thermal insulation of external walls, roofs, floors, and doors are more stringent. Additionally, the type of partition does not matter anymore (multi- or single-layer) as well as building's designation (housing, public use, warehouse, outbuilding and others). For example, the maximum permitted value of the heat transfer coefficient for external walls of heated rooms ($t_i \geq 16$°C) will be 0.20 W/(m^2·K) from 31.12.2020, but for windows, this will be 0.90 W/(m^2·K). It must be stressed that the values of $U_{c(max)}$ and U_{max} do not allow for the impact of heat bridges that exist in the buildings in reality.

Based on the analysis of legal regulations, it can be stated that the amount of heating energy needed for building usage according to its designation may be kept at a reasonably low level; two methods were predicted to be able to fulfil this requirement in new building projects. The first method consists of the design of partitions in the building that enable heat transfer coefficient values of **U** [W/(m^2·K)] for external partitions, windows, doors and installation technology, which conforms to the requirements of thermal isolation. However, heat flow resulting from the occurrence of linear and spatial heat bridges which are created, for example, by the connection of a few partitions in the building must be allowed for in both two-dimensional (2D) and three-dimensional fields (3D).

In this work, an analysis has been undertaken concerning the optimisation of the physical parameters of a building's external partitions and joints of low energy consumption. According to regulation [3], a series of options for the calculation of heat transfer coefficient U was allowed for specific external partitions. Apart from the PN-EN ISO 6946 standard [4], there were other parallel standards mentioned which fundamentally enhance instruments serving for specifying the basic calculation method for the heat transfer coefficient for linear (2D) and spatial (3D) heat bridges. The procedure requires information regarding the linear heat transfer coefficient Ψ [W/(m·K)] in the case of linear heat bridges and spot heat transfer coefficient X [W/K] in the case of spatial heat bridges. Additional heat fluxes resulting from improperly or insufficiently insulated bridges can even exceed the values of basic fluxes present in a continuous (full) partition without heat bridges. The numerical procedure (using computer software) required by the PN-EN ISO 10211 standard [5] causes an essential objection; it is difficult in application and is unclear. The Ψ coefficient gives linear (for 1m of length) heat losses by a bridge, usually assumed in extending the calculation area for distance of two partition widths from the edge of a bridge. It often happens that in a partition a few bridges overlay. In such a case, it is impossible to sum up the Ψ coefficients for this geometry, for instance a pillar: window – adjoining wall.

In studies concerning professional calculation of external partitions, the values of the Ψ coefficient are usually given in relation to a whole joint with a thermal bridge participating in heat transfer [6, 7]. To define the share amount of specific bridges in the total heat loss by a given partition, we need to discriminate the partial values of heat transfer coefficient corresponding to the branches of a joint within the partition in question. This can be achieved through additional numerical calculation by setting the values of a branch heat transfer coefficient. The calculation procedures are described in [8, 9].

2. Physical parameters of external partitions and joints in low-energy buildings

A building is a structure of building partitions and their joints of individual physical character, and it is subjected the external and internal environment influence. In most cases, the analysis of building partitions and joints concerning construction, material and technological aspects does not raise objections at the design stage. However, knowledge of physical parameters associated with heat and humidity exchange allows avoiding numerous mistakes regarding design and execution.

To define the temperature distribution and the additional heat loss resulting from the different materials and structures of partitions and joints, we need to perform detailed professional computer calculations of basic physical parameters regarding:

- heat flux **Φ** [W];
- heat transfer coefficient of the whole partition **U** ($\mathbf{U_{1D}}$) [W/(m^2·K)];
- linear coefficient of thermal coupling $\mathbf{L^{2D}}$ [W/(m·K)];
- linear heat transfer coefficient (defining heat losses resulting from linear thermal bridges) **Ψ** [W/(m·K)], and in the case of heat loss calculations for a single partition, a branch coefficient of heat transfer related to part of a joint (for example, the connection of an external wall with a window – linear heat transfer coefficient for the window part $\mathbf{\Psi_O}$ and linear heat transfer coefficient for an external wall part $\mathbf{\Psi_{sz}}$;
- minimum temperature of an external surface of a partition in the thermal bridge $\mathbf{t_{min.}}$ [°C];
- temperature factor, defined from the minimum temperature of an internal surface of a partition in the thermal bridge $f_{Rsi(2D)}$ [-].

Based on the tests and analyses performed, it is proposed to define the shape coefficient of the thermal bridge [10]. The graphic solution method for two-dimensional issues of stabilized heat flow is determined according to [11] so called the shape coefficient of the system, based on the properties of the orthogonal net of curvilinear squares, consisting of adiabats and isotherms. In a one-dimensional field, for example in a wall with a surface width equal to its thickness, the relation n_2/n_1 is 1. However, in a two-dimensional field, this quantity has, in principle, a value greater than 1. Due to this fact, the shape coefficient of a thermal bridge defines how much the heat flow conducted through a tested two-dimensional area (2D) changes in relation to a one-dimensional field (1D) with other parameters influencing the heat transfer remaining unchanged. Using the above described rule of heat flux comparability, it is proposed to define the shape coefficient of a thermal bridge K for total heat exchange (with allowing for heat interception on the partition surface. Additional heat loss caused by the bridge are referred to the partition width equal to its thickness (or heat bridge range) and calculated on the inside surface of the partition. The K coefficient indicates the number of times the heat flux change which permeates the bridge area, flowing to its internal surface (per 1 m of length) in relation to analogical partition area without a heat bridge. This is a non-unit value which needs to be determined by numerical calculation allowing for various forms of resistance for heat interception on the internal area of bridges. The K coefficient is calculated from the linear heat transfer Ψ and heat transfer coefficient U:

$$K=(\Psi_i+U\cdot g)/(U\cdot g)=\Psi_i/(U\cdot g)+1 \text{ [-]} \quad (1)$$

where:
K – shape coefficient of a thermal bridge, [-]
Ψ_i – linear heat transfer coefficient of a thermal bridge branch within the partition range, [W/(m·K)]
U – heat transfer coefficient for a full partition far from a heat bridge (U_{1D} – heat transfer coefficient in a one-dimension field), [W/(m^2·K)]
g – partition thickness (or heat bridge range), [m].
Detailed calculation procedures for physical parameters of building partition joints are discussed in [8, 9, 10].

3. Calculations of physical parameters of external partition joints of a low-energy building
The calculation of the physical parameters of external partition joints using TRSCO was performed with the following assumptions:
- modelling of joints was performed according to the rules described in the PN-EN ISO 10211 standard [5] and articles [8] and [9];
- heat interception resistances (R_{si}, R_{se}) were assumed according to the PN-EN ISO 6946 standard [4] for the calculation of heat fluxes, and according to the PN-EN ISO 13788 standard [11] for the calculation of heat distribution and temperature factor $f_{Rsi(2D)}$;
- internal air temperature t_i = 20 °C (living-room), external air temperature t_e = -20 °C (zone 3);
- values of heat transfer coefficient for building materials λ [W/(m·K)] assumed based on tables from [9];
Calculations were performed for the following cases:
- corner of two-layer external wall, external wall joint with a window in section through the casing (styrofoam boards with λ=0,04 W/(m·K) – variant I;
- PIR boards with λ=0,021 W/(m·K) of thickness 10, 12, 15, 20 cm – variant II,
- without a window jamb – variant III, IV,
- with a window jamb – variant V, VI.
Figures 1, 2, and 3 show graphic results of a computer simulation of the analysed joints using TRISCO software. The results of physical parameter calculations are presented in Tables 1 and 2.

IOP Conf. Series: Materials Science and Engineering **415** (2018) 012016 doi:10.1088/1757-899X/415/1/012016

Variant I – Styrofoam board insulation, variant II – PIR board insulation

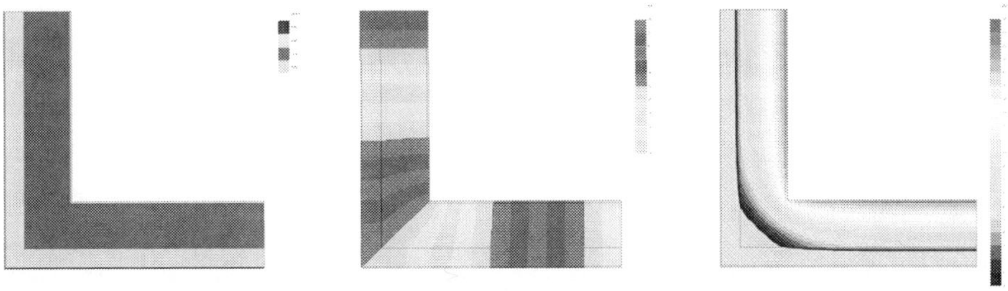

a) calculation model b) heat flux lines c) temperature distribution (isotherms)

Fig. 1. Exemplary graphic representation of computer simulation results for a corner joint of an external two-layer wall – own study

Table 1. Calculation of physical parameters for a corner joint of an external two-layer wall using TRISCO software

	Physical parameters						
	U (U_{1D}) [W/(m²·K)]	**Φ** [W]	**L^{2D}** [W/(m·K)]	**Ψ_i** [W/(m·K)]	K [-]	**$t_{min.}$** [°C]	***$f_{Rsi(2D)}$*** [-]
$I_{(10)}$	0.260	24.11	0.603	0.082 a) 0.041 b)	1.432 c) 1.158 d)	14.08	0.851
$I_{(12)}$	0.230	21.55	0.539	0.078 a) 0.039 b)	1.440 c) 1.170 d)	14.57	0.864
$I_{(15)}$	0.196	18.63	0.466	0.073 a) 0.036 b)	1.449 c) 1.186 d)	15.18	0.879
$I_{(20)}$	0.158	15.25	0.381	0.066 a) 0.033 b)	1.449 c) 1.209 d)	15.91	0.898
$II_{(10)}$	0.164	15.64	0.391	0.063 a) 0.031 b)	1.526 c) 1.192 d)	15.88	0.897
$II_{(12)}$	0.142	13.66	0.342	0.058 a) 0.029 b)	1.530 c) 1.204 d)	16.33	0.908
$II_{(15)}$	0.118	11.51	0.288	0.052 a) 0.026 b)	1.531 c) 1.220 d)	16.84	0.921
$II_{(20)}$	0.092	9.15	0.229	0.045 a) 0.022 b)	1.526 c) 1.245 d)	17.41	0.935

- variant I and PIR boards with λ=0.021 W/(m·K) – variant II of thickness: 10, 12, 15, 20 cm
- linear coefficient of thermal coupling: $\mathbf{L^{2D}=\Phi/(\Delta t \cdot l)}$ [W/(m·K)], according to [8, 9]
- linear heat transfer coefficient: $\mathbf{\Psi= L^{2D}-(\Sigma U_i \cdot l_i)}$ [W/(m·K)], according to [8, 9]
- shape coefficient of a thermal bridge: $\mathbf{K=(\Psi_i+U \cdot g)/(U \cdot g)=\Psi_i/(U \cdot g)+1}$ [-], according to [10]
- temperature factor: $\mathbf{f_{Rsi(2D)} = (t_{min.}- t_e)/(t_i - t_e)}$ [-], according to [12]

a) – value of heat transfer coefficient for overall joint
b) – value of linear heat transfer coefficient for a single corner branch
c) – value of shape coefficient for a heat bridge related to partition thickness
d) – value of shape coefficient for a heat bridge related to 1m bridge branch length

variant III – without a window jamb (styrofoam board insulation), variant IV – without a window jamb (PIR board insulation)

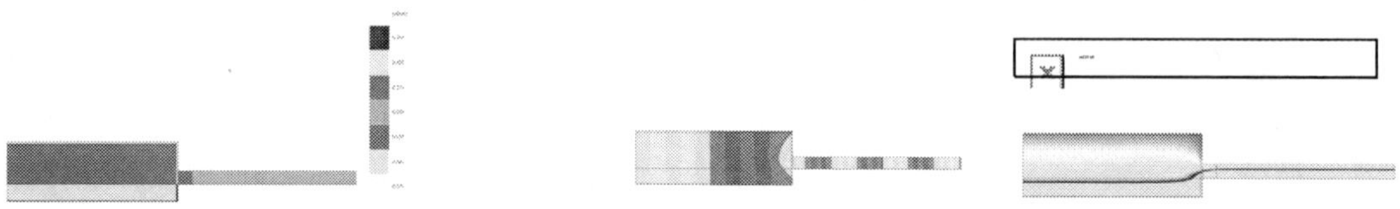

a) calculation model b) heat flux lines c) temperature distribution (isotherms)

Fig. 2. Graphic representation for a joint of an external two-layer wall with a window in section through the casing (without nib) – own study

variant V – with a window jamb (Styrofoam board insulation), variant VI – with a window jamb (PIR board insulation)

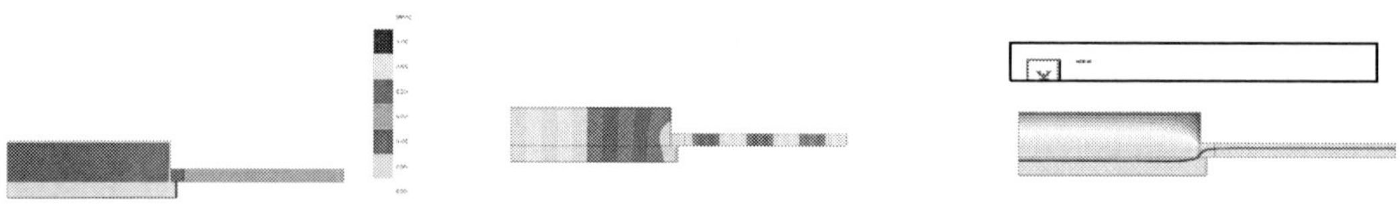

a) calculation model b) heat flux lines c) temperature distribution (isotherms)

Fig. 3. Graphic representation for a joint of an external two-layer wall with a window in section through the casing (with nib) – own study

Table 2. Physical parameters for a joint of an external two-layer wall with a window in section through the casing (TRISCO software)

	Physical parameters						
	U (U_{1D}) [W/(m²·K)]	**Φ** [W]	**L^{2D}** [W/(m·K)]	**Ψ_i** [W/(m·K)]	K [-]	**$t_{min.}$** [°C]	**$f_{Rsi(2D)}$** [-]
$III_{(10)}$	0.260 0.762	45.33	1.133	0.043 a) 0.064b) / -0.021 c)	1.674 d) 0.656 e)	14.72	0.868
$III_{(12)}$	0.230 0.762	44.52	1.113	0.046 a) 0.066b) / -0.021 c)	1.745 d) 0.656 e)	14.83	0.871
$III_{(15)}$	0.196 0.762	43.06	1.077	0.050 a) 0.072b) / -0.022 c)	1.885 d) 0.639 e)	14.90	0.873
$III_{(20)}$	0.158 0.762	41.64	1041	0.053 a) 0.075b) / -0.022 c)	2.021 d) 0.639 e)	15.03	0.876
$IV_{(10)}$	0.164 0.762	41.59	1.040	0.046 a) 0.068b) / -0.022 c)	1.485 d) 0.639 e)	15.08	0.877
$IV_{(12)}$	0.142 0.762	40.81	1.020	0.048 a) 0.070b) / -0.022 c)	1.542 d) 0.639 e)	15.12	0.878
$IV_{(15)}$	0118 0,.762	39.98	1.000	0.051 a) 0.074b) / -0.023 c)	1.627 d) 0.623 e)	15.16	0.879
$IV_{(20)}$	0.092 0762	39.00	0.975	0.053 a) 0.076b) / -0.023 c)	1.721 d) 0.623 e)	15.27	0.882
$V_{(10)}$	0.260 0.762	43.98	1.100	0.034 a) 0.037b) / -0.003 c)	1.390 d) 0.951 e)	15.72	0.893
$V_{(12)}$	0.230 0.762	42.83	1.071	0.037 a) 0.038b) / -0.001 c)	1.429 d) 0.984 e)	15.82	0.896

V(15)	0.196 0.762	41.55	1.039	0,040 [a)] 0.041[b)] / -0.001 [c)]	1.504 [d)] 0.984 [e)]	15.99	0.900
V(20)	0.158 0.762	40.09	1.002	0.044 [a)] 0.043[b)] / -0.001 [c)]	1.585 [d)] 0.984 [e)]	16.14	0904
VI(10)	0.164 0.762	39.94	0.999	0.034 [a)] 0.031[b)] / 0.001 [c)]	1.327 [d)] 1.016 [e)]	16.30	0.908
VI(12)	0.142 0.762	39.10	0.978	0.037 [a)] 0.035[b)] / 0.002 [c)]	1.395 [d)] 1.033 [e)]	16.37	0.909
VI(15)	0.118 0.762	38.21	0.955	0.039 [a)] 0.037[b)] / 0.002 [c)]	1.455 [d)] 1.033 [e)]	16.44	0.911
VI(20)	0.092 0.762	37.22	0.931	0.042 [a)] 0.039[b)] / 0.003 [c)]	1.531 [d)] 1.049 [e)]	16.55	0.914

- variant III, V – styrofoam board insulation with λ=0.04 W/(m·K),
- variant IV, VI – polyurethane board with λ=0.021 W/(m·K)
- linear coefficient of thermal coupling: $\mathbf{L^{2D}=\Phi/(\Delta t \cdot l)}$ [W/(m·K)], according to [8, 9]
- linear heat transfer coefficient: $\mathbf{\Psi = L^{2D} - (\Sigma U_i \cdot l_i)}$ [W/(m·K)], according to [8, 9]
- shape coefficient of a thermal bridge: $\mathbf{K=(\Psi_i+U\cdot g)/(U\cdot g)=\Psi_i/(U\cdot g)+1}$ [-], according to [10]
- temperature factor: $f_{Rsi(2D)} = (t_{min.} - t_e)/(t_i - t_e)$ [-], according to [12]

[a)] – value of heat transfer coefficient for overall joint
[b)] – value of linear (branch) heat transfer coefficient related to external wall
[c)] – value of linear (branch) heat transfer coefficient related to a wall
[d)] – value of shape coefficient for a heat bridge related to external wall
[e)] – value of shape coefficient for a heat bridge related to window

4. Calculation analysis of physical parameters for external partitions and their joints

In choice of material sets for construction of external partitions and their joints for a **low-energy building** using various insulation materials, we must limit additional heat loss. It can be done with the linear heat transfer coefficient Ψ [W/(m·K)] or by the shape coefficient of heat transfer K [-], and the risk of the occurrence of surface humidity (to define the minimal temperature on internal surface of a partition and temperature factor $f_{Rsi(2D)}$), and the risk of the occurrence of inter-layer condensation. The values of physical parameters for external partitions and their joints depend on the type and thickness of insulation material (heat conduction coefficient λ [W/(m·K)]), and the shape of the joint between these two various partitions (for instance, an external wall and window). The location of a window casing in contact of construction layer and insulation with a nib allows lower heat loss (Table 2). It must be noticed that extending the insulation over the casing causes the temperature of the external surface of the partition (in contact with the external wall and the casing) is higher than in the case without insulation on the casing (Table 2). In the case of these types of joints, it is a good practice to define the branch coefficient of heat transfer separately for a part of the external wall $\Psi_{\acute{s}c.}$ [W/(m·K)] and for a part of the window Ψ_O [W/(m·K)] because this enables defining additional heat loss separately for the external wall and the window (Table 2). Using approximate or rough values, for example, based on the PN-EN ISO 14683 standard [7] becomes unfounded because they do not allow for changes of material sets, or insulation type and thickness. However, in heat loss evaluation, we must also analyse other parameters: Φ (amount of heat flux flowing through a joint) [W]; L^{2D} (heat-coupling coefficient) [W/(m·K)] reflecting the heat loss in a joint, or the heat bridge shape K [-].

The fulfilment of the criterion for avoiding surface condensation risk (mildew and fungi development): $f_{Rsi.(2D)} \geq f_{Rsi.(critical)}$, requires defining the value of $f_{Rsi.(2D)}$ based on the minimum temperature on the internal surface of the partition at the location of the heat bridge (2D) $t_{min.}$ [°C] and the value of $f_{Rsi.(critical)}$, accounting for the internal and external air parameters (humidity and air temperature). According to the PN-EN ISO 13788 standard [12], the temperature factor $f_{Rsi(critical)}$ is calculated or assumed, depending on the ventilation type used in the building (for gravity ventilation – dominant in housing buildings, or mechanical ventilation being often a part of conditioning systems allowing in nearly free way to shape properties of interior microclimate). The maximum value obtained over a 12-month relating to a building in Bydgoszcz, assuming humidity class III, amounts to $f_{Rsi.(max)}=f_{Rsi.(critical)}=0.785$ (February). This means that in each month of the year and for any other values of limit temperatures, in order to avoid surface condensation $f_{Rsi.(2D)}$ it should be higher than 0.785. In regulation [3], despite from regarding PN-EN ISO 13788 standard [12] as the binding one in designing, there is a deviation from its requirements consisting in assuming monthly average of internal relative air humidity at a constant value of $\varphi_i=0.50$ (50 %) (point 2.2.2 of appendix No. 2 [3]) for rooms with internal temperature equal at least 20°C. Additionally, it was allowed (without calculation) for these rooms the assumption of factor value $f_{Rsi(critical)}$ = 0.72, which practically means resignation from setting humidity classes of rooms equipped with gravity ventilation. This departure does not allow for humidity calculation of the real localization conditions (climatic) and microclimatic of the examined building, at least in relation to rooms with internal temperature $t_i \geq 20°C$, quite drastically lowering the level of requirements in protection against fungi of buildings located in Poland in stricter climatic zones (zone IV and V). Based on the performed calculations (Table 1 and 2) it can be stated that in the analyzed joints, there is no risk of mildew because the condition $f_{Rsi.(2D)} \geq f_{Rsi.(critical)}$ = 0.785 is fulfilled.

5. Conclusion and final propositions

The quest for fulfilling the heat-humidity requirements for external partitions and joints for low-energy buildings should be based on clear and precise rules resulting from a broad understanding of the internal and external air parameters of the analysed building; this understanding can be acquired by use of modern numerical tools for the analysis of the construction physics involved.

The selection of a system of material layers for external partitions and their joints should consider the following criteria: isolation (Tables 1 and 2); inter-layer surface condensation (Tables 1 and 2); acoustic isolation; fire protection, construction load bearing, and duration capacity. Some sets of material layers fulfil the requirements for thermal insulation ($\mathbf{U_c} \leq \mathbf{U_{c(max)}}$); however, after analyses concerning requirements for humidity, acoustics, or fire protection, the location of heat insulation inside the partition is unacceptable.

It is necessary to restore the calculation of authoritative values of heat transfer coefficients $\mathbf{U_c}$ of singular partitions of a building, regarding two-dimensional fluxes (2D), and for partition ns in contact with ground, with 3D fluxes. The lowering of the limit coefficients for heat transfer $\mathbf{U_{max}}$ without taking heat bridges into account, i.e. heat flows in 2D and 3D fields causes allowing higher heat loss through construction partitions and their joints. The definition of shape coefficients for the heat bridges analysed in this work confirms the need to consider additional heat losses resulting from the connection of a minimum of two partitions of the building (Table 1 and 2). Additionally, it is correct to define limit values of the linear heat transfer $\mathbf{\Psi_{max}}$ at level of 0.05-0.10 W/(m·K) depending on the specifics of the analysed joint.

It is also correct to use the PN-EN ISO 13788 standard [12] in the analysis of partitions and their joints regarding humidity. It is a conclusion, in regulation [3], the paragraph about allowing the application of the limit value of $\boldsymbol{f_{Rsi}}$ at a level of 0.72 should be removed. The author believes that there is a need to develop new design guidelines in regarding humidity, including the definition of critical values for $\boldsymbol{f_{Rsi}}$ and accounting for specific parameters of internal and external air.

References

[1] Regulation by the Council of Ministers of 22 June 2015 about approving "The National Plan for Development of Low Energy Building".

[2] Regulation of 7 July 1994 – *Construction Law* (Dz. U. of 2013, position 1409, with later amendments).

[3] Regulation by the Minister of Transport, Building and Marine Economy amending the Regulation about technical requirements for building and their location (Dz. U. of 2013 position 926). /Announcement by the Minister of Infrastructure and Development of 17 July 2015 about unified text for technical requirements for buildings and their location. (Dz. U. of 2015, position 1422)/.

[4] PN-EN ISO 6946 :2008 *Construction components and building elements. Thermal resistance and heat transfer coefficient. Calculation method.*

[5] PN-EN ISO 10211 :2008 *Heat bridges in buildings. Heat fluxes and surface temperatures. Detailed calculations.*

[6] J.A. Pogorzelski, J. Awksientjuk, *The catalogue of heat bridges* (Wydawnictwo ITB Warszawa, 2003).

IOP Conf. Series: Materials Science and Engineering **415** (2018) 012016 doi:10.1088/1757-899X/415/1/012016

[7] PN-EN ISO 14683:2008 *Heat bridges in buildings. Linear coefficient of heat transfer. Simplified methods and rough values.*
[8] A. Dylla, *The thermal physics of buildings in practice. Thermal and humidity calculations* (Wydawnictwo Naukowe PWN S.A. Warszawa, 2015).
[9] K. Pawłowski, *The design of external partitions in view of current requirements concerning buildings. Thermal and humidity calculations of internal partitions and their joints* (Grupa Wydawnicza Medium Warszawa, 2016).
[10] A. Dylla, K. Pawłowski, *Shape coefficient of a heat bridge.* (Budownictwo Ogólne, Construction, material, thermal and humidity issues in building, Wydawnictwa Uczelniane ATR, Bydgoszcz, 2005).
[11] B. Staniszewski, *Heat exchange* (Wydawnictwo PWN Warszawa, 1963).
[12] PN-EN ISO 13788:2003 *Heat and humidity properties of building components and building elements. Temperature of internal surface allowing to avoid the critical humidity of internal surface condensation. Calculation methods.*

IOP Conf. Series: Materials Science and Engineering **415** (2018) 012017 doi:10.1088/1757-899X/415/1/012017

The characteristics of temperature fluctuations in thermal insulation covered with layer of PCM

A Wieprzkowicz[1] **and D Heim**[1]

[1]Department of Environmental Engineering, Lodz University of Technology Wolczanska 213, 90-924 Lodz, Poland

E-mail: anna.wieprzkowicz@p.lodz.pl

Abstract. Energy performance of opaque elements of building envelope can be improved by the decrease of heat losses by conduction or by effective utilization of solar heat gains. Enhancement of thermal insulation by application a thin layer of phase change material as an external finishing coating contributes to the attenuation and delay of heat flux transferred through partition. Based on the real scale experiment of one-year measurements, dynamics of temperature fluctuations in thermal insulation covered with external layer of PCM was characterized and analysed. It was concluded that outer application of PCM contributes to the attenuation of temperature fluctuations of the whole wall component.

1. Introduction

It can be observed that there is a strong need for the new, smart technologies on thermal insulation industry [1, 2]. The predicted trends of such kind materials development are moving towards application of dynamic thermal insulation materials. It means that, in the contrary to the pursuit of last decades, lowering the thermal conductivity loses its importance in favour for activating the material. There is no doubt that thermal insulation materials of the future will have to be still characterized by excellent insulating properties but simultaneously should reveal the ability of its properties adjustment to dynamic thermal conditions. There are different concepts of achieving those assumptions [3], using e.g. sensible and latent heat accumulation. The comprehensive analyses for the wide range of concretes was done by Garbalińska and Strzałkowski 2016 [4]. The authors analysed and compared the results of internal temperature and internal heat flux fluctuations. One of the alternative and most promising idea is to utilize the effect of latent heat accumulation induced by phase transition of material caused by natural, diurnal temperature fluctuation [5].

Application of phase change materials (PCM) has been considered for last 30 years [6,7,8] as an internal finishing layer, which accumulate excessive internal heat loads during the day and release it back to the internal environment during the night time what contributes to the attenuation of daily temperature fluctuations (Fig. 1a). Another approach, which has been recently investigated concerns different positioning of PCM layer [9] in heavyweight [10] and lightweight [11,12] constructions. Based on previous studies [13,14] in can be assumed that in case of external application of PCM, solar Energy is accumulated in the material and PCM layer plays a role of a thermal shield [15], attenuating the heat flux by conduction and temperature fluctuations in the whole partition (Fig 1b).

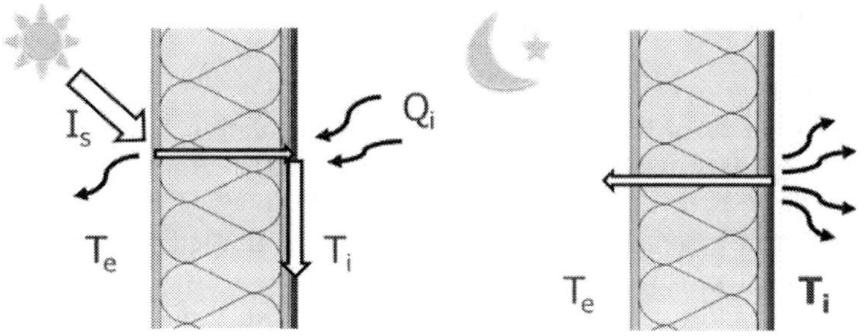

a)

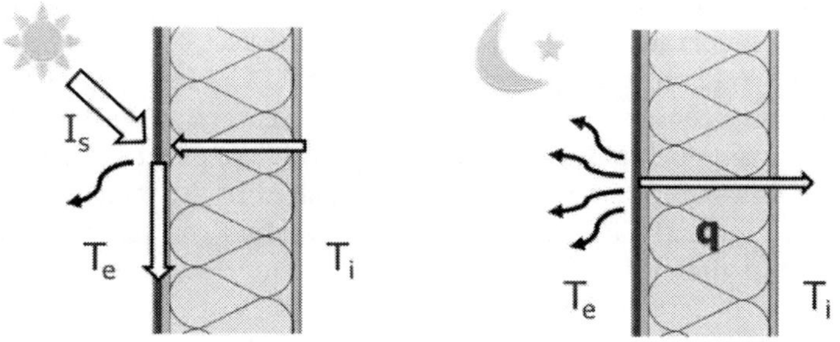

b)

Figure 1. Dominant thermal processes occurring in the lightweight external wall enhanced with PCM layer (marked with the blue line) positioned from: a) internal, b) external side.

Considering PCM integration within external wall, the agent that induce the heat accumulation and release is temperature difference generating heat flux on material surface. Thus, efficiency of the system depends predominately on the occurrence of the process trigger. Depending on the assumption of the specific phase change temperature and position of the material (external, internal or intermediate), different frequency of accumulation and release processes can be observed. It means, that if material was placed in the thermally stable environment (e.g. internal surface of the wall during heating season when system keeps room temperature at lowest required level), there would be no initiator inducing the phase change. On the other hand, highly changeable thermal conditions (e.g. external surface of the wall during the summer) makes problem of proper phase change temperature selection very complicated. Since material properties should match the conditions that trigger the heat accumulation process, selection of material properties can be done only for specific conditions, characteristic for specific period of time.

Different positioning of the PCM in external wall could result in different temperature stabilization effect. Heat accumulation during phase transition is isothermal process, thus gives possibility to keep the temperature on the desired level till melting process is finished. By changing the temperature distribution in the depth of the wall, heat flux by conduction is also affected. In order to limit heat exchange between external and indoor environment, heat flux by conduction should be decreased. One of the possible ways to achieve this is to decrease the temperature difference between external and internal surfaces temperatures, by shifting one of those. In general, to limit heat losses in winter conditions, temperature of external surface should be kept on possibly highest level, while under summer conditions at the lowest.

The purpose of the study was investigation of the dynamics of temperature fluctuations in the thermal insulation covered with PCM layer from the external side of the wall. The analysis was conducted based on the measurement results from real-scale experiment. The thermal effect of PCM layer application on the temperature fluctuations in thermal insulation was evaluated in accordance to the reference case, monitored under the same boundary conditions.

2. Materials and methods

2.1. Experimental set-up

Experimental investigation of the dynamics of thermal insulation enhanced by external layer of PCM was conducted under moderate climatic conditions characteristic for the location of experimental set-up (the city of Lodz, 51°46 N & 19°27 E, Poland, Central Europe). The test stand was designed in such a way that different constructions can be examined under the unified thermal conditions simultaneously. A frame constructed external wall designed with centrally positioned square 1m^2 window and 12 test sections around the window, each with dimensions of 60x60 cm. The basic construction filling the spaces between steel pillars was made of 20 cm of mineral wool and internal finishing lining. The wall was constructed as a ventilated façade, covered from external side with PV panels.

For the purpose of comparative analysis of the thermal effect of PCM application, some of the test section were treated as a reference case, where no additional PCM layer was added to the basic structure.

Each test section was equipped with 6 temperature sensors, placed in the cross section of the component. Since the main goal of the research presented in the paper is characterictic of themperature fluctuations in thermal insulation, the temperature measured in the middle point of thermal insulation layer was analysed. Additionally, temperature in the ventilation gap was registered. The measurements were done with 5-minute time step. Two types of digital temperature sensors were used: TSic 506 and TSic 306. The first one is characterized by very high accuracy of ±0.1 K in a range of 40 K (from 5 to 45°C), but the measurement range of this sensor is −10 to 60°C. Since the temperature in the ventilation cavity could exceed this range, sensors with a lower accuracy (of ±0.3 K in a range from 10 to 90°C) but wider measurement range (−50 to 150°C) were applied. The temperature recorded in the ventilation gap was treated as an external one.

In the paper results from the measurements performed for two test sections of east directed facade were presented and analysed: for the basic – reference case and one enhanced with novel PCM layer.

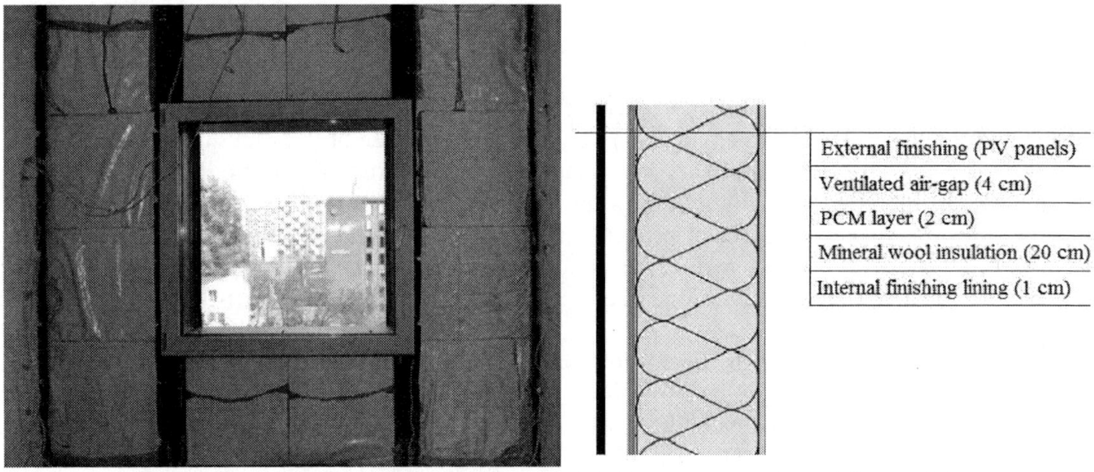

Figure 2. Façade view from inside and detailed cross section of the partition.

2.2. PCM composite

Based on theoretical considerations a novel methods of PCM applications were developed and tested. The basic assumption was to develop cheap and easy way of encapsulation. The first step to achieve this goal was a choice of the material that could work as a matrix that can hold melted material. On the experimental stage of the analysis perlite with different partition size was tested. Furthermore, due to better impregnability, availability and costs, perlite with 1÷6 mm fraction was chosen to be applied in the façade construction.

Furthermore, optimal weight proportion between paraffin and perlite was experimentally investigated. Different proportions were tested and it occurred that maximum content of PCM without further leakage can be obtained with ratio 1:1. The core material was soaked with paraffin wax and coated with liquid foil to additionally prevent from leakage. The minimum weight proportion between perlite and liquid foil, which assured full coverage of the perlite particles surfaces was 1:4. Tested sample was exposed to high temperature (above melting point) and observed for two weeks. After that time no leakage was noted what confirmed proper design.

In order to speed up solidification process after mixing all products perlite and liquid foil were previously cooled down. During the preparation temperature sensors were placed in the component. In the paper, results obtained for the paraffin RT25HC were presented and analysed. Due to the melting temperature of the paraffin around 25°C, it was expected that application of such PCM will be beneficial during the summer period, under high external temperatures and solar radiation. Due to that assumption, measurements conducted during four hottest months were analysed, from May to August.

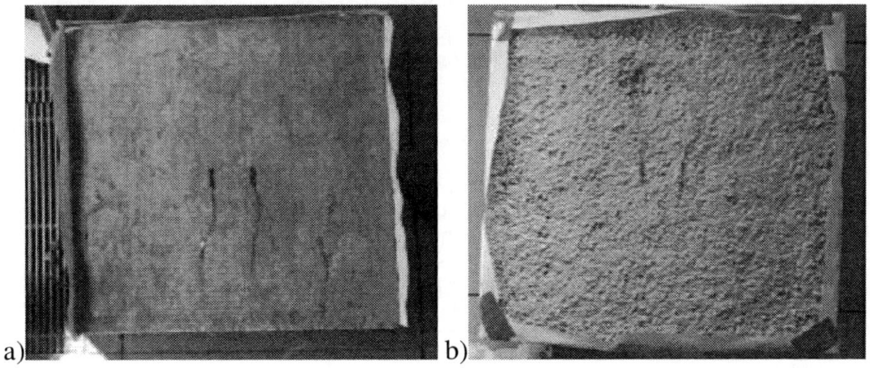

Figure 3. Analysed panels: a) reference, b) modified with PCM layer.

The total thickness of PCM-perlite composite external layer was 2 cm, and the total density 694 kg/m^3. The latent heat capacity of the composite was estimated by DSC measurements. Heating and cooling cycles were proceeded with 1.0 K/min rate. Based on the results presented in Figure 4 calculated value of latent heat capacity of the composite equalled 54 J/g.

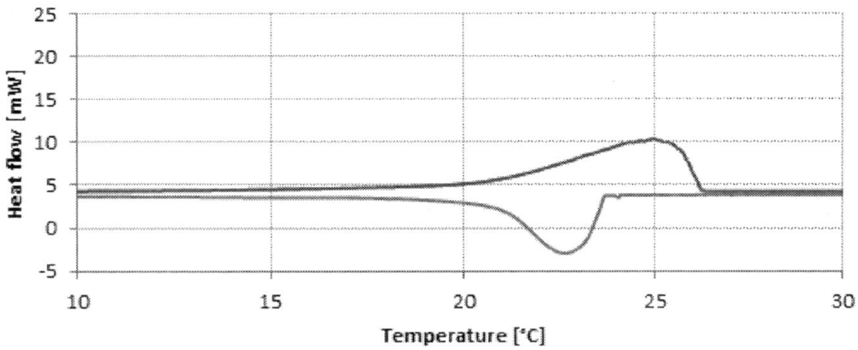

Figure 4. Results of DCS measurement of PCM-perlite composite.

2.3. Evaluation method

2.3.1. Daily temperature fluctuations

Temperature values analysed with 5-minute time step, for modified and reference panel directly reveal the differences in the thermal performance of subsequent test sections. Nevertheless, considering large amount of data only selected, representative period of time can be presented in the paper.

2.3.2. Dynamics of thermal insulation.
In order to describe the dynamics of temperature fluctuations in thermal insulation a novel index was proposed. It describes the speed of temperature increase in thermal insulation layer during the day. Temperature rise rate was calculated taking into account the value of maximum temperature and time when it occurred. This factor was calculated separately for each analysed test section and for each day. Nevertheless, designation of this factor was possible only for days with sinusoidal temperature fluctuations. Dynamics of thermal insulation (Figure 5) was calculated by the following formula:

$$DOTI^n = \frac{\Delta Ti^n}{\Delta t_{Ti}^n} = \frac{Ti_{max}^n - Ti_0^n}{t_{Ti_{max}}^n - t_{Ti_0}^n} \left[\frac{K}{h}\right] \quad (1)$$

where:
Ti_{max}^n - maximum temperature measured in the half of the thermal insulation thickness in panel n
Ti_0^n - temperature measured in the half of the thermal insulation thickness in panel n in time 0
$t_{Ti_{max}}^n$ - hour when temperature Ti_{max}^n occurred
$t_{Ti_0}^n$ - time 0 (beginning of the day)

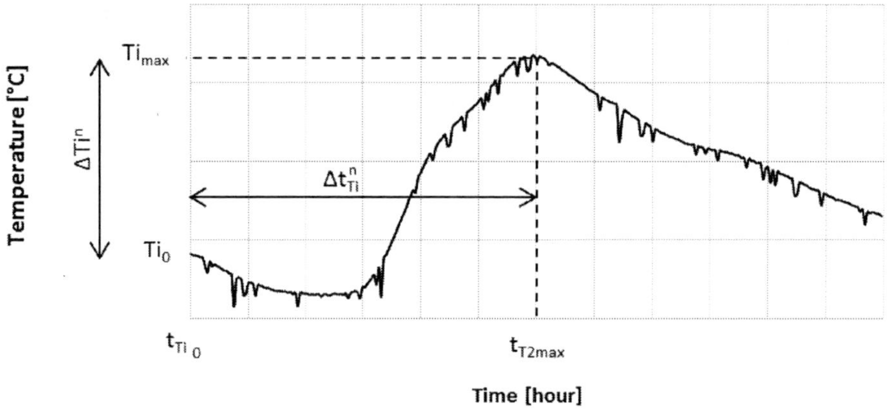

Figure 5. Example values used for dynamics of thermal insulation (DOTI) calculation.

3. Results and analysis

3.1. Daily temperature fluctuations.
Due to large amount of measured data (more than 35 thousands readouts for one test section during analysed period of four months), temperature fluctuations were presented only for selected, representative period of time – one week. Temperature distribution in the middle point of thermal insulation layer in two test sections: reference (REF) and modified (PCM) were presented in Figure 6.

It can be observed that temperature of the insulation covered with PCM is visibly lower during the day and higher during the night-time, comparing to the reference panel. It can be stated that application of PCM reduced daily amplitudes of insulation temperature. It can be also observed that due to highly intense and sudden occurrence of solar radiation during the accumulation of latent heat by PCM, temperature rise can be observed, but slower than for reference panel. Moreover, when temperature dropped to solidification temperature of paraffin, release of accumulated heat caused very slow decrease of temperature.

The temperature in the mineral wool varied from 20°C to 32°C for test section with PCM while for reference panel from 16°C to 40°C. It can be concluded that for panels with paraffin temperature was kept in the phase transition range and maximum temperature peak was shifted in time.

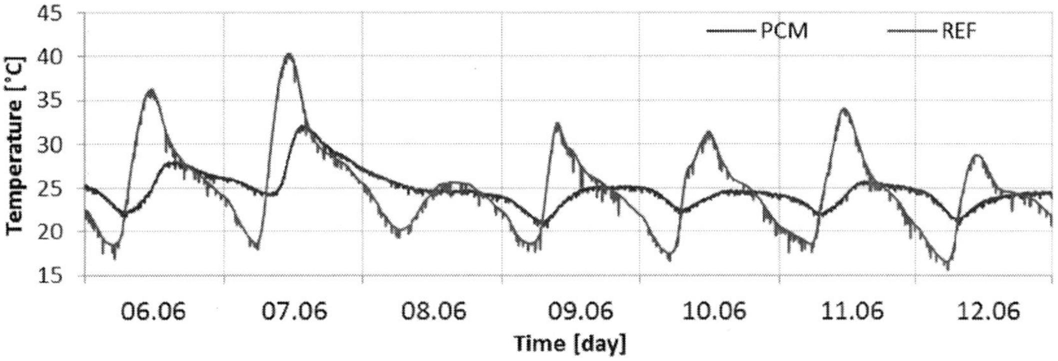

Figure 6. Temperature fluctuations in the middle point of mineral wool measured for reference and modified test section.

3.2. Dynamics of thermal insulation (DOTI).

To assess the attenuation of the influence of external temperature fluctuations on the thermal performance of insulation layer, another parameter - DOTI was introduced. It reveals the speed of temperature rise in the material, so reflects the ability of the material to attenuate the influence of thermal fluctuations of the surrounding conditions. Based on the results presented in Figure 7 it can be observed that the rate of temperature increase for the case with PCM do not exceeds 0.5 K/h for the most of analysed days, while for reference case mostly varies between 0.5 and 1.5 K/h.

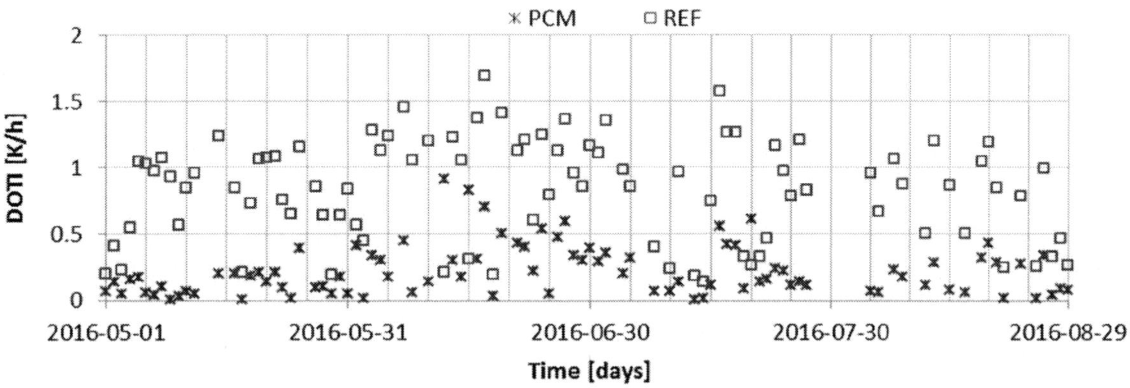

Figure 7. Values of DOTI parameter calculated for modified and reference test sections, in the analysed period of time.

It can be found obvious that for the lower external temperature fluctuation the lower temperature fluctuations should be noted in the inside layers of the wall. Nevertheless, based on the results presented in Figure 8 it can be observed that in case of PCM application values of DOTI index remain almost constant, independently to daily amplitude of external temperature. On the contrary, for the reference case it can be visibly noted that insulation temperature increases with the increase of external temperature fluctuations. It confirms that application of an additional layer of PCM contributes to the attenuation of temperature fluctuations in the layer characterized by low thermal mass even under highly variable external thermal conditions.

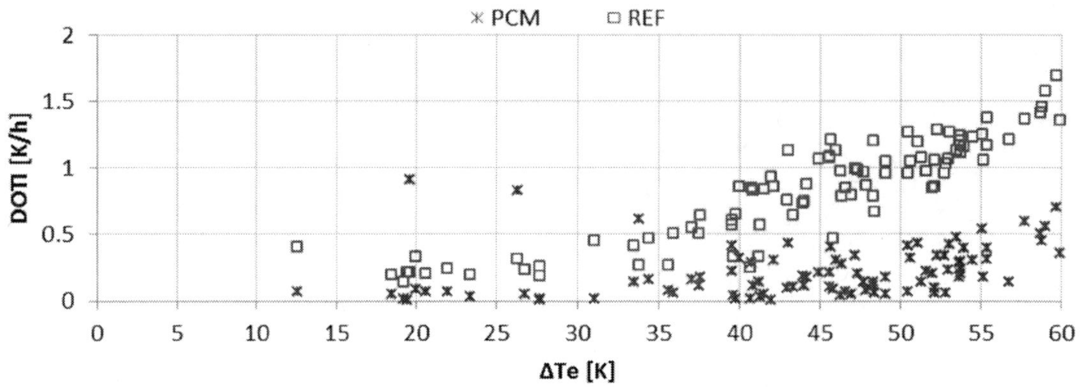

Figure 8. Values of DOTI parameter calculated for modified and reference test sections, in accordance to daily external temperature fluctuations.

4. Conclusions

The technical solution described and analysed in that paper is devoted to increase thermal inertia of external surface of thermal insulation. It is expected that such concept of external wall will stabilize daily temperature fluctuations and effect on the solar heat gains through opaque partition. Usually, this type of gains are omitted in standard buildings but in passive and zero-energy ones can play a significant role in energy cooling demand.

The technical solution presented in the paper is based on monolithic, external plaster pasted to the surface of thermal insulation. That solution was originally developed by authors and further tested in laboratory and real scale application. The observation and analyses showed that composite material is tight and thermally stable during long term period.

Analysing experimental results obtained for selected 7 days of June, characterised by high solar radiation it can be stated that daily fluctuations of external surface temperature can be even four times higher for standard insulation than in case of PCM modified material. Moreover, the temperature surface with PCM did not exceed 27°C, while for standard partition it reached up to 40°C. The dynamics of thermal insulation expressed by the DOTI index is strongly changeable an depends on daily external temperature fluctuations.

References

[1] Jelle B P 2011 *Energy and Buildings* Traditional, state-of-the-art and future thermal building insulation materials and solutions – Properties, requirements and possibilities **43** 2549–63

[2] Aditya L, Mahlia T M I, Rismanchi B, Ng H M, Hasan M H, Metselaar H S C, Muraza O, Aditiya H B 2017 *Renewable and Sustainable Energy Reviews* A review on insulation materials for energy conservation in buildings **73** 1352–65

[3] Hoes P, Hensen J L M 2016 *Energy and Buildings* Thepotential of lightweight low-energy houses with hybrid adaptable thermal storage: Comparing the performance of promising concepts **110** 79–93

[4] Garbalińska H, Strzałkowski J 2016 *Technical Transactions – Civil Engineering* Wave amplitudes of temperature and heat flux in surroundings of external walls, **3-B/2016** 65-73

[5] Zastawna-Rumin A, Nowak K 2015 *Procedia Engineering* Experimental Thermal Performance Analysis of Building Components Containing Phase Change Material (PCM) **108** 428-435.

[6] Drake J B 1987 *Report ORNL – 6359. Oak Ridge National Laboratory* A study of the optimal transition temperature of PCM wallboard for solar energy storage

[7] Heim D, Clarke J. 2004 *Energy and Buildings* Numerical modelling and thermal simulation of

PCM-gypsum composites with ESP-r **36(8)** 795-805

[8] Rhafiki T El, Kousksou T, Allouhi A, Benomar W, Zennouhi H, Jamil A, Zeraouli Y 2017 *Journal of Building Engineering* Numerical analysis of a micro-encapsulated PCM wallboard: Fluxmeter applications **14** 127-33

[9] Chwieduk D A, 2013 *Energy* Dynamics of external wall structures with a PCM (phase change materials) in high latitude countries **59** 301-13

[10] Shi X, Memon S A, Tang W, Cui H, Xing F 2014 *Energy and Buildings* Experimental assessment of position of macro encapsulated phase change material in concrete walls on indoor temperatures and humidity levels **71** 80–87

[11] Jin X, Medina M A, Zhang X 2013 *Applied Energy* On the importance of the location of PCMs in building walls for enhanced thermal performance **106** 72–78

[12] Kosny J, Fallahi A, Shukla N, Kossecka E, Ahbari R 2014 *Solar Energy* Thermal load mitigation and passive cooling in residential attics containing PCM-enhanced insulations **108** 164–77

[13] Heim D, Wieprzkowicz A 2016 *Journal of Building Performance and Simulation* Positioning of an isothermal heat storage layer in a building wall exposed to the external environment **9** 542–554

[14] Heim D, Wieprzkowicz A 2018 *Applied Sciences* Attenuation of temperature fluctuations on external surface of the wall by phase change material activated layer **8(11)**

[15] Wieprzkowicz A, Heim D 2016 *Environmental Management and Quality, An International Journal* Energy performance of dynamic thermal insulation built in the experimental façade system **27(6)** 681-694

Tests of the innovative building materials used for external walls in a case-study construction objects

B Pietruszka[1], **K Oberda**[1], **M Piasecki**[1]

[1]The Thermal Physics, Acoustics and Environment Department, The Building Research Institute, ITB, 1 Filtrowa Street, 00-611 Warsaw, Poland
b.pietruszka@itb.pl

Abstract. The article discuss performance of the innovative building materials used for external walls in a case study construction objects. Following the targets of EPBD directive all new constructions shall be nearly zero-energy buildings (NZEB). In this framework the goal of façade elements proposed is therefore to achieve or undercut a thermal transmittance of 0.20 $W/(m^2 \cdot K)$. An assessment of the thermal behaviour of the innovative composites elements was carried out considering the physical and thermal properties. The goal of investigation was to confirm that the results obtained in a laboratory can be successfully transferred to demo-case buildings taking into account thermal and mechanical behaviour of the façade elements. The results are presented in the paper.

1. Introduction

The materials and building components of modern buildings systems used in sustainable buildings to a society where environmental awareness and a high degree of living comfort are required. This is an actual research and development challenge faced by research centers from around the world including research projects done by ITB. Of particular importance are technical aspects such as mechanical performance, protection against moisture [1], heat loss in winter, overheating in summer and noise protection [1-2]. Components for the interior should be able to buffer heat and humidity peaks and also prevent pollutants and noise. The solutions currently in use are precast concrete wall panels [3]. Wall panels are structural elements subjected to in-plane action having negligible thickness compared to their length and breadth. In order to have sufficient strength to take in-plane loads, they may be reinforced with mild steel, steel fibers or steel fabric mesh. They are economical, not only from the structural design point, but also from the point of view overall construction. [4]. In this group of materials occur various types of concrete like Normal Strength Concrete (NSC), High Strength Concrete (HSC), Steel Fiber Reinforced Concrete (SFRC), Self-Compacting Concrete (SCC), Steel Fiber Reinforced Self Compacting Concrete (SFRSCC) Geo Polymer Concrete (GPC), Steel Fiber Light Weight Concrete (SFLWC), Hygrothermally Treated Ultra-High Performance Concrete (UHPC) in combination with autoclaved aerated concrete (AAC), textile reinforced concrete (TRC) or cellular lightweight concrete (CLC) [1-2, 4-5]. UHCP and other materials are promising alternatives with advantages such as lower energy consumption and reduced environmental impact. The materials used in the partitions shouldn't additionally affect negatively the thermal comfort of users and the quality of the indoor air of buildings [6]. Predictions suggest sandwich elements for building envelopes have other benefits such as an increased service life, optimized use of building area due to thinner elements, and minimized maintenance due to non-corrosive reinforcing materials. Typical precast concrete sandwich wall panels consist of two concrete layers and a layer of insulation between them. The concrete layers may be conventionally reinforced or prestressed and the insulation layer may be made of extruded polystyrene (XPS) or expanded polystyrene (EPS) rigid foam insulation to provide high thermal resistance (R-value) for the wall panels [3].

The case study solutions are presented in the paper. The innovative sustainable façades are based on type of concrete as UHPC, TRF, AAC and FC, while partition walls are based on earthen materials, optimized cementitious materials with modified surfaces and wooden/cellulose materials. The demonstrator buildings made for H-House project (ID 608893, founded under FP7-NMP), applying available technics, i.e. heat-flow meter technique, were investigated during the winter time.

2. The object of the research.

The final exposure of project developed external and internal walls made of new materials were assembled at demonstration site at ITB premises at Warsaw, Poland, and at Buzzi-Unicem premises at Trino, Italy.

The Object of Research (demo-cases) at Warsaw were two separate buildings (5 m x 5 m) made of prefabricated TRC-FC and UHPC-AAC composite elements (Figure 1). Both buildings show different propositions for the developed systems, i.e. for construction of new buildings and renovation, as well as different element of assemblies. The UHPC-AAC demonstrator building consists of eight prefabricated elements (four wall elements and four corner elements, Figure 1). One of the UHPC-AAC wall element (its external side) has a super hydrophobic surface, while the other three UHPC-AAC wall elements have photocatalytic surfaces. The developed UHPC-AAC elements may be used both for a new building construction and a building renovation.

Figure 1. UHPC-AAC demonstrator building; left: view on super hydrophobic surface; right: view photocatalytic surface.

The TRC-FC demonstrator building is composed of five composite elements (Figure 2). Four of these are sandwich elements, consisting of external TRC layer, FC (Foam Concrete) insulation layer and internal load-bearing TRC layer, to be used for new construction. The fifth element is a half sandwich element, consisting of external TRC layer and internal FC insulation layer, to be used for renovation (element is non load-bearing and applied as curtain to the walls of an existing structure). The interior surface of the TRC-FC sandwich element is presented in figure 2.

Figure 2. TRC-FC demonstrator building; left: external view on TRC-FC sandwich elements; right: interior FC layer.

The development of internal walls was based on natural materials (wood, wood fibre boards, straw boards and cellulose boards, earth plasters, etc.) which are exhibited in both demonstrator buildings, as shown in figure 3. In the UHPC-AAC demonstrator building the partition wall constructed (Figure 3, left) was based on timber studs anchored to floor and roof construction. Wood fibre insulation was placed between the timber studs. The stud structure was covered with wood fibre boards that are finished with an earth adhesive with integrated flax fibre and earth plaster final coat. The TRC-FC demonstrator building includes a partition wall made out of solid straw boards covered with cellulose boards (Figure 3, right). The finishing layer consists of an earth filler, reinforced with a glass fibre reinforcement mesh. The interior surface of the TRC-FC sandwich element is presented in finished with an earth plaster final coat with integrated flax fibre reinforcement mesh (Figure 3, right).

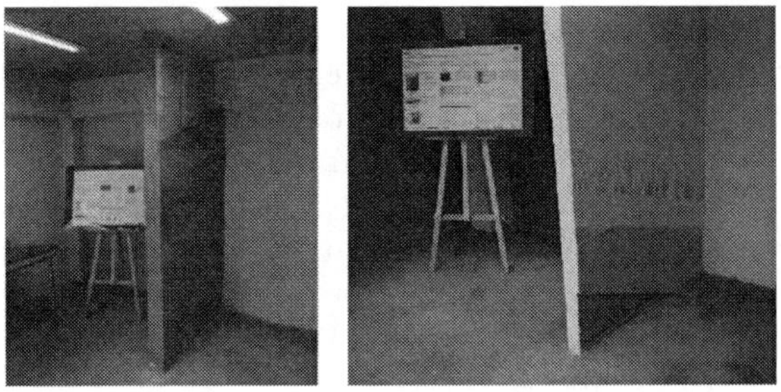

Figure 3. Internal walls in the UHPC-ACC (left) and TRC-FC (right) case-study buildings. Components and materials of the external and internal walls

3. Methods- technical performance assessment

3.1 Thermal performance.

Determination of thermal conductivity λ of the developed in the project materials was done in steady-state conditions by one–sample heat flow meter apparatus type FOX 314 or FOX 50 with horizontal orientation and hot plate situated on underside of sample, according to PN-EN 12667:2002 [8] and PN-EN 12664:2002 [9], respectively in ITB. Measurements were performed in mean specimen temperature of 10°C, temperature difference over sample thickness 20 K, heat movement vertically up, in environment temperature about 20°C. Before measurements the samples were conditioned at 23°C and 50% relatively air humidity till the stable masses of the samples were obtained. For each material, at least 5 samples have been tested and the mean thermal conductivity value was determined. On the basis of the measurement results the thermal transmittance in steady-state conditions of developed external walls was done applying 3D numerical modelling (Physibel, Solido) according to PN-EN ISO 6946:2008 [10] and PN-EN ISO 10211:2008 [11].

For the case-study objects during the winter period (2017/2018) the thermal performance tests were done applying the heat-flow meter technique for on-site measurements of thermal resistance, R, according to the standards of ISO 9869:1994 [7]. The measurements of R were based on simultaneous time averaged measurement of heat flux Φ and differential temperature, ΔT, (using two temperature sensors on each on a different side of the wall). During these tests the infrared thermal mapping was also performed to inspect the developed panels construction and locate in the building defects such as thermal linear and point bridges, missing insulation, delaminating render or condensation problems. For each of the two separate buildings (Figures 1and 2) the thermal imaging mapping were performed using FLIR-T62101 thermal imaging camera.

3.2 Mechanical tests.

The durability of the developed products i.e. TRC-FC and UHPC-AAC panels intended to be used in the building envelope was assessed by testing them in various sequences following the EOTA protocol i.e. ETAG 004, *Guideline for European Technical Approval of External thermal Insulation Composite Systems (ETICS) with Rendering* [12], applied directly and/or modified in ITB. The components were tested in the temperature range between -20° C and +70° C. Deterioration, cracking, blistering etc. of the components during the tests were closely observed.

The following tests of durability performance were performed:

3.2.1 Heat - rain cycles

Before the tests, samples were conditioned for 4 weeks at the temperature between 10°C and 25°C and relative humidity (RH) of about 50%. The tested sample was subjected to a series of 80 cycles, containing the three following stages:

1st - heating the sample surface up to 70°C (temperature rising within 1 hour) and maintaining at (70 + 5)° C and at RH 10- 30% for 2 hours (total heating of 3 hours),

2nd – spraying-on water for 1 hour (water temperature (+15 ± 5)° C, water flow rate of 1 l/m² min),

3rd – sample drying for 2 hours.

After every 4 cycles the sample surface was investigated, i.e. any deterioration, cracking, blistering, declaration, etc. of the sample surface occurring during the tests were closely observed. To illustrate the changes (if visible/noticed) photographs of the exposed sample surface were taken.

3.2.2 Heat-cold cycles

Before the tests the sample was conditioned, at least 48 hours, at temperatures between 10°C and 25° C and RH ≥ 50%. The tested sample was exposed to 5 heat/cold cycles, each of 24 hours, including two following stages:

1st – sample exposure to temperature (50 ± 5)° C (temp. rising within 1 hour) at maximum RH of 30% for 7 hours (total of 8 hours),

2nd – sample exposure to temperature (-20 ± 5)° C (temp. falling within 2 hours) at RH of (50 ± 5)% for 14 hours (total of 16 hours). At every cycle during the heat/cold cycles, observations relating to a change in characteristics or performance (blistering, detachment, crazing, loss of adhesion, formation of cracks, deceleration, etc., of the sample were recorded.

For the heat-rain cycles and heat-cold cycles tests the dimension of the UHPC-AAC panel was 0,6 m x 1,6 m × 0,38 m (width× height× thickness) and of the TRC-FC panel of a size 0,6 m× 1,3 m× 0,23 m. To avoid wetting of the panel sides during the heat-rain cycles its inner layers were protected by a waterproof emulsion as shown in figure 4.

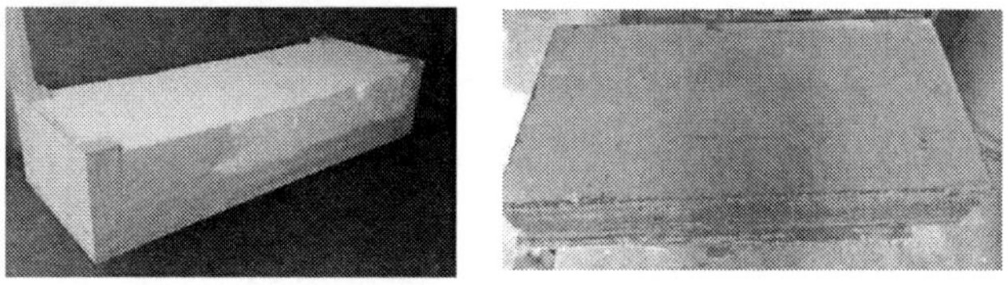

Figure 4. The view of the UHPC-AAC panel (left) and TRF-FC panel (right).

3.2.3 Freeze-thaw behavior.

Before the tests the samples were conditioned for 28 days at (23 + 2)° C and (50 + 5)% RH. The samples were handled manually, the size was about 200 mm × 200 mm and 300 mm × 300 mm.

The samples were subjected to a series of 30 cycles including the following stages:

1st – sample exposure to water for 8 hours at initial temperature of (23 ± 2)° C by immersion of the sample in a water bath (max. immersion till 8 mm)

2nd – sample freezing at temperature (- 20 ± 2)° C for 16 hours.

When the test was interrupted, i.e. because the samples were handled manually and there were stops during weekends or holidays, the samples were always maintained immersed in water between the cycles. At the end of the tests, observations relating to a change in characteristics of the sample surface i.e. distortion, deceleration, cracking etc. were recorded.

4. Tests results

4.1. Thermal performance.

The thermal conductivity of tested materials are presented in Table 1 (uncertainty of thermal conductivity measurement is 3%). The values of thermal transmittance of the façade element calculated according to PN-EN ISO 6946:2008 [10] and PN-EN ISO 10211:2008 [11] and measured on-site on the demo-cases are reported in Table 2 (uncertainty of heat flux measurement is 5%). The temperature distribution (isotherms) for the façade element obtained during the calculation method and during in-situ infrared measurements are presented in Figure 5.

Table 1. The thermal conductivity of developed materials for external walls

Component	Density (kgm^{-3})	Thermal conductivity ($Wm^{-1}K^{-1}$)
UHPC - Ultra-High Performance Concrete	2200	1.5
AAC- autoclaved aerated concrete	120	0.042
TRC- textile reinforced concrete	1800	1.3
FC – foamed concrete	200	0.040

Table 2. The thermal transmittance of developed materials for external walls

Component	Calculated U value in the central part of the panel ($Wm^{-2}K^{-1}$)	Measured on-site U value in the central part of the panel ($Wm^{-2}K^{-1}$)
UHPC-AAC-Ultra-High Performance Concrete (UHPC) in combination with autoclaved aerated concrete (AAC)	0.15	0.13
TRC-FC Textile Reinforced Concrete (TRC) with a combination with low density foamed concrete (FC)	0.30	0.29

(heat flux - uncertainty of measurement 5%)

The temperature distribution (isotherms) of the UHPC-AAC panel as 3D model (bottom) and the infrared mapping of the tested samples are presented in figure 5

IOP Conf. Series: Materials Science and Engineering **415** (2018) 012018 doi:10.1088/1757-899X/415/1/012018

Figure 5. The view of the UHPC-AAC panel of the demo side (left) and infrared mapping (right) and 3D model of temperature distribution (bottom).

4.2 Durability result results.

After the mechanical testing "no damage" i.e. distortion or cracking on the surface of the tested panels was recorded. Only small deceleration was observed, mainly near, along edges of the sample what was probably due to insufficient samples sides protection, as shown in figure 6.

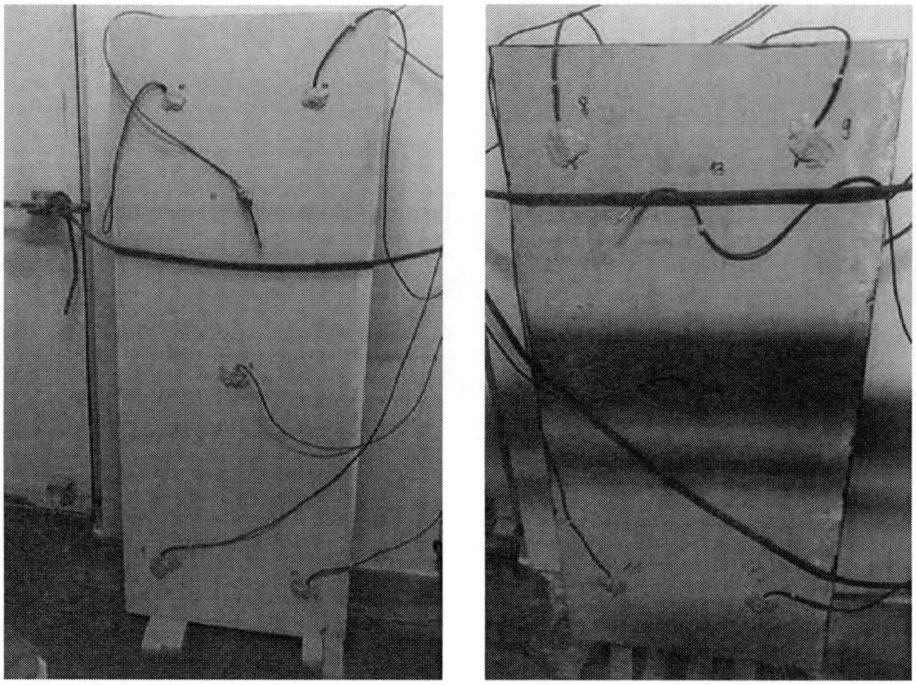

Figure 6. The view of the UHPC-AAC panel (left) and TRF-FC panel (right) after the durability tests.

Summary
Durability and thermal tests of prefabricated elements built on the basis of innovative material solutions have been carried out. The obtained test results presented in tables and figures indicate that developed within the H-House project two innovative panels i.e. UHPC-AAC and TRC-FC can be proposed as solutions for building external walls. Thermal and durability analyses made for the developed within the H-House project demo-cases supported the idea that the prefabricated TRC-FC and UHPC-AAC composite elements can be used in climate zone as it is in Poland. Performed thermal analyses showed that the installed panels achieve the thermal transmittance values and infrared imaging supported didn't show any changes in the structure during the tests.

Acknowledgments
H-HOUSE is a research project with a budget near to 6,5 millions of Euro, co-funded for about 4,75 million of Euro by the European Commission under the 7th Framework Programme (2007–2013). Grant Agreement No. 608893.

References

[1] Miccoli L., Fontana P., Silva N., Klinge A., Cederqvist C., Kreft O., Qvaeschning D., Sjöström C., *Composite UHPC-AAC/CLC facade elements with modified interior plaster for new buildings and refurbishment. Materials and production technology*, Journal of Facade Design and Engineering, vol. 3, no. 1, pp. 91-102, 2015.

[2] Miccoli L., Fontana P., Johansson G., Zandi K., Williams Portal N., Müller U., *Numerical modelling of UHPC and TRC sandwich elements for building envelopes*, IABSE Conference – Structural Engineering: Providing Solutions to Global Challenges, 23-25 2015, Geneva, Switzerland

[3] Qing Z. and Zhengxing G., *Experimental evaluation of precast concrete sandwich wall panels with steel–glass fiber–reinforced polymer shear connectors*, Advances in Structural Engineering 2017, Vol. 20(10) 1476–1492.

[4] Himasree P. et al 2017 IOP Conf. Ser.: Earth Environ. Sci. 80 012041

[5] Miccoli L., Fontana P., Kreft O., Pietruszka B., Łukaszewska A., Klinge A., *Design of UHPC-AAC light-weight composite façade elements for refurbishment*, 2017 Proc. Int. Conf. on Durability of Building Materials and Components (Belgium) Vol 29-31 (Ghent University, Belgium) p 95

[6] Piasecki M., Kostyrko K., Pykacz S., *Indoor environmental quality assessment, part I: choice of the indoor environmental quality sub-component models.* Journal of Building Physics 41: (3) p. 294–289, 2017.

[7] PN-EN ISO 9869:1994. Thermal insulation - Building elements - In-situ measurement of thermal resistance and thermal transmittance - Part 1: Heat flow meter method.

[8] PN-EN 12664:2002 Thermal performance of building materials and products. Determination of thermal resistance by means of guarded hot plate and heat flow meter methods. Dry and moist products of medium and low thermal resistance.

[9] PN-EN 12667:2002. Thermal performance of building materials and products. Determination of thermal resistance by means of guarded hot plate and heat flow meter methods. Products of high and medium thermal resistance.

[10] PN-EN ISO 6946:2008 Building components and building elements. Thermal resistance and thermal transmittance. Calculation method.

[11] PN-EN ISO 10211:2008. Thermal bridges in building construction. Heat flows and surface temperatures. Detailed calculations.

[12] ETAG 004, Edition 2011, Guideline for European Technical Approval of External thermal Insulation Composite Systems (ETICS) with Rendering.

IOP Conf. Series: Materials Science and Engineering **415** (2018) 012019 doi:10.1088/1757-899X/415/1/012019

Polycarbonate multi-wall panels integrated in multi-layer solar façade concepts

M Čekon[1] and K Struhala[1]

[1] Centre AdMaS, Faculty of Civil Engineering, Brno University of Technology, Czechia

cekon.m@fce.vutbr.cz

Abstract. Recent technological advances in transparent insulation materials´ (TIMs) production may have opened ways for new integration of these materials in buildings. This paper presents a study analysing several variants of façades incorporating TIMs on solar façade principles. The analysis is based on ongoing long-term full-scale experiments in Brno, Czech Republic. The paper introduces the test platform which is used for evaluation of specific aspects of integrated polycarbonate TIMs. The paper describes the experiment results, especially long-term thermal response and passive solar gain data. These data are applied for analysis of: influence of various polycarbonate-based TIMs on the real performance of the façades; influence of implementation of different solar absorbers on the performance of the proposed facades; effectivity of application of latent thermal energy storage (based on PCMs) as a part of heat accumulation layer and coupling of TIM with prismatic glass to enhance optical selectivity aspect. Presented results demonstrate for example significant influence of the type of solar absorber on the thermal performance of tested solar facades: the difference is up to 35% to 54%. Also the integration of prismatic glass coupled with simple two-wall polycarbonate panel can reduce solar penetration through components at the level comparable with the most complex six-wall panel.

1. Introduction

Commonly, the efficient way to improve energy consumption of buildings is based on reduction of the heating and ventilation energy losses. This can be achieved through addition of thermal insulation on the building envelope and installation of efficient HVAC system with heat recovery [1, 2]. However the potential for further savings in future building designs is limited by increasing investment costs, embodied energy and other environmental impact [3, 4, 5]. Therefore the building sector is looking for solutions that would not only minimize the energy losses [6], but also utilize the energy gains available on-site. These solutions integrate renewable energy sources (RES) such as photovoltaics (PV) or solar thermal collectors for on-site energy generation [7, 8, 9]. Although the field of RES is already well covered in literature, there are still some directions that need to be carefully investigated. Therefore this paper focuses on one of more straightforward ways for exploitation of available renewable energy gains: solar (heat) energy gains through TIMs based on Polycarbonate multi-wall panels integrated in building envelopes, particularly in multi-layer solar façade.

Integration of TIMs is already proven in case of solar thermal collectors [10, 11]. Existing TIMs [12] could be also integrated into solar thermal facades providing both transparency and thermal insulation. Several studies in the last two decades for example applied TIMs as replacement of the original glazing elements in solar wall concepts [13]. However, their involvement in building envelope is so specific that application in contemporary building practice is rather rare, even though commercial products already exist [14]. One of the main drawbacks is that specialized TIMs are rather expensive. Therefore application of cheaper materials with similar properties (such as polycarbonate sheets and

panels) is studied. Various types of multiwall panels with improved thermal properties (with optical efficiency remaining the same) are being developed. Their application in building envelopes (facades or roofs) can provide cost-effective way for achieving passive energy gains in building design [15].
This paper introduces eight experimental polycarbonate solar façade (PSF) concepts utilizing different types of TIMs (based on multiwall co-extruded polycarbonate). Components representing these PSFs are investigated as a part of long-term experiment in Brno, Czechia. Especially the influence of varying structure of TIMs (and related thermal and optical parameters) on the performance of the components is studied in detail. However, other issues like selection of proper solar absorber or incorporation of phase-change materials (PCMs) for better heat accumulation are also investigated. The paper presents comparative analysis of thermal, optical and solar properties of individual materials and illustrates performance of the whole PSF concepts on selected summer and winter days (with clear sky) measurements.

2. Experimental polycarbonate solar façade concepts

At first the properties of materials applied in the proposed concepts were tested. This was followed by full-scale testing of eight experimental PSF components in real climate conditions (see Figure 1). The full-scale tests are taking place at AdMaS research centre belonging to Brno University of Technology, Czechia [16]. The PSF components are based on four polycarbonate TIMs. The TIMs are installed in pairs of components opposite or next to each other. This limits the number of tested materials. On the other hand, it allows comparative measurements and increased accuracy of the experimental results. The test cell's longitudinal axis is oriented towards South-East / North-West. One of the walls with the test components is oriented to the South-East to receive as much solar radiation as possible. The other (North-Western) wall is adjacent to one of the AdMaS center's buildings. Therefore, the test components there are not exposed to the solar radiation directly. The space inside the test cell is divided into eight compensating boxes (TCBs) made of particleboard. They are constructed around individual PSF components of tested materials – four components on each side. The TCBs separate individual testing components from each other and the internal climate of the test cell to dampen the temperature amplitudes of the internal air. Internal climate in the test cell (not in the individual test boxes) is controlled by an air-conditioning unit.

Figure 1. Outside view of the test cell (left), inside view of the test cell with individual TCBs (middle), test cell cross section (right).

Eight full-scale PSF components (A to H) are long-term tested in real climate conditions. Components A to D are oriented to South-East, while components E to H are oriented to North-West. Thermal and optical performance of all components are described in sections 3 and 4, while real outdoor performance in section 5 is described only for components A to D oriented to the South-East (due to direct solar radiation). Schemes of the components are in Figures 2 and 3. Each tested component fulfils several different functions thanks to four functional layers. First layer of tested components is made of clear polycarbonate multi-wall panel in the role of TIM. All the TIMs are installed with air cavities parallel to the longer side of the components. Components A and H integrate simple 10mm thick two-wall panel (PC10). In case of component A the PC10 panel is covered by prismatic glass pane rayWall90 to improve its overall optical performance.

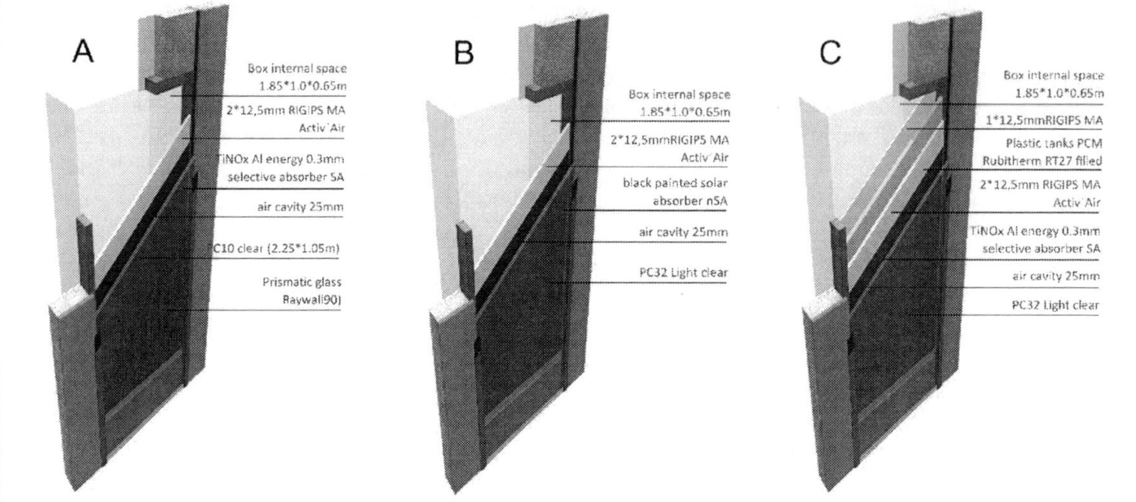

Figure 2. Test components A, B and C, key material layers, TIM and solar absorber types.

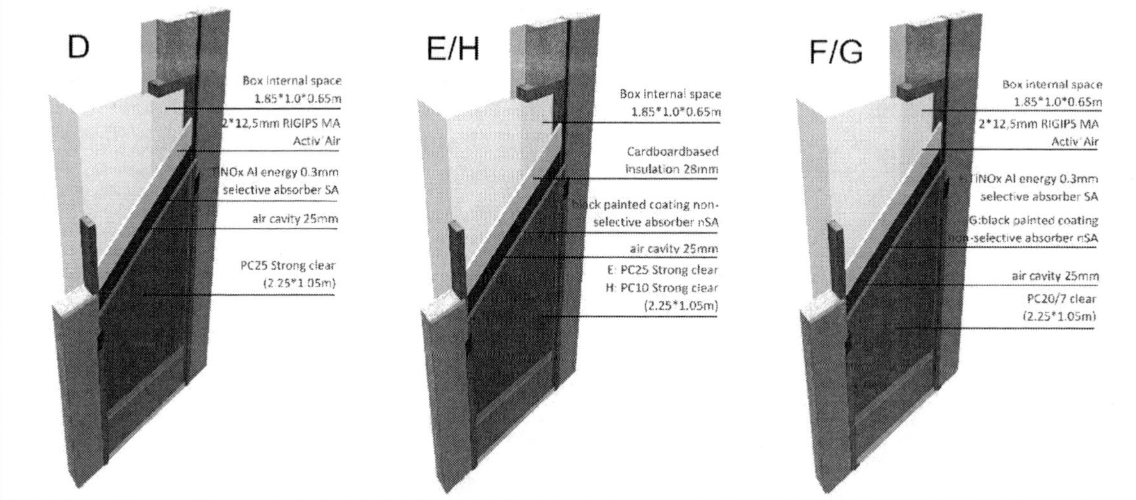

Figure 3. Test components D, E, F, G and H, key material layers, TIM and solar absorber types.

Components B and C integrate the most complex 32mm thick six-walls system with combined parallel and diagonal partitioning (PC32). Components D and E integrate 25mm thick three-wall panel with combined partitioning (PC25). Components F and G integrate 20 mm thick seven-wall panel (PC20). Basic parameters of the panels are described in Table 1. Second layer of the tested components is 25mm non-ventilated air gap. Third layer is made of a solar absorber. Two types of solar absorber are incorporated in the experimental components: selective absorber (SA) with low-e coating in components A, C, D and F and non-selective absorber (nSA) in components B, E, G and H. Reason for this is difference in the absorber's properties. The absorbers have similar solar absorbance, approx. 0.94 (SA) and 0.96 (nSA), however they significantly differ in thermal emissivity (0.06 and 0.94, respectively). Final layer of the components is the base layer for application of solar absorber. In real façade this layer would also fulfil the role of heat accumulator. For the purpose of the described experiment this layer is simplified to a 25mm plasterboard panel in components A, B, D, F and G. 28mm corrugated fiberboard panel is incorporated in the same role in components E and H for the purpose of performance comparison. Component C is used for evaluation of PCM application. Therefore the heat accumulating layer in this component is made of Rubitherm RT27 in plastic tanks enclosed between plasterboards.

3. Thermal performance of experimental PSFs

This section presents the results of standardized calculations employed to reproduce the theoretical thermal performance of presented PSF concepts. Calculated parameters consider the influence of different thermal and optical properties. In addition, the results of the PSF concepts with and without a selective absorber are also calculated. Firstly, measured thermal conductivities of Polycarbonate multi-wall systems are presented. Secondly, the thermal parameters of the whole concepts are calculated.
Thermal properties of the TIMs were evaluated with TLP 300-DTX-1P apparatus. It utilizes guarded hot plate (GHP) method to measure heat resistance and calculate equivalent thermal conductivity in compliance with ISO 8302 [17] and EN 12667 [18]. Results of the evaluation are summed in Table 1. Temperature dependency of the thermal conductivity is taken into account as well [19].

Table 1. Key material and measured thermal parameters of evaluated polycarbonate TIMs. [19]

	PC type	*Thickness* [mm]	*Weight* [$kg.m^{-2}$]	*Internal cells´ structure*	Equivalent thermal conductivity at mean temperature [$W{\cdot}m^{-1}{\cdot}K^{-1}$]		
					10°C	20°C	30°C
PC10	Clear 2walls	10.3	1.7	Parallel	0.0668	0.0712	0.0763
PC20	Clear 7walls	20.7	3.0	Parallel	0.0527	0.0551	0.0582
PC25	Clear 3walls	25.1	3.4	Parallel/Diagonal	0.0665	0.0702	0.0752
PC32	Clear 6walls	32.5	3.6	Parallel/Diagonal	0.0601	0.0634	0.0684

Thermal resistances and transmittances of individual parts as well as whole experimental components were calculated according to ISO 6946 [20] using thermal conductivity values described in Table 1. Results of these calculations are summed in Table 2. The table shows values representing individual "layers" (R_t is for TIM, R_g for air gap and solar absorber, R_i heat accumulation material) or parts (R_{te} for TIM and air gap coupled with solar absorber together and their calculated equivalent thermal conductivity λ_{te}) as well as whole components (R_c). For this study, dynamic effect of phase change integrated in test component C is not considered, thus only standard calculation is provided.

Table 2. Thermal resistance for individual layers and whole components.

[$m^2{\cdot}K{\cdot}W^{-1}$]	R_t	R_g	R_i	R_{te}	λ_{te}	R_c	U_c
A	0.145	0.57	0.125	0.795	0.044	1.050	0.95
B	0.513	0.18	0.125	0.733	0.078	0.988	1.01
C	0.513	0.57	0.275	1.123	0.051	1.528	0.65
D	0.357	0.57	0.125	0.968	0.052	1.223	0.82
E	0.357	0.18	0.444	0.578	0.086	1.152	0.87
F	0.376	0.57	0.125	0.986	0.046	1.241	0.81
G	0.376	0.18	0.125	0.596	0.075	0.851	1.17
H	0.145	0.18	0.444	0.365	0.096	0.939	1.06

The different performance of both solar absorber types is clearly visible in thermal resistance R_{te} and equivalent thermal conductivity values λ_{te}. The difference in those parameters of components A and H is 54% (with PC10), B and C is 35% (PC32), D and E represents 40% (PC25) and F with G corresponds to 40% (PC20). This difference means that for example the overall *U*-value of component A (with simplest PC10 TIM) bests the *U*-value of component B (with most complex PC32 TIM).

Component B with nSA has thermal resistance of 0.988 $m^2 \cdot K \cdot W^{-1}$. Compare this with the thermal resistance of 1.528 $m^2 \cdot K \cdot W^{-1}$ of component C with the same PC32 TIM combined with SA. The equivalent thermal conductivity of all components integrating SA from 0.044 (PC10) to 0.052 (PC25) approaches to recent conventional thermal insulations (e.g. expanded polystyrene reaches the value around 0.04).

4. Optical performance of polycarbonate TIM panels

Solar transmittance of polycarbonate TIMs presented in this paper is based on previous work [19]. Table 3 presents solar transmittance values based on laboratory spectrophotometer measurements using a Perkin Lambda 1050 UV/VIS/NIR spectrophotometer with a 150mm Spectralon integrating sphere. These spectral data are registered ranging from 200 nm to 3300 nm representing all spectral regions ($\tau_{\lambda\ SRS}UV$ for ultraviolet, $\tau_{\lambda\ SRS}\ VIS$ for visible and $\tau_{\lambda\ SRS}\ NIR$ for near infrared region respectively). Each sample was measured in the centre position of vertically oriented chambers. Spectral values and integrated Total Solar Transmittance (TST - τ_λ) parameters from 280 nm to 2 500 nm and declared values τ_{decl} are shown as well. The lowest value represents the system PC32 with solar transmittance of 43%, while the system PC10 has transmittance of 74%. It should be noted that solar transmittance obtained by laboratory spectral measurements does not correspond with producer-declared values. The difference is more than 10%.

Table 3. Solar transmittance of tested polycarbonate TIM panels [19].

	τ_λ ASTM G 173 [21]	$\tau_{\lambda\ SRS}UV$ 0.2 – 0.3 μm	$\tau_{\lambda\ SRS}\ VIS$ 0.3 – 0.8 μm	$\tau_{\lambda\ SRS}\ NIR$ 0.8 – 2.5 μm	τ_{decl}
PC10	0.74	0.08	0.79	0.77	0.84
PC20	0.50	0.04	0.54	0.52	0.62
PC25	0.48	0.07	0.50	0.50	0.63
PC32	0.43	0.03	0.46	0.45	0.53

5. Real outdoor performance and thermal response

As the proposed concept is based on various solar active materials, real outdoor testing may determine the maximum temperature that can be achieved in these test solar components and thus determine their potentials. There are many specific issues to take into account, such as inclined angular dependence, fluctuations of solar irradiation, overall solar distribution or orientation to the cardinal points. These issues could be solved all together with real-life monitoring of solar thermal performance. Results of full-size experiments in real climate condition demonstrate influence of different absorber efficiencies and effect of various geometrical and physical characteristics on the performance of individual PSF components. This part of the paper focuses on components A, B, C and D oriented to the South-East to analyze the performance of the components in direct solar exposure using contact digital thermometers Maxim DS18B20. Performance of the components is analyzed during representative clear sky day (24 hours) in winter and summer, with maximum sun height above the horizon of 31° and 59° respectively. The indoor climate was not controlled during the winter day, while during summer day the indoor air temperature in the test cell was maintained approx. at 21 °C. Average outdoor air temperature was T_e = -8.63 °C in winter (24.53 °C in summer), while interior temperature was T_{ai} = 5.16 °C in winter (21.11 °C in summer). Average indoor air temperature in individual TCBs in winter (summer) day reached: T_A = 6.99 °C (27.40 °C), T_B = 8.20 °C (26.77 °C), T_C = 6.98 °C (26.16 °C) and T_D = 10.20 °C (29.36 °C). Detailed temperature measurement results are shown in Figure 4. Overall, the best efficiency (represented by the highest indoor TCB temperature) was achieved with component D. It should be highlighted that test components A and B show similar summer performance (similarly to section 3) even though their composition is significantly different. This time, the reason is likely the application of rayWall90 prismatic glass in component A, which improves summer performance of this component. It should be also noted, that winter results were influenced by the fact that indoor conditions were not controlled. This means that PCM in component

C was not fully activated and the component could not perform at peak efficiency. Therefore, the effect of C in winter is represented by more stable temperature progression comparing to those with no PCM. When looking at minimum temperatures for summer test period, test component C shows the best results due to PCM integration. Its peak indoor temperature is more than 10 K lower compared to other components.

Figure 5 presents maximum temperatures achieved on solar absorber surfaces (T_{abs}) of particular components. Interestingly, component A exhibits almost the same maximum temperatures for both winter and summer days. Component B has maximum temperatures approx. 65 °C in winter and 80 °C in summer. Here the specific optical raster applied on the model A influence the internal temperature at the level comparable with the component B. The test component C has the highest surface temperature in both winter and summer. The winter temperature is beneficial, however in summer its overall performance could cause unnecessary overheating.

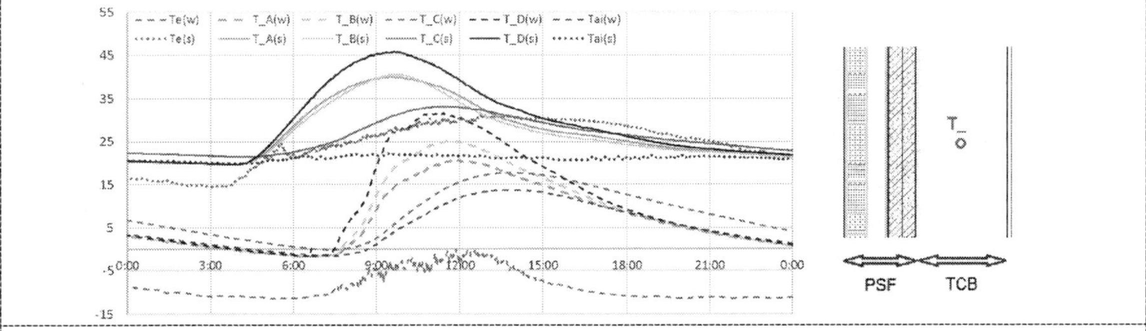

Figure 4. Internal test compensating box (TCB) temperatures; (w) winter day; (s) summer day.

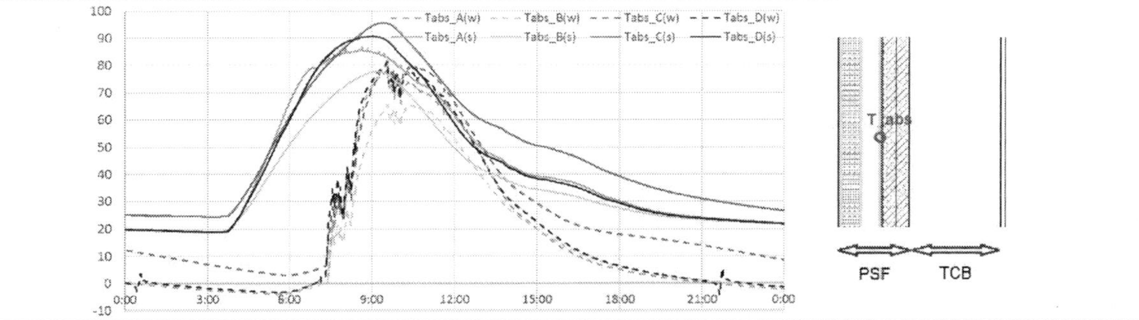

Figure 5. Absorber temperatures; (w) winter; (s) summer.

This is the reason for integration of PCM as a latent heat storage that would eliminate internal temperature fluctuations and limit the overheating. Last component D has the same absorber response in winter as C, however in summer it reaches the lower absorber temperature due to lower thermal capacity of plasterboard without PCM. Concerning only the polycarbonate TIMs, the temperature difference $\Delta T pc\_$ is a key element when temperature gradient needed to be considered for thermal calculations. Figure 6 presents results of maximum temperature difference obtained on polycarbonate systems in presented days. The maximum is achieved in winter case approaching 70K difference. When taking one particular type of PC32 into account, both test components B and C are in a strong contrast. The type of solar absorber significantly affects this aspect and the difference is almost 20K. In addition, measuring solar irradiance rates I_r and dynamic solar transmittance based on evaluating the amount of solar radiation penetrating through the multi-wall structure of the polycarbonate TIMs is provided as well (Figure 7). A specific methodical approach is applied in this relation to provide real data. During clear sky conditions, huge variation of penetrated part of solar irradiance can be observed. Depending on the particular TIM, the values significantly vary, which influences hourly solar transmittance values. Activation of the prismatic glass integrated in A is clearly demonstrated in Figure 7, however the internal structure of C indicates the same response.

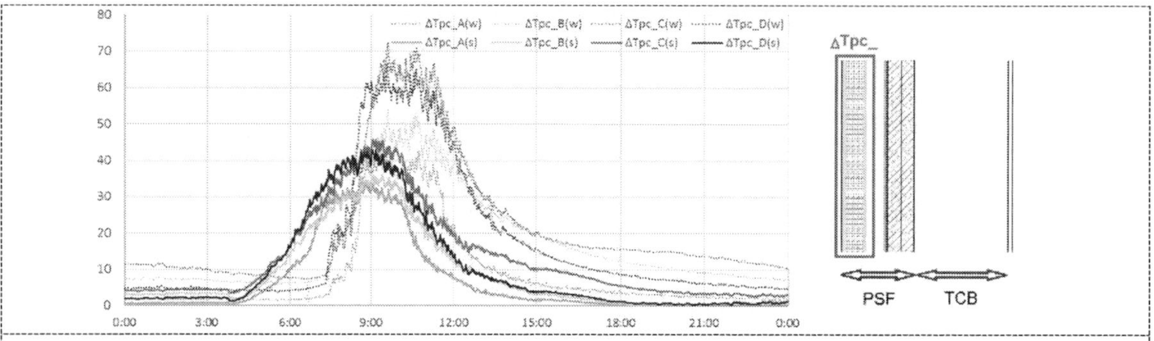

Figure 6. Temperature difference of polycarbonate systems; (w) winter; (s) summer

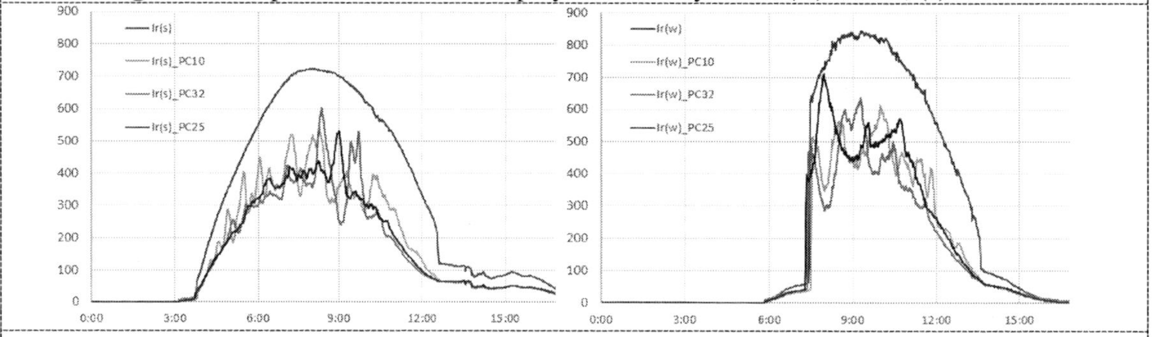

Figure 7. Solar irradiance rates penetrated through PC systems; (w) winter; (s) summer.

Conclusion

The paper presents the results of the solar and thermal characterization of proposed polycarbonate solar façade (PSF) concepts. The concepts evaluated impacts of different TIMs, selective and non-selective solar absorbers, as well as and latent thermal energy storage. Experimental monitoring of full-scale PSF components focused on the dynamic annual performance of TIMs coupled with solar absorbers as a fundamental step towards understanding overall interactions influencing the heat transfer of the presented concepts. Experimental results and calculations indicate that application of SA in proposed PSFs significantly improves their thermal parameters compared to application of nSA. Results show that when thermal resistance or equivalent thermal conductivity is considered, proposed concepts implementing the most complex PC32 TIM with SA (component C) provide comparable performance to conventional thermal insulation. When comparing the temperature response of the experimental PSF components invoked by solar radiation at its peak, a significant effect of particular solar absorber type and type of polycarbonate TIM is observed. Further research will be focused on detailed experimental analysis of long-term outdoor experiments and numerical modelling as well as the optimization of the described PSF concepts.

Acknowledgments

This research was supported by the project GJ 16-02430Y "Contemporary concepts of climatically active solar façades integrating advanced material solutions" supported by Czech Science Foundation.

References

[1] X. Liang, Z. Wang, M. Royapoor, Q. Wu, T. Roskilly (2017), Comparison of building

performance between Conventional House and Passive House in the UK, *En. Proc.* **142** 1823
[2] S. Guillén-Lambea, B. Rodríguez-Soria, J. M. Marín (2016), Review of European ventilation strategies to meet the cooling and heating demands of nearly zero energy buildings (nZEB)/Passivhaus. Comparison with the USA, *Renew Sust Energ Rev* **62** 561
[3] I. Zacà, D. D'Agostino, P. M. Congedo, C. Baglivo (2015), Assessment of cost-optimality and technical solutions in high performance multi-residential buildings in the Mediterranean area, *Energ Buildings* **102** 250
[4] T. Ekström, R. Bernardo, Å. Blomsterberg (2018), Cost-effective passive house renovation package for Swedish single-family houses from the 1960s and 1970s, *En. Buildings* **161** 89
[5] W. Ott, R. Bolliger (2015), Pitfalls in the Economic and Ecological Evaluation of Energy Related Building Renovation Strategies and Measures, *Energy Procedia* **78** 2340
[6] P. Ďurica, P. Juráš, R. Ponechal, D. Štaffenová (2016), Long-Time evaluation of thermal-Technical parameters of experimental wooden walls. In *CESB 2016 - Central Europe Towards Sustainable Building 2016*: *Innovations for Sustainable Future*, pp. 1063-1070.
[7] I. Visa, M. Moldovan, M. Comsit, M. Neagoe, A. Duta (2017), Facades Integrated Solar-thermal Collectors – Challenges and Solutions, *Energy Procedia* **112** 176
[8] R. Bolliger, W. Ott, S. von Grünigen (2015), Finding the Balance between Energy Efficiency Measures and Renewable Energy Measures in Building Renovation: An Assessment Based on Generic Calculations in 8 European Countries, *Energy Procedia* **78** 2372
[9] F. Favoino, F. Goia, M. Perino, V. Serra (2014), Experimental assessment of the energy performance of an advanced responsive multifunct. façade module, *Energ Buildings* **68** 647
[10] J. Cadafalch, R. Cònsul (2014), Detailed modelling of flat plate solar thermal collectors with honeycomb-like transparent insulation, *Solar Energy* **107** 202
[11] J. D. Osorio, A. Rivera-Alvarez, P. Girurugwiro, S. Yang, R. Hovsapian, J. C. Ordonez (2017), Integration of transparent insulation materials into solar collector devices, *Sol Energy* **147** 8
[12] N. D. Kaushika, K. Sumathy (2003) Solar transparent insulation materials: a review. *Renew Sust Energ Rev*. **7**(4) 317
[13] I. L. Wong, P. C. Eames, R. S. Perrera (2007), A review of transparent insulation systems and the evaluation of payback period for building applications, *Solar Energy* **81** 1058
[14] M. Casini (2016), Smart Buildings: Advanced Materials and Nanotechnology to Improve Energy-Efficiency and Environmental Performance, Elsevier: Cambridge, 2016, 363 p.
[15] E. Moretti, M. Zinzi, E. Belloni (2014) Polycarbonate panels for buildings: Experimental investigation of thermal and optical performance, *En Buildings* **70** 23
[16] M. Čekon, R. Slávik, K. Struhala, M. Formánek (2017) Experimental Full-Scale Test Cell Optimizing for Research of Novel Concepts towards Climatically Active Solar Façade Design, *Applied Mechanics and Materials* **861** 213
[17] ISO (1991), *Thermal insulation - Determination of steady-state thermal resistance and related properties - Guarded hot plate apparatus*, ISO 8302, International Organization for Standardization (ISO): Geneva, 1991, 47 p.
[18] ČSN EN (2001) *Thermal perforance of building materials and products – Deterination of thermal resistance by means of guarded hot plate and heat flow method*, ČSN EN 12667, Czech normalization institute (CNI): Prague, 2001, 60 p.
[19] M. Čekon, R. Slávik, J.Zach (2017) Experimental Analysis of Transparent Insulation Based on Polycarbonate Multi-Wall Systems: Thermal and Optical Performance, *Energy Procedia* **132** 502
[20] ISO (2017), *Building components and building elements — Thermal resistance and thermal transmittance — Calculation method*, ISO 6946, International Organization for Standardization (ISO): Geneva, 2017, 40 p.
[21] ASTM (2012), *Standard Tables for Reference Solar Spectral Irradiances: Direct Normal and Hemispherical on 37_ Tilted Surface;* ASTM G173-03*;* American Society for Testing and Materials (ASTM): West Conshohocken, PA, USA, 2012.

IOP Conf. Series: Materials Science and Engineering **415** (2018) 012020 doi:10.1088/1757-899X/415/1/012020

Comparison of triple glazed windows based on long-term measurement

P Juras[1]

[1] Department of Building Engineering and Urban Planning, Faculty of Civil Engineering, University of Zilina, Univerzitna 8215/1, 010 26 Zilina, Slovakia

peter.juras@fstav.uniza.sk

Abstract. This paper deals with measurement of three windows suitable for low-energy or passive houses. Windows are evaluated since 2011 in the laboratory of pavilion type with constant indoor boundary conditions and real outdoor climate. There are three windows, one wooden and two plastic which differ from each other by use of thermo modules inside the frame and center seal. Temperatures and heat fluxes on the glass, sash and frame of the windows are measured continuously. Outdoor climate is measured by the mobile weather station. During the measured period - winter 2017, windows were equipped with more sensors for temperature measuring than before. Sensors were added on the glazing, on the jamb, into the space between frame and sash and also on the montage gap filled with polyurethane foam. Temperatures on the same places of different windows are compared in this paper.

1. Introduction

With the advance in the thermal protection of the buildings and energy performance it is still important to build in better windows. The Slovak standard [1] dealing with thermal protection (STN 73 0540:2012) followed the EC directive about Energy performance of building (EPBD) 2010/31/EU [2]. Valid values for windows are since 2016 $U_w = 1.0$ W/(m^2.K) and from 2021 will be on the level 0.6. Current values are suitable for ultra-low energy buildings. The window not only protects the indoor environment against the outdoor climate like a non-transparent part of building envelope, but it has more function. Very important is the insolation of the space, views, contact with outdoor, ventilation, etc. With the more strict requirements of thermal properties it is necessary to increase the thermal properties of used materials, both for glazings (triple, gas fillings, heat mirrors and low emissivity surfaces) and frames (wooden, aluminium, plastic with thermo modules, more chambers etc.). Nonetheless is the advance with incorporating the photovoltaics into the glazing or usage of transparent thermal insulations [3]. Overcome of overheating is also a problem which has to be dealt with in the low-energy buildings [4]. Another problem is the minimizing of thermal loss through the thermal bridges around the windows [5].

The use of triple glazing helps with avoiding the cool radiation from the glass surface and also the vapor condensation at the bottom of the glazing.

The three windows (Figure 1), which are analyzed in this paper are in the laboratory of pavilion type evaluated since 2011 (Figure 2). The windows are back there suitable for low-energy and nearly-zero buildings. Several articles about measurements, analyzing the winter and summer periods, measuring the transmission etc., were published [6-7].

IOP Conf. Series: Materials Science and Engineering **415** (2018) 012020 doi:10.1088/1757-899X/415/1/012020

Direct comparison of heat fluxes measured during winter 2017 [7] and 2018 is made in this article. Since December 2017 were also added themocouples to the new positions, which enable possibility to make deeper analysis of the regime during freezing nights and sunny winter days.

Figure 1. Outside view on the tested windows in laboratory. Visible sensors on surfaces and AC chiller.

Figure 2. Inside view in the room with test windows. Visible positions of sensors and heat flux plates.

2. Laboratory and test setup

The analyzed windows are built into the testing laboratory of the Department since 2011. This is a so-called "laboratory of the pavilion type", which nowadays consists of three different rooms with aim on the nearly zero building envelopes [8].

The windows are oriented to the south with slight declination to the west (15°). From outside, windows are exposed to the real climate boundary conditions. The outdoor climate is monitored and recorded by the mobile experimental weather station placed on the nearby building`s roof [8]. Exact monitoring of outdoor boundary conditions in-situ enables the possibility for more precise measurement and simulations. The indoor climate is controlled by the AC unit, which is set to maintain the Slovak standard indoor boundary conditions: 20 °C, 50 %. The indoor air temperature is recorded and showed small amplitude about ± 1 °C, during the very cold night about - 2 ~ 3 °C, which is taken into the account for calculation of the thermal transmittance coefficient.

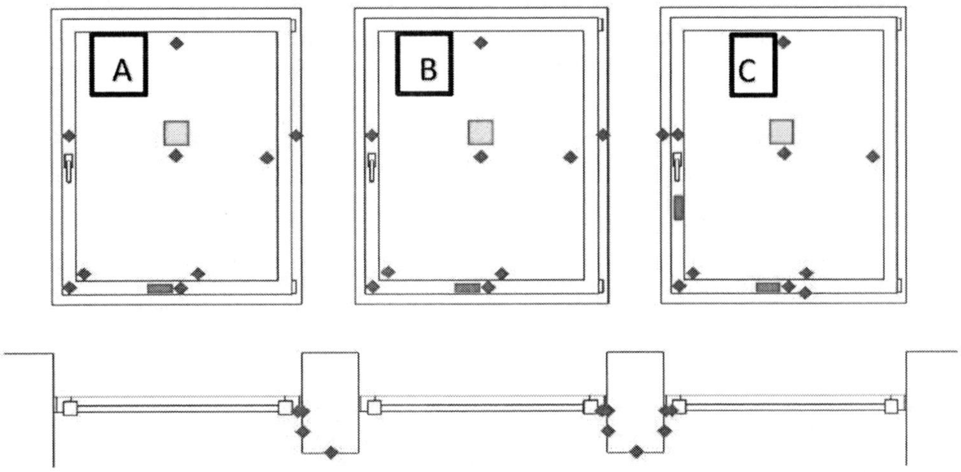

Figure 3. Positions of thermocouples and heat flux plates on the windows. View from the interior side.

Sensor used for this measurement consist of NiCR-Ni thermocouples [10] and heat flux plates (HFP) equipped also with thermocouple with correction (standard ones (120 x 120 mm) and half-sized for window sashes and frames.) [11]. Monitoring positions are frame, sash and glazing. Some sensors are also in the cavity between frame and sash. Totally from previous set up were added 21 thermocouples. The pattern of the positions is the same on all three windows to make possibility of direct comparison. Totally are 57 thermocouples used, there are 6 positions on each window on both sides, 4 positions on each window only inside and 2 more position on two windows from inside. Some thermocouples are also on the pillar between the windows and on the joint between pillar and frame, but these were not utilized in this article. Detailed position of the thermocouples and HFP are in Figure 3. The datalogger and both sensor types are from Ahlborn. For this measurements the recording interval was five minutes.

Evolution of the laboratory equipment contain: since 2013 the outdoor climate has been monitored more precise by the experimental weather station [9], from autumn 2015 there are also the heat flux plates (HFP) fixed on the inner surfaces, which creates the possibility to measure the heat flux and thermal conductance coefficient of the glass and frame, depending on the position of the plates and since December 2017 were added 21 thermocouples to new positions.

For needs of this article, seven HFP were used. Three of them have a standard dimensions, placed in the center of glazing, three small ones placed on the windows sashes and another one on the frame (wooden window C). Positions are marked in Figure 3. Each HFP measures the thermal flow and also the surface temperature.

The windows properties based on the manufacturers data are summarized in Table 1. It is good to mention, that "paper" based values comply nowadays standard values several years ago.

Table 1. Specified properties of measured windows based on the manufacturers data

Window	Material	No. of chambers	Glazing	Gas	Heat gain [%]	U_w calculated	U_f	U_g
						[W/(m².K)]		
A	plastic	6	triple	Ar	36	0.80	1.0	0.5
B	plastic with thermo modules	6	triple	Kr	47	0.78	0.85	0.5
C	wood	-	triple	Ar	48	0.79	0.8	0.6

3. Heat fluxes

The thermal transmittance coefficient (*U*-value) of a structural element describes the quantity of thermal energy that passes through it from one side to another per second and per square meter surface at a constant difference in ambient temperature of 1 °K. With measured heat flow density using heat flux plates, knowing the outdoor and indoor air temperatures, the thermal transmittance coefficient can be calculated according the equation (1):

$$U = \frac{q}{\theta_{ai} - \theta_{ae}} \tag{1}$$

U – thermal transmittance coefficient [W/(m².K)]
q – heat flow density [W/(m².K)]
θ_{ai} – temperature of indoor air [°C]
θ_{ae} – temperature of outdoor air [°C]

Thermal surface transfer coefficient can be calculated according to Equation 2 and 3:

$$h_i = \frac{q}{\theta_{si} - \theta_{ai}} \tag{2}$$

$$h_e = \frac{q}{\theta_{se} - \theta_{ae}} \tag{3}$$

h_i – surface coefficient of heat transfer, inside [W/(m^2.K)]
h_e – surface coefficient of heat transfer, outside [W/(m^2.K)]
θ_{si} – temperature of inner surface [°C]
θ_{se} – temperature of outer surface [°C]

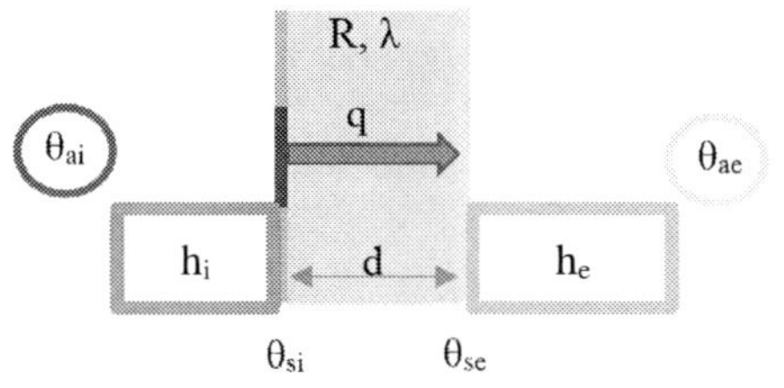

Figure 4. Illustration of heat flow from warmer indoor side to outdoor and the coefficients and physical properties which influence it.

Calculated heat fluxes represented in form of thermal conductance coefficients are summarized in Table 2. There is also the comparison with measured values from January 2017 [7]. The measured differences are accounted to the uncertainty during the measurement period, air flow in the room, not perfectly contact between windows and HFP. There was also different average outdoor air temperature during the measurement period. It was selected five days and the time periods from 20:00 to 5:00, with 5 minutes interval, which creates enough measured values.

The differences can be also accounted on the fact that the thermal conductance coefficient is not constant, but slightly varies depending on the temperature differences (outdoor air temperature). This is showed in the Figures 5 and 6. The influence of temperature on the thermal conductance and quantification cannot be determined in this case. The measured values are very different from the manufacturers` values, but it has to be taken the difference compared to the laboratory measurement. The glazing values for window A and B are significantly higher than window C.

Table 2. Measured heat fluxes represented in form of thermal conductance coefficient [W/(m^2.K)]. Values from January 2017[7] - 300 values, average θ_{ae} = -11.0 °C, February 2018 - 682 values, θ_{ae} = -12.5 °C

Window	Frame [W/(m^2.K)]		Glazing [W/(m^2.K)]	
	January 2017	February 2018	January 2017	February 2018
A	0.76	0.78	0.90	0.92
B	0.73	0.80	1.05	1.01
C	0.65	0.67	0.67	0.69

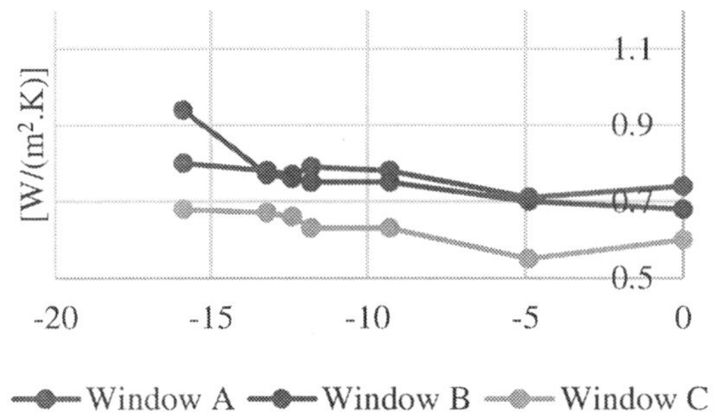

Figure 5. Dependence between temperature of outdoor air and thermal conductance for the window sash.

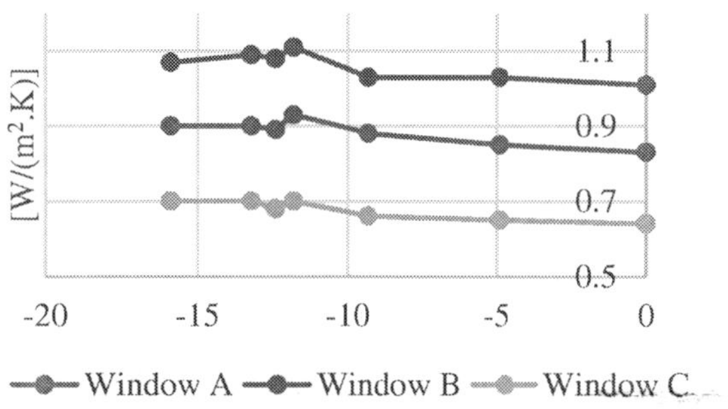

Figure 6. Dependence between temperature of outdoor air and thermal conductance for the glazing.

The surface temperatures are influenced by the surface coefficient of heat transfer. These coefficients are stated in the Slovak Standard [1] for various surfaces and indoor and outdoor surfaces. For the windows are valid different values than for walls, or non-transparent structures. It differs also by the heat flow orientation. Calculated values for glazing and window´s sash are in Table 3. For calculations were used equations (2 and 3). For h_i were used surface temperatures measured with the sensors integrated into the heat flux plates.

Values for glazing for interior were above the Standard value, for sash lower. If averaged, there will be a good match. For exterior is the trend opposite, glazing has lower values than sash. These surface heat transfer coefficients are probably influenced by the night sky cool radiation. This is planned to be compared to the laboratory measurement in the climate chamber in the future.

Table 3. Calculated surface coefficient of heat transfer based on the measured heat fluxed with Equation (2 and 3) and defined values in Slovak Standard [1]

Surface coefficient of heat transfer [W/(m².K)]	STN 73 0540	A glazing	B glazing	C glazing	A sash	B sash	C sash
h_i **(HFP)**	7.62	10.5	9.53	13.2	5.1	5.2	6.3
h_e	25	10.3	8.1	10.3	12.7	18.4	11.7

4. Temperatures

All measured courses are for the time period 25th of February - 1st of March 2018. Analyzed were positions in the cavity between frame and sash (Figure 7 and 8), on the frame (Figure 9), on the glazing with position on the bottom (Figure 10) and on the top (Figure 11). The window B has additional sensor on the contact between glazing and sash. Standardly, the thermocouple is about 2 cm above this position. Difference between this two temperatures are in Figure 12. Because of the amplitude of the indoor air temperature during the night, there is also added required surface temperature from Slovak standard for 18 °C. Temperatures for the glazing of Window C for all five positions are presented in Figure 13.

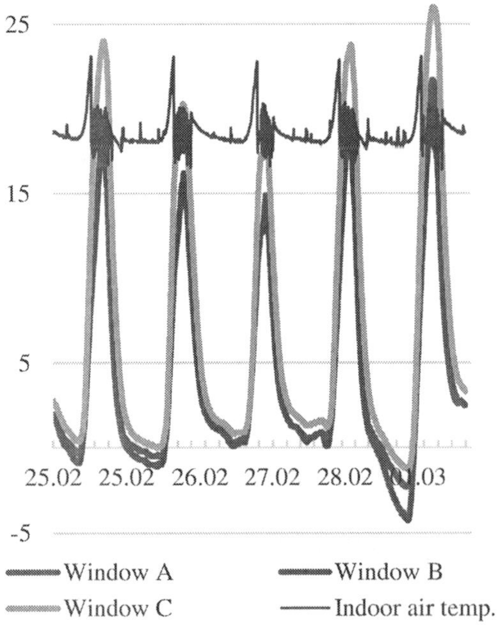

Figure 7. Temperature courses for cavity between frame and sash at the bottom.

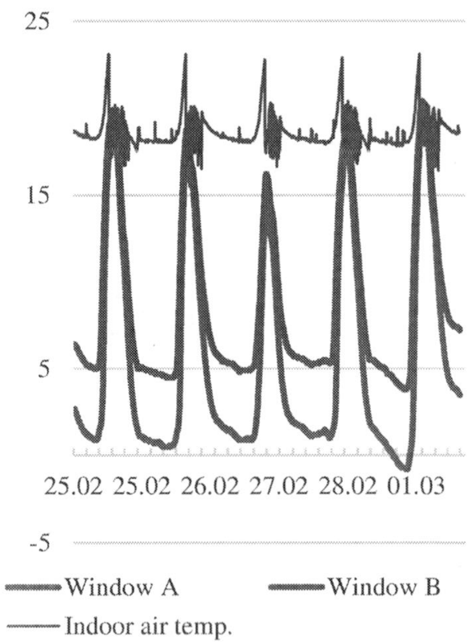

Figure 8. Temperature courses for cavity between frame and sash on the right side.

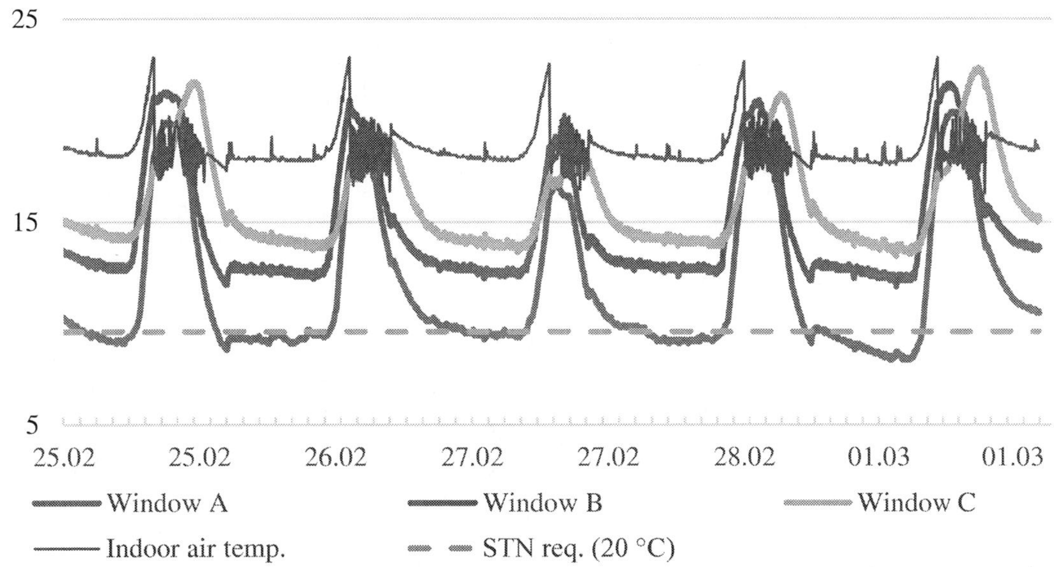

Figure 9. Temperature courses for position on the right side of frame.

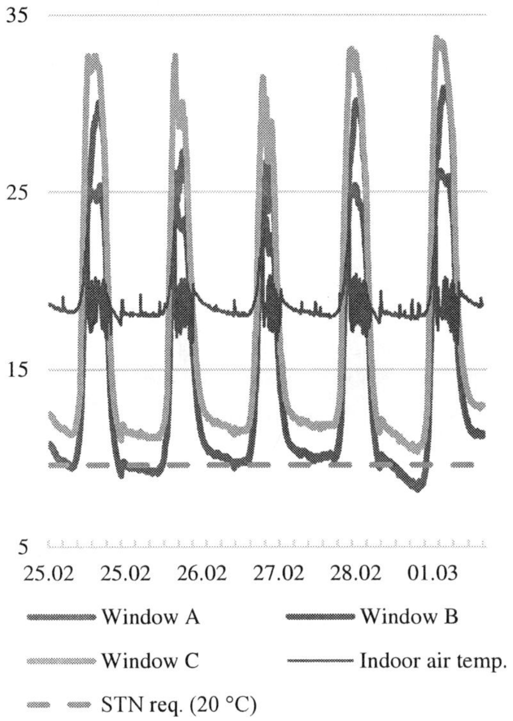

Figure 10. Temperature courses at the bottom of the glazing.

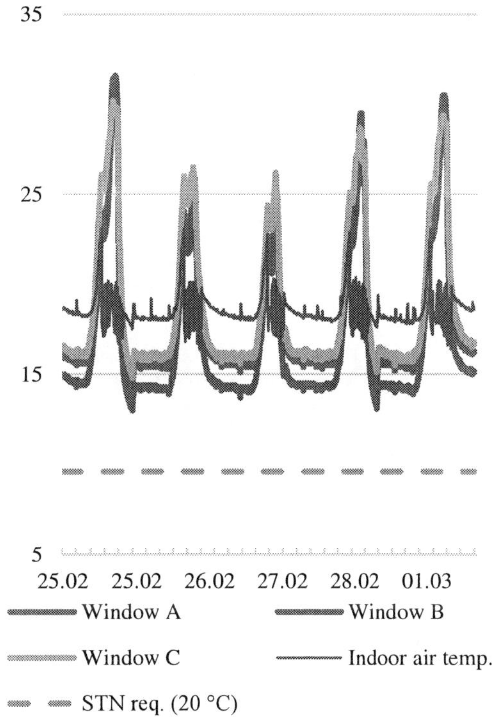

Figure 11. Temperature courses at the top of the glazing.

5. Discussion

Temperatures in the cavity were compared for the position at the bottom (all three windows) and on the right side (two windows). There are big differences between this positions for the windows B (Figure 7 and 8). Differences between Window A and B is not only in use of thermomodules but also Window B has additional sealing in the middle. At the bottom there could be problem with the contact, because there is a minimal difference. On the other hand, there is a significant difference up to 5°C in the cavity between windows A and B. Window C has no thermocouple in this position.

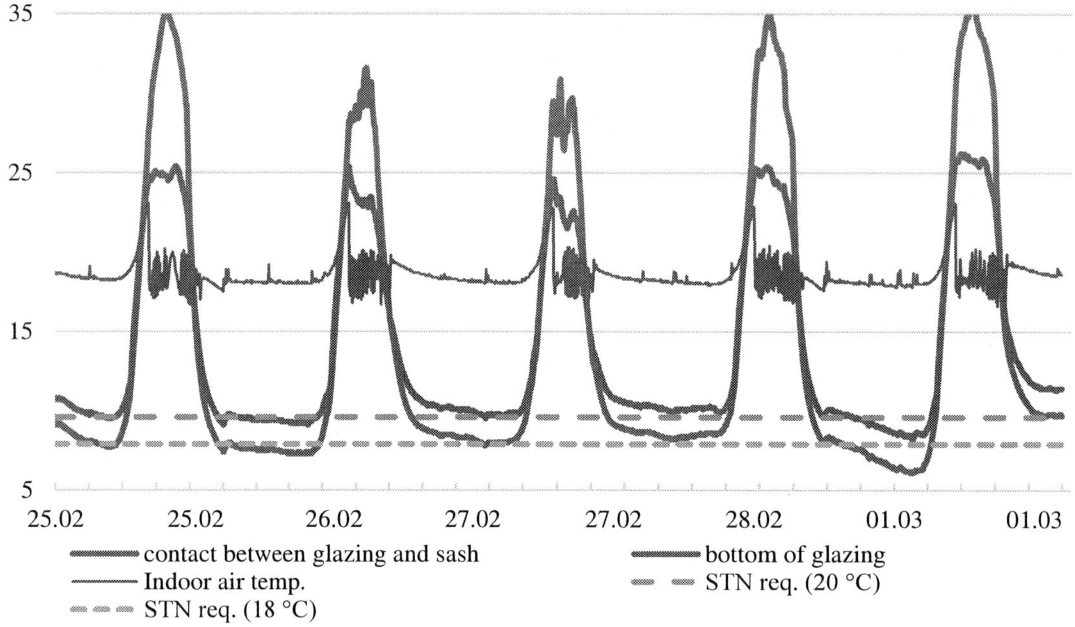

Figure 12. Temperature courses on the Window B glazing with lowest temperatures - bottom of the glazing (about 20 mm from the bottom) and direct contact between sash and glazing at the bottom.

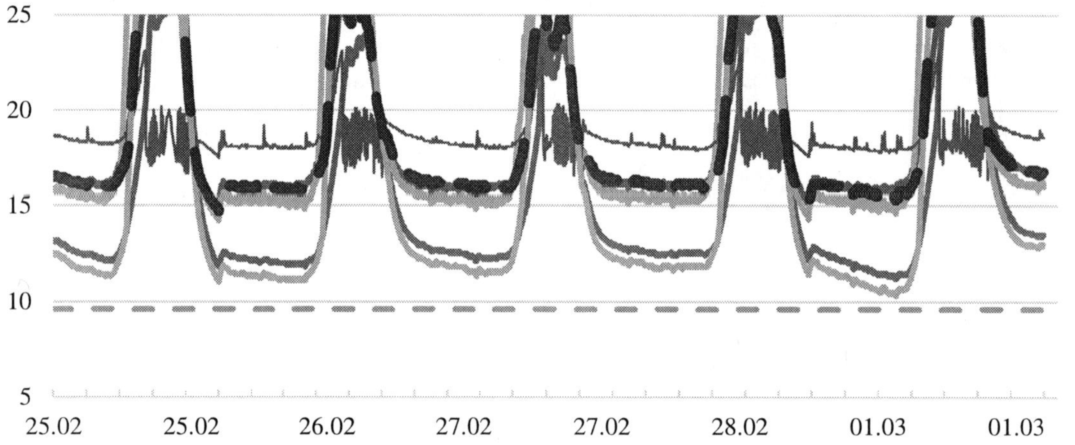

Figure 13. Temperature courses on the window C for all position on the glazing - during night.

Differences on the position at the bottom and top of the glazing are shown in Figures 10 and 11. During the night, temperatures on all windows are about 5 °C higher than at the bottom. Interesting are the daily peaks, which are lower on the top, so the daily amplitude is lower.

According to the previous results [6, 7] and also this, the critical position on the glazing is not the corner but the bottom of the glazing. The thermocouples are fixed on the windows approximately 20 mm from the bottom. For the window B was added thermocouple on the direct joint of glazing and sash. These courses are in Figure 12. The graph showed that the difference between these two positions is up to 2 °C. For comparison are stated two surface temperatures from the Slovak standard, one for 20 °C and second for 18 °C. This is because the indoor air temperature during the night was about 18 °C. The measured temperature on the contact was lower than both required values, bottom of the glazing was above the required temperature for 18 °C. Critical temperature for the outdoor air temperature is the same as for the frame (Figure 9).

From the comparison it is visible that the wooden Window C has the best results. Comparison of surface temperatures on the glazing for all five positions is on the Figure 13. This is only for the night representative because of the solar insolation on the additional sensors. Bottom and corner values are much lower than the middle or center of glazing, the top and right side courses are similar to the middle position.

6. Conclusion

Three windows in the pavilion type laboratory are constantly measured with the real outdoor climate. In this paper were monitored temperatures with additional thermocouples mounted in December 2017. This deals with the temperature courses in the frame/sash cavity, more detailed temperature distribution on the glazing and comparison between sash and frame. Detailed analysis confirms that the best window is with wooden frame, which complies the Standard requirements and based on the heat flux measurements has stable values over the years.

The window C has also the best measured values from the heat flux measurement compared to the other two windows. Comparison with the laboratory measurements cannot be made in this type of measurement because of the use of real windows, changing outdoor climate and other uncertainties such as air conditioning unit working interval etc

Acknowledgments
The research was done with the support of grant project VEGA no. 1/0945/16.

References

[1] STN 73 0540:2012: Thermal protection of buidings. Thermal performance of buildings and components. (In Slovak)
[2] Directive 2010/31/EU Energy performance of buildings
[3] Cekon M and Slavik R 2017 A Non-Ventilated Solar Façade Concept Based on Selective and Transparent Insulation Material Integration: An Experimental Study J. Energies **10**:815
[4] Iringova A 2016 Impact of external shading on light comfort and energy efficiency in apartment buildings J. Advanced Materials Research: enviBUILD vol. **861** pp. 485-492
[5] Ingeli R et al. 2014 Thermal bridges minimizing through window jamb in low energy buildings. Advanced Materials Research vol. **899** pp. 66-69
[6] Durica P et al. 2013 Long time testing of temperature parameters in selected windows J. Advanced Materials Research: enviBUILD vol. **855** pp. 81-84
[7] Juras P et al. 2017 Comparison of different windows for low-energy houses In MATEC web of conf. XXVI R-S-P seminar vol. **117**
[8] Durica P et al. 2016 Lightweight wood-based wall: the long-time evaluation of heat-air-moisture transport J. Communications vol. **14** no. 4 pp. 68-76
[9] Staffenova D et al. 2014 Climate Data Processing for Needs of Energy Analysis. J. Advanced Materials Research: enviBUILD vol. **1041** pp. 129-134
[10] NiCr-Ni thermocouple T190-0, website < http://ahlborn.com/getfile.php?2052.pdf>
[11] Thermal Flux Plates, website <http://www.ahlborn.com/download/pdfs/kap13/WflPlatten.pdf>

IOP Conf. Series: Materials Science and Engineering **415** (2018) 012021 doi:10.1088/1757-899X/415/1/012021

Energy Certification Label vs. Passive Discomfort Index for existing dwellings

S A Magalhães, V P de Freitas, J L Alexandre

LFC-Construct Rua Roberto Frias
silvia.magalhaes@fe.up.pt

Abstract. Several studies highlight the significant differences between theoretical energy consumptions, - considered in the Energy Certification Schemes (ECS), following the Energy Performance of Buildings Directive (EPBD) -, and the actual energy household consumptions, obtained from extended surveys and monitoring campaigns, especially in Southern European countries. These differences are referred in the literature as *"heating gaps"*. Actual consumptions are significantly unpredictable (depending on outdoor climate, income, energy prices, cultural habits, etc.), but, in most cases, lower than permanent heating/cooling assumptions, present on regulations. This inaccuracy may lead to the suggestion of inappropriate measures on ECS, for low energy-consumption dwellings. Inappropriate because the impact of the same measure on energy consumptions and thermal comfort is different for cases with a permanent heating behavior and cases without it, and because it may lead to hygrothermal pathologies and, finally, an economic mistake, as the real payback time of that measure will be higher than the one suggested, misleading economically more vulnerable households. A complementary approach for the evaluation of thermal performance in existing dwellings, considering intermittent heating scenarios is proposed, by the definition of the "Passive Discomfort Index" (PDI), complementary to the energy label. This index is quantified by the calculation of the temperatures outside the comfort range within the building, in realistic use conditions. A 19th century building, located in Porto, was used to perform a monitoring campaign of temperature and HR to calibrate a numerical model, developed using an advanced simulation tool. It was then performed a sensivity analysis, comparing the energy label with the PDI value, for different retrofitting scenarios, mainly with different insulation thicknesses. The results show that there is no obvious relation between the energy label and the comfort conditions, for intermittent heating scenarios, among others, reinforcing the need of a complementary approach. This is especially important for existent buildings and for Southern European countries, where the "*heating gap*" is higher.

1. Introduction

Reducing household energy consumption is a fundamental European goal, since 40% of the total energy consumption is spent inside buildings [1]. It is very important, although, to identify where and how it is spent, so that policies are drawn right to the target.

Energy Performance of Buildings Directive (EPBD)[1] was designed assuming that there is a permanent (24h/7d) heating/cooling practice inside buildings, to achieve permanent comfort conditions, being space heating the higher end-use consumption. The evolution of the Directives, from 2002 to its recast at 2010, and of the National Regulations following it, presents ever-decreasing U impositions, only achievable with large thicknesses of insulation. However, several studies focusing on actual consumptions, provided by energy suppliers databases and household surveys, point to the fact that actual household heating consumption are lower (intermittent and not in 100% of the floor area) than theoretical assumptions from regulations, especially in the less efficient buildings, what is referred on literature review as "*rebound effect*" [2-4]. The "rebound effect" means people tend to spend more energy on heating, as more efficient the house is and vice versa, as mentioned by Hens [2] and Sorrel [5], among others .

Laurent et al. [4] compared theoretical with actual heating consumptions, in four european countries (United Kingdom, France, The Netherlands and Germany) with the higher differences marked with a dashed array (Figure 1.a). The variability and unpredictability of occupant's behavior is also often referred to as a major factor to explain these differences. Guerra Santin [3] concluded the same for a Dutch Residential Stock sample (Figure 1.b)).

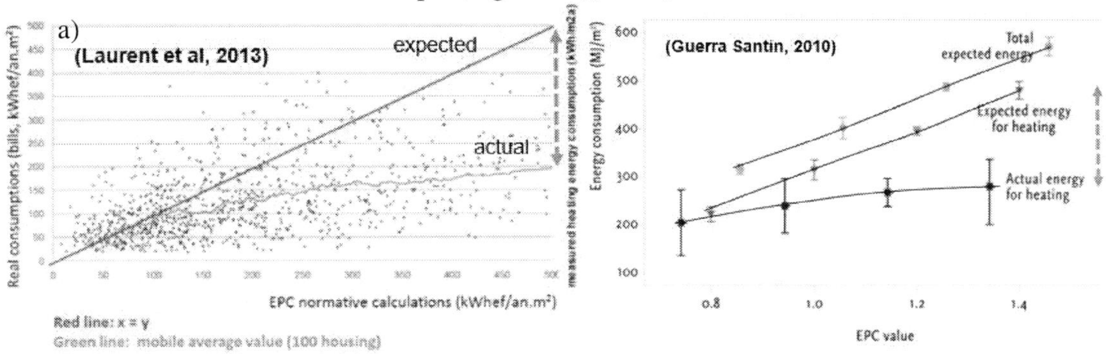

Figure 1. Heating energy consumptions (theoretical vs. actual) for European samples

This discrepancy is even higher in southern/ Mediterranean Europe, which present a moderate climate, compared to the central/northern countries. In fact, and observing the Portuguese example, only 21% [6] of the annual household energy bill is spent with space heating (kitchen represents 39%, water heating 23%, beside other end-uses), what is a much lower value than 67%, for the European average (for space heating weight) [7]. Magalhães [8] found an "*heating gap*" of 95% in Portugal, which means only 5% of the regulation assumption heating needs are actually consumed (considering all dwelling at 20ºC all winter period) .

Though, despite the lower energy consumptions, these southern/Mediterranean countries still present conditions of discomfort in the winter for most of the time [9]. This means that these countries don't need to heat so tightly, as the central and northern countries, but, when they need to do it, they just can't afford it, most of the time [8, 10]. This is explained by the phenomenon of "fuel" or "energy poverty" and in a simple way, results from the reunion of two main conditions: low incomes and high energy costs [11].

The improvement measures suggested by the Portuguese Regulation methodology [12], following EPBD, point mainly to the introduction of strong insulation thicknesses in the opaque envelope. There are several studies with Life Cycle Cost analysis to evaluate the impact of implementing these measures, but always with the assumption of permanent heating. However, there are very few studies evaluating the impact of these same measures, but with intermittent heating profiles, which is the realistic scenario for Portugal and other vulnerable countries. And these few studies point to some interesting conclusions. If, on the one hand, the introduction of insulation does not have a significant impact for comfort conditions improvement in winter, when there is an intermittent heating behaviour [10, 13], in summer it can even lead to an overheating risk, when inappropriately placed (on the internal side, reducing building inertia) [14]. Moreover, other disadvantages are identified, like increasing pathology risks (interstitial condensations), air quality prejudice, effects that can be worsened with the wrong using behaviours. Some of these buildings, were built in a totally different perspective, considering natural ventilation most of the times, and the increasing tightness demanded on the regulations distorts its thermal performance. Adding to the previous aspects, it is also important to mention the strong investment that some measures imply for the owners, not having the expected payback, presented in the Energy Certificates, that are based on theoretical savings. It's impossible to save what is not spent. Finally, the architectural value of some existing buildings that are often incompatible with the introduction of insulation.

In a macro perspective view, national and European policy targets for reducing energy consumption, based on the theoretical consumptions, can be also misleading, with all the economic consequences.

It is, therefore, clear that EPBD is a very useful tool for new buildings and when there is permanent heating, but it may reveal inappropriate for countries where there is no such practice and for existing buildings, where improvement measures have different impacts and implications compared to new ones.

The present work proposes an alternative methodology for evaluating the thermal performance of existing buildings, based on the quantification of the discomfort, which exists and it is a problem, not the heating consumptions, which are already low many times, especially in Portugal.

2. Energy poverty and heating consumptions in Portugal and southern Europe

Portugal is one of the European countries that most suffers from "fuel poverty" or "energy poverty" [15], defined in literature as the *'lack of affordability of energy for heating'* [11]. Another Mediterranean or East countries, are also the most energy poverty vulnerable. This incurs in low heating habits for these countries.

2.1. Energy poverty – Portugal vs. Europe

Portuguese median disposable income[1][16] was, in 2016, 9916€, a 39% lower value than EU-27 European average of 16351€. On the other hand, electricity price was, in 2015, 9% more expensive (0.23€/kwh) than EU-27 average (0.21€/kwh) and gas price, 38% more expensive than EU-27 average (0.098 and 0.071€/kwh, respectively), as shown in Figure 2 (a). Both conditions: - income and energy costs - lead to an energy poverty vulnerability context and low heating consumptions. This phenomenon is not exclusive from Portugal, despite this is a paradigmatic case.

Hills [17] proposed in his research concerning Fuel Poverty in Europe, the index "Low income – high cost" (LIHC). This index was applied at a cross-national level, in the present study, using the same statistic data from Figure 2 (a). Results are presented on Figure 2 (b), highlighting the fuel poverty vulnerable area, where four countries stand out: Portugal, Spain, Greece and Cyprus.

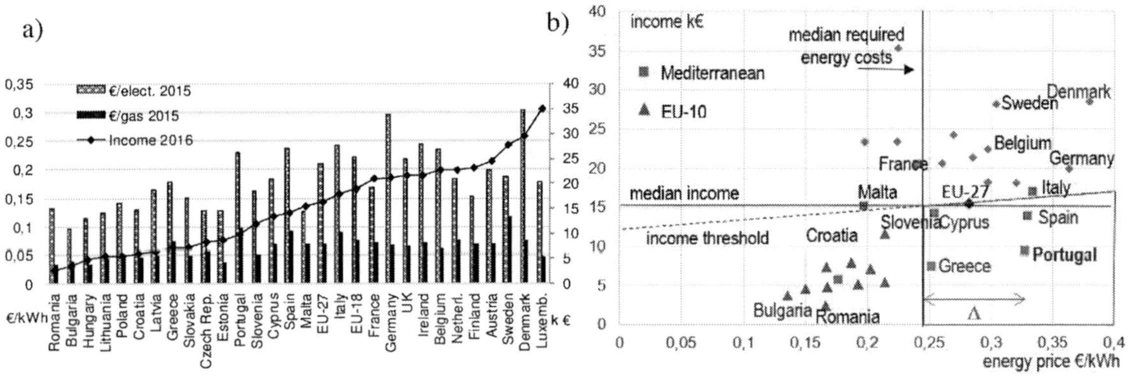

Figure 2. Median disposable income, electricity and gas prices in EU-28 (a) and Hills LIHC index adaptation

It's interesting to notice, moreover, that it is in the Mediterranean area that there are more winter deaths. A study from Oxford University [18], calculated the EWDI (Excess Winter Deaths Index), defined by the ratio of winter deaths for total deaths, from 2005 to 2014, for 31 European countries. Malta presented the highest value (28.3%) followed by Portugal (25.9%), Spain and Cyprus, confirming the previous findings from Figure 2 (b).

2.2. Household heating consumptions – Portugal vs. Europe

[1] For upper secondary and post-secondary non-tertiary education (levels 3 and 4), from 18 to 64 years old.

Previous findings regarding energy poverty phenomenon - namely high energy costs and low-income levels – are present all across Europe, but with more expression in Mediterranean area and some East countries, being Portugal one of the most evident case. These aspects, justify, itself, the low heating habits for these countries, but there is one more obvious for Mediterranean area, that is climate. Portugal has 1155ºCd, 63% less "heating degrees days" (HDD) than EU-28 average of 2817ºCd and only 19% of the Finland value [19]. So, it's not surprising that Portugal, as other southern/Mediterranean countries, heat much less than EU average and, especially, northern countries.

Indeed, space heating weight in Portuguese average household energy bill is 21% of the total consumption and 11% of the total cost [6]. In contrast, this end-use weigh rises considerably for EU average being the main energy end-use with 67% of total consumption [7].

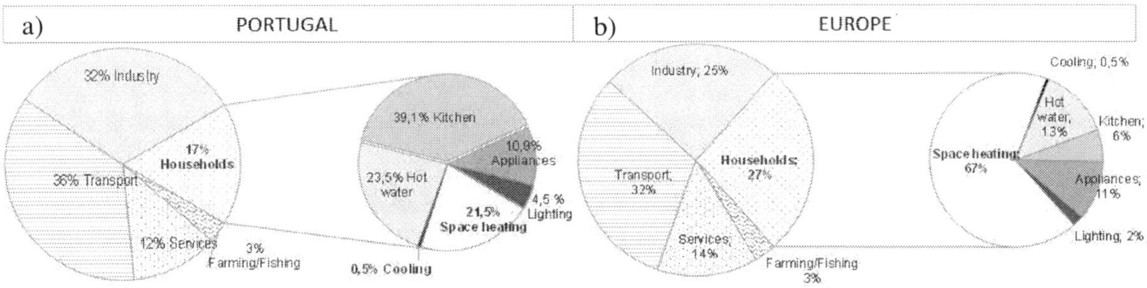

Figure 3. Energy total consumptions by end-use and household detail, for Portugal (a) and EU (b)

This way, being heating consumptions already low (and cooling, negligible), discomfort minimization should the target for these countries, more than energy efficiency for heating or cooling. That is the motivation and main purpose for this study.

3. Methodology

3.1. Experimental campaign and numerical model

A century building located in Porto, northern Portugal (Figure 4 a)), was used to validate the numerical model. The building presents a typical construction from the 19th century, with stone heavy external walls, wooden floors and roof, naturally ventilated, which was refurbished maintaining the main characteristics. Both apartments, north (N) and south (S) oriented, are located on the 1st floor and were monitored (temperature and relative humidity) during 2016. A geometric model for both apartments was built on ©WUFIPlus software (Figure 4 b)). This has been calibrated by experimental data (Figure 4c)), with 0.5ºC maximum temperature difference and 3.5% relative maximum humidity difference in the 5 rooms/2 dwellings.

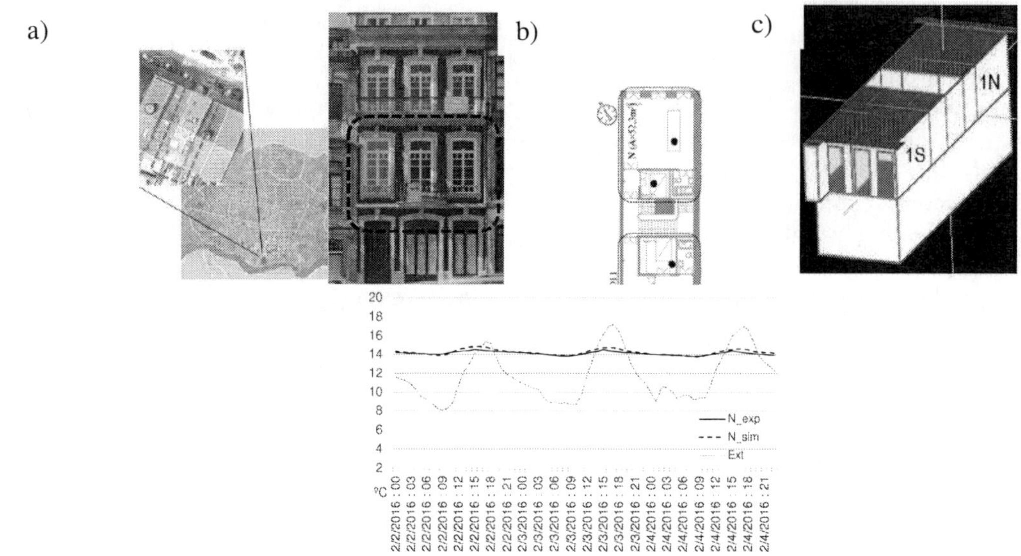

Figure 4. Case study building and experimental campaign (a), numerical model (b) and temperature validation (num. vs. exp.) (c)

3.2. *"Passive Discomfort Index" (PDI) definition*

The present study main purpose is to develop a new approach for evaluating the thermal performance of existing dwellings in low heating countries. This new approach consists in quantifying the discomfort outside a defined comfort range, that is, for the present study, [20-25] ºC or [18-25] ºC, this last case the Portuguese regulation "definition" of comfort (steady state assumption). This "discomfort" is quantified by the PDI that is numerical defined by formula (1) and in a graphic way by Figure 5.

$$PDI = PDI_{win} + PDI_{sum} = \int_{year} {}^{\circ}C*h_{\ win} + \int_{year} {}^{\circ}C*h_{\ sum} = \sum_{h=1}^{8760}(20-Ti) + \sum_{h=1}^{8760}(Ti-25) \qquad (1)$$

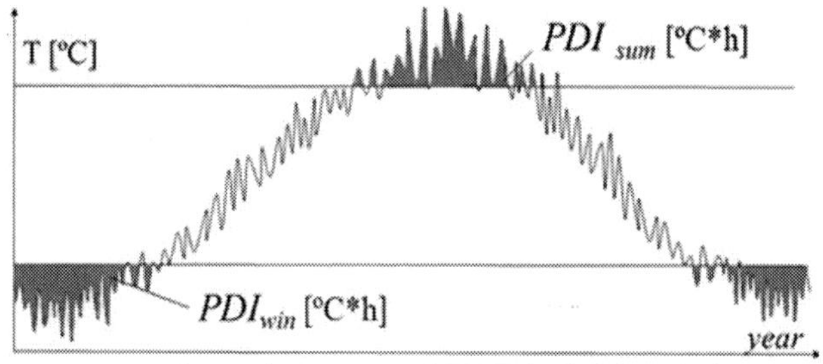

Figure 5. Free-floating temperature along the year and PDI definition (winter and summer)

3.3. *Energy Label – Portuguese Regulation (REH)*

REH [12] is the Portuguese thermal regulation for residential buildings, being the current transposition of the recast EPBD. It is based on the comparison of the nominal primary energy needs (N_{tc}) and the same needs for a reference building (N_t), summing different components as the heating needs, the cooling needs, the ventilation losses, solar gains, among others.

The Certification Scheme assigns an energy label to the dwelling, depending on the value of $R_{NT} = N_{tc} * N_t^{-1}$, from **A** (when $\mathbf{R_{NT} \leq 0,25}$) to **F** (when $\mathbf{R_{NT}}$ > 2,5). It's important to notice that REH´s building energy calculation models consider a permanent, in space and time, heating behavior inside dwelling (if the temperature is above 18ºC, in the current version).

3.4. Sensitivity analysis – scenarios definition

Variables considered to perform a sensitivity analysis include: three insulation thicknesses ("I" 4/6/8cm, with U=0.63/0.48/0.39 W/m2°C, respectively) by the interior side of the wall, and a non-insulated or "base" scenario ("B", U=2 W/m2°C), reflecting the real building; five heating profiles (0/4/6/8/24h) and three technical heating systems, to evaluate its influence on discomfort hours and energy label assigned to the building. Table 1 presents the several options.

Table 1. Scenarios definition [d]

[Comfort range]	A - PDI_{win} vs. Heating Energy [20-25] °C												
	B - PDI vs. Energy Label [18-25] °C [a]												
Scenario Code [b]	**B0h_N/S**	**B4h_N/S**	**B6h_N/S**	**B8h_N/S**	**B24h_N/S**	**I#_0h_N/S**	**I#_4h_N/S**	**I#_6h_N/S**	**I#_8h_N/S**	**I#_24h_N/S**	**BS2_0h_N/S**	**BS3_0h_N/S**	**I4S3_0h_N/S**
Envelope	"B" (non insulated)					"I#" (insulated with #4/6/8 cm)					"B"		I4
Heating (h)	0h	Intermittent			24h	0h	Intermittent			24h	0h		
Cooling (h)	0h				24h	0h				24h	0h		
Systems [c]	S1 ("Default system")										S2	S3	

a [18-25] °C is the Comfort range considered on Portuguese Regulation, REH
b Scenario code means: Insulation option_Hours of heating_North/South orientated; In the case of B evaluation, after Insulation option, it's mentioned also the alternative technical system considered.
c S1 means Electrical appliances (default system, with efficiency factor η = 1); S2 means Gas Boiler for heating (η = 0,85); S3 means Heat pump+Solar Panels (COP = 3,5 / EER=2,5)
d Other parameters: Ventilation = $0.5h^{-1}$; no occupation; A_{win}/A_{floor} = 20%; Openings: U_w=2.71/Solar factor=0.65; adjacent buildings and dwellings with the same inner climate.

Two approaches are considered. First, in "A" evaluation, it's compared the "passive discomfort index" (PDI), considering the comfort range [20-25] °C and the total energy consumption, obtained from the numerical advanced model in a dynamic state. In "B" evaluation, on the other hand, it's compared the PDI with the energy label, obtained from the Portuguese Regulation REH calculation model (where it's assumed the comfort range [18-25] °C in a steady state approach). So that there is a common basis, for B evaluation, PDI is calculated considering also the comfort range [18-25] °C.

4. Results and discussion

In the present chapter, the results for the two separated approaches mentioned on §3, are presented. Results for A are presented in Figure 6, in an energy consumption ranking, for all scenarios. Still for A approach, it's compared, in Figure 7, the PDI values with the heating consumptions, both only for winter, comparing the non-insulated scenario with the 8cm insulated scenario, and north and south orientated, for different intermittent profiles. Figure 8 presents, on the other hand, the annual energy consumptions, obtained from the dynamic model, with the energy label obtained from the ECS, for all scenarios.

4.1. A: PDI vs. Actual heating consumption

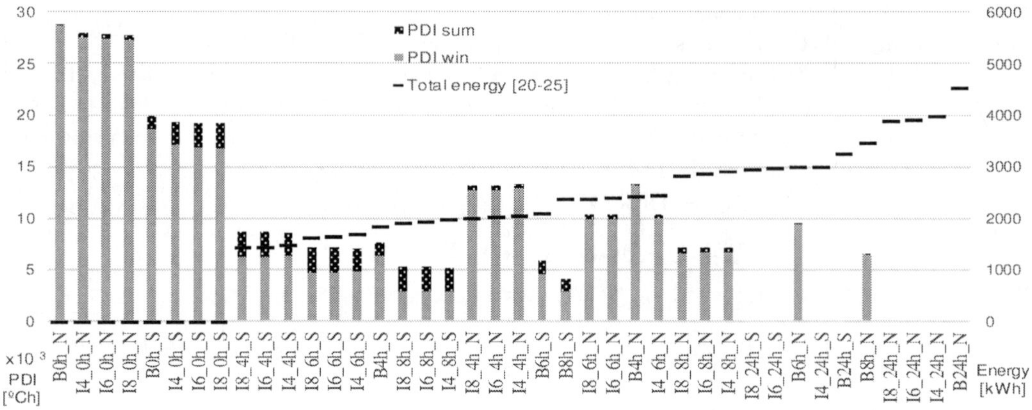

Figure 6. Ranking of the total energy consumption (dynamic) and relative PDI_{win} and PDI_{sum}

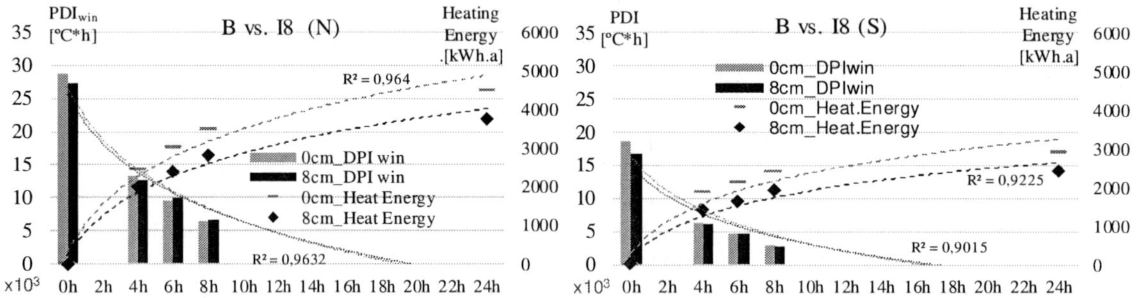

Figure 7. PDI_{win} vs. Heating energy consumption for different heating scenarios, without or with 8cm insulation, for North (a) and South (S) orientated dwellings

It's possible to observe, in Figure 6, that there isn't a clear inversely proportional relation between discomfort and energy consumption, considering all variables. There are scenarios with more energy consumptions and more discomfort than others, even for the same orientation. Also, the insulation thickness has a very small impact on the annual discomfort, as the total discomfort is very similar, for scenarios I8/6/4, regardless the heating profile or orientation. The summer discomfort for north orientated dwelling is negligible, but for south orientated it can have a significant impact. From Figure 7 a) and b) it's possible to observe, only for winter, the important role of orientation on total heating energy/PDI_{win}. Also, it's clear, once again, the low influence of insulation, considering the extreme cases: without insulation or with 8cm, on total heating energy/$PDI_{win.}$ Finally, it is possible to see that the major decrease of discomfort happens with the first hours heating and this decrease is increasingly lower with the number of heating hours. It is evident, in Figure 7, the non-linear (logarithm) relation between energy/discomfort and the heating profile.

4.2. *B: PDI vs. Energy Label*

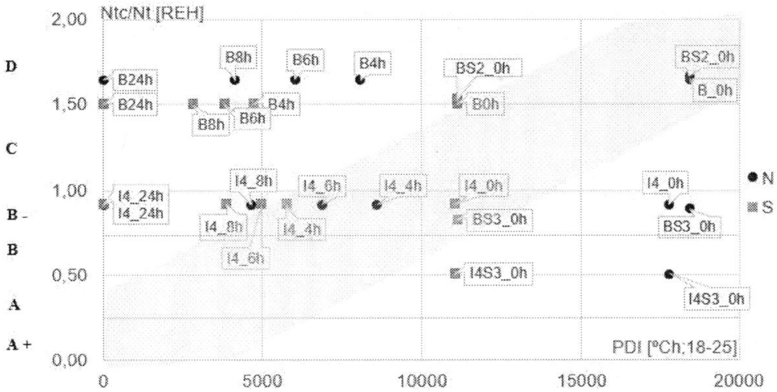

Figure 8. PDI vs. Energy Label (REH)

In B approach, it's possible to observe, in Figure 8 now considering all scenarios, there isn't any visible relation between the energy label and the annual discomfort (the expected relation should be in the grey area). Also, the technical system plays an important role on the label assigned, regardless the discomfort of that scenario. And a better energy label does not always mean a lower discomfort. For instance, I4S3_0h (north) is an A label, but presents a higher discomfort than I4_0h (south), that is a B, both non- heated. Considering extreme cases, the same north orientated dwelling, can present totally different discomfort levels, from 0 to almost 20000ºCh (B24h and BS2_0h), despite always being a D label. If 4cm of insulation are introduced, the label is improved for an A, but the discomfort remains the same if it is not heated.

5. Conclusions

From the results presented on the previous chapter, it is possible to conclude:

- In intermittent heating scenarios - the most common on southern Europe -, insulation thickness (when placed by the interior side) has a negligible impact on winter discomfort hours, for Porto climate; for free-floating scenarios, it is irrelevant to place insulation at all; this raises the issue of the real interest of the ever-increasing insulation thicknesses imposed by the regulations, for these climates and for existing buildings;
- There are multiple variables that have a much higher impact on annual/winter discomfort hours, than insulation thickness, such as orientation and the heating profile;
- There is a logarithm relation between discomfort/energy and the heating profile; this means the number of heating hours have an ever-decreasing impact on winter discomfort reduction (or energy increase);
- There is no evident relationship between the energy label, obtained from the national regulation transposed by EPBD, and the annual discomfort level, considering intermittent heating scenarios on winter;
- For this reason, a complementary approach to energy is needed to evaluate the thermal performance of existent residential buildings, especially when located in temperate climates, where intermittent heating habits prevail.

Acknowledgements

This work was financially supported by: Project POCI-01-0145-FEDER-007457 - CONSTRUCT - Institute of R&D In Structures and Construction funded by FEDER funds through COMPETE2020 - Programa Operacional Competitividade e Internacionalização (POCI) – and by national funds through FCT - Fundação para a Ciência e a Tecnologia; SFRH/BD/96068/2013.

COMPETE 2020 PORTUGAL 2020 UNIÃO EUROPEIA Fundo Europeu de Desenvolvimento Regional FCT Fundação para a Ciência e a Tecnologia

References

1. DIRECTIVE 2010/31/UE, *19-05-2010, Buildings Energy Performance (recast)*. 2010.
2. Hens, H., W. Parijs, and M. Deurinck, *Energy consumption for heating and rebound effects.* Energy and Buildings, 2010. **42**(1): p. 105-110.
3. Guerra Santin, O., *Actual energy consumption in dwellings: The effect of energy performance regulations and occupant behaviour*, in *Housing*. 2010, OTB Research Institute: IOS Press.
4. Laurent, M.-H., et al. *Back to reality: How domestic energy efficiency policies in four European countries can be improved by using empirical data instead of normative calculation.* . in *ECEEE Summer Study Proceedings. Pages 2057-2070.* 2013.
5. Sorrell, S., *The Rebound Effect: an assessment of the evidence for economy-wide energy savings from improved energy efficiency*. 2007, UK Energy Research Centre London.
6. INE/DGEG, *Household Energy Consumption National Survey* 2010: Instituto Nacional de Estatística, I.P. Direção Geral de Energia e Geologia.
7. Lapillonne, B., et al., *Energy Efficiency Trends for households in the EU - Odyssee Project*, ENERDATA, Editor. 2015, Inteligent Energy Europe.
8. Magalhães, S., *The relationship between heating energy use, indoor temperature and heating energy demand under reference conditions in residential buildings. PhD Thesis.*, in *INEGI*. 2016, Universidade do Porto.
9. Silva, P.C.P.d., et al., *Portuguese building stock indoor environmental quality "in-situ" assessment.* 2010
10. Magalhães, S.A. and V.P. de Freitas, *A complementary approach for energy efficiency and comfort evaluation of renovated dwellings in Southern Europe.* Energy Procedia, 2017. **132**: p. 909-914.
11. Boardman, B., *Fuel poverty: from cold homes to affordable warmth.* 1991: Pinter Pub Limited.
12. Portaria 379-A, *Regulamento de Desempenho Térmico de Edifícios de Habitação (REH) - Requisitos de conceção para edifícios novos e intervenções.* 2015: Ministério do Ambiente, Ordenamento, do Território e Energia.
13. Jerónimo, R., *Avaliação do desempenho higrotérmico e do conforto de edifícios rurais reabilitados. PhD Thesis.*, in *FEUP - Civil Engineering Department*. 2014, Oporto University.
14. Chvatal, K.M.S. and H. Corvacho, *The impact of increasing the building envelope insulation upon the risk of overheating in summer and an increased energy consumption.* Journal of Building Performance Simulation, 2009. **2**(4): p. 267-282.
15. Bouzarovski, S., *Energy poverty in the EU: a review of the evidence.* 2011.
16. EUROSTAT. *Living Conditions Statistics - Average Income.* 2018 15-02-2018 [cited 2018 12-03-2018]; Available from: http://ec.europa.eu/eurostat/web/gdp-and-beyond/quality-of-life/median-income.
17. Hills, J., *Fuel Poverty - The problem and its measurements* 2011, CASE - Centre of Analysis of Social Exclusion.
18. Tom Fowler, R.J.S., Thomas Waite, Ruth Harrell, Sari Kovats, and Y.D. Angie Bone, Virginia Murray, *Excess winter deaths in Europe: a multi-country descriptive analysis.* European Journal of Public Health, 2014. **25, nº 2**: p. 339–345.
19. EUROSTAT. *Heating Degree Days in Europe Countries.* 2018 19-02-2018 12/03/2018]; Available from: http://appsso.eurostat.ec.europa.eu/nui/setupDownloads.do.

Assessment of the impact of spatial development changes on thermal comfort experienced by man in the external environment

A D Bochenek and K Klemm

Institute of Environmental Engineering and Building Installations, Faculty of Civil Engineering, Architecture and Environmental Engineering, Lodz University of Technology, Politechniki 6 Avenue, Lodz, Poland

katarzyna.klemm@p.lodz.pl

Abstract. Progressive urbanization has a direct influence on both the local and consequently global climate. The spatial development methods of the area require the use of modern computer technologies that support planning decisions by key decision makers. Evaluation of the prevailing microclimate conditions in the external environment is possible using CFD software. In this paper were presented scenarios of the impact of changes in spatial development of the area covered by revitalization process in the most urbanized zone of Lodz. The changes proposed by the authors included two types of simulation input data, i.e. values determined for a typical meteorological year and values of meteorological parameters from the nearest measuring station – Lodz-Lublinek. PMV index enabled to evaluate the thermal comfort of human in an external environment. Conducted research has shown that the more effective solutions were scenarios based on a growth of the green area rather than introduction of water elements within the analyzed public space.

1. Introduction

Dynamically progressing urbanization process contributes to disadvantageous changes in spatial development. The creation of new buildings, increase in land cover with impermeable materials and decrease in the share of green areas result in emergence of urban heat island phenomenon in urban areas. As a result, there is an increase in energy consumption necessary to cool buildings, water is used to lower the temperature in external environment (e.g. in a form of water curtains), air quality deteriorates, thermal discomfort is created, as well as morbidity and mortality caused by thermal stress are increased [1].

One of possibilities to limit the negative effects of phenomena caused by transformations of the spatial development method is to undertake mitigation actions. The collection of mitigation strategies was published by the Environmental Protection Agency (EPA) in 2008. The compendium describes, among others the use of traditional and vertical forms of vegetation. High greenery enables to reduce the amount of sunlight reaching the earth's surface (70-90%). It provides a shadow and contributes to reduce the air temperature in the vicinity of trees. It also modifies the airflow in cities. Green roofs and walls, limiting the amount of radiation reaching the surface of building partitions, are responsible for temperature reduction inside the buildings. By capturing rainwater, they contribute to the increase of relative air humidity. The next steps concern the use of building materials characterized by high albedo values, absorbing less heat during the daytime and reflecting a significant amount of solar radiation back into the atmosphere [2]. However, these are only selected mitigation strategies. According to empirical research, the introduction of water elements into the environment causes modification of microclimatic parameters. This is confirmed by research carried out by Chatzidimitriou and Bruse. The use of water elements in a form of fountains and water curtains reduces the air temperature [3]. According to Barakat et al., water resources cause a cooling effect (they modify the value of temperature and relative humidity) [4]. Tests performed by Chan et al. confirm that the presence of water surfaces increases the acceptance of prevailing microclimatic conditions in the external environment [5].

The authors took into account only the summer period due to the fact that the highest temperature values occur during this period (overheating conditions), and thus the most uncomfortable conditions prevailing in the external environment. In addition, it was associated with the highest intensity of use of public spaces in the city. Nine scenarios of the impact of changes in spatial development on perceived thermal comfort were considered in this work. They concerned transformations of selected public space in terms of vegetation and water elements. Simulations were conducted using the Envi-met program. The input parameters of models were data of Typical Meteorological Year obtained from databases of the European Commission and information obtained from the Lodz-Lublinek meteorological station. Finally, the assessment of thermal comfort was carried out by comparing the values of thermal comfort index – PMV determined for individual scenarios with its value for the baseline scenario (determined on the basis of current state of spatial development of the area).

2. Area of research

One of the oldest public spaces in the city, located in so-called Greater City Zone of Lodz, has been selected as a research area. This area – The Old Market was indicated as a priority area for the revitalization process. Multi-family residential building is the dominant type of development. The objects constituting the frontage of square have been entered into the municipal register of monuments in Lodz. As a result, the potential change in land development related to the geometry of objects has been significantly hampered. All transformations require arrangements with the monument conservator, urban planners and municipal officials. However, modifications within the public space may occur by changing the surface types, introduction of additional plantings and water elements.

3. Thermal comfort

Thermal comfort is defined as a state of mind expressing satisfaction from the thermal environment (ASHRAE 55) [6]. It is dependent on physical condition of the man, resulting from the pace of metabolism, clothing and physical activity. What is more, it is felt in a subjective way. Its assessment is related to psychological factors, i.e. expectations, experiences, acceptance of external conditions, mobility and thermal adaptation by changing the exposure to microclimate parameters [7].

Thermal conditions can be estimated using the Predicted Mean Vote (PMV) index, which is the average rating of a group of people who determine their thermal impressions. At first, it was used to determine indoor conditions. Modification of the Fanger's thermal balance equation by Jendritzky and Nubler enabled to use it in external environment research. The PMV index can be calculated for any combination of metabolism, clothing, air temperature, average temperature radiation, air-flow velocity and humidity. The evaluation of thermal sensations is carried out on a seven-stage scale from [-3] to [+3]. The conditions described as comfortable are for the range [-0.5, +0.5] [8].

The PMV index is commonly used in studies of highly urbanized areas, including public spaces. This fact is confirmed by the research of Coisson et al. (assessment of thermal comfort in urban areas in Italy), Djekic et al. (assessment of thermal comfort in public spaces – Serbia), Wang et al. (assessment of thermal comfort depending on spatial development method – Netherlands).

4. Simulation of thermal comfort

The assessment of thermal comfort was conducted with the use of CFD type software, i.e. Envi-met program. This tool enables creation of three-dimensional, non-hydrostatic microclimatic models for urbanized areas. Simulations of meteorological parameters take into account the air-flows between buildings, processes of vertical and horizontal heat exchange of surfaces, turbulence, transpiration, vegetation parameters, as well as dispersion of pollutants [9]. The program enables to carry out simulate of the relation between substrate – vegetation – air in the twenty-four hour cycle (from 24 to 48 hours) [10-11]. In order to estimate meteorological parameters, it is necessary to define the initial conditions in the external environment. One way to determine them is to adopt data of Typical Meteorological Year. Finally, there is an option to calculate the basic thermal comfort indicators, i.e. PMV (Predicted Mean Vote), PET (Physiological Equivalent Temperature) and UTCI (Universal Thermal Climat Index) [12-13].

In work, the assessment of thermal comfort was conducted for the warmest day of Typical Meteorological Year, determined on the basis of data from a 10-year period (from 2006 to 2015). In order to estimate the PMV index, it was necessary to determine the individual characteristics of the man, his clothing and type of performed physical activity. The thermal comfort was estimated for a 35-year-old man with a height of 1.75 m and weight of 75 kg, with a clothing insulation of 0.50 clo [14]. The authors considered two cases, i.e. for a person not performing physical activity and moving at a speed of 1.21 m/s (average speed of the man's movement – walk). In the latter case it should be expected that PMV values will be lower due to the formula of the index.

5. Typical meteorological year

Over the past decades, methods have been developed to determine average meteorological conditions. The most commonly used are: Weather Year for Energy Calculation (WYEC2) developed for ASHRAE by Watsun Simulation Laboratory, Typical Meteorological Year (TMY2) updated by the National Renewable Energy Laboratory, the Canadian Weather for Energy Calculation (CWEC) established by the WATSUN Simulation Laboratory and the Test Reference Year (TRY) proposed by the Technical University of Denmark. Usually, climate data are developed from actual measurements of at least 30 years of observation per location. They concern parameters such as: dry bulb temperature, incident radiation, wind direction and speed, as well as dew point temperature [15].

The data set for selected weather parameters for Poland is defined by EN ISO 15927-4:2005 – Hygrothermal performance of buildings – Calculation and presentation of climatic data – Part 4: Hourly data for assessing the annual energy use for heating and cooling. Typical Meteorological Year is created on the basis of hourly values of weather parameters, i.e. dry bulb temperature, incident radiation and relative humidity over a period of at least 10 years.

Standardized data – Typical Meteorological Year reflect long-term conditions ruling in the external environment. Therefore, they can be used as input data of the simulation for atmospheric processes.

6. Input data

The analysis of thermal comfort in the external environment was carried out on the basis of average atmospheric conditions characterized by the warmest day of Typical Meteorological Year. At first, information on hourly values of selected meteorological parameters was obtained from archival databases of the European Commission for a period of 10 years (from 2006 to 2015). The warmest day of Typical Meteorological Year is July 2, 2012. Information on relative humidity and wind direction has been used as an input data to simulate atmospheric processes. The specificity of used software required to determine hourly values of the air temperature, wind speed, atmospheric pressure, dew point temperature and specific humidity at a height of 2.500 m. The data necessary for their calculation were collected from databases of the Lodz-Lublinek measurement station located in the suburban zone, in the south-western part of the city. Parameter values measured in an open area do not fully correspond to conditions prevailing in the city centre. Therefore, it was necessary to calculate the wind speed for the centre area of Lodz. For this purpose, a modified version of the logarithmic equation (1) was used [16], i.e.:

$$\overline{U}_{(z)} = \frac{u_*}{\mathrm{K}} * \ln\frac{(z-z_\mathrm{d})}{z_0} \tag{1}$$

where:
$\overline{U}_{(z)}$ – average wind speed at a height of 10 m [m/s],
u_* – friction speed [m/s],
K – Von Kármán constant (0.4),
z – altitude above the area level [m],
$(z - z_d)$ – effective height [m],
z_d – displacement of zero plane [m],
z_0 – roughness coefficient (also called the roughness parameter or surface roughness level) [m].

At first, the friction speed was calculated for an open area, where the Lodz-Lublinek meteorological station was located.

$$u_* = \frac{\overline{U}_{(z)}}{\ln\frac{(z-z_d)}{z_0}} * \mathrm{K} \tag{2}$$

The use of modified version of the logarithmic formula is justified from a level of $z \cong (z_d+10)$, below the value of wind speed should be considered as constant and equal to $\overline{U}(z_d + 10)$ [17]. In this study, the friction speed was calculated using a simplified form of equation (3), because the test was carried out at a height not exceeding 10 m.

$$u_* = \frac{\overline{U}_{(z)}}{\ln\frac{10}{z_0}} * \mathrm{K} \tag{3}$$

Then it was necessary to determine the friction speed for an area of the city centre of Lodz. The relation proposed by Simiu provides a link between the speed profiles of an open area and inner city [18], i.e.:

$$\frac{u_*}{u_{*1}} = \left(\frac{z_0}{z_{01}}\right)^{0,0706} \tag{4}$$

where:
u_* – friction speed on the Lublinek area [m/s],
u_{*1} – friction speed in the central zone of Lodz [m/s],
z_0 – roughness coefficient for the Lublinek area [m],
z_{01} – roughness coefficient for the central zone of Lodz [m].

$$q = q_1 + (25 * \Delta q_{100}) \tag{5}$$

where:
q – specific air humidity at a height of 2500 m [g/kg],
q_1 – specific humidity by ground surface [g/kg],
Δq_{100} – vertical gradient of specific humidity at a height of 100 m.

Table 1. Values of the simulation input parameters.

Input parameters	User input during simulation	Input parameters	User input during simulation
Simulation days	2 July 2012 – 3 July 2012	Dew point [°C]	18.5
Simulation time [h]	25 hours	Specific humidity	7.14
Wind speed [m/s]	1.79	Temperature [°C]	20.55 – 32.27
Wind direction [°]	234.10	Relative humidity [%]	39.04 – 84.64
Air pressure [hPa]	989		

As a result, the value of air-flow velocity was determined at a height of 10 m in the central part of Lodz. Another necessary input data of simulation is the value of specific humidity at 2500 m. Knowing its value at the ground level and vertical humidity gradient, it is possible to determine it using the equation (5) [19].

The values of model input parameters determined on the basis of data from the Typical Meteorological Year and calculated on the basis of information obtained from the Lodz-Lublinek meteorological station were presented in table 1.

7. Scenarios of spatial development changes

The way of spatial development affects the microclimate parameters, i.e. air temperature, solar radiation reaching the ground atmospheric layers, air-flow and relative humidity. Through appropriate modifications within the public space, it is possible to provide thermal comfort to people living in the external environment.

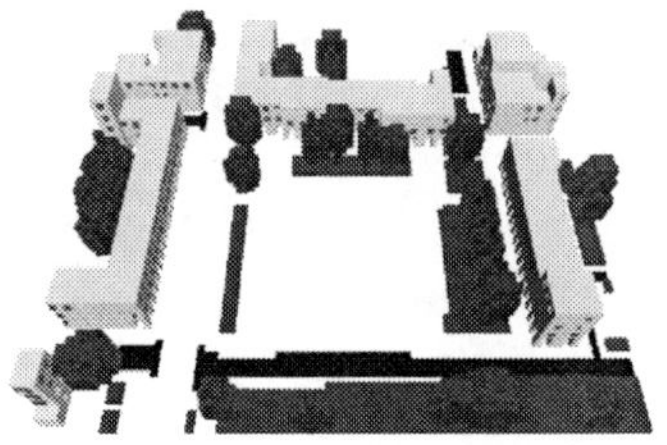

Figure 1. Base Case (BC).
Green cover – 16%

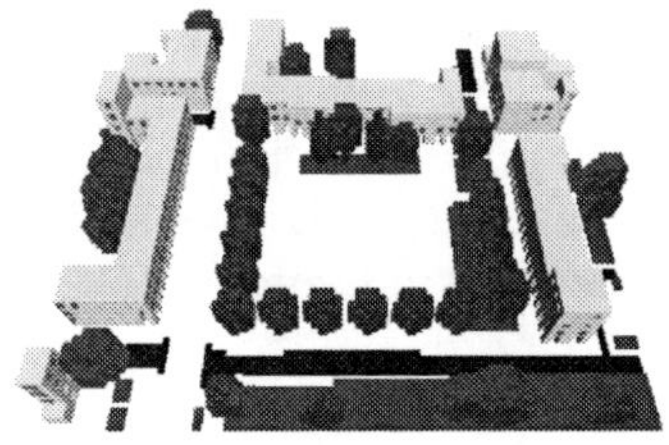

Figure 2. Additional Trees (AT).
Green cover – 22%

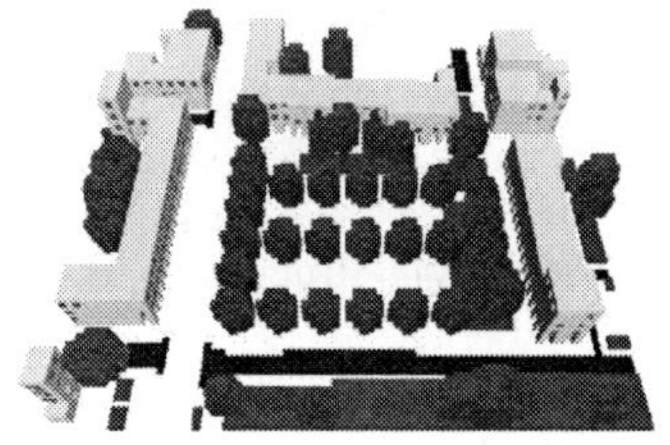

Figure 3. Additional Plantings (AP).
Green cover – 30%

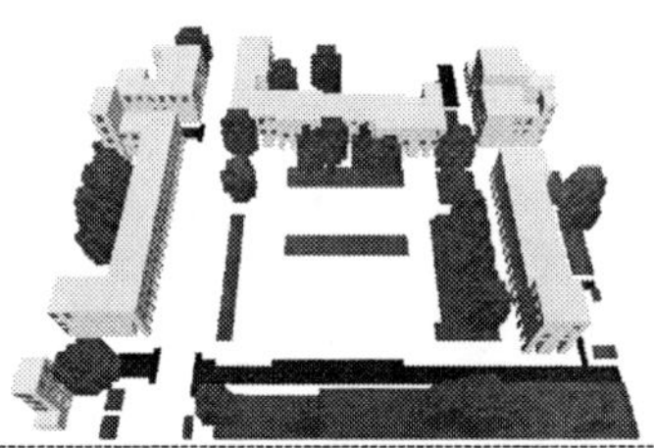

Figure 4. Additional Water (AW).
Green cover – 16%
Water surface – 3%

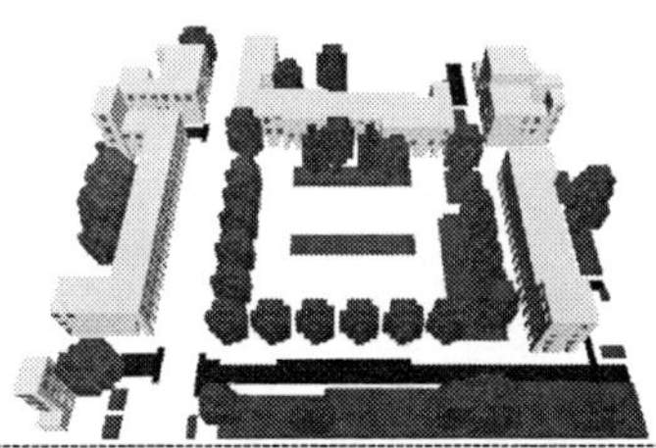

Figure 5. Water – Trees (WT).
Green cover – 22%
Water surface – 3%

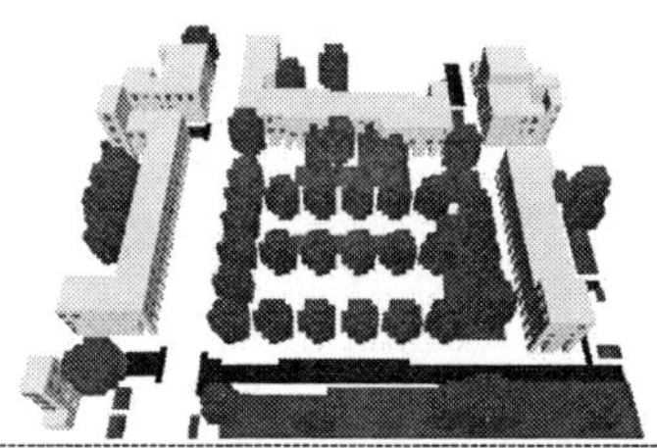

Figure 6. Water – Plantings (WP).
Green cover – 30%
Water surface – 3%

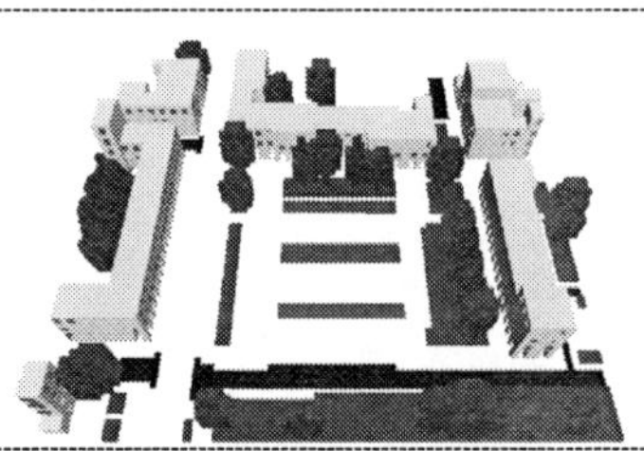

Figure 7. Water Surface (WS).
Green cover – 16%
Water surface – 8%

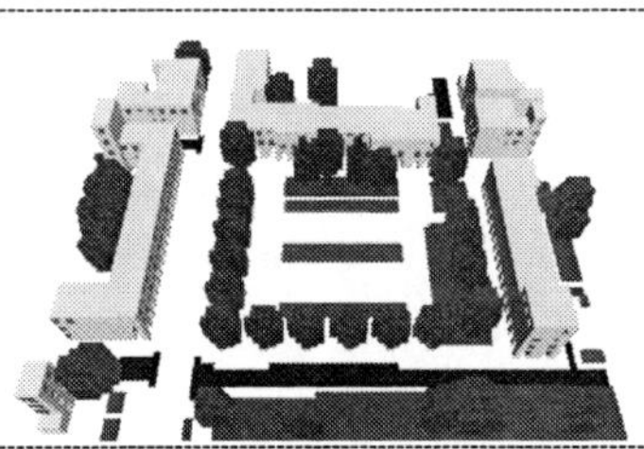

Figure 8. Water Surface – Trees (WST).
Green cover – 22%
Water surface – 8%

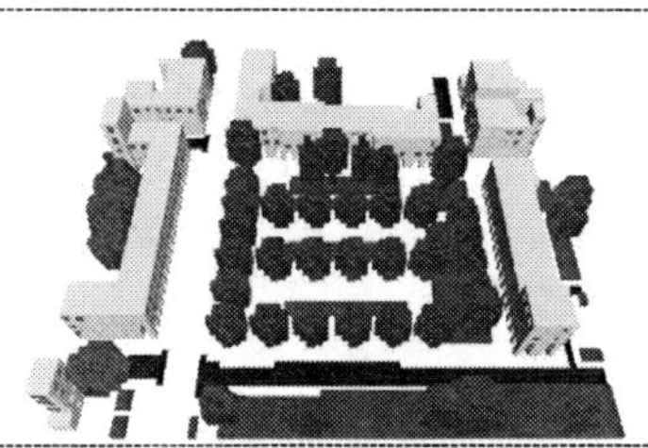

Figure 9. Water Surface – Plantings (WSP).
Green cover – 30%
Water surface – 8%

The impact of spatial development changes on thermal comfort was estimated by comparing the PMV index value calculated on the basis of nine scenarios related to introduction of greenery and water elements within the main square plate.

Three-dimensional model of public space – Base Case (BC) – shows the current state of land development. The size of area is 80 x 80 x 30 cells with a resolution of dx, dy = 2m and dz = 1m. The basic materials used in this model are: concrete (building structures – 20%), impermeable surface (concrete cube – 63.94%, asphalt – 0.06%), natural surface (land with grass – 16%). In addition, high greenery was included, which is the native species occurring in the area of study, i.e.: acer pseudoplatanus, ash (fraxinus excelsior), lime tree (tilia).

Simulations take into account two scenarios based on introduction of additional vegetation in the square. Additional Trees (AT) assumes addition of two tree frontages along the communication routes (south and west). The plantings were provided at a distance of 10 m. In this case, the rate of green cover of the square was increased from 16% to 22%. Additional Plantings (AP) include maintenance of plantings along communication routes and the addition of plant cover on the main square plate. Percentage of greenery on the land has increased to 30%. The study uses native tree species, i.e. acer pseudoplatanus, as new plantings.

Next variants are related to change of the type of existing surface. Additional Water (AW) assumes introduction of water elements in the central part of square. Its area is 8 m x 40 m – 3% of the land cover. As a result, the impermeable part of surface – concrete slab – is eliminated. Water Surface (WS) consists in adding three water elements with dimensions of 6 m x 40 m (reservoir in the northern and southern part of the square) and 8 m x 40 m (reservoir in the central part of the square).

Subsequent scenarios take into account the addition of both greenery and water elements. An overview of individual variants was presented in Figure 1-9.

8. Analysis of obtained results

Comparison of the average value of PMV index for a typical warmest day of the year, depending on the spatial development method, was presented in Figure 10-11.

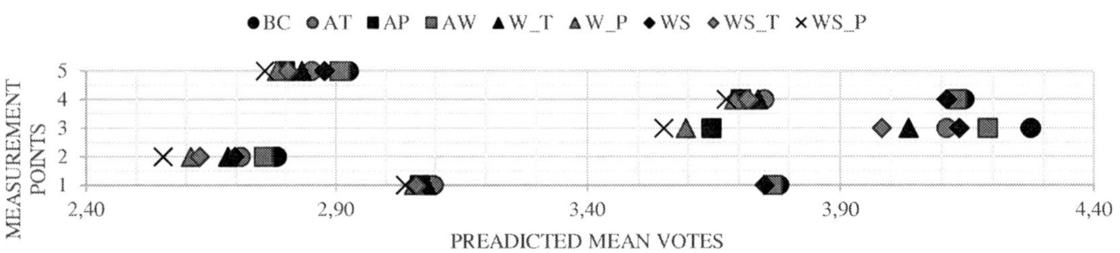

Figure 10. PMV values depending on the spatial changes scenario (standing person).

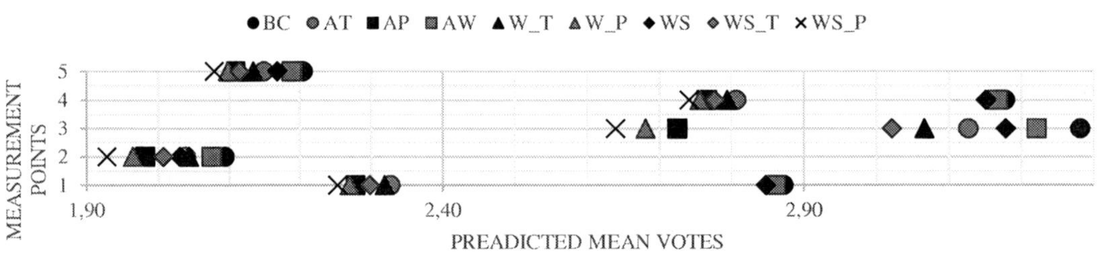

Figure 11. PMV values depending on the spatial changes scenario (walking person).

The analysis showed that regardless of the type of performed physical activity, the nature of changes in PMV index is analogous. Variants related to physical activity (slow walking) are characterized by lower PMV values. It results from the fact that PMV value is dependent on the relative air velocity. The research conducted by Sugiono and Hardiningtyas confirmed that the higher relative air velocity the lower PMV value [20]. In all the measurement points, the WSP scenario proved to be the best, consisting in introduction of additional plantings (30%) and water elements (8%) within the main square plate. The value of index has been reduced from 3.54 to 3.08 (sitting person) and from 2.72 to 2.33 (walking person). This means that thermal sensations of people staying in the public space for a long period of time have changed from “extreme heat stress” to “hot”. In case of people performing physical activity, they were reduced to “warm” level.

The greatest variation of PMV values was recorded within the third site. It was located in the middle of square’s main plate. Thus, prevailing there conditions were not affected by the land developed method. This enabled the assessment of extreme thermal conditions within analyzed space. In this case, the maximum change in PMV was 0.72 for a person not performing physical activity and 0.64 for a person with moderate physical activity (walk). Strategies to mitigate thermal discomfort enabled to reduce the average PMV up to 3.09, which meant that there were “hot” conditions in the third site. The lower daily variations in PMV values were observed in points 2 and 5. Measurement sites were located along the main facades of buildings (point 2 – northern and point 5 – western frontage). Therefore, the objects contributed to reduction of the amount of solar radiation reaching the earth surface. Thus, lower PMV values were observed in their vicinity. Difference between the extreme values of PMV for a person not performing physical activity was 0.23 (point 2), 0.17 (point 5) and for moving person 0.16 (point 2) and 0.13 (point 5). Thermal conditions within buildings have been reduced to “warm” conditions (PMV from 2.47 to 2.33).

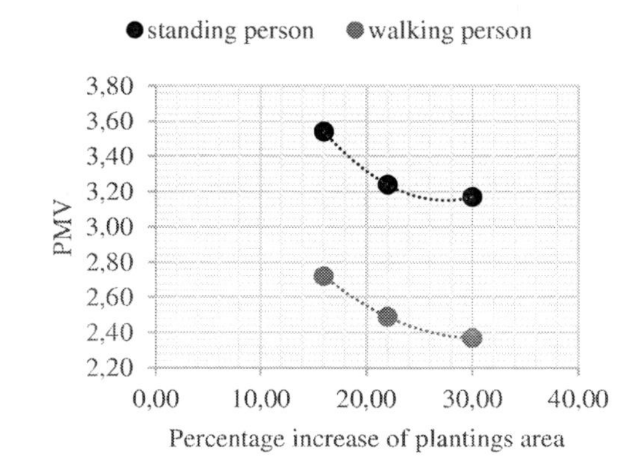

Figure 12. Dependence PMV values on plantings area.

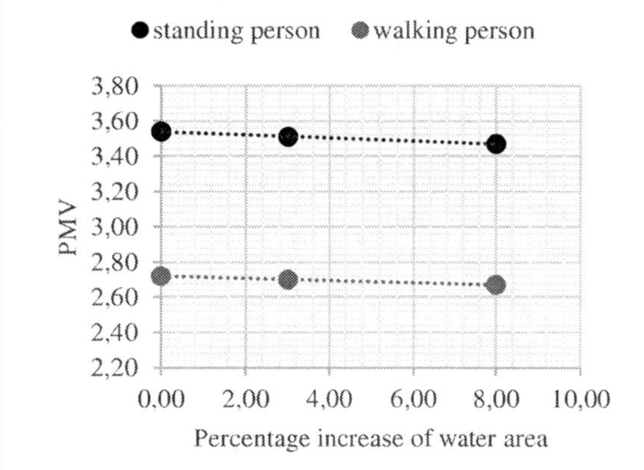

Figure 13. Dependence PMV values on water area.

Variants of changes related to introduction of greenery proved to be more effective in reducing the thermal discomfort experienced by man compared to scenarios assuming the change of surface type through introduction of water elements (Fig. 12-13). Studies show that an increase in the planting surface by 1% reduces thermal discomfort by an average of 0.038 (person not performing physical activity), 0.033 (moving person). Also, the removal of impermeable (concrete) surfaces for introduction of water elements contributes to reduction of PMV index. An increase of 1% in the natural surface (water) results in an average decrease of 0.009 (person not performing physical activity), 0.007 (walk).

9. Conclusions

The prevailing climate in highly urbanized areas depends on the land developed method. Persons responsible for spatial planning have the opportunity to transform the urban tissue through modification of the buildings, introduction and liquidation of construction works. However, transformation of historical areas of cities requires agreement with the conservators of monuments. Sometimes it is not possible to modify the structure of buildings, in particular those entered in the register of monuments. Then, the only solutions are transformations within the existing surface (their replacement), as well as creation of plantings in vertical and horizontal form. Changes in the way of spatial development translate into the modification of microclimatic parameters. The result is formation of thermal environment.

Nine scenarios of changes in the development of one of the oldest squares in the city – the Old Market Square – were considered in this work. The possibility of transforming urban structure has been limited to changing the surface types (introduction of water elements) and increasing the surface area of plantations. This was due to the fact that objects forming the frontage of square were entered in the municipal register of monuments.

Conducted research has shown that the more effective solutions were scenarios based on a growth of the green area rather than introduction of water elements within the analyzed public space. The authors showed that increase in the planting area by 1% contributed to a decrease in PMV index by an average of 0.036. The difference in case of water elements was 0.008. Scenarios consisting in introduction of both greenery and water elements resulted in the resultant of above mentioned variants.

Change of the land development method can have a positive impact on the thermal sensations of people in the external environment. Introduction of new plantings and water elements into the strongly urbanized environment modifies microclimatic parameters. In this way, it has an impact on thermal comfort. However, these strategies do not significantly reduce thermal discomfort. They can only be a part of mitigation strategies. Adaptation of urban spaces to changing climatic conditions should primarily take into account changes in the geometry of construction works and in their arrangement in relation to each other. In areas where the historical urban tissue is present, such activities are significantly hampered. An increase in the planting surface and change in the type of surface can only have so-called cooling effect.

10. References

[1] Middel A, Chhetri N and Quay R 2015 *Urban Forestry and Urban Greening* **14** 178-186
[2] United States Environmental Protection Agency 2008 *Reducing Urban Heat Islands: Compendium of Strategies* **1-6**
[3] Chatzidimitriou A, Liveris P, Bruse M and Topli L 2013 *Proc. 29th Int. Con. Of the Passive and Low Energy Architecture. Sustainable Architecture for a Renewable Future (Munich)*
[4] Barakat A, Ayad H and El-Sayed Z 2017 *Alexandria Engineering Journal* **56** 533-543
[5] Chan S Y, Chau T M and Leung T M 2017 *Building and Environment* **122** 171-183
[6] American Society of Heating, Refrigerating and Air-Conditioning Engineers (ASHRAE) 2010 *Thermal environment conditions for human occupancy (America)* p 62
[7] Thorsson S, Lindqvist M and Lindqvist S 2004 *International Journal of Biometeorology* **48** 149-156
[8] Bohojło A 2010 *Budownictwo i Inżynieria Środowiska* **1** 11-16
[9] Elnabawi M H, Hamza N and Dudek S 2015 *HBRC Journal* **11** 246-251
[10] Huttner S, Bruse M and Dostal P 2008 Using ENVI-met to simulate the impact of global warming on the microclimate in central European cities *5th Japanese-German Meeting on Urban Climatology (Freburg, Germany)* ed H Mayer and A Matzarakis p 307-312
[11] Srivanit M and Hokao K 2013 *Building and Environment* **66** 158-172
[12] Salata F, Golasi I and De Lieto Vollaro R 2016 *Sustainable City and Society* **26** 318-343
[13] Huang Q, Meng X, Yang X, Jin L, Liu X and Hu W 2016 *Energy Procedia* **104** 177-182
[14] Gaspari J and Fabbri K 2017 *Energy Procedia* **111** 500-509
[15] Crawley D B 1998 *ASHRAE Transactions* **104** 498-515
[16] Żurański J A 2005 Wpływ warunków klimatycznych i terenowych na obciążenie wiatrem konstrukcji budowlanych. *Prace Naukowe Instytutu Techniki Budowlanej* (Warszawa) p 128
[17] Flaga A 2008 Inżynieria wiatrowa. Podstawy i zastosowania. *Arkady* p 720
[18] Simiu E 1975 *Proc. of the 4-th International Conference on Wind Effects on Buildings and Structures* (Heathrow) p 721-735
[19] Stopa-Boryczka M and Boryczka J 2016 Atlas współzależności parametrów meteorologicznych i geograficznych w Polsce *Uniwersytet Warszawski* p 416
[20] Sugiono S and Hardiningtyas D 2014 International Science Index **8** 612-618

IOP Conf. Series: Materials Science and Engineering **415** (2018) 012023 doi:10.1088/1757-899X/415/1/012023

Analysis of energy efficiency and thermal comfort for an office building complex located in Poland – a case study

M Zygmunt [1], D Gawin [1]

[1] Lodz University of Technology, Faculty of Civil Engineering, Architecture and Environmental Engineering, Lodz, Poland.

* Corresponding author e-mail: marcin.zygmunt@p.lodz.pl

Abstract. This article presents an analysis of energy demand for an office buildings complex located in Gdansk, Poland. The analyses were performed with advanced computer simulations using the Energy Plus software. Simulations were validated by comparing their results with the actual energy consumption data for one of the buildings belonging to the complex and information gathered from complex owner and technical office. The paper includes the effectiveness analysis of using various sources and distribution systems for heating/cooling, use of the phase–change material as internal ceiling layer, reflective glass application and use solar energy produced by photovoltaic glass and PV panels located on the roofs. For each of the variants, thermal comfort analysis was carried out. The aim of the analysis is to examine the time variation and peak load reductions of power demand for the considered variants, indicated as possible to be introduced by the complex owner.

1. Introduction

In Poland the electricity demand is growing each year [1,2]. The growing electricity consumption is also associated with the increasing power peak demand. This is mainly due to the economic development of Poland, as well as due to increasingly higher requirements of building users regarding quality of the internal environment.

One of the most important aspects of the proper and effective functioning of individual buildings as well as large enterprises is proper energy management. Important tasks of such an energy management are rational energy consumption, reduction of power peak demand and choosing appropriate modernization tasks. Nowadays, due to global trends towards sustainable development and increasing requirements of regulations concerning installations of a building, it is necessary to use renewable energy sources, among which solar energy is the most popular in urban areas.

Moreover, an architect should create favorable indoor conditions for people inside every building, among others by providing the thermal comfort conditions [3]. Diversified types of heating, ventilation and air-conditioning (HVAC) installations, as well as development of smart building management systems (BMS) allow dynamic control of a building internal environment.

This article presents an analysis of power peak demand and energy consumption by the existing office building complex located in Gdansk, Poland. The analysis is based on some available real energy consumption data and technical documentation for one of the buildings belonging to the analyzed complex. The research is based on the results of computer simulations for the whole office buildings complex, performed by means the Energy Plus software [4]. A number of simulations is made in order to predict effects of application of phase–change materials (PCM), reflective glass, photovoltaic glass, PV panels as well as different source and distribution systems of chill in air

IOP Conf. Series: Materials Science and Engineering **415** (2018) 012023 doi:10.1088/1757-899X/415/1/012023

conditioning system. Analysis focused on thermal comfort, cooling power demand and heating energy consumption.

2. Description of the analysis

In Poland, the current methodology for determining the energy performance of a building [5] requires monthly balance method, based on average values of parameters related to external climate, heat gains and losses, as well as the installations' efficiency.

The analysis was carried out using advanced energy simulations, performed with a minimum one hour calculation step. The computer model of a real office buildings' complex located in Gdansk, Poland, defined by means of the Energy Plus software, is analyzed. Such type of simulations fully reflects the dynamics of changes in external climate conditions and the response of building installations to the environmental thermal loads. All simulations were performed for a full year, using weather data of Typical Meteorological Year (TMY) for Gdansk, Poland [6].

2.1. Subject of the analysis

The subject of the analysis is the existing complex of office buildings located in Gdansk (figure 1). The complex consists of 4 buildings, one of which is under construction (planned commission date – end of 2019). The buildings are characterized by a large office open space areas and significant area of external glass façades. The total office space of the analyzed complex is 98 314,71 m^2.

Figure 1. Visualization of the completed office buildings complex in Gdansk

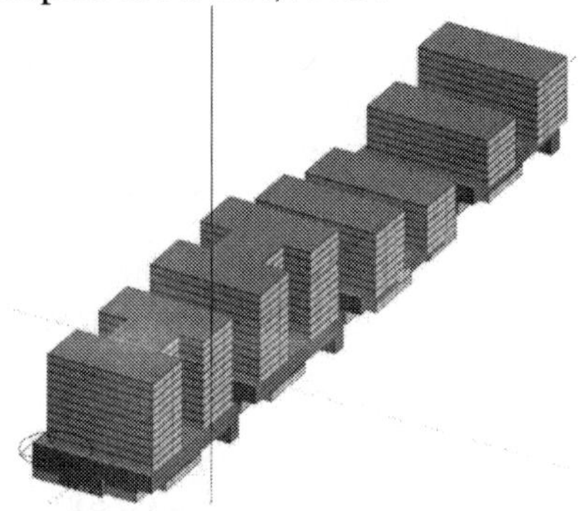

Figure 2. A computer model of the analyzed office buildings complex

The study is based on the real parameters (including building project) and data for one of the analyzed complex buildings. The data included time history of electricity consumption for different purposes, total energy consumption from district heating system, work schedules of individual devices and working hours of the facility. Moreover, air flow of the ventilation system was set according to parameters gathered from technical office of analyzed complex.

Table 1. Comparison of the results of computer simulation of energy consumption, made with Energy Plus software with real values.

	Actual use [kWh]	Simulation [kWh]	Absolute difference [kWh]	Relative difference [%]
Cooling	168 568.36	167 354.67	- 1 213.69	- 0.72
Int. Lighting	350 830.63	360 022.39	9 191.76	2.62
Int. Equipment	564 306.17	560 694.61	- 3 611.56	- 0.64
Fans	813 125.94	748 726.37	- 64 399.57	- 7.92
Pumps	57 102.08	57 907.22	805.14	1.41
Humidification	145 107.47	132 968.01	- 12 139.46	- 8.37
Sum (electricity)	2 099 040.65	2 027 673.27	- 71 367.38	- 3.40
Heating	1 211 413.24	1 154 476.82	- 56 936.42	- 4.70

Firstly, a calibration of the computer model (figure 2) was performed, comparing simulation results to the real energy consumption data. Electricity consumption was calibrated on the basis of hourly readings, while heating energy consumption was compared to monthly energy bills. Table 1 presents comparison of the annual electricity and heat consumption obtained from simulations (variant 0) with the actual values. Worth mentioning fact is that the computer model of the analyzed building was accepted by the complex owner. It was assumed that other buildings of the complex are characterized by the same parameters of the energy model (thermal insulation, installations, working hours schedule etc.).

The research does not cover the first floors of the analyzed buildings, which are used for different purposes (space types): car parking area, restaurants, gyms and swimming pool. This is due to the lack of data concerning the actual energy consumption for these parts of analyzed buildings.

2.2. Description of simulation variants

The following variants were analyzed in the study:

- variant 1: application of phase–change material (PCM) as internal layer of ceilings

The use of PCM as internal ceilings layer has been analyzed. It was assumed, that PCM was encapsulated in plasterboards. The material phase change temperature range was between 22 and 24 °C.

- variant 2: use of the reflective glass [7] on the external façade

It is assumed use of reflective glass on the exterior façade of the buildings. Reflective glass is coated with a metal oxide in order to obtain solar control and reduce solar energy penetration. The glass assumed in the simulations [8] is characterized by a lower value of light transmittance (L_T) and lower solar heat gain coefficient (SGHC) value than real elevation glass of the analyzed complex.

- variant 3: application of photovoltaic glass on the front elevation

It is assumed use of photovoltaic glass [9], for electricity production (figure 3). It was assumed that the photovoltaic modules constitute 15 % of front façade, (oriented to the southwest) glazing area for the all buildings of complex. The nearest area of the analyzed complex as well as its individual buildings do not generate shadings on the front façade. The total area of photovoltaic modules is 917,42 m^2. Analysis takes into account the lower natural light transmittance (L_T) and lower solar heat gain coefficient (SGHC) value..

Figure 3. An example of the existing photovoltaic glass use.

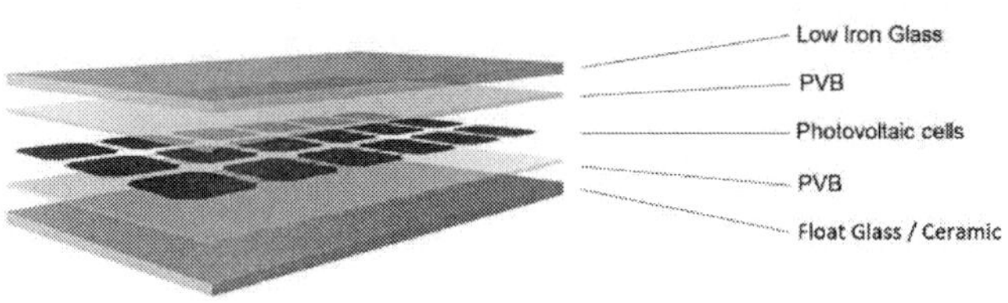

Figure 4. Typical photovoltaic glass construction.

- variant 4: use of photovoltaic panels on the roof of the analyzed complex

It is assumed use of photovoltaic panels installation on the available roof area of the complex. The available roof area for solar installation is 4 600.68 m^2, which is 39.67% of total roof area of the complex. PV panels are oriented directly to the south, inclined at 30 degrees to maximize electricity production during summer. The arrangement of the solar installation assumed shading minimization generated by individual panels. Total area of used PV panels [10] is 1 155.29 m^2.

- variant 5: change of chill distribution system from fan coils to cooling beams [11]

This case was analyzed in order to compare those two most commonly used systems. It was assumed that priority of the system is cooling, but room heating is also possible (4 pipe installation system). In the assumed installation scheme, the heating elements were water convectors, and not hot air supply as in variant 0.

- variant 6: change of chill source from air–cooled chillers to water cooled

This variant assumed chill source change from air–cooled aggregates to dry liquid coolers, which were connected to the water–glycol mixture installation. The variant implied significant changes in the hydraulic system of the HVAC installation.

3. Simulation results

In subsection 3.1 the energy demand and cooling power for the analyzed variants of buildings complex were analyzed, while in subsection 3.1 thermal comfort was considered.

The results concerning time variation of power demand and thermal comfort are presented for extreme summer and winter week periods. In each figure (from figure 5 to figure 8) working hours were shown by grey background. The extreme weeks were selected from the TMY weather data for Gdansk, Poland. The Extreme Summer Week (ESW) was chosen on the basis of the highest average temperature of external air. Similarly, an Extreme Winter Week (EWW) was selected considering the lowest average temperature of external air. ESW lasts from 12th to 18th July, while EWW – from 10th to 16th December. Calculations of the electricity yield by means solar installation were performed in accordance with [12].

3.1. Energy demand results

The energy demand for different purposes obtained from the energy simulations, is summarized in Table 2. Each of the considered variants was compared with the base one (variant 0).

Table 2. Comparison of energy consumption results obtained for the analyzed variants.

	Variant 0 [kWh]	Variant 1 [kWh]	Variant 2 [kWh]	Variant 3 [kWh]	Variant 4 [kWh]	Variant 5 [kWh]	Variant 6 [kWh]
Cooling	927 592.87	853 453.28	785 238.58	823 030.74[a]	704 088.68[b]	788 232.74	719 594.09
Int. Lighting	2 046 267.06	2 046 267.06	2 107 119.83	2 049 193.15	2 046 267.06	2 046 267.06	2 046 267.06
Int. Equip.	2 990 732.74	2 990 732.74	2 990 732.74	3 014 576.02	3 070 210.35	2 990 732.74	2 990 732.74
Fans	3 908 200.41	3 884 049.47	3 660 744.94	3 894 364.83	3 519 681.49	2 712 113.35	3 908 200.41
Pumps	298 816.86	296 500.16	273 448.22	297 309.11	258 986.68	327 281.80	583 737.53
Humidif.	715 784.42	554 465.82	682 702.71	712 332.94	663 844.29	636 090.83	715 784.42
Sum (electricity)	10 887 394.36	10 625 468.53	10 499 987.02	10 790 806.79	10 263 078.55	9 500 718.52	10 964 316.25
Heating	6 160 098.76	6 139 094.12	6 626 522.81	6 179 328.85	6 160 098.76	6 635 229.02	6 160 098.76

[a] The value of cooling energy demand reduced by energy yield from photovoltaic glass.
[b] The value of cooling energy demand reduced by energy yield from photovoltaic panels.

The most advantageous, in terms of electricity consumption, was variant 5, i.e. cooling beams as a chill distribution source. This variant results in reduction of cooling energy demand by 15.02% and lower electric energy consumption by fans by 30.60%. Total annual electricity consumption has been reduced by 12.74%.

Variant 6, i.e. change of chill source from air–cooled chillers to water cooled ones, has lower cooling energy demand than variant 5. Modification in the installation HVAC schema led to a change in energy consumption structure (see Table 2). Total annual electricity consumption is higher than in the base variant 0.

The lowest value of cooling energy demand gave variant 4, i.e. use of photovoltaic panels on the roof of the analyzed complex. This variant assumes use of energy produced by photovoltaic panels for cooling the buildings. The PV panels produced 223 504.19 kWh of electricity per year. Variant 4 reduced cooling energy demand by 24.10%.

Variant 3, i.e. the photovoltaic glass applied on the front elevation, produced electricity more evenly throughout the year than variant 4. The total amount of the produced electricity was equal to 95 534.28 kWh per year. This variant only slightly affected the interior lighting conditions in the rooms of the analyzed buildings.

Most of the analyzed variants had similar energy demand for heating. The most advantageous is variant 1, i.e. application of PCM as internal layer of ceilings. However, the achieved reduction is very small (by 0.34%) that this variant cannot be considered as profitable. The highest heating demand was obtained for variant 2, i.e. use of reflective glass. The increase of heating demand by 7.57 % was caused by a reduction of solar gains in the heating period.

The internal equipment energy consumption for the analyzed variants, except of variants 3 and 4, were equal. These variants' results take into account the additional electricity consumption by auxiliary drivers of the solar installation. In addition, for variants 2 and 3 the energy consumption for internal lighting was increased due to lower light transmission through windows.

The hourly cooling power demand history for ESW period is shown in figure 5. In addition, the power production output by means of photovoltaic glass (sol_3) and PV panels (sol_4) has been shown.

The analysis showed that the all considered variants applied in order to reduce cooling peak demand were effective, but to different extent.

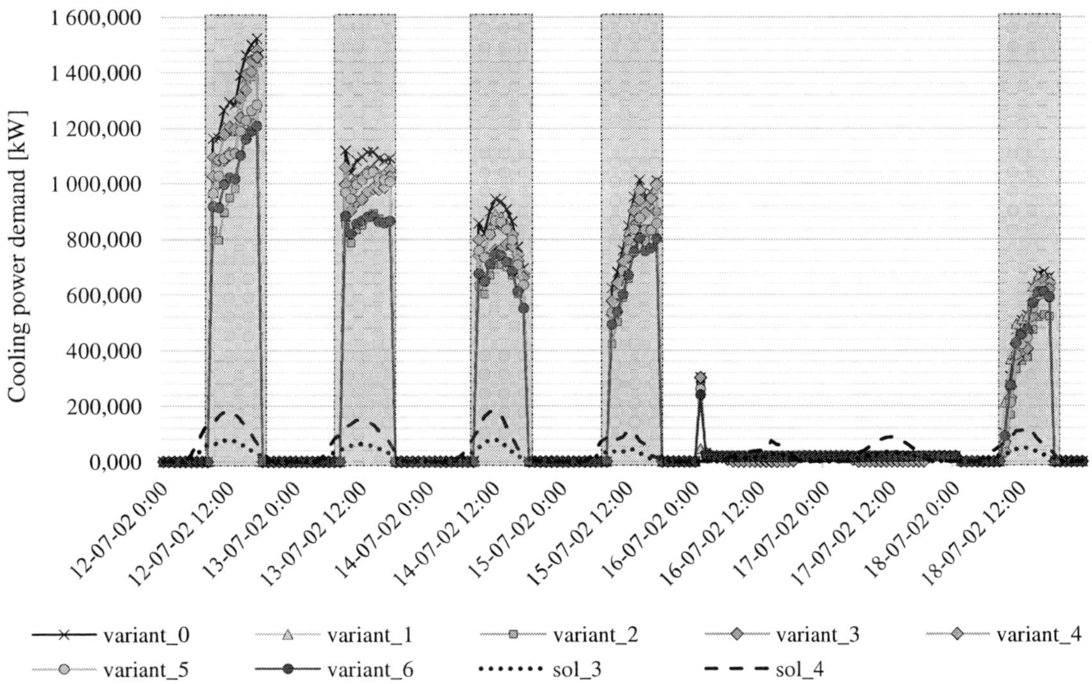

Figure 5. Cooling power demand distribution of the analyzed variants for the ESW period.

Table 3 presents the obtained cooling power demand and resulting power reduction for the analyzed variants during the highest cooling demand period (i.e. between 4 p.m. and 5 p.m. on 12th July). The biggest reduction, by 313.91 kW (20.62%), was obtained for variant 6.

Table 3. Comparison of peak power cooling demand of the analyzed variants.

	Variant 0	Variant 1	Variant 2	Variant 3	Variant 4	Variant 5	Variant 6

Cooling power demand [kW]	1 522.64	1 440.77	1 272.43	1 480.81	1 456.04	1 284.56	1 208.72
Absolute difference [kW]	-	- 81.86	- 250.21	- 41.83	- 66.60	- 238.08	- 313.91
Relative difference [%]	-	- 5.38	- 16.43	- 2.75	- 4.37	- 15.64	- 20.62

3.2. Thermal comfort analysis

For each of the considered variants thermal comfort conditions were analyzed, based on Predicted Mean Vote values (PMV index). There are four internal environment categories [13]. For the analyzed buildings complex the 2nd category of internal environment was assumed (standard expectations, recommended for new and modernized buildings). The value of PMV index should be in the range of:

$$-0.5 < PMV < +0.5$$

Within a few hours of building operation time it is acceptable to lower the category of internal environment to the 3rd category.

The time histories of the PMV index are shown for the ESW period in figure 6 and for the EWW period in figure 7. The presented results concern the corner room, for which, due to glazing façades of both outer walls, it is the most difficult to maintain thermal comfort. The complex does not have working weekends, hence the analysis of PMV index distribution does not concern 16th and 17th July (ESW period) and 10th and 11th December (EWW period). Simulations by means the Energy Plus software performed thermal comfort calculations accordance with [14].

In figure 6 it can be seen that during the working hours the PMV index values oscillate within the recommended limits of thermal comfort. The best PMV index values were obtained for variant 2, i.e. use of reflective glass. It may be observed that the assumed internal air temperature for the summer period (24 °C) is correctly selected. Solar radiation, which is reduced in variant 2, has a large influence on the room thermal comfort..

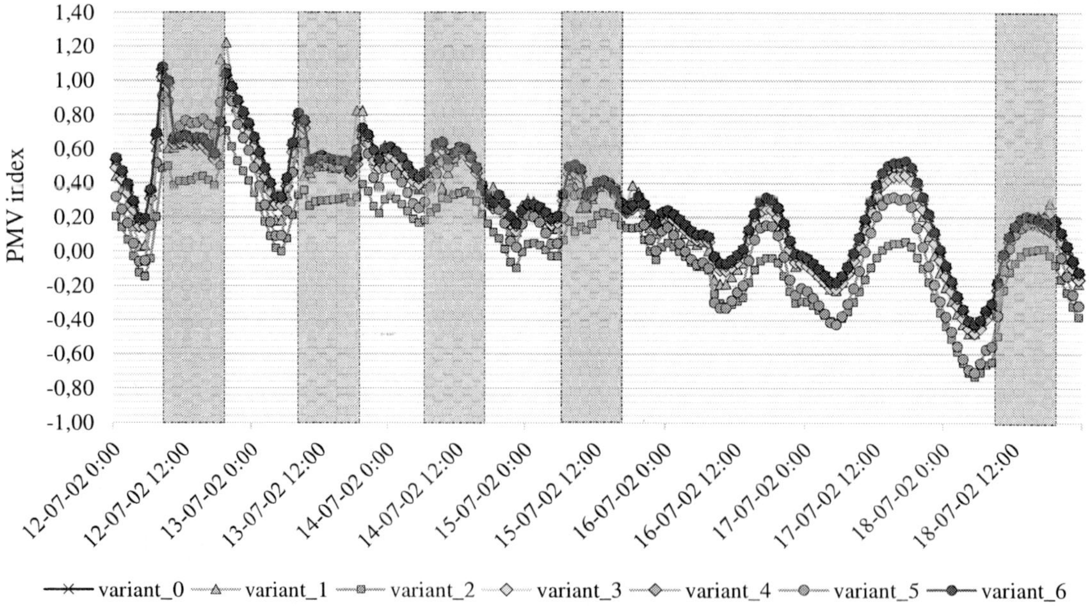

Figure 6. The time histories of the PMV index for ESW period.

Figure 7 shows that the all considered variants have similar PMV index values for the EWW period. During buildings working hours, the obtained PMV index values are below recommended values of thermal comfort. The analysis included the following parameters related to human physiology: typical office work (M= 70 W/m^2) and common clothing set during winter period (I_{cl} = 1.0 clo). This led to the conclusion that indoor air temperature in the winter period (assumed as 22 °C, which was given by the building administrator) is too low.

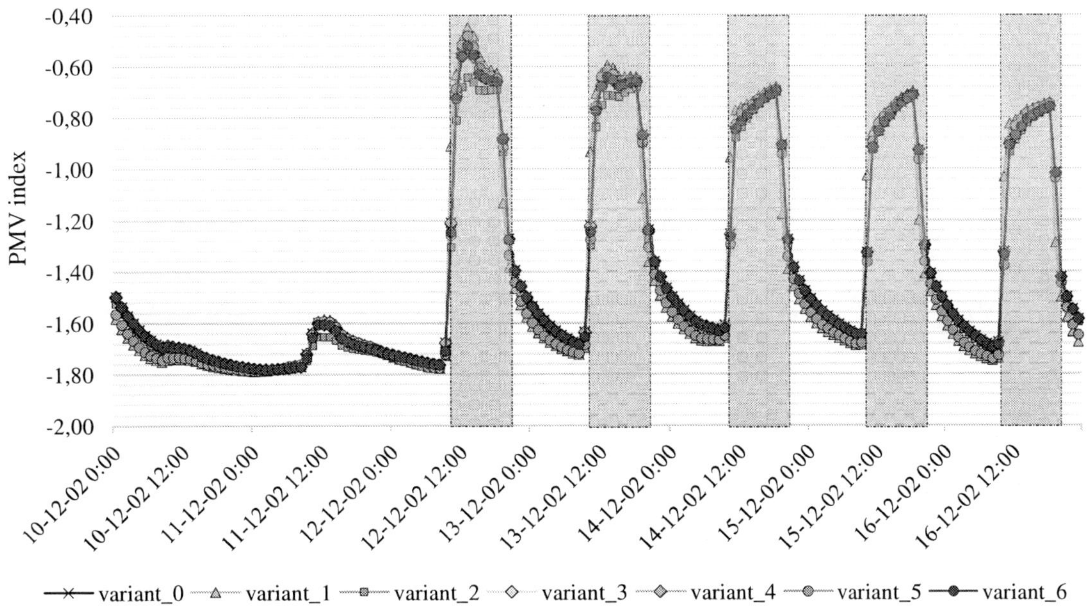

Figure 7. The time histories of the PMV index for EWW period.

Hence additional simulations were done for the increased air temperature (24 °C) to analyze thermal comfort during the EWW period. In figure 8 it can be seen that, for each variants, thermal comfort has improved comparing to the previous analysis (see figure 7). It can be concluded that higher internal air temperature is required in order to meet thermal comfort requirements for buildings with glass exterior façades than for ones with masonry walls. Mean radiant temperature (which, in this case, is much lower than air temperature) had significant influence in the PMV index results. The total heating demand, for the EWW period, increased from 4.77 to 6.99 % (depending on the variant).

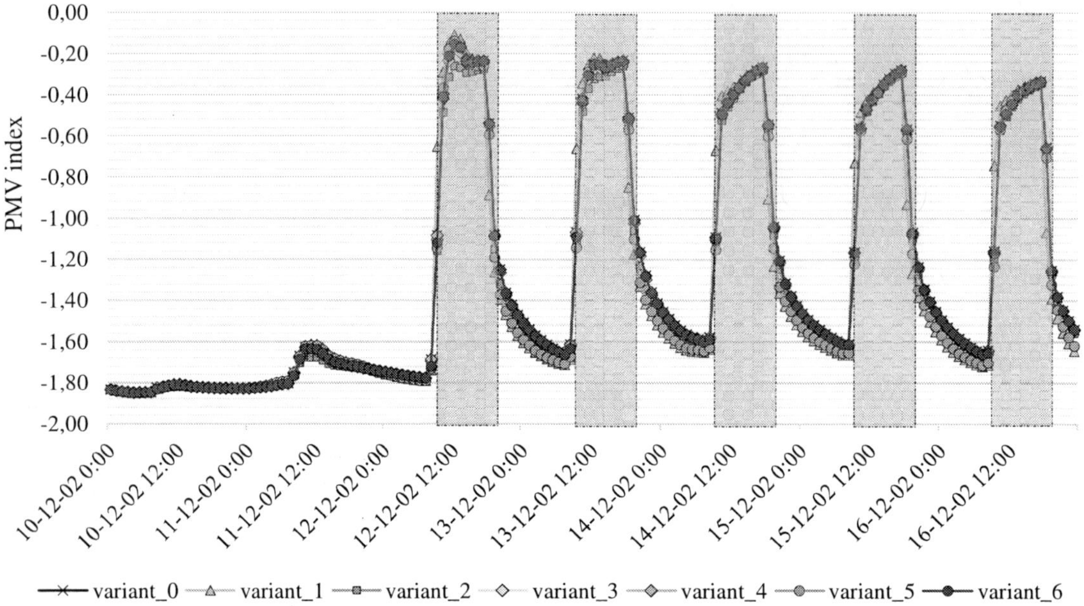

Figure 8. The time histories of the PMV index for EWW period for higher internal air temperature adjustment.

4. Conclusions

The results of the analysis performed lead to the following conclusions:

- The use of advanced computer simulations for large public buildings has been demonstrated in order to select the most appropriate construction and HVAC installation solutions.
- The computer simulation is an effective tool for selecting the proper solution of HVAC system at the designing stage of building project.
- In order to check the profitability of modernization variant for an existing building, the analysis should be extended by appropriate economic analyses.
- The analysis showed a potential for using solar energy in urban areas, using both photovoltaic glass façade and PV panels on the building roof.
- All considered modernization variants resulted in cooling power demand reduction during summer period. Reduction of power peaks is one of the most important issues for proper building energy management.
- The need of limiting solar radiation causing uncontrolled inner heat gains during summer has been demonstrated. Reducing the adverse effects of solar radiation leads to improvement of thermal comfort and reduction of energy consumption during summer.
- Using cooling beams instead of fan coils in chill distribution system (variant 5) was shown to be the most advantageous in the study performed. This variant results in reduction of cooling energy demand by 15.02%, while total electricity consumption was reduced by 12.74%. However, it should be noted that cooling beams may perform well only up to a certain limit value of inner heat gains.
- Any proposed modernization variant should be assessed in terms of meeting the required internal environment conditions, because providing thermal comfort is a priority for spaces intended for permanent stay/work of people.

5. References

IOP Conf. Series: Materials Science and Engineering **415** (2018) 012023 doi:10.1088/1757-899X/415/1/012023

[1] Polskie Sieci Elektroenergetyczne, *Plan rozwoju w zakresie zaspokojenia obecnego i przyszłego zapotrzebowania na moc elektryczną na lata 2016–2025*, Konstancin – Jeziorna, 2014.
[2] Polskie Sieci Elektroenergetyczne – www.pse.pl {access: 10.03.2018r.}.
[3] Fanger P O, *Komfort cieplny*, Arkady, 1974
[4] EnergyPlus Engineering Reference, Illinois 2014
[5] Rozporządzenie Ministra Infrastruktury i Rozwoju z dnia 27 lutego 2015 r., w sprawie metodologii wyznaczania charakterystyki energetycznej budynku lub części budynku oraz świadectw charakterystyki energetycznej.
[6] Energy Plus – www.energyplus.net {access: 01.03.2018r.}.
[7] SunGuard Glass – www.sunguardglass.com {access: 18.03.2018r.}.
[8] Press Glass – www.pressglass.com {access: 10.03.2018r.}.
[9] Energy Glass – www.energyglass.eu {access: 18.03.2018r.}.
[10] EnergySage – www.energysage.com {access: 18.03.2018r.}.
[11] Swegon – www.swegon.com {access: 22.03.2018r.}.
[12] Chwieduk D, *Solar Energy in Buildings*, Elsevier Science, 2014
[13] prEN 15251:2006 – *Indoor environmental input parameters for design and assessment of energy performance of buildings – addressing indoor air quality, thermal environment, lighting and acoustics.*
[14] ISO 7730:2005 – *Ergonomics of the thermal environment – Analytical determination and interpretation of thermal comfort using calculation of the PMV and PPD indices and local thermal comfort criteria.*

IOP Conf. Series: Materials Science and Engineering **415** (2018) 012024 doi:10.1088/1757-899X/415/1/012024

Thermal comfort of the sport facilities on the example of indoor tennis court

S Bugaj, P Kosiński

Institute of Building Engineering, University of Warmia and Mazury in Poland, ul. Jana Heweliusza 4, 10-724 Olsztyn, Poland

piotr.kosinski@uwm.edu.pl

Abstract. An opinion about the difficulties with maintaining appropriate climate conditions in the sports facilities is quite popular amongst sportsman of various disciplines. The problem can result in health and economic problems. First of all, the users of facilities are not provided with appropriate conditions for sport practicing and thus it may lead to contusions. Secondly the cost of using the object rises. Taking into account the increasing number of sports facilities, maintaining of indoor thermal comfort of them, while ensuring low maintenance costs, gains in importance. The paper presents the research conducted in a pneumatic tennis hall, a detailed analysis of maintaining thermal comfort in the hall and cognitive and utilitarian conclusions.

1. Introduction

The rules governing the construction of sports facilities do not differ from the implementation of other construction projects. However, the architecture of sport facilities is governed by its own rules. A good quality sport facility should meet the construction, functional and economical requirements. Functional features are particularly important for athletes, because apart from the right selection of materials, surface or shape of the facility, the microclimate of the interior affects the achievements of the players and safety of the sport (injuries). Regardless of the sports discipline, an important feature of any indoor sports facility is to provide athletes with comfortable conditions for practicing their disciplines. The independent factors, like: air temperature, humidity and flows play an important role here.

In the top ten most popular sports of the Biggest Global Sports ranking [1] there are four disciplines that are played in the indoor facilities or whose specificity requires training in the indoor facilities. Tennis is at the 3rd place of this ranking.

Thermal comfort of indoor sports facilities is a rarely discussed issue. World or national federations and associations of sports disciplines recommend to maintain specific thermal parameters in sports facilities. It should be noticed, that these requirements are primarily dedicated to new buildings. There are some works that discuss the fact that designers sometimes forget that the requirements appropriate for athletes do not need to be the same for the audience or service staff. This applies for example to swimming pools, which are a great challenge for designers [2, 3]. There are also works that describe difficulties with maintaining the thermal comfort of new sports facilities, especially in the aspect of overheating [4]. Overheating is one of the more frequent problems that affects sports halls. It can be a bit decreased by using efficient ventilation causing air movement and heat removal, but at the same time, the air speed cannot be high to lower the risk of draft or irritation due to air movement [5, 6].

2. A case study of thermal comfort in the indoor tennis court

2.1. Description of the facility

The pneumatic hall covering two full-size clay courts is located in the center of Olsztyn. External dimensions 36 x 36 m, height 9 m. The cover of the hall was made of multilayer PVC, supported on two columns, reinforced with cables. Entrance to the hall through revolving doors. The hall is supplied with air by the pump which maintain the overpressure inside. The same pump is also used for heating, but the heater furnace stops when the indoor air temperature exceeds 15 °C. The hall is operated in autumn and winter according to the planned training cycle, in the spring and summer season the cover is removed.

2.2. Measurements of thermal state of the facility

2.2.1. Method

The measurements were carried out in the autumn of 2015, in October from the 23th (at 8.17 pm) to the 25th (at 1.00 pm). At that time, the hall was used on the 1st day until 8.30 pm, the 2nd between 9.00 am and 9.00 pm and the 3rd day from 9.00 am.

There was used the set of the Ahlborn thermal comfort sensors connected with a data recorder. The used sensors: Globe thermometer FPA805GTS (accuracy ±0.15°C), Thermo-hygrometer FHA646E1 (accuracy 0.1°C, ±2%RH), Thermometer FPA32L0250 (accuracy 0.1°C) Thermo-anemometer FVA605TA10U (accuracy ±1%), Pressure difference meter FDA602S1K (accuracy ±0.5%). The sensors were calibrated before use. A tripod with sensors was placed between two courts in the central part of the hall. The outdoor air temperature sensor and the pressure tube were moved through the inspection hole in the hall cover. Outdoor air relative humidity was also recorded, but the sensor moved from the original place and thus it measured the parameters of exhaust air from the hall. Data were recorded every 3 minutes.

2.2.2. Results

Figure 1a presents the registered outdoor and indoor temperature and air speed. In the analyzed period, the outdoor temperature fluctuated in the range 6.0-10.6°C. During this time, the indoor air temperature fluctuated in the range 6.77-17.44°C. The average temperature for the period of use of the hall was 14.08°C. During the night, when the heating system is switched off, the indoor temperature drops significantly and approaches the outside temperature (1-2°C difference). The maximal measured temperature difference during the day was 9.46°C. Indoor temperature fluctuations are clearly visible during the heating of the hall. It could be related to the thermostat regulation of the heating system and also the use of the hall – entering and leaving the hall. The increases of indoor temperature correspond to the decreases of the relative humidity of the air and the difference of the pressure between the hall and the surroundings.

Figure 1b presents the registered indoor relative humidity of air and the pressure difference. The relative indoor humidity was at the level 90-95% in the night and 50-83% during the day. The overpressure in the hall was maintained at the level 150-175 Pa in the night and 75-175 Pa in the period of use. Fluctuations in the pressure difference can also be related to the heating system as well as entering and leaving the hall.

The measured air speed was in the range 0,03 – 0,3 m/s. On the second day of measurements at 15.00 the sensor has turned off and since then air speed in the hall was no further registered. Revel and Arnesano conducted the sensitivity analysis of the air speed and confirmed that it is a low-sensitive parameter and it could be supposed to be constant for values <0.2 m/s [7].

The variability of the air speed could be caused by air movement caused by the air supply system and the movement of the players. Despite the revolving doors, frequent entering and leaving the hall were the reason for drafts and thus for pressure peaks. The climatic data registered in the hall were within the limit values in ISO 7730:2006 [8], and thus enable the use of PMV and PPD indicators to determine the thermal comfort of the hall.

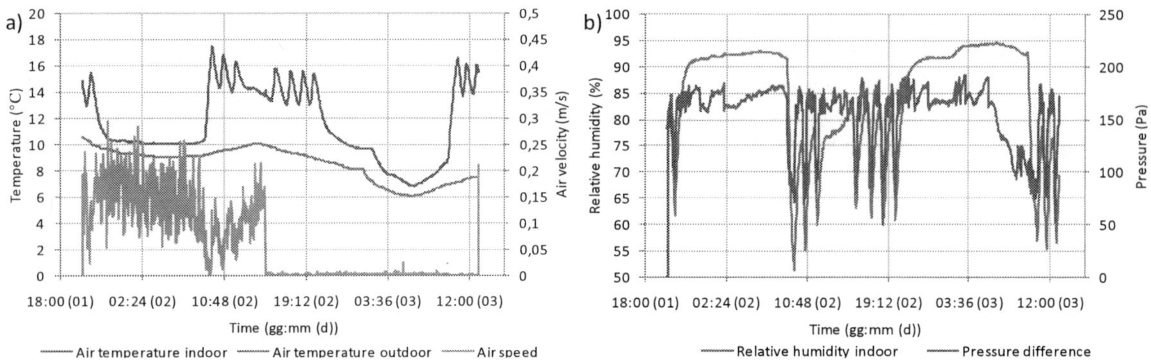

Figure 1. The measured values: a) air temperature indoor and outdoor and air speed, b) relative humidity indoor and pressure difference.

2.3. *PMV and PPD prediction*

2.3.1. *Input data and assumptions.*

In order to determine the thermal comfort indicators (PMV, PPD), the athlete's level of metabolism was assumed to 4 met (232,8 Wm^{-2}). The total thermal resistance of clothing was set in two variants: basic clothing (t-shirt, shorts, underwear and socks) and warm (additional sweatshirt and light trousers instead of shorts). Calculated thermal resistance of the basic variant is 0.27 clo (0.042 m^2KW^{-1}), while the warm variant 0.57 clo (0.088 m^2KW^{-1}). The PMV indicator is the most reliable for predefined conditions, but it can also be used for slow changes in the value of individual microclimate parameters of interiors in specific intervals [9].

2.3.2. *Results*

Figure 2a presents the PMV indicator determined on the basis of registered measurement data, metabolism and clothes of the athlete. It can be clearly seen that except the night period, PMV assumes positive values. In the later part of the paper, the PMV values will be considered only for the hall usage period. PMV values are definitely higher for the case of warm clothing where the average PMV is 1.45, while for basic clothing it is 0.84. PMV fluctuations correspond to temperature and humidity fluctuations. The maximum value of PMV for warm clothing is 1.91, minimum 0.47. The maximum value of PMV for basic clothing is 1.44, minimum -0.42. The average PPD values (Figure 2b) determined on the basis of PMV were 48.7 for warm clothing and 22.6 for basic clothing.

Comparison the PMV results with the 7-point scale of thermal sensations has been assumed that there is warm indoor environment in the tennis hall. Therefore, these are not comfortable conditions for sports. Athletes wearing warmth can feel more discomfort, what is in agreement with the users' opinions. The PMV characterizing the comfortable environment (not too warm, not too cold) should be found in the range $-0.5<PMV<0.5$. Such values were met in the hall only during the night, but the temperature drops below 13°C – the minimum indoor temperature for playing tennis (by International Tennis Federation).

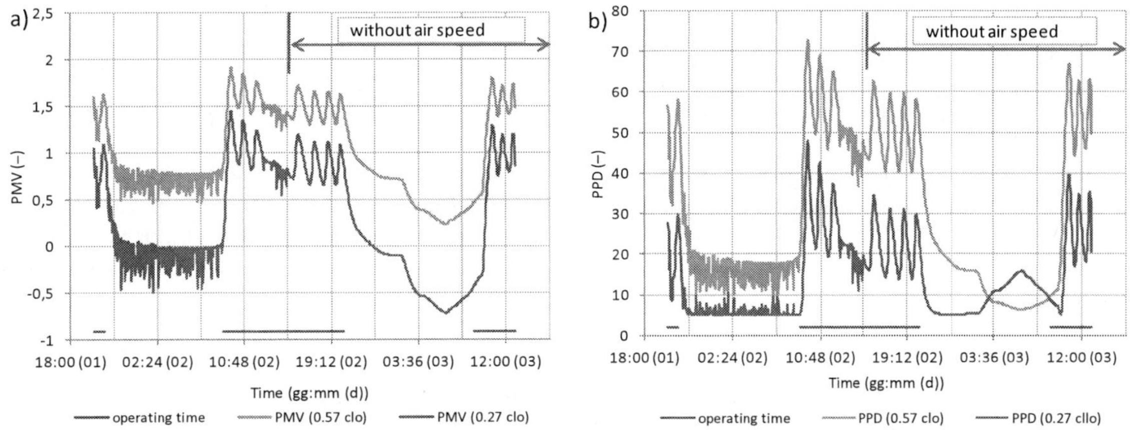

Figure 2. a) PMV and b) PPD predicted for both variants of sportswear.

2.4. The study on possible improvement of the thermal comfort in the examined tennis pneumatic hall

At the beginning of the analyze the linear dependence between the PMV components was checked: radiant temperature, air temperature, air relative humidity and air speed. Afterwards, the linear dependence between these components and PMV was examined. A strong linear correlation was found between radiant temperature, air temperature and relative humidity of air (0.788÷0.929). The linear correlation between these three and air speed is much smaller 2·E-2÷4.6·E-2 for correlation with temperature and 9·E-6 for correlation with air relative humidity.

Although, the PMV value depends on each of the measured components, a linear correlation was examined to obtain information about the proportion of individual values in PMV formation. The strongest is the correlation with air temperature 0.976÷0.977, with radiant temperature 0.947÷0.948 and air relative humidity 0.873÷0.876. The correlation between PMV and air speed 6.1·E-3÷8.3·E-3 reveals by far the lowest dependence, which may result from low air speed.

Table 1. The correlation between PMV and its components.

Linear correlation	Radiant temperature	Air temperature	Air speed	Relative humidity	PMV (0.57 clo)	PMV (0.27 clo)
Radiant temperature	x	0.929	4.6·E-2	0.788	0.947	0.948
Air temperature	0.929	x	2·E-2	0.917	0.976	0.977
Air speed	4.6·E-2	2·E-2	x	9·E-6	6.1·E-3	8.1·E-3
Relative humidity	0.788	0.917	9·E-6	x	0.876	0.873
PMV (0.57 clo)	0.947	0.976	6.1·E-3	0.876	x	-
PMV (0.27 clo)	0.948	0.977	8.1·E-3	0.909	-	x

The second step of the investigation was to simulate the PMV change due to the change of the measurable components. The reduction of radiant and air temperature, and air relative humidity as well as the increase of air velocity were analyzed. It was assumed that a satisfactory change will occur when the PMV values during the hall utility time will be in the range -0.5÷0.5.

Tables 2 and 3 present the effect of proposed changes of the PMV components on its value. Presented are the averaged values, standard deviation, maximal and minimal values of PMV. The table presents only changes that correspond to the adopted boundary conditions: maintenance of the average temperature in the hall above 13.0°C, not exceeding the air speed of 0.9 m/s, maintaining the relative humidity at the level no less than 55-60%. The temperature criterion is fundamental to protect the health of athletes, below this temperature the risk of injury is very high. The air speed criterion is related to the risk of drafts, and interfering the ball's flight. Lawn Tennis Association [10] recommends that in traditional constructions of tennis halls air speed should not exceed 0.5 m/s, but there is no such requirement for pneumatic halls. The air distribution in tennis hall should consider ball movement requirements as well as human thermal comfort requirements [7].

The order of presented solutions corresponds to the positions of variants selected on the basis of a multicriterial rating: from the most favourable one. The following optimization parameters were adopted: average PMV value close to 0.00 (weight 0.65) and as low as possible standard deviation (weight 0.35).

Table 2. Conception of the improvement of the thermal comfort in the case of light sportswear (0.27 clo) by changing the measurable parameters of PMV. T – indoor radiant and air temperature, RH – indoor relative humidity, v – indoor air speed.

PMV (0.27 clo)	Base value	v+0.4 m/s	v+0.4m/s; T-1°C	v+0.4 m/s; T-1°C; RH-20%	v+0.2 m/s; T-1°C	v+0.2 m/s	T-1°C
Average	0,84	0,08	-0,17	-0,23	0,28	0,51	0,63
Standard deviation	0,37	0,41	0,41	0,41	0,37	0,37	0,37
Maximum	1,44	0,76	0,51	0,44	0,85	1,08	1,23
Minimum	-0,42	-1,3	-1,54	-1,58	-0,86	-0,65	-0,63

Table 3. Conception of the improvement of the thermal comfort in the case of warm sportswear (0.57 clo) by changing the measurable parameters of PMV. T – indoor radiant and air temperature, RH – indoor relative humidity, v – indoor air speed.

PMV (0.57 clo)	Base value	v+0.6 m/s; T-1°C; RH-20%	v+0.4 m/s; T-1°C; RH-20%	v+0.4 m/s; T-1°C	v+0.6 m/s	v+0.2 m/s; T-1°C	v+0.4 m/s
Average	1,45	0,57	0,73	0,79	0,83	1,04	0,98
Standard deviation	0,29	0,33	0,31	0,31	0,34	0,28	0,32
Maximum	1,91	1,12	1,24	1,31	1,4	1,49	1,51
Minimum	0,47	-0,55	-0,31	-0,27	-0,31	0,11	-0,08

The results of the analysis show that it is possible to increase the thermal comfort in the pneumatic tennis hall. In the case of a light sportswear (0.27 clo), it seems easier to achieve. For both cases of clothing it is clearly visible that reduction of temperature and relative humidity only, does not bring the desired effect. Only increasing the air speed in the hall brings the expected improvement. Among the analyzed variants of the component parameter changes for light clothing, the most favourable is the increase of the air speed by 0.4 m/s, which will give an average PMV 0.08 with a standard deviation of 0.41, and thus ideal values. The average air speed in the hall during the period of use will be increased to 0.50 m/s and the maximum values would be at the level of 0.65 m/s what is still acceptable in pneumatic halls.

In the case of warmer clothing (0.57 clo), increasing the thermal comfort is much more difficult. It was assumed no to increase the air speed by more than 0.6 m/s. Among the analyzed variants of the component parameter changes for warm clothing, the most favourable is to simultaneously: increasing the air speed by 0.6 m/s (average 0.70 m/s, maximum 0.85 m/s), reducing the air and radiant temperature by 1°C (average air temperature 13.17°C) and reducing the air relative humidity by 20% (average RH 56%).

For both clothing standards, for most favourable cases, the PPD values were determined. For light sportswear the average PPD is 8.72 with a standard deviation 7.47. For warm sportswear the average PPD is 14.63 with standard deviation 5.52.

3. Discussion and conclusions

The research revealed that sports facilities are characterized with different specification to other buildings where the state of thermal comfort was analyzed. Thermal comfort of the sports hall should be primarily provided to athletes. It is assumed that the level o metabolism of tennis players is 4 met. This means that it is over 3 times more than for someone doing sedentary activity. Thus, it means that in sports facilities it is not easy to achieve a thermal comfort conditions both for athletes and viewers or service staff. The analysis showed that the slight difference in the sportswear (long sleeve sweathsirt and light trousers instead of shorts, 0.3 clo) caused significant differences in the values of PMV and PPD.

The research carried out in the pneumatic hall revealed that the indoor thermal sensation is too warm. This is particularly interesting because the temperature in the hall was kept at the level 14÷16°C with occasional peaks up to 17.44°C . The specificity of the pneumatic hall - continuous drafts caused by the air supply system do not provide indoor thermal comfort for athletes, what is in accordance with their opinion.

Although the linear correlation of PMV with air speed was small, the analysis revealed that increasing the air speed in the hall was the fastest way of improving indoor thermal comfort there. This is also the easier way of receiving heat from tennis players. On the other hand, in all of the interval exercises, like tennis playing, the high physical effort is interrupted with rest. This is why the air speed in indoor tennis courts should not be too high. It is also connected with the tennis ball movement [11].

It can be concluded that achieving a satisfactory level of thermal comfort is not always achievable. Clothing plays the major role in thermal sensitivity of someone involved in the activity with high metabolism. It was possible to achieve the almost ideal PMV level for light sporstwear by increasing the air speed in the hall by 0.4 m/s. Increasing the air speed in the pneumatic hall is a technical challenge. It may be necessary, for example, to install confusors in the air supply system of the hall. In the case of warm sporstwear (only 0.3 clo difference) the ideal level of PMV was not achieved, although the temperature was reduced by 1°C, the relative humidity was reduced by 20% and the air speed was increased by 0.6 m/s. These were limit values or even almost a bit exceeded (air speed). Despite these changes, the thermal conditions were still warm and it might be one of the reasons that modern indoor tennis courts are constructed in other techniques than pneumatic hall. Nowadays, the most popular are the halls, in which the roof construction is supported on girders made of light plastics, glued wood or truss. This type of construction enable the side curtains being opened ensuring a constant supply of fresh air and adequate humidity, what increases the comfort of use of the object.

References

[1] http://www.biggestglobalsports.com/

[2] Rajagopalan P and Jamei E 2015 Thermal comfort of multiple user groups in indoor aquatic centres *Energy and Build.* **150**, 129-38.

[3] Kampel W and Aas B and Bruland A 2013 Energy-use in Norwegian swimming halls *Energy and Build.* **59,** 181–86.

[4] Kisielewicz T and Dudzińska A 2015 Summer overheating of a passive sports hall building *Arch. of Civil and Mech. Eng.* **15** 1193-1201

[5] Veselý M and Zeiler W 2014 Personalized conditioning and its impact on thermal comfort and energy performance – A review *Renewable and Sustainable Energy Reviews* **34(C)** 401-08.

[6] Sui X and Han G and Chen F 2014 Numerical simulation on air distribution of a tennis hall in

winter and evaluation on indoor thermal environment *Eng. Rev.* 34(2) 109-18.
[7] Revel G and Arnesano M 2014 Measuring overall thermal comfort to balance energy use in sports facilities *Meas.* **55** 382-93.
[8] ISO 7730:2006 Ergonomics of the thermal environment - Analytical determination and interpretation of thermal comfort using calculation of the PMV and PPD indices and local thermal comfort criteria Geneva: International Organization for Standardization.
[9] Fanger P 1970 *Thermal comfort-analysis and applications in environmental engineering.* 1970, (Danish Technical Press: Copenhagen).
[10] https://www.lta.org.uk/
[11] Chadwick S 2003 *The aerodynamic properties of tennis balls* PhD thesis University of Sheffield.

IOP Conf. Series: Materials Science and Engineering **415** (2018) 012025 doi:10.1088/1757-899X/415/1/012025

Numerical analysis of outdoor climate conditions based on the 3-years measurement for Zilina

D Jurasova[1]

[1]Department of Building Engineering and Urban Planning, University of Zilina, Univerzitna 8215/1, 01026 Zilina, Slovakia

daniela.jurasova@fstav.uniza.sk

Abstract. The need for reference files resulted from the development of advanced simulation models. Few methodologies for creation of test reference years were developed in last decades. Most common are following two methods: a simple selection of worst year from recommended period or the numerical analysis of months and then the combination of months into one reference file, which in is not real, because the months are selected from various years. It is necessary to temper the connection of individual months. With the outdoor climate change and therefore the synthetically (combined) based reference years for many areas can not satisfy more and more advanced simulation models. Reference years are created from values measured mostly by weather stations belonging to the national weather forecast service, which is not very dense and the resolution and measured climate data are insufficient. Problems of urban heat islands in the cities is also widely discussed. In this paper, the outdoor climate characteristics measured by the mobile weather station located in the University of Zilina campus are analyzed. The station records most of the climate parameters required for HAM simulations since 2014. Measured data are also compared to data obtained from national weather forecast service stations located nearby.

1. Introduction

During last years it is to observe that the climate as a long-term weather regime is visibly changing [1, 2, 3]. It has its prior role and strongly influences the indoor environment of buildings. Nowadays, the climate change is becoming crucial. It is not just about predictions, the result is visible right now. The analysis presented in this article should emphasize the importance of use as accurate boundary conditions for HAM simulations as possible. Building simulation assumes dynamic boundary conditions (continuous in time). From this fact it follows that creating suitable reference file is one of the most challenging parts how to obtain meaningful results. Area of Slovakia has not its actual reference complex files usable for building simulations, HAM and wind-driven rain simulations. Test reference years (TRYs) for several places, based on the long-term measurement exist [4, 5]. These were created as average ones, without severe years. The existing TRYs for Slovakia need to be updated due to changing climate conditions and resulting new requirements at the user side. They are basically based on the 30-year climate normal period 1961-1990 [13]. Researchers often tend to use files of similar climate – not really measured right at the experiment site. Actual trends are focused on obtaining not only random numbers [6, 7, 8]. So which data file is the right exterior boundary condition for numerical simulation? Possibilities in mentioned area of research are wide. This study is based on the need of precise integration of boundary condition for analysis of hygrothermal behavior e. g. of building envelope structures [9, 10, 11]. Choosing them in the right way is a key aspect for relevant results [12, 13]. Role of the building envelope in area of sustainable building and energy assessment is vital. Hygrothermal conditions (humidity and temperature) inside the building depend on the outside climate [9, 14, 15].

For needs of energy simulations in area of Civil engineering, relevant data about exterior boundary conditions are mostly needed in shorter time periods. This highlights the advantage of having own experimental weather station. Building simulation tools need the right interval of data recording. The data should be complete, consistent and logical. Typical weather data files contain incorrect or missing entries caused mostly by malfunction. Correction and adding of optimal values are crucial aspects of climate data preparation especially for numerical simulations. The used external boundary conditions should be representative of the building location [5, 11, 16]. According to the territorial study of Slovakia about climate change, the global warming could reflect in our area by increase of temperature averages till the year 2075 at about 2 – 4 °C [2, 3].

Gridded climate datasets are required for numerous purposes in applied and theoretical climate and environmental sciences [17]. Often is also combination of observed weather (temperature, humidity, wind, …) data with modeled solar radiation data, transforming these into Actual Meteorological Year [16].

2. Input data

Measurement of possible meteorological parameters runs with a 10-minute recording interval. For needs of this article, air temperature and relative humidity were chosen, as two important characteristics for hygrothermal simulations. Measuring interval of experimental weather station is adjustable as required. Results of measurement were elaborated in a way of hourly means and described.

Exterior air temperature changes relatively slow. Mentioning solar radiation, it is possible to come to big and quick changes of values within a few minutes. Simplifying of climate factor course to hourly values can be misrepresentative in this case. Experimental weather station (Figure 2, left) is a benefit.

Figure 1. Location of weather stations (Dolny Hricov, DH (3) and Zilina, ZA, EX (2, 1), Slovakia).

For the analysis in this article, years 2014 (partly), 2015, 2016 and 2017 were selected. Using the reference file from Dolny Hricov (DH) and Zilina (ZA), based on measuring range 1961-1990 (Table 1), it could be determined, e. g. for evaluation of individual years, if these years were above or below the reference value. The files were obtained from Slovak Hydrometeorological Institute (SHMI).

Figure 2. Measuring devices (from left: experimental weather station – EX, DH, ZA).

Table. 1 Reference file, weather station DH and ZA – Air temperatures [°C]

Time period	I	II	III	IV	V	VI	VII	VIII	IX	X	XI	XII
Monthly min. DH	-9.2	-7.4	-2.8	4.9	10.2	13.1	14.8	14.1	10.2	5.0	-2.0	-5.6
Monthly min. ZA	-9.1	-6.8	-2.3	5.4	10.4	13.1	15.2	14.0	10.4	5.3	-1.8	-5.3
Monthly max. DH	1.2	4.2	5.9	10.7	15.0	18.8	19.3	18.1	15.8	12.2	7.2	2,.2
Monthly max. ZA	1.5	4.0	6.3	11.1	15.5	18.6	19.4	18.0	15.8	12.4	7.3	2.4
Monthly av. DH	-3.6	-1.4	2.5	7.7	12.8	15.6	17.0	16.3	12.7	8.1	3.1	-1.7
Monthly av. ZA	-3.4	-1.1	2.9	8.1	13.1	15.6	17.0	16.3	12.8	8.2	3.2	-1.5

January is usually the coldest and July the warmest month of the year. It is visible that values for both weather stations are closely similar. The same can be said about seasons, summer and winter periods and years. Data were obtained from a long-time period (30 years). Measuring devices are located in a distance of about 20 km from each other (Figure 1). The experimental meteorological station is surrounded by buildings and forest – it is placed on the roof nearby the laboratory. The meteorological station Zilina-Borik is situated in urban area, but surrounded by low trees and flat terrain (Figure 2, right). Dolny Hricov is paced near the airport runway in flatland surrounded by meadow. General face of surroundings is absolutely free (Figure 2, middle).

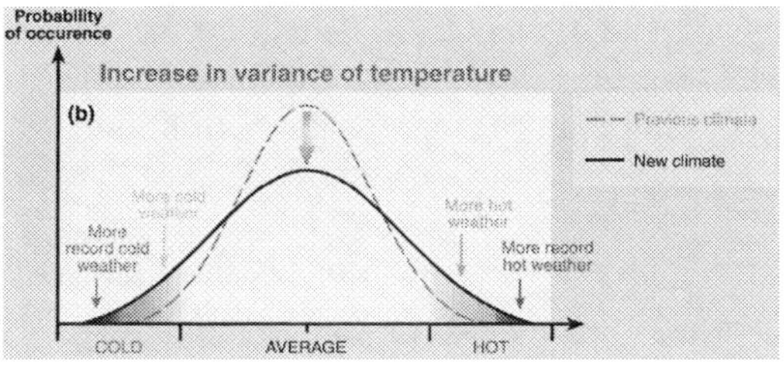

Figure 3. Diagram showing the climate change effect on extreme temperatures [3].

2.1. Air temperature

Year 2017 reached average monthly values closest to the average from all 3 evaluated experimental yearly data (Figure 4). The coldest month should be January and the warmest one should be July (according to long-term past measurements and knowledge). Yearly course of the year 2015 shows February as the coldest and August of the warmest month. Year 2016 is represented by January as the coldest and June as the warmest month. Watching the year 2017 it can be seen that June and August were warmer than July. January 2017 reached the lowest value of average temperature. As it is clear from the graph (Figure 4, left), the biggest differences among individual summer periods are for June, July and August. These months can be marked as critical for building indoor environment (focused on the overheating during summer). Summer 2016 seems to be closest to the average. The biggest differences among measured values and its averages during winter period were recorded in January (Figure 4, left), which is critical for energy use and heating.

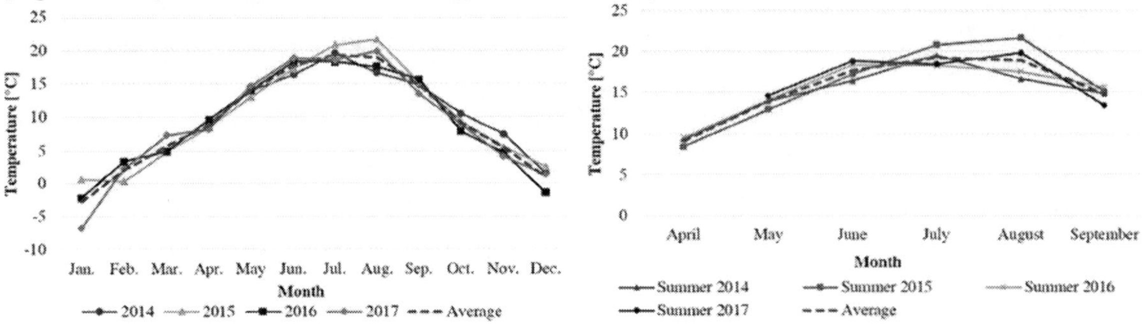

Figure 4. Annual (left), summer (right) air temperature, experimental weather station (EX).

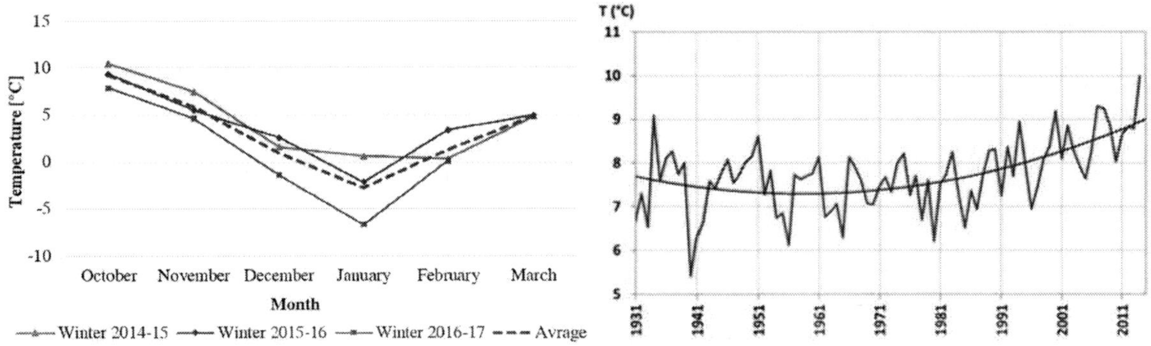

Figure 5. Exterior air temperature, winter period, experimental weather station (EX, left), spatial average annual air temperature since 1931, the area of Slovakia (right) [2]

Highlighting the trends, January becomes colder from year to year (according to monthly averages). December measurement shows a trend of changes. The coldest was December 2016, subsequently 2014 and later on 2015. Winter 2015-16 seems to suit best to the average from assessed period. Winter 2016-17 forms the lower boundary of temperature courses, the bottom one is a combination of winter 2014-15 and 2015-16.

Table 2. Monthly means [°C] for various locations (from EnergyPlus) vs. measured by EX.

	Jan.	**Feb.**	**Mar.**	**Apr.**	**May**	**Jun.**	**Jul.**	**Aug.**	**Sep.**	**Oct.**	**Nov.**	**Dec.**
Bratislava	0	0	5	10	15	18	21	20	15	9	5	1
Kosice	-2	-1	4	9	15	17	19	19	14	9	3	-1
Ostrava	-1	-1	3	8	13	16	18	17	13	9	3	0
Zilina	**-2**	**1**	**5**	**8**	**13**	**18**	**19**	**19**	**14**	**8**	**4**	**1**

Monthly temperature means available through IWEC files for EnergyPlus [19] are quite similar e. g. to Ostrava (Table 2, deviation by around 1 °C), but focusing on extremes in individual days, the differences are more visible and influencing. Analyzing Table 3 it is visible that monthly temperature averages of selected years 2015, 2016 and 2017 recorded by experimental weather station (EX) are closer to values obtained in Zilina-Borik (ZA), which was expected. These weather stations are at a 3-km distance.

Direct temperature readings dating back to the 19th century show that the last 10 years had 8 of the 10 warmest years on record [3]. In Hurbanovo (Slovakia), from 10 highest values of yearly air temperature means since 1901, 8 were recorded since 2000 [2].

Table 3. Monthly temperature averages, weather stations DH, ZA, EX.

	2015 EX	**2015** DH	**2015** ZA	**2016** EX	**2016** DH	**2016** ZA	**2017** EX	**2017** DH	**2017** ZA
Jan.	0.58	-0.18	0.6	-2.17	-2.59	-2.43	-6.72	-3.88	-7.12
Feb.	0.33	-0.78	0.16	3.40	3.51	3.77	2.33	0.71	1.47
Mar.	4.85	4.27	4.35	4.93	4.54	4.68	7.33	6.08	6.55
Apr.	8.50	7.89	8.2	9.60	9.13	9.48	8.28	7.24	7.58
May	12.97	12.18	12.87	14.07	13.84	13.76	14.66	14.08	14.66
Jun.	17.22	17.36	17.27	18,36	18,17	17.93	18.83	18.57	19.16
Jul.	20.80	20.46	20.06	18.28	19.06	18.8	18.49	18.63	19.2
Aug.	21.65	21.09	20.82	17.55	17.27	16.65	19.85	19.52	19.89
Sep.	15.09	13.17	14.54	15.69	15.49	15.15	13.45	13.21	13.53
Oct.	9.34	8.72	8.82	7.90	7.78	7.86	8.95	8.72	8.98
Nov.	5.51	4.55	4.9	4.63	4.10	4.58	4.13	3.79	4.26
Dec.	2.60	2.28	2.58	-1.39	-1.45	-1.42	1.36	0.94	1.27

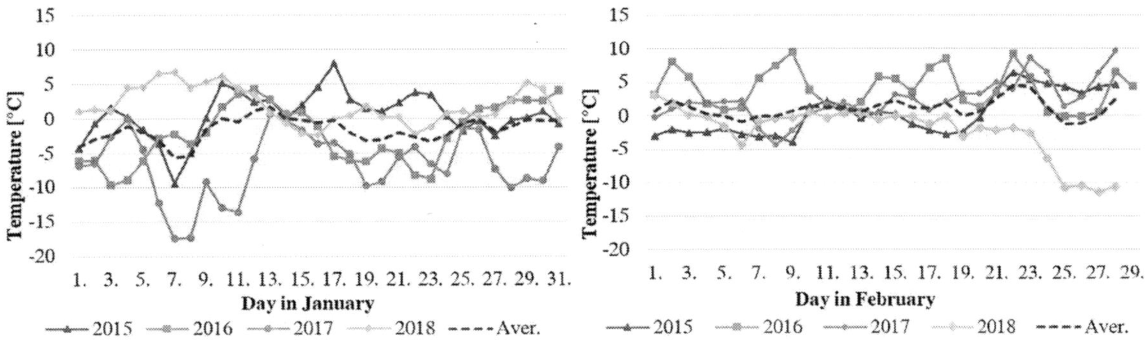

Figure 6. Daily averages, experimental weather station (EX), January (left) and February (right).

In winter period, 12-14 January and 11-13 February (middle of the month), all the curves have closely identical courses (Figure 6, left, right). But there are also visible significant differences, which is interesting e. g. from the point of view of building simulations. Noticeably coldest February from selected years was in 2018. The biggest temperature difference was 20.32 °C. First 11 days of January 2016 and 2017 (Figure 8) show reversed temperature courses. The same is visible during last 5 days of the month. 7 January shows maximum contrast of recorded daily averages among selected years. The difference reached up to 24.14 °C. A lot of differences but also similarities in individual courses are evident. The lowest yearly temperature was measured in 2017. The difference reached up to 8.63 °C in yearly average (2017, 2018).

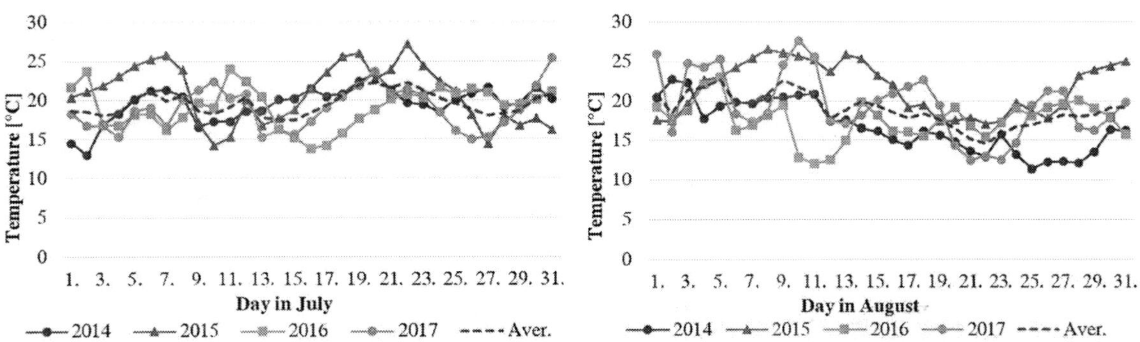

Figure 7. Daily averages, experimental weather station (EX), July (left) and August (right).

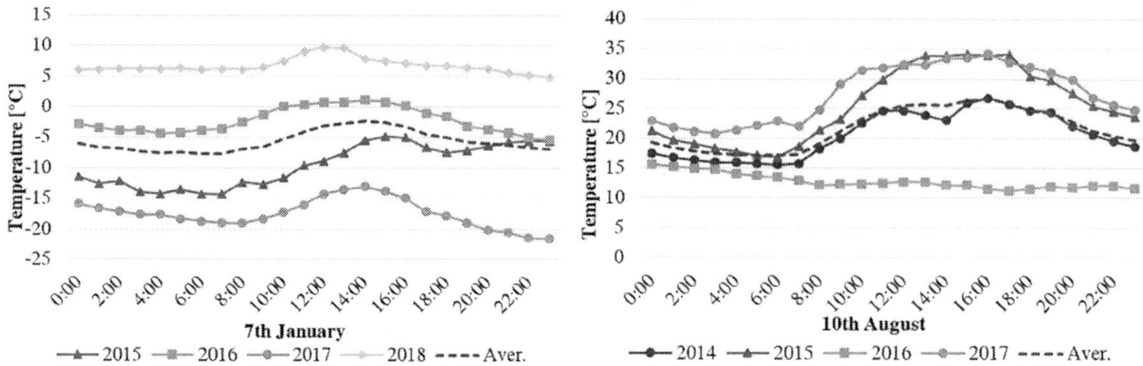

Figure 8. Hourly averages, experimental weather station (EX), selected days (cold-left, warm-right).

2.2. Relative air humidity

Relative humidity is a more changeable and unexpected parameter than the temperature. It shows wider differences among individual years (Figure 9, left, right). To illustrate some facts, annual, summer and winter periods were selected. Year 2016 seems to be closest to average (Figure 9, left).

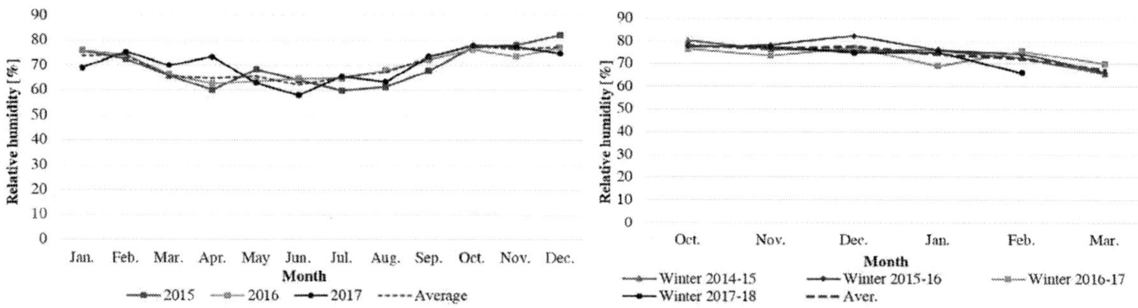

Figure 9. Monthly means, exterior air humidity (left), exterior air humidity, summer period (right).

Implementing building envelopes into real climate conditions is a current trend of research in area of building simulation. In hygrothermal simulations, the initial boundary condition of temperature and water content in the structure is representative and crucial [13, 14, 18].

Winter 2014-15 puts well to the average of evaluated period (Figure 9, right).

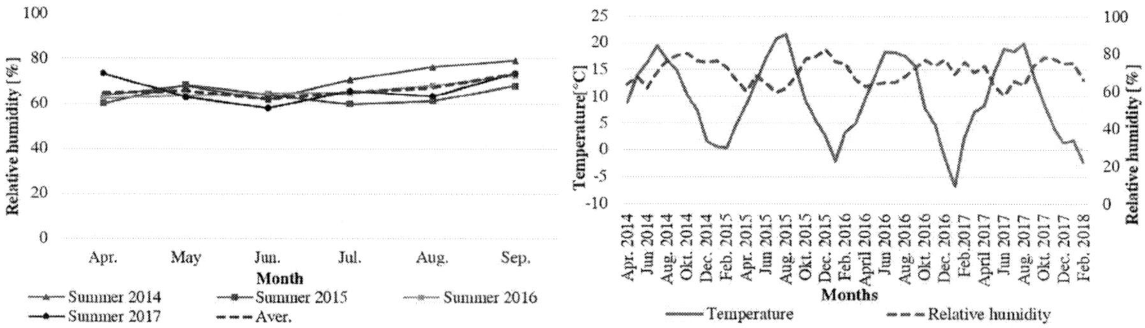

Figure 10. Relative humidity-summer (left), temperature and relative humidity-whole period (right).

Calculation of monthly mean was processed according to [20]. Hourly data were used for the monthly mean calculation.

Table 4. Measured air temperature (T) and relative humidity (RH), vs. reference values.

Year	2015	2016	2017	Average
T EX [°C]	9.95	9.24	9.25	**9.48**
T DH [°C]	9.25	9.07	8.97	**9.10**
T ZA [°C]	9.60	9.07	9.12	**9.26**
RH EX [%]	69.42	69.92	70.07	69.80
RH DH [%]	78.39	79.56	78.84	78.93

Reference file	
RF1 DH [°C]	**7.4**
RF1 ZA [°C]	**7.6**
RF2 DH [°C]	**8.1**
RF2 ZA [°C]	**8.1**

As Table 4 presents, increase of yearly average temperature is visible (in comparison to the reference file). Use of the reference data sets for example 1961-1990 (RF1) or 1981-2010 (RF2) could not be suitable enough for numerical simulations during these and future days. Relative air humidity is a strongly influencing parameter for building simulations (e. g. WUFI) [9, 14]. This parameter is more depending on the location (Table 4, 5) closely related to wind driven rain.

Table 5. Monthly means of relative humidity for various locations (fromEnergy Plus files).

	Jan.	Feb.	Mar.	Apr.	May	Jun.	Jul.	Aug.	Sep.	Oct.	Nov.	Dec.
Bratislava	85	84	67	61	63	72	62	65	75	76	85	81
Kosice	82	81	67	64	70	69	70	65	73	84	82	84
Ostrava	83	83	73	69	69	71	76	68	70	76	82	87
Zilina	**73**	**74**	**66**	**65**	**66**	**62**	**65**	**67**	**73**	**78**	**76**	**77**

As precisely the relative air humidity will be implemented, so exact will be the results [12, 13]. Table 6 shows variability of measured relative humidity for various locations in comparison to the near-real time experimental measurement in Zilina.

3. Results and discussion

The WMO has set a 30-year period for a meteorological standard, which should cover all extremes at a given interval. Analysis for Zilina was done for time period April 2014 – March 2018. Over time the database becomes more extensive. Purpose of this analysis for the future is to suggest some optimum yearly reference file for computer simulations, as well as an extreme reference file, which would allow monitoring of selected parameters of building envelope under non-standard conditions. These should be the most critical ones. The deeper analysis of measured data requires a wider range of measured data, which will be at disposal by ending each following year. Most suitable way is construction of a year from 12 months of real measured weather data. Generated data are usable, but they cannot predict all extremes, which locally occur in last decades also in Slovakia. Existing reference files need the reanalysis. In a long-time point of view, there are smaller differences, but analyzing monthly, daily or hourly values it is visible that the differences are bigger. December, January and February often reach some extremes, which are crucial from the point of view of variable measured data. In 2017, the temperature for 8 January was registered as the lowest for last 20 years (-24.6 °C).

Slovakia is orographically a very ragged country. Climate character has its typical changes according to the altitude and characteristic daily and yearly regime of meteorological variables. Region of Slovakia is missing actual reference files suitable for building simulations. All these problems are complicated and complex. Crucial are also winter and summer period with their extremes. Some years can be exceptionally wet, which more influences the hygrothermal behavior of a building envelope. In general, it is usual to use the arithmetic mean to give a picture of more measured values with one number.

Future predictions are also possible to do, but it is dependent on a measurement quality, methodology of used computing tools and character of monitored variable. First of all, data have to be professionally assessed. The next step after evaluation of recorded parameters will be creation of characteristic reference file for Zilina, which would serve best for numerical simulations, e. g. WUFI, ESP-r, Delphin.

The wider database of measurement, the better. Having a period of 3 complete years at the disposal enabled first analysis, which is planned to be continuously updated and repeatedly analyzed.

Acknowledgements
This article was created with support of the grant project VEGA No. 1/0683/16 and also with provided data from Slovak Hydrometeorological Institute in Bratislava (SHMI) for comparison and analysis as well as consultation.

References

[1] Climate Change Reference Guide, Worldwatch Institute, 2009.
[2] Climate Atlas of Slovakia, Slovak Hydrometeorological Institure, Bratislava, 2015.
[3] Climate Change 2001, 2014: Synthesis report, An Assessment of the Intergovernmental Panel on Climate Change.
[4] Levermore, G. – Hoffmann, S. – Rabenseifer, R.: Comparison of TRYs for Bratislava, CESB Conference, Prague, 2007.
[5] Bielek, M. at. al.: Research report No. 09, Outdoor climate model – reference year for 1st temperature range of Slovakia, Bratislava, 1993 (in Slovak).
[6] Schoner, T. – Zirkelbach, D.: Erstellung Hygrothermischer Referenzjahre (HRY) in Deutschland, Neue Forschungsergebnisse, Kurz Gefasst, IBP-Mitteilung 547, 2016.
[7] Hygrothermische Referenzjahre und Lokalklimamodelle, cited online: <https://www.ibp.fraunhofer.de/de/Kompetenzen/hygrothermik/projekte/hygrothermische-referenzjahre-und-lokalklimamodelle.html>.
[8] Mesinger, F.: North American Regional Reanalysis, American Meteorological Society, 2006.
[9] Slavik, R. – Cekon, M.: Study of Surface Temperature Monitoring in the Field of Buildings, World Multidisciplinary Civil Engineering-Architecture-Urban Planning Symposium, WMCAUS 2016.
[10] Pernigotto, G. – Prada, A. – Costola, D. – Gasparella, A. – Hensen, J. L. M: Multi-year and reference year weather data for building energy labelling in North Italy climates, Energy and Buildings, vol. 72, 2014.
[11] Juras, P. – Zilinsky, P.: Comparison of HAM Simulation with Different Reference Years for Bratislava, Advanced Building Construction and Materials 855, p. 149-153, 2014.
[12] Ortsgenaue Testreferenzjahre von Deutschland für mittlere, extreme und zukünftige Witterungsverhältnisse, Bundesinstitut für Bau-, Stadt- und Raumforschung, Deutscher Wetterdienst, Offenbach, Juli 2017.
[13] Vertal, M. – Zozulak, M. – Vaskova, A. – Korjenic, A.: Hygrothermal initial condition for simulation of green building construction, Energy and Buildings, In press, Accepted Manuscript, 2018.
[14] Sanders, C.: IEA ANNEX 24, Final report, Vol. 2, Task 2: Environmental conditions, Leuven, 1996.
[15] Maile, T. – Fischer, M. – Bazjanac, V.: A Method to Compare Measured and Simulated Data to Assess Building Energy Performance, Standford University, CIFE Working Paper, August, 2010.
[16] Lundstrom, L.: Weather data for building simulation, New actual weather files for North Europe combining observed weather and modelled solar radiation, Vasteras, Sweden, 2012.
[17] Krahnemann, S. at. al.: High-resolution grids of hourly meteorological variables for Germany, DOI 10.1007/s00704-016-2003-7.
[18] STN EN 15026, Hygrothermal performance of building components and building elements – Assessment of moisture transfer by numerical simulation, 2007.
[19] EnergyPlus, cited online: <https://energyplus.net/>.
[20] STN EN ISO 15927-1, Hygrothermal performance of buildings – Calculation and presentation of climatic data – Part 1: Monthly means of meteorological elements, 2004.

Influence of microclimate control scenarios on energy consumption in the Gallery of the 19th-Century Polish Art in the Sukiennice (the former Cloth Hall) of The National Museum in Krakow

A Sadłowska-Sałęga[1], J Radoń[1], J Sobczyk[2], K Wąs[1]

[1] Faculty of Environmental Engineering and Land Surveying, University of Agriculture in Krakow, al. A. Mickiewicza 24-28, 30-059 Krakow, Poland
[2] Laboratory of Analysis and Non-Destructive Investigation of Heritage Objects (LANBOZ), National Museum in Krakow, al. 3 Maja 1, 30-062 Krakow, Poland

Abstract. The main aim of preventive conservation, ensuring protection of historic collections, is stabilization of relative humidity, and for some materials also temperature, in the interiors of historic buildings or museums. The improper microclimatic parameters may cause irreversible damages. On the other hand, strict control of indoor climate is expensive. Definition of optimal microclimate control in terms of energy use, taking into account most important, sometimes even contradictory factors, requires interdisciplinary cooperation, among others in the field of conservation, collections' care, architecture, building physics and HVAC systems. Due to the type of historic objects and specific microclimate requirements, hygrothermal simulations of whole historical buildings are essential for the most part of a case study. The paper presents an analysis of energy consumption according to various microclimate control scenarios for The National Museum in Kraków: The Gallery of 19th-Century Polish Art - located in Renaissance Cloth Hall in Kraków (UNESCO World Heritage Site). The analysis is based on the results of simulations performed with WUFI®PLUS software and in situ measurements. The microclimate control variants were developed, among others, on the basis of ASHRAE and ICOM-IIC guidelines. The research was supported by The Polish National Centre for Research and Development within "HERIVERDE" project.

1. Introduction

The main activity of preventive conservation, ensuring protection of historic collections, is stabilization of relative humidity, and for some materials also temperature, in the interiors of historic buildings or museums. The improper microclimatic parameters may cause irreversible damages. On the other hand, strict control of indoor climate is expensive both in the stage of investment in the installation of air conditioning systems as well as the current operation of these devices. The microclimate inside museum should also ensure, if possible, thermal comfort for the people.

Relative humidity is, according to scientific publications and international standards [1, 2], the most important parameter responsible for the formation of physical damage, especially in objects composed of complex hygroscopic organic materials, i.e. paintings, furniture, etc. Fluctuations of relative humidity cause a dimensional response of organic hygroscopic materials which can lead to the formation of looses, cracks, etc. Relative humidity accelerates the natural aging processes of objects. Higher values of relative humidity significantly affect the acceleration of metal corrosion processes. They also are conducive to the growth of microorganisms. It can be assumed that reduction of relative humidity by half reduces the rate of chemical degradation by more than half [3].

The temperature also influences the acceleration of aging processes. It can be assumed that a 5°C increase in temperature doubles the rate of chemical degradation [3]. Temperature in the range of 20 ÷ 30°C favors the growth of microorganisms [4].

On the basis of many years of research, institutions dealing with heritage protection have developed guidelines regarding permissible temperature and relative humidity changes in museums. The most restrictive approach to microclimate control in museums, archives and libraries presents American Society of Heating, Refrigerating and Air-Conditioning Engineers (ASHRAE). The maximum fluctuations and gradients are divided into 5 basic control classes for general collections. In addition, separate control classes have been distinguished for collections containing chemically unstable materials and metal collections [5]. In 2014, International Council of Museums (ICOM-CC) and International Institute for Conservation of Historic and Artistic Works (IIC) announced a joint declaration regarding the requirements for microclimate control in museum facilities. Two ranges of fluctuations in temperature and relative air humidity were developed depending on the type of collections: general collections and containing hygroscopic materials [6]. Some museum institutions, for example the National Museum in Krakow (MNK), developed their own guidelines for the specifity of stored exhibits or the constructions of buildings in which museums are located. Examples of guidelines of these institutions are presented in the Table 1.

Present paper is a step forward into an idea of "green museum" which comprises of safety of heritage objects, thermal comfort of visitors and staff, environment protection and energy saving [7].

The Sukiennice (the former Cloth Hall) is a large market hall erected in the 13th century in the middle of the Market Square, extended in the 14th century in the Gothic style and remodelled in the mid-16th century after Renaissance fashion. Following its restoration in the years 1875 ÷ 1879, it began to serve a representative function and became a venue for grand balls and patriotic celebrations. Due to their location and character, the halls on the first floor of the Cloth Hall building became the first seat of the National Museum in Krakow. In the years 2006 ÷ 2010, the building was thoroughly renovated.

The museum has eight air-conditioning units which implement four scenarios of microclimate control (depending on the way the room is used). The building also has central heating and a floor heating system (in exhibition halls), which is switched on if necessary.
The exhibition in the Cloth Hall is presented in the following four rooms: The Bacciarelli Room, The Michałowski Room, The Siemiradzki Room, The Chełmoński Room.

2. WUFI®plus model of the building

Calculations of microclimate and energy demand in the Cloth Hall in Krakow were made using the WUFI®plus software. Program developed by the Fraunhofer Institute for Building Physics, allows realistic calculation of the transient coupled one- and two-dimensional heat and moisture transport in walls and other multi-layer building components exposed to natural weather. The program has been widely validated on the basis of many years experimental research including historical buildings [8, 9].

The Cloth Hall model was created in the Sketchup program, based on the plans made available by the National Museum in Krakow, and then imported into the WUFI®plus program (Figure 1). Due to the size of the building, the complex layout of the rooms and the complexity of the construction of partitions, the model underwent some simplifications, which included unification of the thickness of partitions and simplification of their geometry (e.g. replacement of arches with straight lines). These simplifications were made so as not to change the parameters affecting the heat and moisture exchange through the partitions, i.e. the surfaces and the total mass of the partitions were preserved.

IOP Conf. Series: Materials Science and Engineering **415** (2018) 012026 doi:10.1088/1757-899X/415/1/012026

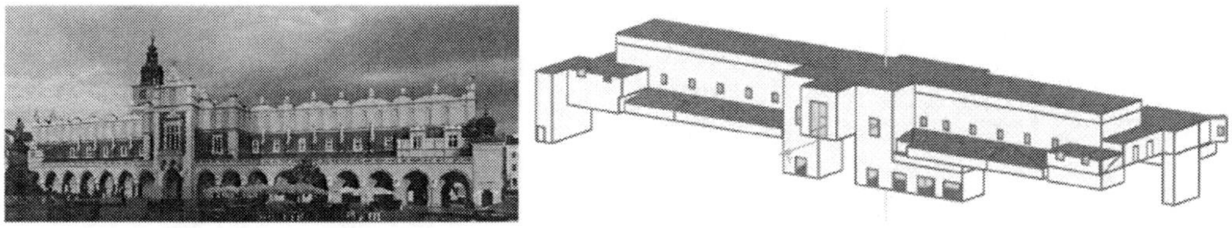

Figure 1. The Cloth Hall: western façade (a), WUFI®plus model (b)

The building's model covers all rooms belonging to the Gallery of Polish Art of the 19th century of the National Museum in Krakow. These rooms are divided, depending on the method of microclimate control, into nine calculation zones (exhibition rooms, heated rooms, a room for variable exhibitions and educational rooms, offices and storage rooms, a multimedia room, 3 air-conditioned zones and an attic). The main zone is the combined four exhibition rooms with a total area of 1 144 m^2 and a volume of 6 942 m^3. The remaining rooms of the building (such as restaurants, souvenir shops) are, so called, attached zones.

The basic construction material of the cloth hall is full brick. In the past it was usually used for building external and internal walls and vaults. The thickness of external walls in the modeled exhibition halls ranges from 0.7 to 1.0 meter. In the calculation model, the assumed value is 0.87 m. The floors of the exhibition halls are a combination of historical brick and modern reinforced concrete. The ceiling of the halls is made of steel structure covered with lighting panels. It separates the hall space from the attic of the cloth hall. The roof of the building is made of insulated partitions covered with copper sheets. The windows of the exhibition halls have been mostly walled with brick during one of the renovations; the remaining ones are made of wood equipped with a double, composite glazing units. The material data necessary for calculations, including historical bricks, was adopted from the WUFI®plus database.

3. Model validation

Parameters of the external climate (temperature, relative humidity, wind speed and direction, total and diffused solar radiation) were recorded on the roof of the Main Building of the National Museum in Krakow. The distance to the station from the object was approximately 1 km in a straight line. The use of an external climate measured at a certain distance from the object may lead to a slight error in the simulation calculations [9]. Based on the measured values (for 2015), a *.WAC file (WUFI ASCII CLIMAT) was created.

The parameters of the internal air were measured as part of the microclimate monitoring in the exhibition halls. Due to the fact that all exhibition rooms are interconnected (the doors between rooms are not closed), the average temperature and relative humidity of air were calculated based on the theory of mixing air streams due to the volume of individual rooms.

On the basis of ticket sales in 2015, the average hourly number of visitors was estimated (depending on the month). The adopted unit heat gains (by convection – 101 W and by radiation – 69 W) and moisture (98 $g \cdot h^{-1}$) correspond to the metabolic activity at the level of 1.6 met (adult standing person). For this activity, the unitary CO_2 emission is 59 $g \cdot h^{-1}$.

The calculations also include heat gains from lighting. Average hourly gains heat from the light is assumed to be 17 160 $W \cdot h^{-1}$ [10].

Exhibition halls are supported by four air handling units (Figure 2a). The air from the intake extends goes through the rotary heat exchanger, then: in the summer it is cooled and dried in the wet cooler, and then heated to the supply air temperature in the heater; in winter it is heated in the heater and then moistened with a steam generator (Figure 2b). The air cooler located before outlet is used for heat recovery for chillers.

IOP Conf. Series: Materials Science and Engineering **415** (2018) 012026 doi:10.1088/1757-899X/415/1/012026

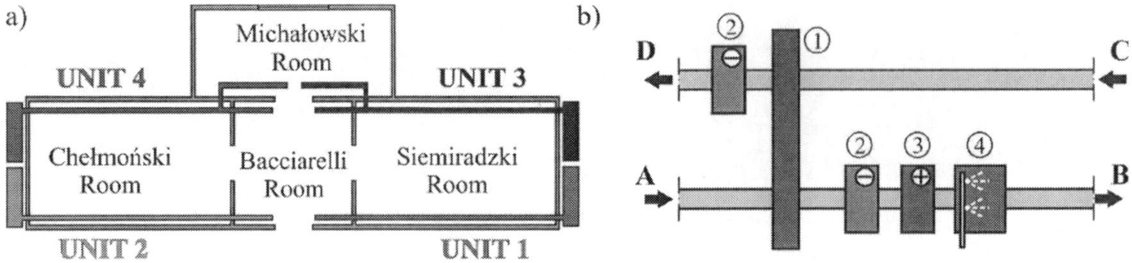

Figure 2. Separation of air from air-conditioning units to individual exhibition rooms (a) and operating diagram of air-conditioning units (b): A – intake air, B – supply air, C – exhaust air, D – outlet air; 1 – rotary heat exchanger, 2 – wet cooler, 3 – heater, 4 – steam generator

Both the temperature and relative humidity of the supply air to the exhibition rooms, as well as the airflow of individual air-conditioning units differ from each other. In the model, the exhibition halls constitute one zone serviced by one virtual unit. The averaged values of temperature and relative humidity of the air as well as the total efficiency of units were used for the calculations.

3.1 Microclimate parameters

Simulations were carried out assuming mentioned internal heat gains and supply air with averaged parameters. As a result of these simulations, annual variations in the temperature and relative humidity of the indoor air were obtained. These parameters are summarized by calculated averaged air parameters in exhibition halls.

Maximum absolute error for temperature is 3.7 K and mean one is 0.7 K. Errors below 1 K constitute, on a yearly basis, 72.7%. For relative humidity, the maximum error and mean are 13.2% and 2.8%, respectively. Errors below 5% constitute annually up to 85%. Over the year, the correlation, both for temperature and relative humidity, is significant – the correlation coefficient exceeds 0.6 (Spearman's test). It should be remembered that at each hour in a given month, the same number of visitors was established. The actual number of people for each hour of museum use is difficult to determine. As a result, there may be slight underestimates or overestimation of internal heat and moisture gains at individual times. This affects the course of parameter variability and, consequently, the value of the correlation coefficient. As shown in the monthly analysis from May to September, the correlation coefficients, both for temperature and relative humidity, exceed 0.8.

3.2 Energy consumption

To determine the accuracy of the calculation of energy consumption the differences in the enthalpy of external and supply air were calculated. Based on these difference, the hourly energy consumption for heating and cooling for air conditioning units serving the exhibition rooms was calculated. In 2015, air handling units delivered 663 990 kWh of energy for heating and 73 023 kWh of energy for cooling. It should be mentioned that these calculations do not include heat recovery.

The energy demands for individual air conditioning processes were calculated in WUFI®plus. In the calculations, as boundary conditions measured values of external and internal air parameters were assumed. Internal heat and humidity are also taken into account. According to simulation, the energy demand for space heating in 2015 amounted to 705 kWh. This value is 106% of the value calculated on the basis of the enthalpy difference in the air-handling unit. The Figure 3 shows the course of variation for energy consumption for heating based on the difference of enthalpy at the units and based on the simulation in the WUFI®plus program.

IOP Conf. Series: Materials Science and Engineering **415** (2018) 012026 doi:10.1088/1757-899X/415/1/012026

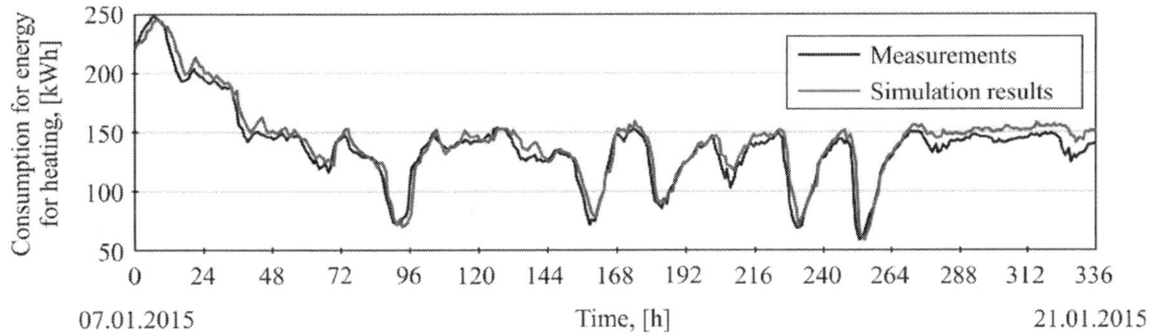

Figure 3. Comparison of calculated and actual heating energy consumption for the period 07 ÷ 21.01.2015

4. Theoretical calculation

To determine the impact of climate control scenarios on energy consumption in the calculations, three scenarios of full climate control including full air conditioning developed on the basis of ASHRAE, ICOM-IIC and MNK guidelines (Table 1), as well as scenarios considering only selected processes and their total absence have been included.

Table 1. Microclimate control option

	Temperature [°C]		Relative humidity [%]	
	Warm season	Cold season	Warm season	Cold season
MNK	18 ÷ 25	18 ÷ 21	40 ÷ 60	35 ÷ 55
ASHRAE AA	19 ÷ 23		42 ÷ 52[a]	
ICOM-CC - IIC	18 ÷ 25		40 ÷ 60	

[a] 47% is a long-term average from 2012 to 2016

At present, due to the lack of connection between the exhausting transfer grille and the exhaust duct, the air-conditioned zone (exhibition halls) is connected to the attic zone. The calculations additionally include variants assuming the separation of these zones by sealing the connection grille - exhaust duct. Calculations were carried out for two cases: with a fixed exchange rate equal to 3.0 h^{-1} (average airflow of air-conditioning units measured in 2015 related to the volume of exhibition halls) and with constant average airflow of units measured in 2015.

Air handling units serving exhibition halls work only on the outside air (due to the lack of space in the technical space it is impossible to install the air recirculation section). Hypothetically, however, it is possible to control ventilation due to the permissible concentration of CO_2 (the maximum permissible level of CO_2 concentration was assumed at 1 000 ppm; the minimum fresh air quantity for hygienic reasons is 20%).

The calculation was made taking into account heat recovery or its absence. Taking into account all assumptions resulted in the development of 29 variants of simulation.

4.1 Microclimate shaping

As a result of the calculations annual courses of variability of temperature and relative humidity of air in exhibition halls were obtained. To determine the "zero state", calculations were made assuming passive shaping of the microclimate, followed by heating, humidification and cooling with dehumidification.

If microclimate is shaped passively and with the air change rate (*ACR*) at 0.1 h^{-1}, annual fluctuations of temperature in the range 0.8 ÷ 28.4°C (with annual average 13.5°C) and relative humidity in the range

50.6 ÷ 91.7% (with annual average 76.5%) were obtained. With only infiltration, the CO_2 concentration exceeded the permissible value of 1 000 ppm. The maximum concentration was 1990 ppm and the average was 834 ppm. Based on the calculations, it was found that the *ACR* at which the concentration of CO_2 does not exceed 1000 ppm is 0.5 h^{-1}. The maximum concentration decreased to 882 ppm, and the average to 447 ppm. A change in *ACR* resulted in a change in the parameters of the indoor air. The minimum air temperature was -0.2°C, maximum 28.5°C, and the average 13.0°C. Relative humidity ranged from 37.8 ÷ 95.5% with an average of 69.6%.

To meet the requirements of the National Museum in Krakow it is necessary to introduce heating in the winter as well as year-round dehumidification of the air. Increasing the air temperature reduces its relative humidity, so in the next stage a simulation including heating was carried out. In the heating season, the minimum relative humidity decreased to 17.5%. The period in which relative humidity does not exceed 30% covers 927 hours, which is 10.6% of the year. In the summer, relative humidity still exceeds 60% (2138 hours, which is 24.4% per annum). In the summer there is little overheating (maximum temperature reached 28.6°C). The time of exceeding the recommended temperature of 25°C was 424 hours, which is 4.8% of the year. However, 45% of this period are exceeded only 1 K. According to the above analysis, to maintain even the least restrictive requirements, it is necessary to use full air conditioning.

4.2 *Energy consumption*

Given the current usage (compliance with the requirements of the National Museum in Krakow and the connection of exhibition halls with an attic, 85% heat recovery) as a result of the calculations annual energy demand for heating equal to 269 987 kWh were obtained. The maximum hourly energy demand amounted to 11 393 kW. The period in which heating was necessary comprised 5972 hours, which is 68% per annum. In general, the heating period covered the period from January to the end of March and from November to December. However, from April to June there were periods in which heating was required, a similar situation took place in September and October. In April there was a need for heating in almost 91% of the month, in May in 63%, in September in 80%. Cooling is required only in 8% per annum. The energy demand for this process is 11393 kWh. In order to dry the air to the desired level of relative humidity during the year, 226 522 kg of water should be drained out of it. The energy demand for dehumidification was therefore 430 391 kWh [11]. The period when dehumidification is required is 2 920 hours, which is 33% of the year. In order to maintain the minimum required humidity, it is necessary to bring to it, annually, 233 192 kg of water [11]. To evaporate this amount of water, it is necessary to bring 161 939 kW of energy. The period in which air humidification was required amounted to 4 106 hours, which is 47% of the year. Depending on the calculation variants, the share of energy demand for individual air treatment processes may differ from each other (Figure 4).

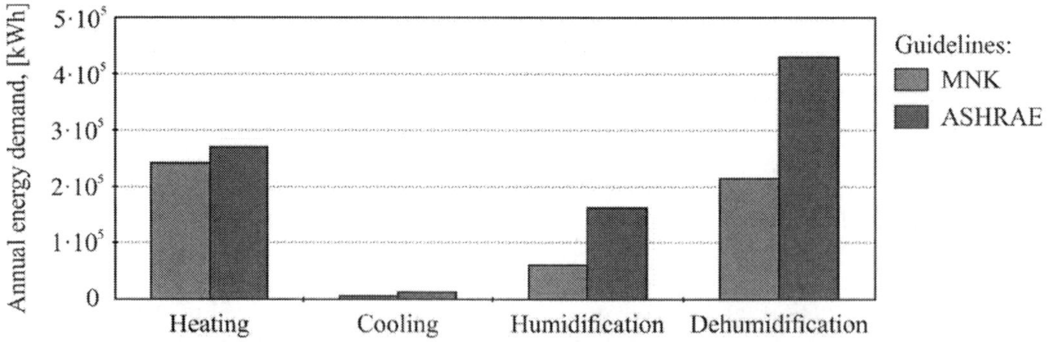

Figure 4. Annual energy consumption for individual processes for ASHRAE and the National Museum in Krakow (MNK) guidelines

Separation of the exhibition rooms from the attic results in a reduction of the heating energy demand, in the case of work only in fresh air by 22% and by 2% in the case of ventilation control based on the permissible CO_2 concentration. Introduction of ventilation control based on CO_2 concentration results in a significant reduction in the demand for heating by about 76%.

For variants assuming work only on outdoor air, separating the attic gives savings by an average of 20%. In the case of ventilation control based on CO_2 concentration, this separation results in an increase in cooling demand by up to 50%. The cooling demand decreases in case of change to ventilation control based on the CO_2 concentration (savings amounting to 80% in the case of a connection with an attic and 60% without this connection). The narrowing of the permissible variations of temperature and relative humidity of the air resulted in an approximate 59% increase in the energy demand for cooling.

The change of the requirements regarding the formation of the microclimate to the recommended by ICOM-CC and IIC resulted only in the increase of the energy demand for humidification (65% for calculations taking into account work only on fresh air and 140% for calculations with CO_2 concentration control) and slight reduction for dehumidification (about 9%). In the case of changes to the ASHRAE guidelines energy demand for heating increased by an average of 12%, for cooling of 185% and for humidification and dehumidification of 244% and 112%, respectively. In the case of variants in which CO_2 control was taken into account, almost 3.5 times more energy demand for humidification was induced.

To estimate the actual benefits or losses resulting from the change of use, all processes (heating, cooling, humidification, dehumidification) should be analysed all together. In the total energy demand, the energy demand for cooling was omitted. This is due to the fact that in the case of the Cloth Hall, air cooling is carried out together with dehumidification (in a wet cooler). By separating the exhibition rooms from the attic, a 15 ÷ 20% reduction in energy demand was obtained (depending on the ventilation capacity) – Figure 5. Controlling ventilation based on the permissible CO_2 concentration using air recirculation is of course the most economically advantageous variant. If such a solution was technically possible, 70% savings could be made for the Cloth Hall. The application of the ICOM-CC - IIC guidelines does not significantly increase the energy demand. The total increase in the amount of energy required in this case results from the higher relative humidity in the winter than in the MNK guidelines. In order to meet the requirements of ASHRAE, an increase of 65% of the energy demand should be expected.

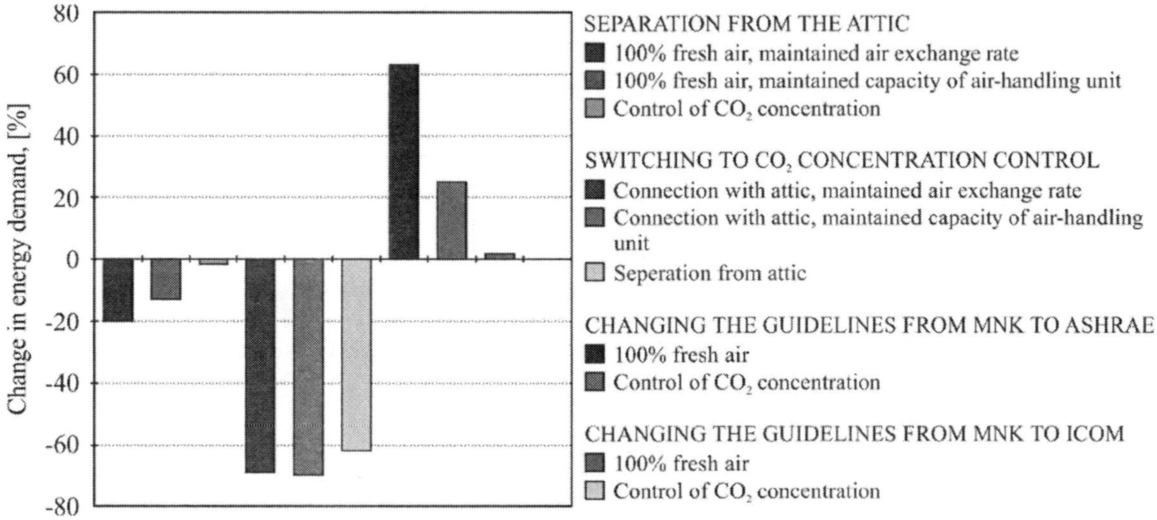

Figure 5. The average change in the total energy demand for air conditioning

5. Conclusions

Due to the specificity of historical building, reduction of the energy consumption to "0" is practically impossible. The construction of the building and the conservation requirements limit the thermal insulation of the partitions. Maintaining appropriate, for the museum collections, microclimatic conditions requires full air-conditioning (heating, cooling, humidifying, dehumidifying). As shown in the above analysis, savings should be sought in the correct technique of microclimate control, understanding of the technology used as well as the manner in which a given installation is designed, executed and operated. Where possible, museums that are housed in historical buildings should be treated as a case study. Wherever applicable, the requirements regarding the microclimate parameters should be relaxed to the highest extent possible.

6. Acknowledgments

Measurements and computational analysis were made as part of the "Heriverde" research project funded by The Polish National Centre for Research and Development.

7. Literature

[1] Italian Standad UNI 10969:2002 Cultural Heritage. General principles for the choice and the control of the microclimate to preserve cultural heritage in indoor environments

[2] EN 15757:2010 Conservation of cultural property - specifications for temperature and relative humidity to limit climate-induced mechanical damage in organic hygroscopic materials

[3] Michalski S 2002 Double the life for each five-degree drop, more than double the life for each halving of relative humidity *13th ICOM-CC Triennial Meeting Rio Janeiro* (James & James, London)

[4] Sedlbauer K 2001 Prediction of Mould Fungus Formation on the Surface of and Inside Building Components, *Fraunhofer Institute for Building Physics*

[5] American Society of Heating, Refrigerating and Air-Conditioning Engineers 2007 *Ashrae Handbook: Fundamentals chap. 21 Museums. libraries and archives* (Atlanta, Ga.: Ashrae)

[6] http://www.icom-cc.org/332/-icom-cc-documents/declaration-on-environmental-guidelines/#.WrPqZ5fkU2w (29.03.2018)

[7] Bratasz Ł (ed) 2012 Zarządzanie klimatem w muzeach: ochrona zbiorów i energooszczędność *Ochrona Zbiorów* **02** Seria wydawnicza Narodowego Instytutu Muzealnictwa i Ochrony Zbiorów oraz Muzeum Narodowego w Krakowie

[8] Antretter F, Sauer F, Schöpfer T and Holm A 2011 Validation of a hygrothermal whole building simulation software Proc. Int. Conf. *12th Conference of International Building Performance Simulation Association* (Sydney) (Preprint https://wufi.de/literatur)

[9] Sadłowska-Sałęga A 2014 Możliwości i ograniczenia obliczeniowego wyznaczenia cieplno-wilgotnościowych parametrów powietrza w budynkach historycznych (ang. Possibilities and limitations of computational designation of the temperature and humidity in historical building) *Dissertation Cracow University of Technology*

[10] PN-84/E-02033: Oświetlenie wnętrz światłem elektrycznym (Interior lighting with electric light)

[11] Kupczak A, Sadłowska-Sałęga A, Krzemień L, Sobczyk J, Radoń J and Kozłowski R 2018 Impact of paper and wooden collections on humidity stability and energy consumption in museums and libraries *Energy and Buildings* vol **158** pp. 77–85

IOP Conf. Series: Materials Science and Engineering **415** (2018) 012027 doi:10.1088/1757-899X/415/1/012027

Effect of the variable air volume on energy consumption - case study

P Kapalo[1] and N Spodyniuk[2]

[1]Technical University of Kosice, Faculty of Civil Engineering, Kosice, Slovakia
[2]Lviv Polytechnic National University, Department of Heat, Gas Supply and Ventilation, Lviv, Ukraine

E-mail: peter.kapalo@tuke.sk

Abstract. At present, administrative buildings consume more energy than in the past, and they also contribute to environmental pollution. It is an effort to build buildings with minimal energy consumption. The energy consumed to operate the ventilation air is not insignificant. Techniques and control systems allow different ways of saving energy. However, all the savings measures should also take into account the quality of the internal environment for the health of people. This article compares a ventilation system with a constant flow of ventilating air with an air-conditioning system with varying air flow rates ensuring the required air quality.

1. Introduction

At present, administrative buildings are large energy consumers. From the study [1], about 12% of the world carbon dioxide (CO_2) releases in 2010 are generated, directly or indirectly, by producing the needed energy for buildings in order to ensure their adequate operation. Energy use in buildings and the desire to reduce the undesirable CO_2 production from polluting atmospheres is also a motive for making energy use more efficient and reducing overall energy consumption. When reducing the energy intensity of a building, it is also necessary to take into account the quality of the indoor environment, the reduction of which can cause various health problems as well as reduce the performance of employees [2].

One of the options for reducing energy consumption is the use of a variable air volume (VAV) system. A similar issue was dealt with by Eskin and Turkmen [3], who analyzed the annual energy requirements needed to heat and cool the office building in the Turkish climate. Their analysis shows that 48% of the energy was used for heating and 52% for cooling the building. Different studies can see similar proportions in Central Europe as well - depending on the building envelope and climatic conditions that are different.

Despite the fact that the natural advantages of the VAV system in terms of its energy efficiency are well understood, the designer is often subjected to complications due to different building characteristics, different climatic conditions and several zones. In the hot and humid environment, it is often found that the benefits of VAV systems are not fully realized [4]. Therefore, Sekhar [4] examined five different buildings in an environment whose internal load remained constant, but the thermal load caused by the envelope of the building and the orientation differs between the five buildings. This examines the presence of a variety of cooling loads, which then leads to the study of potential energy savings by the VAV system.

In the area of reducing energy consumption, Aunur et al. [5] compared VAV with Variable Refrigerant Flow (VRF) system. In their work they found that a large amount of energy is consumed by the VAV itself. They found that the VRF system is more economical than the VAV system. Huang et al. [6] examined the simulation of the air conditioning control system, where they presented energy savings of 17%. According to Liang et al. [7] it is possible to monitor the ventilation air and optimize the VAV system to achieve energy savings about 28%.

In our article, we compare ventilation system with constant flow of ventilation air with ventilation system with varying ventilation air flow, which is controlled on the basis of the CO_2 concentration at the locations of the persons in the building.

2. Description of the building and ventilation system

2.1 Description of the building

The administrative building is rectangular, without protruding and receding parts. The building has a total of six above-ground floors. On the first floor there is a porter, a room for visits, a sanitary facility, a dining room and a food dispenser. From the second up to the fifth floor there are 44 offices. On the last above-ground floor there is a boiler room and an air-conditioning machine room.

Figure 1. Southeast (left) and northeast (right) view of the building [Source: Jankura D.]

The building should be located in Poprad city, where the external calculation temperature is -16°C and the outside summer calculation temperature is +32°C. The heat transfer coefficient for the peripheral wall U = 0.11 W/(m^2·K), roof U = 0.09 W/(m^2·K), window U = 1.1 W/(m^2·K), door U = 1.9 W/(m^2·K) and thermal resistance for the floor on the ground R = 6.64 (m^2·K)/W. The total design heat capacity is 192 kW. The total amount of heat is 131 kW. Heat gains were calculated separately for months for each building's hours of operation from 7:00 am to 5:00 pm and are listed in Table 1.

Table 1. Heat gains after months

Time	March Q (kW)	April Q (kW)	May Q (kW)	June Q (kW)	July Q (kW)	August Q (kW)	September Q (kW)	October Q (kW)
07:00	67	79	89	86	84	82	60	34
08:00	85	89	95	92	91	93	79	63
09:00	86	89	93	90	91	93	84	75
10:00	77	80	83	82	83	84	78	71
11:00	63	66	70	71	71	71	66	60
12:00	53	58	61	63	64	62	58	52
13:00	59	63	68	68	69	67	63	57
14:00	68	71	75	75	76	75	70	64
15:00	73	76	80	80	80	80	73	66
16:00	71	75	81	80	79	79	70	57
17:00	59	68	76	76	74	72	56	38

2.2 Description of ventilation system

Equipment for air conditioning in the building is divided into five independent air-conditioning systems from an operational and hygienic point of view:

- ventilation and cooling of office space,
- ventilation and cooling of the dining room,
- ventilation of food dispensers,
- ventilation of sanitary facilities.

For the entire building, the heat produced in the boiler room is located on the 6th floor. Cooling is ensured by using chillers placed on the roof.

In the building there is dominant office ventilation, for which a separate air-conditioning unit is used, located in the technical room on the top floor of the building. In all offices, are used comfortable slot connections for the air supply, which are located in the overhead construction. The hoses have a long range of jets and the possibility of changing the flow (horizontal, vertical). To remove dirty air, shock absorbers are built into the roof structure.

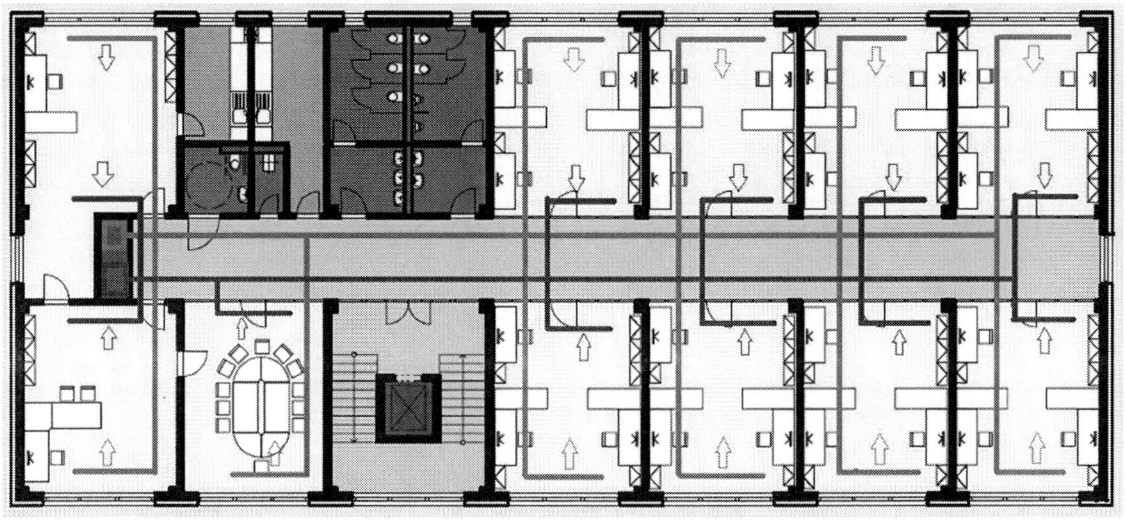

Figure 2. Typical floor plan of the administrative part of the building

3. Calculation of the need heat and cold for ventilation air

The need for heating and cooling for the heat treatment of ventilation air is designed for two versions of the air conditioning system.

3.1 First version of ventilation systems

The first air-conditioning system has a given constant flow of ventilating air during working hours (7:00 - 17:00), which is installed on the total projected number of people in the building. The total projected number of people is 105 people. Calculations are based on the volume flow of air V = 1.31 m^3/s, air density ρ = 1.27 kg/m^3 in winter and air density ρ = 1.18 kg/m^3 in summer, with a thermal capacity c_{pv} = 1.005 kJ/(kg·K) in winter and c_{pv} = 1.01 kJ/(kg·K) in summer. Number of hours of daily operation $n_{h,o}$ = 10 h/d. Air temperature control is carried out in the engine room of the air conditioner in accordance with the temperature of the outside air.

3.2 Second version of ventilation systems

The second air-conditioning system has a variable air flow rate during the working hours (7:00 - 17:00), which is controlled by a CO_2 sensor located in each room. The Variable Air Flow (VAV) regulator is installed in front of each room at the inlet and outlet air ducts. If the CO_2 concentration increases in the room, the VAV valve will open to the necessary position. If the CO_2 concentration decreases, the VAV valves will close. The central air handling unit senses the air in the air duct and adjusts the airflow as required. Thus, the reduction of heat consumption for air heating (cooling), as well as the reduction of electricity consumption for fans and pumps is achieved. For the second variant, the calculations take into account the air parameters, as in the first variant.

The maximum estimated number of people is 105 people. Figure 3 shows the occupancy diagram of the building by persons.

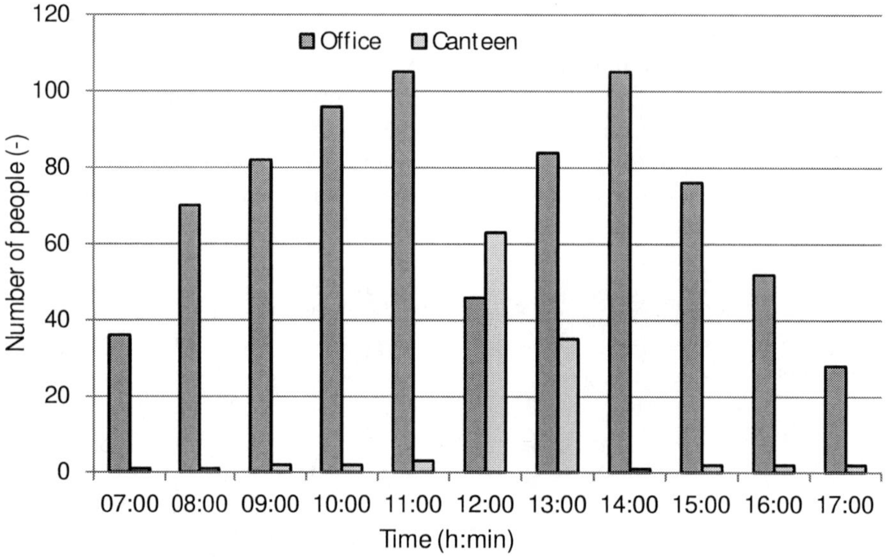

Figure 3. Building occupancy diagram

Due to the varying number of people in the building, it is necessary to consider varying air flow rates. Dosage of fresh air is set at 45 $m^3/(h.pers)$. Predicted air flow rates are shown in Table 2. Average volume air flow rate is 0.89 m^3/s.

Table 2. Ventilation air flow rates

Hour (h:min)	Number of people (-)	Volume air flow V (m^3/h)	Volume air flow V (m^3/s)
07:00	36	1,620	0.45
08:00	70	3,150	0.88
09:00	82	3,690	1.03
10:00	96	4,320	1.20
11:00	105	4,725	1.31
12:00	46	2,070	0.58
13:00	84	3,780	1.05
14:00	105	4,725	1.31
15:00	76	3,420	0.95
16:00	52	2,340	0.65
17:00	28	1,260	0.35

3.3 The heat demand for heating the ventilation air

The heat demand for air heating $Q_{t,\,vent}$ (kWh/year) is calculated according to the following formula:

$$Q_{t,\,vent} = V_e \cdot \rho_e \cdot c_{pv} \cdot n_h \cdot D_{vent} \qquad (1)$$

where: V_e - volume of outside air flow (m^3/s); ρ_e - outdoor air density in winter months (kg/m^3); c_{pv} - specific heat capacity of the ventilation air at constant winter pressure (kJ/(kg·K)); $n_{h,h}$ - hours of operation of the air conditioner (h); D_{vet} - the number of degree days in the period required to heat (K·d) [8].
The received heat demand for air heating for the first and second variants is given in the table. 3

Table 3. The heat demand for heating the ventilation air

Month	First version Without VAV valve		Second version With VAV valve	
	D_{vent} (K·d)	$Q_{t,vent}$ (kWh)	D_{vent} (K·d)	$Q_{t,vent}$ (kWh)
January	725	12.410	725	8.438
February	607	10.390	607	7.064
March	517	8.849	517	6.017
April	336	5.751	336	3.910
May	177	3.030	177	2.060
June	84	1.438	84	978
July	28	479	28	326
August	56	959	56	652
September	174	2.978	174	2.025
October	350	5.991	350	4.073
November	498	8.524	498	5.796
December	648	11.092	648	7.541
Total (kWh/year)	Σ	71.891	Σ	48.879

3.4 The cool demand for cooling the ventilation air
The heat demand for air heating $Q_{c,\,vent}$ (kWh/year) is calculated according to the following formula:

$$Q_{c,\,vent} = V_e \cdot \rho_e \cdot c_{pv} \cdot H_c \qquad (2)$$

where: V_e - volume flow of outdoor air in the summer months (m^3/s); ρ_e - density of outdoor air in the summer months (kg/m^3); c_{pv} - specific heat capacity of the ventilation air at constant summer pressure (kJ/(kg·K)); H_c - number of cooling stages (K·h) [8].
The resulting cooling air requirements for the first and second versions are in Table 4.

Table 4. The cool demand for cooling the ventilation air

Month	First version Without VAV valve		Second version With VAV valve	
	H_c (K·d)	$Q_{c,vent}$ (kWh)	H_c (K·d)	$Q_{c,vent}$ (kWh)
March	17.5	25	17.5	18
April	34.6	50	34.6	35
May	315.2	456	315.2	317
June	574.2	832	574.2	577
July	768.4	1.113	768.4	772
August	554.8	803	554.8	557
September	241.4	350	241.4	242
October	11.0	16	11.0	11
Total (kWh/year)	Σ	3.645	Σ	2.528

4. Results and discussion

In order to evaluate the systems, it is also necessary to know the cool demand to remove heat flows from internal heat sources (people, computers, lighting, food etc.), from external heat sources (from radiation, heat transfer through building structures) and the need for electric power to drive fans and pumps. The individual partial calculations are not processed in this article but the results are shown in Table 5.

Table 5. Comparison of energy consumption

	First variant Without VAV valve (kWh/year)	Second variant With VAV valve (kWh/year)
The heat demand for heating the ventilation air	71.891	48.879
The cool demand for cooling the ventilation air	3.645	2.528
The cool demand in the allocation of warm streams from internal heat sources	8.496	5.735
The cool demand in the allocation of warm streams from external heat sources	116.851	116.851
Electricity required for fans and pumps	11.412	5.005

From Table 5, we can see that the heat demand for air heating lower by 32%. Also, the cool demand for air cooling is lower by 31%. The heat demand from internal heat sources is 33% lower. The power requirement for fan and pump drive is 56% lower. The need for cooling to reduce the heat load from outdoor heat generators, which, in the first variant, is 55% remains unchanged and strongly influences the need for energy.

Despite the fact that the second variant has much less energy needs, the total can be calculated after counting the losses for individual energies with a total reduction of only about 4%.

In the second variant, it is necessary to purchase and install a CO_2 sensor in each ventilated room, VAV valve and control valve pair. After a full review of investment and operating costs (investment costs is approximately 25,230 and operating costs for system with VAV valves is 17,108 Eur/year, systems without VAV valves is 19,533 Eur/year), it can be assumed that after 11 years, CO_2 sensors will be paid, VAV valves and control valves, and after 11 years the VAV valve system will be profitable.

5. Conclusion

Based on the analysis of the ventilation system it can be stated that it is appropriate to install a system with VAV valves for all buildings. Although this system is beneficial for the first period, it is not always beneficial for economic reasons.

However, it is also necessary to look at the environment. By regulating the airflow of the VAV valves, we save about 18% of the energy that needs to be produced and often air pollution. Since, due to 4% annual financial savings, it is not beneficial for an investor to increase both investment and operational costs, so it's beneficial to save the environment. Therefore, it would be necessary for the state to contribute to subsidies to environment-friendly systems.

Acknowledgments
This article was elaborated within the framework of the project VEGA 1/0697/17 "Hygienic water audit platform as transition tool to Legionella free water and HVAC systems in hospitals".

References

[1] Pachauri R K, Meyer L A et al. 2014 Climate Change 2014 *Synthesis Report Contribution of Working Groups I, II and III to the Fifth Assessment Report of the Intergovernmental Panel on Climate Change* (IPCC Geneva Switzerland) p 151
[2] Wargocki P 2017 Improving indoor air quality improves the performance of office work and school work and provides economic benefits *Healthy and Sustainable Buildings Conference* (IAQ 2007 Baltimore)
[3] Eskin N and Türkmen H 2008 Analysis of annual heating and cooling energy requirements for office buildings in different climates in Turkey *Energy and Buildings* **40(5)** 763-773
[4] Sekhar S C 1997 A critical evaluation of variable air volume system in hot and humid climates *Energy and Buildings* **26(2)** 223-232
[5] Aunur T N, Hwang Y and Rademarch R 2008 Simulation comparison of VAV and VRF air conditioning systems in an existing building for the cooling season *Energy and Buildings* **41(11)** 1143-50
[6] Huang W Z, Zaheeruddin M and Cho S H 2006 Dynamic simulation of energy management control functions for HVAC systems in buildings *Energy Conversion and Management* **47(7)** 926-943
[7] Liang W, Quinte R, Jia X and Sun J-Q 2015 MPC control for improving energy efficiency of a building air handler for multi-zone VAVs. *Building and Environment (Elsevier)* **92** 256-268
[8] Dahlsveen T, Petras D et al. 2008 *Energy auditing and building certification* (JAGA GROUP Bratislava, Slovak) p 166

Analysis of ventilation air exchange rate and indoor air quality in the office room using metabolically generated CO_2.

Nowak K., Nowak-Dzieszko K., Marcinowski A.

Cracow University of Technology

Abstract. The article presents measurements carried out in an office room using metabolically generated carbon dioxide. The measurements were conducted using concentration decay gas tracer method. Increase of the concentration of CO_2 in the room occupied by workers allowed for evaluation of internal air quality and thermal work conditions. Analysis of the CO_2 concentration decay conditions after leaving the space by users was used to determine the efficiency of ventilation system and room envelope airtightness. The obtained results were used to calculate specific airflow rate of natural ventilation system being used in the analysed room.

1. Introduction

Due to the increasing concerns regarding building energy demand reduction of energy for ventilation is one of the main targets in the designing process. Ventilation cannot be analyzed without influence of infiltration (uncontrolled and unintentional airflows) which also affects total energy losses. On the other hand, ventilation system provides the fresh and removes contaminated and humid air to and from the interior environment which is crucial for the internal air quality. Indoor air quality is significantly associated with ventilation. Low ventilation air exchange can affect the inside air quality and going further the people's health negatively.

The efficiency of ventilation system can be evaluated using gas tracing method. One of the gases which may be used in those measurements is carbon dioxide. The CO_2 generated by users can be treated as the tracer gas. It makes the tests cheaper as no mechanical gas emitters are required. Gas tracing measurement with CO_2 allows for simultaneous evaluation of ventilation efficiency, building airtightness and air quality. This approach to the problem was suggested by Persily [1] who presented some techniques to evaluate the building ventilation efficiency and indoor air quality analyzing carbon dioxide concentration within the internal space. Similar problems were described by Bulinska [2] who used metabolically generated carbon dioxide to determined efficiency of ventilation systems in the buildings. Zhang [3] has used the same method testing the ventilation rates of dormitories and offices. Analysis of increasing concentration of CO_2 when buildings are occupied by users allow for evaluation of indoor air quality. Such measurements were described by Benedettelli [4] and Cichowicz [5]. In both cases quality of internal air in the university facilities were analyzed.

2. Carbon dioxide in the air – level adjustments and restrictions

Polish legal acts do not specify admissible concentrations of carbon dioxide in atmospheric air and in rooms intended for permanent residence, for example for dwellings. Because of that, the standards and recommendations of other European countries and the United States (ASHRAE) [6], [7] are most commonly used to determine acceptable concentrations. Per those requirements the upper level of carbon dioxide concentration for permanent residence of people should not exceed 1000 ppm. To keep this requirement about 27 m^3/h of fresh air per person should be supplied per hour to the internal leaving space.

Polish regulations define the permissible concentrations and intensities of harmful factors in the working environment [8]:

- Permissible exposure limit (NDS) – the weighted mean value of concentration, the impact on an employee of 8 hours per day and the average weekly working time during his/her working activity should not result in negative changes in his/her health and health of his/her future generations;
- Short term permissible exposure limit (NDSCh) – mean value of concentration which should not cause negative changes in the worker's health if it is present in the work environment for not more than 15 minutes and not more than 2 times during the work shift, with the a time interval not shorter than 1 hour.
- Ceiling exposure limit (NDSP) – value of concentration which due to health or life threat of the worker, cannot be exceeded in the work environment at any time.

According to this regulation NDS of carbon dioxide is 5000 ppm (9000 mg/m^3), and NDSCh cannot exceed 15 000 ppm (27 000 mg/m^3).

3. Air quality of the working and leaving internal environment

In every enclosed spaces where a human resides, the concentration of carbon dioxide increases because of depletion of the oxygen due to process of breathing. The carbon dioxide concentration in exhaled air is about 40 000 ppm [7]. The amount of CO_2 produced by human may differ depending on the body weight and level of its metabolic activity [6][7].

The CO_2 exhaled stream is related to the amount of oxygen consumed. The amount of oxygen consumed by one person in m^3/h can be determined from the equation [7]:

$$V_{O2}=3.6\,\frac{0.00276*A_D*M}{0.23R_q+0.7}$$

Where: A_D is DuBois surface (surface of human body) in m^2, M is a metabolic rate, met (1 met is equal to 58.2 W/m^2, R_q is a respiratory quotient which can be assumed depending on the intensity of activity performed by the human, from 0.83-1. The DuBois surface area can be taken as 1.8 m^2 for an average adult.

Permissible levels of carbon dioxide in non-residential spaces are regulated by the standard PN-EN13779 [9]. It introduces four acceptable levels of carbon dioxide concentration depending on the selected level of indoor air quality. The table 1 shows the permissible values for the carbon dioxide concentration according to the air quality category.

Table 1. Permissible values for carbon dioxide concentration to the air quality category per standard PN-EN 13799 [9]

Category ofair quality in the room	Increment of CO_2 concentration, above CO_2 level in external air, ppm
High quality of indoor air	350
Medium quality of indoor air	500
Moderate quality of indoor air	800
Low quality of indoor air	1200

IOP Conf. Series: Materials Science and Engineering **415** (2018) 012028 doi:10.1088/1757-899X/415/1/012028

4. Experimental method

The measurements were conducted using gas tracer concentration decay method. The methodology of measurements is described in standard [10]. Method was also described by Sherman [11]. This method is based on the analysis of decay of concentration of a tracer gas mixed with the air. Metabolically generated carbon dioxide was used as the tracer gas in the measurements. Carbon dioxide is one of the natural compound of atmospheric air. It is an odourless, colourless non-flammable gas, heavier than the air. The density of carbon dioxide is almost 1.5 higher than that of air [7]. Emission of gas to the atmosphere is usually caused because of combustion of chemical compounds containing carbon atoms. Carbon dioxide is also released into the atmosphere as a result of human and animal activity.

Measurements with the gas tracing method were conducted using 3 sensors Rotronic CP11 allowing the continuous measurements of CO_2 concentration, temperature changes and relative humidity (fig.1).

Figure 1. Sensor Rotronic CP11 used for measurements

5. Measurements

The analysed office room is located in the five-storey office building in Cracow at Oleandry street (figure 2). Office room is located at second story of the building and does not have direct contact with outside corridor. The room is designated to perform light computer office work for four working people at the same time. The measuring zone (figure 3) has one wall with direct contact with the external environment with two 1.95m width on 1.90m height windows. Three other walls have contact with other rooms. Door 2.00m height and 0.85 width dividing two office rooms are located on the south wall of the room. There is natural ventilation system with one ventilation duct 18cm on 18cm located 2.35m above floor level shown in figure 2b.

Considered zone is constantly occupied by four people working from 8 am to 4 pm with one half an hour break at 12 am. Volume of the office room is about $58m^3$.

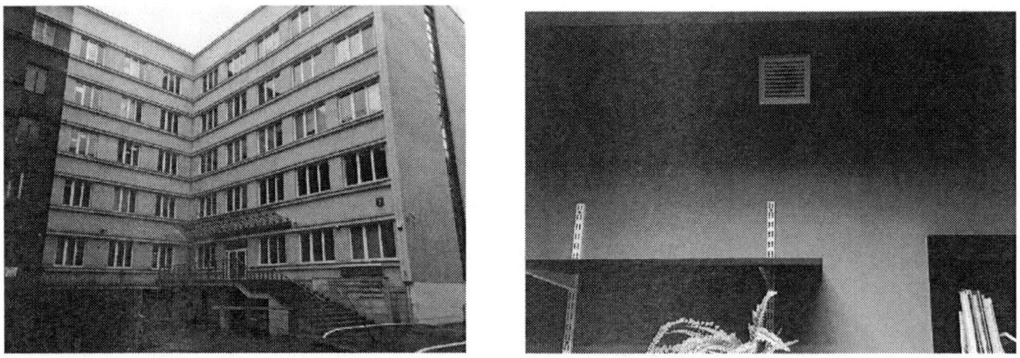

Figure 2. a) office building at Oleandry street, b) ventilation duct in the office room

Figure 3. View of measuring zone

Detectors were located in three different places of the zone in a 2 meters distance from workers to avoid direct contact of the exhaled air with the sensors. Such arrangement ensured appropriate results of measurements, because high concentration of air exhaled gas by people. Standards [6] suggest to keep at least distance of 2m from the occupant. Per standard [10] the concentration of carbon dioxide in a representative measurement place should not differ by more than 10% from the mean concentration value in the measurement zone. Before performing the test all sensor were calibrated in accordance with the manufacturer's instructions.

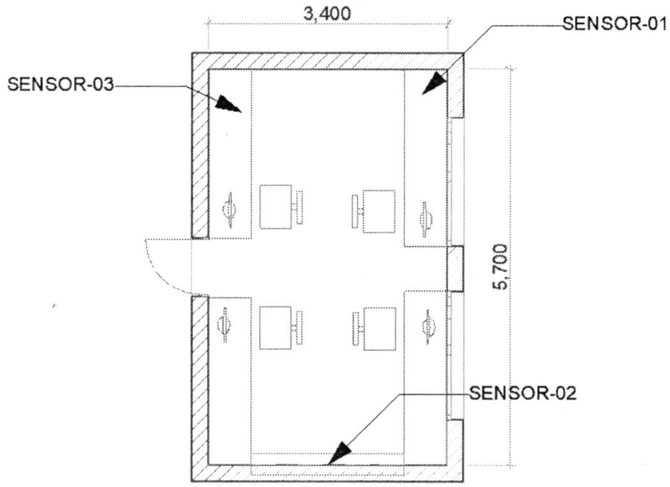

Figure 4. Location of sensors in the measuring zone

All measuring devices were located inside considered room. Sensors 1 and 3 were placed on desks 0.75 m above floor level, sensor 2 was placed on a shelf mounted along the wall located 1.75 m above floor level (figure 3).

Measurements were conducted for four working days between 19.09.2017 and 22.09.2017. Data were collected continuously and recorded every 60 seconds. Analysis below will be divided on two parts, first one while room was occupied by workers and the second part after leaving the room by workers (analysis of CO_2 concentration decay).

5.1. Office room occupied by workers, 8 am to 4 pm

While the office room was occupied by workers the measurements allowed to analyze the internal changes of metabolically generated CO_2 concentration, temperature and relative humidity.

Figure 5 presents changes of CO_2 concentration during one working day from three sensors. The external CO_2 concentration at 8 am, measured by separate sensor was equal to 389 ppm. The initial concentration inside at this time is at the level around 1000 ppm due to the presence of people in the room before the beginning of measurements.

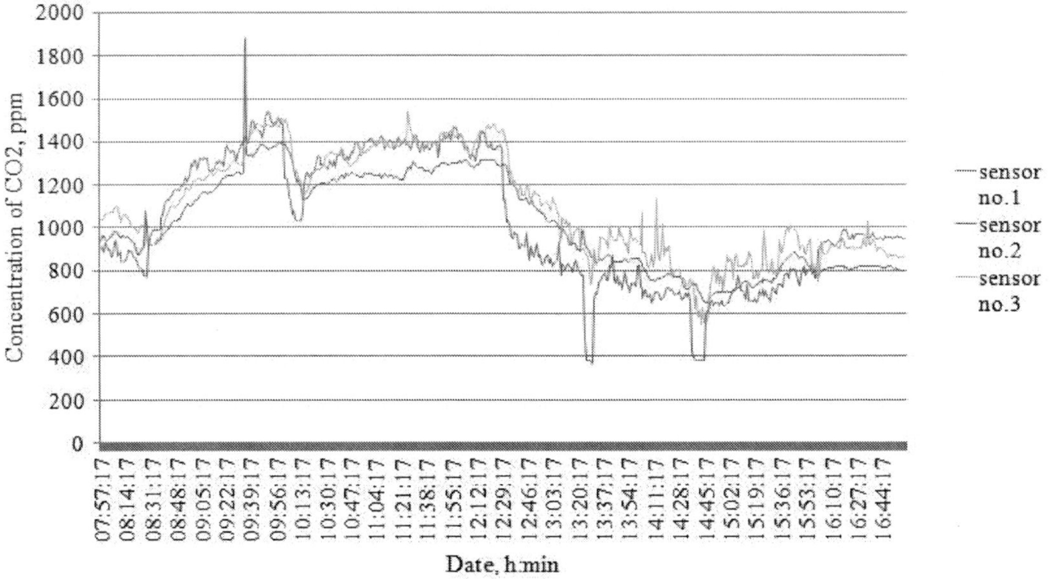

Figure. 5. Carbon dioxide concentration inside the room – 19.09.2017

The sharp increase of carbon dioxide concentration can be observed up to 10 am can be observed. Around 10 am workers opened the window as they started to feel uncomfortable. The CO_2 concentration was at level of 1500 ppm. After opening the window the concentration drop can be observed to about 1200 ppm. At 12 am the workers left the office on lunch break and it caused the decrease of CO_2 concentration. After the break till the end of working day windows inside were partially open so level of CO_2 is between 600 and 1000 ppm. Values recorded by sensor 1 are much lower comparing to other two as this sensor is located closest to the window. When the workers left the room due to the end of the working day the window was completely closed.

Based on the received data the quality of internal air was classified. The external concentration of CO_2 during entire measurement time was assumed at the level of 400 ppm. The result of the internal concentration of carbon dioxide was taken as the average value of all three sensors at 10:03 when the level of CO_2 was the highest. Results were compared with the referenced values found in the table 1, showing the increase in carbon dioxide concentration above the CO_2 level in the outdoor air.

Per the classification of standard [9] the indoor air can be qualified between moderate and low quality, which is very unsatisfactory result. The employees spent about 40% of their working time in conditions qualified by the standard as moderate quality of indoor air. Taking into account the ASHRAE standard [6], the concentration level determining conditions as comfortable are exceeded for more than 50% of the working time.

What is very important those unsatisfactory conditions were noticed despite of the fact that windows were opened during the measurements. Conditions would be much worse if the windows would be closed during entire working day.

Sensor used in the measurements allow for measuring of temperature and humidity changes. Results are presented in figure 6 and 7.

No sensor was used to measure the outdoor temperature, so the data concerning average external temperature were taken from measuring station Vaisala WXT520 located at the roof of Faculty of Physics and Applied Computer Science belonging to AGH University [12].

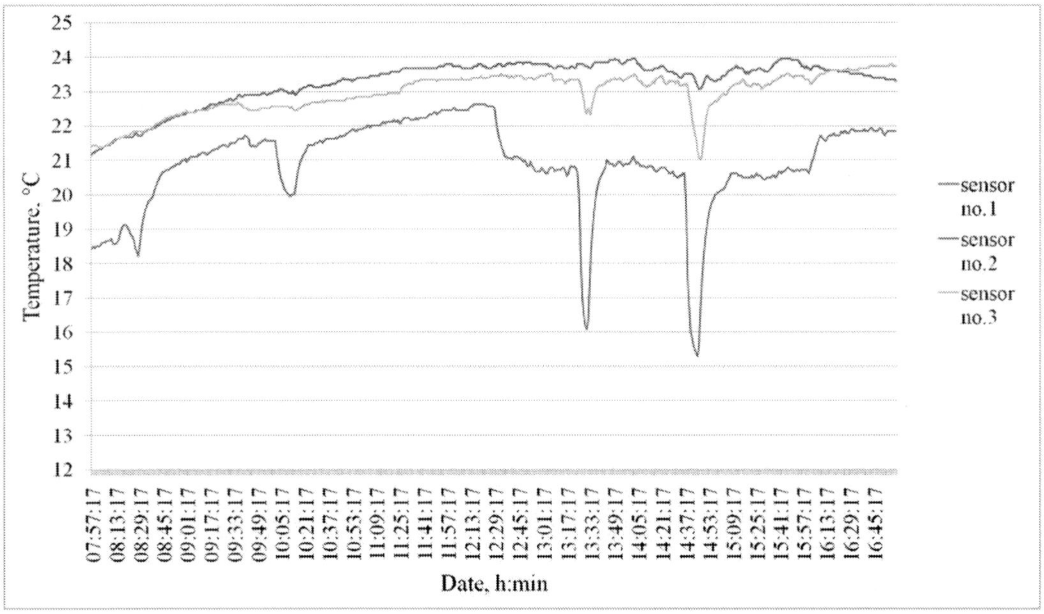

Figure 6. Indoor temperature recorded during measurements in the zone during first day of measurements

Figure 6 presents temperature readings from 3 sensors. The highest values of internal temperature have been indicated throughout the course of the tests by the sensor number 3 located farthest from the windows. Similar values were indicated by sensor number 2 which is located in the corner of the room, but some small temperature drops may already be noticed. Sudden drops of temperature are connected with opening of windows and local movements of cold outside air.

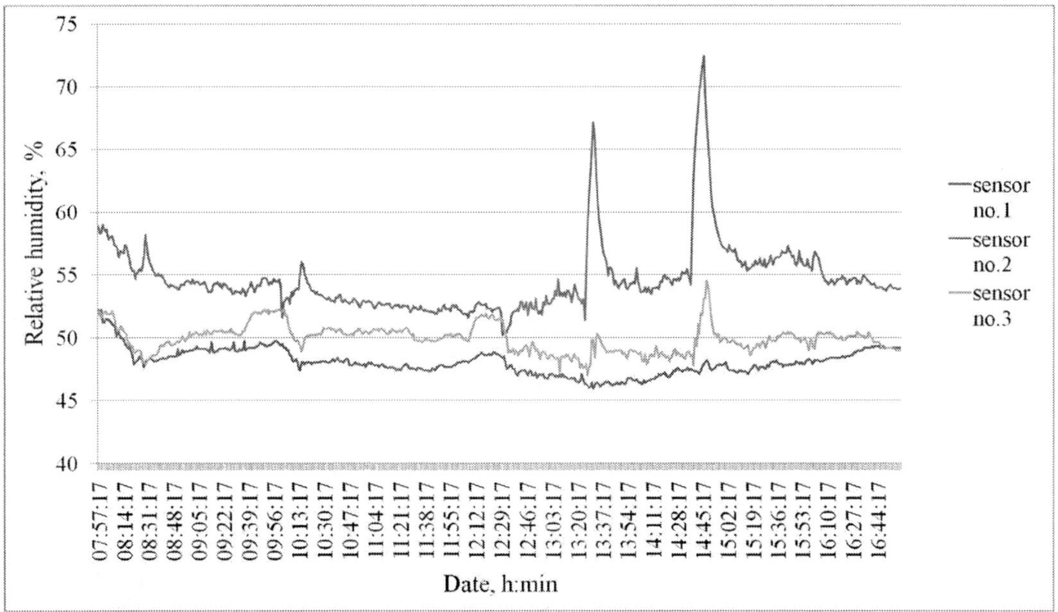

Figure 7. Indoor relative humidity recorded during measurements in the zone during first day of measurements

There are big differences in the values of temperature which is uncomfortable from the users point of view. People sitting in different parts of the room are affected by significantly different temperatures.

Figure 7 presents changes of relative humidity for the sensors. Again significant changes between values can be observed, for the senor located next to the window values are much higher due to much lower temperature.

5.2. Measurements of empty room.

The tests carried out on 19.09.2017 were continued when the employees left the measuring zone. Similar tests were carried in the next four days.

Figures 8-9 present concentration decay diagrams during four days after leaving the space by users. In some days users stayed longer but the data used for calculation of air flow were always taken when

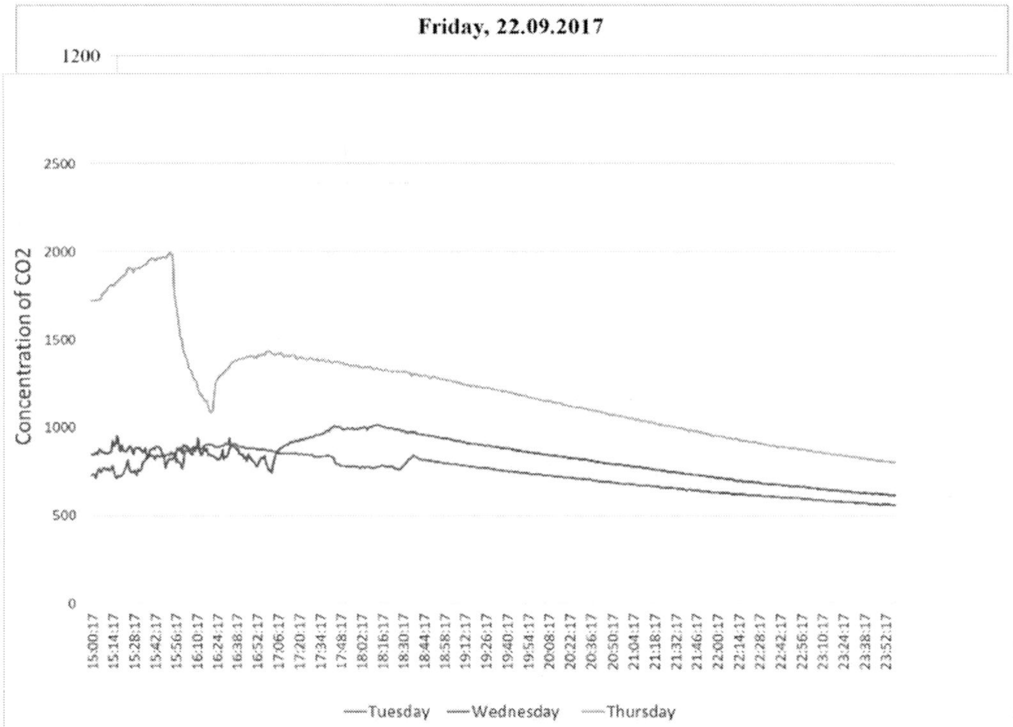

the room was empty. First three days are shown on one diagram, the Friday data are listed on the separate one as the measurement step was different.

Figures 8 thru 9. Concentration decay diagrams during four days of measurements.

The specific airflow rates were calculated separately for each day using two point decay method. As the decay period time is not strictly determined in the standard [10] different periods were analysed: one, three, six and eight hours from initial concentration levels. In order to calculate the air exchange rate, equation concerning 2 point decay was used.

$$\overline{N} = \frac{1}{t_2 - t_1} log_e \frac{C(t_1)}{C(t_2)}$$

where:

$\overline{N}$- time-mean specific airflow rate (1/h)

t – time (t_1: measurement start point, t_2: measurement end point) (h);

Table 2 presents initial and final values of carbon dioxide concentration at specific time ranges.

Table 2. Specific airflow rate calculations for each day

	Concentration of CO_2, ppm			
	Tuesday, 19.09.2017	Wednesday, 20.09.2017	Thursday, 21.09.2017	Friday, 22.09.2017
Initial concentration of CO_2	834	1014	1273	1038
1 hour decay period	753	913	1166	940
3 hours decay period	646	768	986	776
6 hours decay period	512	607	768	599
8 hours decay period	490	535	652	536

The specific values of air flows were calculated and presented in table 3.

Table 3. Table represents concentration of CO_2 in particular time intervals

	SPECIFIC AIRFLOW RATE CALCULATION [1/h]			
	Tuesday, 19.09.2017	Wednesday, 20.09.2017	Thursday, 21.09.2017	Friday, 22.09.2017
1 hour decay period	0.102	0.105	0.088	0.099
3 hours decay period	0.085	0.092	0.085	0.097
6 hours decay period	0.081	0.086	0.084	0.092
8 hours decay period	0.066	0.080	0.084	0.083
mean value	0.078	0.086	0.084	0.090
inside;outside temp	21;12	21;12	21;10.5	20.5;11
ΔT [K]	9.0	9.0	10.5	9.5

The mean values of specific airflow rate do not differ more than 5% between each others. The external temperature conditions were similar in the analysed periods of time. The average temperatures in the analysed days were calculated based on [12].

For further considerations of the ventilation efficiency of the office room, the average airflow rate value n=0.085 [1/h] was assumed. This value allows for calculation of total ventilation air change per hours. The volume of analysed office room is $V=58m^3$ which gives less than 5 [m^3/h] air change rate.

$$V * n = 0{,}085\frac{1}{h} * 58m^3 = 4{,}92\frac{m^3}{h}$$

Assuming 4 people working in the office room the stream of the air per one person is 1,23 m^3/hour per person.

As it can be noticed the obtained result does not meet the requirement of minimum $20m^3/h$ per person described in the standard [13].

6. Conclusions

Metabolically generated carbon dioxide used as a tracer gas can be a source of useful information about the quality of internal air within the space. The decay of this tracer gas allows for calculation and evaluation of ventilation system efficiency. The measurements are time consuming but did not require very complicated equipment and the results can be a source of information about specific ventilation systems.

The conduced measurements and calculations in the office room prove that the efficiency of ventilation system in is not sufficient and the air quality is poor. The basis of the gravitational ventilation is a temperature difference between the internal and external environment. Measurements were conducted in September when temperature difference was only about 10K. The efficiency of ventilation system could be higher during the winter months. The similar tests will be conducted in winter conditions to compare the results.

Low efficiency of ventilation causes high concentration of carbon dioxide in the office which can affect the employees in the room, reducing their productivity and affecting their well-being. In the tested room, the gas concentration often exceeded the recommended values, especially when the windows were closed.

The standards determine the maximum level of CO_2 concentration which in this particular room was exceeded during the working time. Comparing the results with the standards it appears that the air is low quality.

References

[1] Persily A. K. Evaluating Building IAQ and ventilation with indoor carbon dioxide. Ashrae transactions, 103:1-12, 1997

[2] Bulińska A., *Analiza wykorzystania generowanego metabolicznie ditlenku węgla jako znacznika gazowego do określania wymiany powietrza w pomieszczeniach*, Rozprawa Doktorska, Gliwice 2011

[3] Zhang W., Wang L., Ji Z., Hui Y., Test on ventilation Rates of Dormitories and Offices in University by the CO2 Tracer Gas Method, Procedia Engineering 121 (2015) 662-666

[4] Benedettelli M., Naticchia B., Carbonari A., Pascucci M.: Testing of Tracer Gas Based Measurement Procedure to Asses Air Change Rates in Buildings

[5] Cichowicz R., Gawron-Skarbek A., Godala M., Zimna-Walendzik E., Sabiniak H., Szatko F., Assesment of carbon dioxide concentration in the air of selected university facilities, Probl Hig Epidemiol 2014, 95(2): 287-291

[6] ASHRAE 62-1989. *Ventilation for acceptable indoor air quality*, 1989. American Society of Heating, Refrigerating and Air-Conditioning Engineers.

[7] ASTM Standard D 6245-98. *Standard guide for using indoor air quality*, 1989. American Society of Heating, Refrigerating and Air-Conditioning Engineers.

[8] Rozporządzenie Ministra Pracy i Polityki Społecznej z dnia 29 listopada 2002 r. w sprawie najwyższych dopuszczalnych stężeń i natężeń czynników szkodliwych dla zdrowia w środowisku pracy, 2002

[9] PN-EN 13779. Ventilation for non-residential buildings - Performance requirements for ventilation and room-conditioning systems, 2008.

[10] ISO/FDIS 12569:2012 *Thermal performance of buildings and materials – Determination of specific airflow rate in buildings – Tracer gas dilution method*, 2012

[11] Sherman M. H., Tracer gas techniques for measuring ventilation in a single zone, Building and Environment, Vol 25, No 4, pages 365-374, 1990.

[12] http://meteo.ftj.agh.edu.pl/meteo/archiwalneWykresyMeteo, website containing archived weather data measured by the Vaisala WXT520 station belonging to AGH University

[13] PN-83/B-03430. *Wentylacja w budynkach mieszkalnych, zamieszkania zbiorowego i użyteczności publicznej*. Wymagania, 1983. Polski Komitet Normalizacyjny.

[14] PN-EN 15251 - Indoor environmental input parameters for design and assessment of energy performance of buildings addressing indoor air quality, thermal environment, lighting and acoustics.

IOP Conf. Series: Materials Science and Engineering **415** (2018) 012029 doi:10.1088/1757-899X/415/1/012029

Energy use optimization in the building of National Library

J Radoń[1], A Sadłowska-Sałęga[1], K Wąs[1], A Gryc[1], A Kupczak[2],

[1] Faculty of Environmental Engineering and Land Surveying, University of Agriculture in Krakow, al. A. Mickiewicza 24-28, 30-059 Krakow, Poland

[2] Jerzy Haber Institute of Catalysis and Surface Chemistry, Polish Academy of Sciences, Krakow, Poland

Abstract: The analysis of inner climate parameters and energy use was carried out in the Krasinski Palace in Warsaw: seat of The Polish National Library. The building includes books stores, conservation workshops, reading rooms and offices. Central hydronic heating system and sporadically used local air dryers are the only active systems. The purpose of the analysis was to check at which extent actual microclimate and energy management can be improved. Several alternative scenarios, not compromising climatic protection level of valuable collections, to optimize energy use were elaborated. It turned out, that only improving of heating control and doubling of thermal insulation thickness of ceiling to attic allows for up to 30% of energy savings. Application of extended HVAC systems is not necessary due to high hygrothermal inertia of the building. WUFI®PLUS software was used for the building simulations. The calculation model was first validated against energy use for heating and inner climate measurement results. A special module was used to account for moisture exchange between books and inner air.

1 Introduction

The building, called Krasinski Palace in Warsaw, was built 1677 ÷ 1682 by polish magnate family in the Baroque style. In 1765 the Government of the Republic of Poland repurchased the Palace and placed in it some state institutions and offices. Before the II WW it was the seat of The Supreme Court. During the Warsaw Uprising, the building was destroyed in about 80%. Its reconstruction works began in 1948. In 1961 the building was handed over to the needs of the National Library, where the remaining library collections were placed. On July 1, 1965, the Palace was entered into the register of monuments. In the years 2014 ÷ 2016 the building underwent a thorough revitalization. Currently, the Palace houses collections of manuscripts and old prints of the National Library. Beside book storage rooms there are also conservation workshops, reading rooms and offices [1].

Extended research on the building was carried out in the years 2015 ÷ 2017. The main aim was to find solutions for reduction of energy use while preserving appropriate microclimate for valuable books and manuscripts. According to conservator's suggestion, relative humidity of inner air should be maintained in the range 30 ÷ 50%. This is in line with the ASHRAE recommendations [2].

Long term measurement results were used for validation of calculation model, then several, alternative control scenarios were considered to improve inner climate and to optimize energy use. WUFI®PLUS software was used for the building simulations. The software had been extensively validated for hygrothermal building simulations [3] and already applied to calculations of microclimate development and energy aspects in archive depots [4].

IOP Conf. Series: Materials Science and Engineering **415** (2018) 012029 doi:10.1088/1757-899X/415/1/012029

2 Material and methods

2.1 Building

The Krasinski Palace (Figure 1a) has a massive brick construction. It has 3 floors, cellar and hip roof. Roof is covered with painted metal sheets. The heating consists of central hydronic system powered from the municipal heating network. Heating elements are traditional cast, iron radiators. Cellar and attic are not heated.

Outer walls are 95 cm thick (ground floor, cellar) and 70 cm (remaining floors), covered by 2 cm lime-cement plaster. Thickness of inner walls varies between 12 ÷ 140 cm depending on type (partition or supporting). There is cross vault (bricks and sandstone) over cellar. Remaining ceilings consist of ceramic plates reinforced by steel flat bars (so called Klein's construction), approximately 24 cm thick, supported by steel girders. Concrete screed (5 ÷ 8 cm thick) and boards on joists is basic floor structure. Ceiling to not heated attic is poorly insulated by only 10 cm EPS or mineral wool. Windows consists of double glass box with one additional glass pane, put at inner side, to reduce heat loss.

The building comprises rooms of diverse functions and dimensions (Table 1). Almost 40% of total floor area is taken by book magazines (Figure 1b). The total length of book shelfs in storage rooms is estimated to 27 km, the volume of books amounts to about 12% of magazine space. Local air driers are sporadically applied to avoid, hazardous for paper collections, too high air humidity.

The most representative element is so called "Wilanow room" in the central part of the building. It is an open space, starting with its own staircase at the ground floor, stretching from the first to the second floor. The interior is rented for external and internal events, like readings, concerts etc.

About 60 people (researches, conservators, office workers and other stuff) occupy the building at working time. National Library is closed to the public. Catalogue and reading room are irregularly visited by small groups of students and researchers.

a) b)

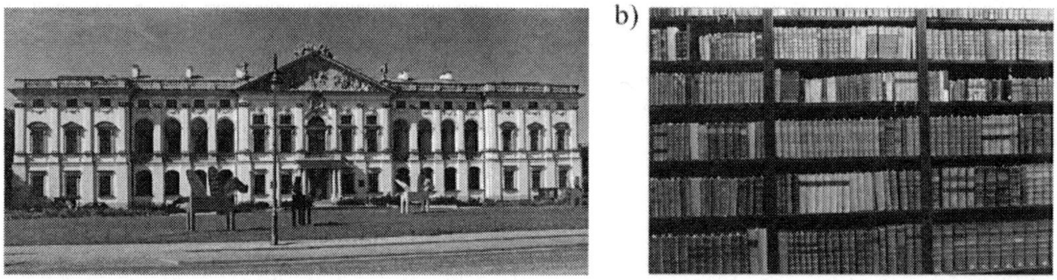

Figure 1. Krasinski Palace: south-west façade (a), collections of antique books in the building (b)

2.2 Measurements

In most of the rooms temperature and relative humidity are constantly monitored. Figure 2 shows yearly pattern (hourly values, averaged for rooms of magazines and offices) of these parameters in book depots, Wilanow room and office rooms in the period Mai 2015 to April 2016.

Outside air temperature and RH were also measured near the building. Data of solar radiation (global and diffuse) was obtained from the nearest weather station Bielany, located about 4,6 km away.

Energy used for heating was determined by monthly readings from heat meter. The heat consumption by the whole building, in the heating period 2015 ÷ 2016, was 1678 GJ, i.e. 466111 kWh. Electricity use was determined from monthly bills.

To define air infiltration, several measurements of CO_2 decay was made. On this basis, the average air exchange rate for the building was estimated to $0.13 \pm 0.06\ h^{-1}$. The measurements were taken after working hours, after employees left the building.

Table 1. Rooms in Krasinski Palace

Room/destiny	Floor area, [m^2]	Net volume, [m^3]
Book magazines		
- ground floor	573.5	2752.8
- 1st floor	220.3	991.4
- 2nd floor	482.1	1470.4
Conservation workshops		
- ground floor	84.2	404.0
- 1st floor	97.0	519.0
- 2nd floor	153.2	467.2
Office rooms		
- ground floor	263.1	1177.3
- 1st floor	307.5	1441.2
- 2nd floor	307.3	1224.6
Catalog and reading room. 1st floor	220.3	1178.6
Representative "Wilanow room"	636.6	4873.7
Corridors. stair cases	176.4	846.0
Attic	1569.0	2982.0
Cellar	1873.8	4506.6
Total without attic and cellar	3521.5	13503.1

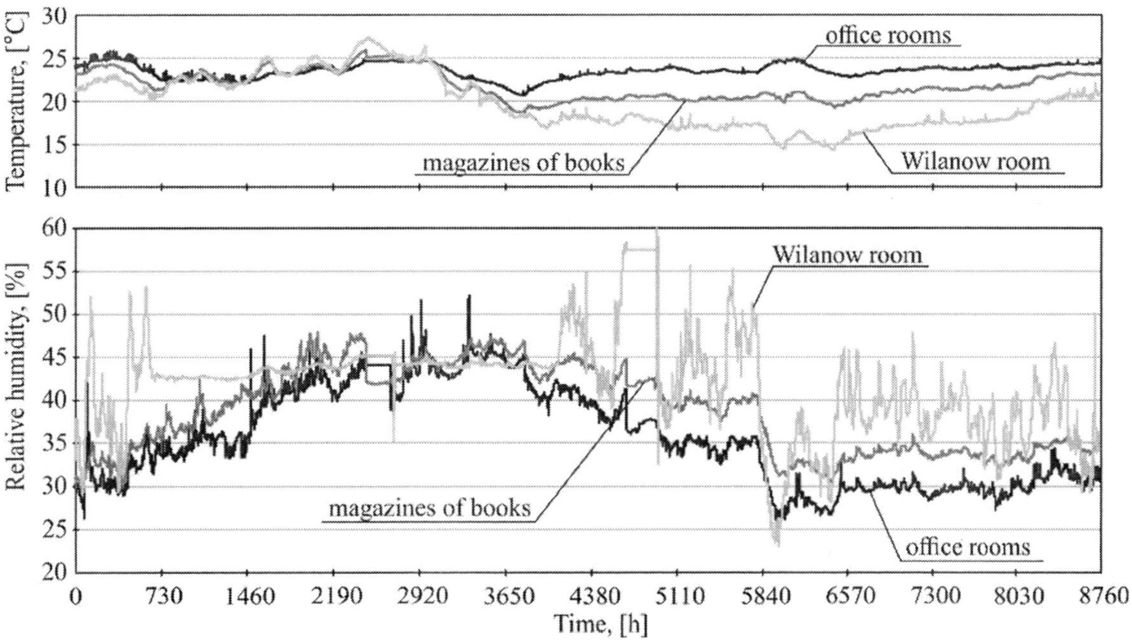

Figure 2. Measured average air temperature and RH in book magazines. office rooms and Wilanow room

2.3 Calculation model

Based on architectural plans. building 3D model was prepared in the SkechUp© software. and then imported into the WUFI®PLUS program. The model includes relevant opaque and transparent assemblies. which have impact on transient microclimate formation. The cubature of individual rooms and area of partitions had been exactly reflected. Based on specific usage of rooms and façade orientation. the building was divided into 18 zones including attic and cellar. Free floating air temperature and RH were assumed in the attic and sinusoidal yearly pattern of this parameters in the cellar. Figure 3. shows the visualization of whole building (all zones) and book magazines.

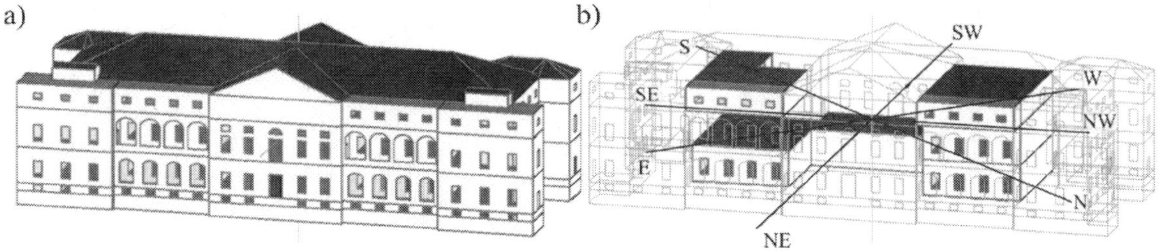

Figure 3. Building visualization in WUFI®PLUS software: a – whole building (all zones), b – book magazines.

Magazines are filled with books up to 12% of inner space. Special attention was then paid to calculation of buffering of heat and moisture in books. WUFI®PLUS can calculate 3D objects only thermally. Development of an algorithm estimating the dynamics of water vapor sorption and consequently the impact of library collections on climatic conditions was necessary. For this purpose. a statistical model of a book standing on a shelf next to other books was developed. First. the size of 384 old prints stored in one of the National Library's magazines was measured. Based on these measurements. an equivalent cuboidal block of paper cards was created assuming that the volume of this block is equal to the sum of the volumes of individual books. and the surface is equal to the sum of the surface of the respective sides of the books standing side by side. An additional condition imposed on the dimensions of the model block was the height-to-width ratio of the books. which has a good approximation of 0.71. The height and width of the statistical book included in the model block obtained on the basis of such accepted criteria are 261 and 186 mm respectively. The simulation of water vapor sorption by a row of books implies the diffusion of the vapor through the two sides of the book - upper and lateral. Access of water vapor to the other lateral side is usually blocked by a firmly glued spine. Based on this diffusion model. the statistical implementation of the book to the WUFI®PLUS module has been adjusted. In the model. the statistical book was changed into cuboid. with the same volume. but with one side diffusion open. which is the sum of the two diffusion-open face surfaces in the output book. The height and width of the modified statistical book are finally 447 and 109 mm, respectively (see Figure 4). In order to obtain agreement between the simulation results for the original (2D moisture flow) and modified statistical books (1D flow). a parametric adjustment of the diffusion coefficient was carried out. The best fit was obtained by division of the original diffusion coefficient by 1.5. COMSOL Multiphysics software was used for 2D moisture flow analysis [5]. Similar algorithm was also used in adjustment for other museum collections (e.g. wooden sculptures) [6]. Consequently. WUFI®PLUS calculated heat exchange with cuboid of books using 3D thermal calculation. and moisture exchange using geometrical and diffusive adjustment to 1D calculation. Material data for books is shown in Figure 5. Moisture storage function has been determined experimentally for several paper types and averaged to best reflect of the paper properties of book collections in the National Library [7]. Remaining data was assumed based on literature study and WUFI®PLUS database.

Assemblies of walls. ceilings as well as roof structures. in particular; layer types and thickness. were assumed based on available building documentation and own observations. Material parameters were defined from material database of WUFI®PLUS. U value of windows was estimated to 1.1 $W \cdot m^{-2} \cdot K^{-1}$ and average solar radiation transmittance (g-value) to 0.6. Windows in book magazines were curtained to reduce potentially harmful effect of intensive light on the state of books. Hence. average g-value of windows in storage rooms was reduced to 0.1.

IOP Conf. Series: Materials Science and Engineering **415** (2018) 012029 doi:10.1088/1757-899X/415/1/012029

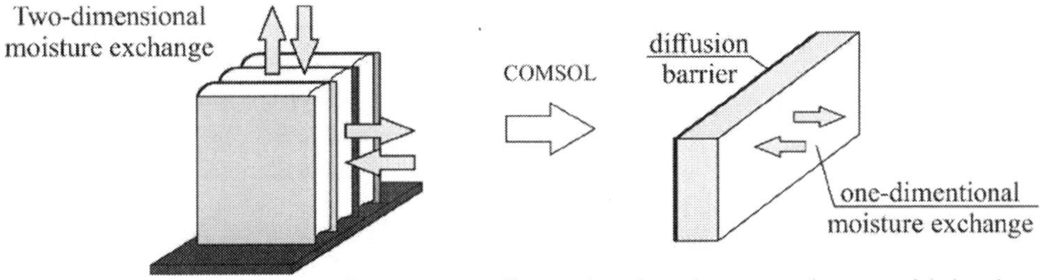

Figure 4. Adjustment of two to one dimensional moisture exchange with books

Bulk desity	[kg·m^{-3}]	618
Porosity	[m^3·m^{-3}]	0,6
Specific heat capasity	[J·kg^{-1}·K^{-1}]	850
Thermal conductivity, dry	[W·m^{-1}·K^{-1}]	0.73
Water vapor diffusion resistance factor	[-]	3.6

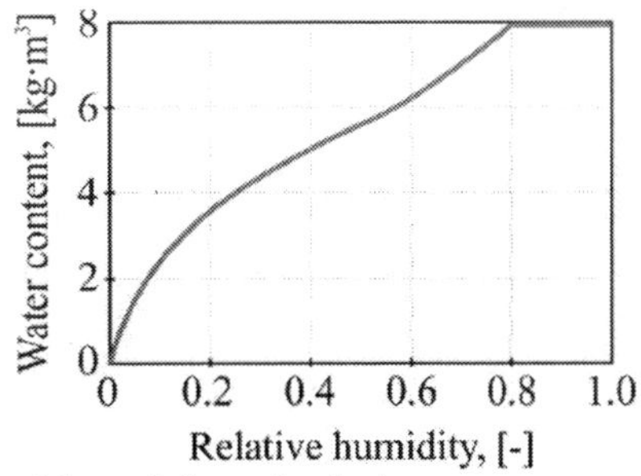

Figure 5. Material properties of books cuboid used for calculation

During work time. for all rooms excluding book magazines. air change rate (ACH) was assumed to 1 h^{-1}. Outer work time. ACH in all rooms. was assumed to 0.13 h^{-1} (based on CO_2 decay measurements). Windows in book magazines are not opened. so ACH in this rooms was assumed also 0.13 h^{-1} for the whole calculation time.

Heat gains from 60 people (convection 4800 W and radiation 2460 W. corresponding to human activity at 1.2 met – an adult performing work in a sitting position) were distributed proportionally to the floor area in all rooms except magazines and corridors. The source of moisture (3540 g·h^{-1}) were similarly assumed and distributed. Heat and moisture gains from visitors were omitted.

Electricity use was also added to heat gains proportionally to floor area. according to monthly power consumption. It was assumed that 75% of the total electricity consumption is a heat gain. of this 90% is consumed during the day at working days and 10% at night. On days off. the heat gain from lighting is constant and amounts to 500 W. It has been assumed that the radiative and convective part of electricity heat source is both 50%.

3 Calculative analysis

3.1 Model validation

Due to the lack of real data of instantaneous heat and moisture sources in individual rooms. poor heating control and uncertainty of air exchange during working time. it is very difficult to make calculations for direct comparison of air temperature and humidity patterns with measured data. Therefore. the model adequacy check was made indirectly. performing calculations with the assumption of measured air temperature and humidity in the zones. The program calculates the values of heat and moisture sources from active systems necessary to maintain the assumed air parameters. There is only heating in the zones. hence the amount of cooling. humidification and drying should be zero. Non-zero values are a measure of the (un)accuracy of the model.

Due to the valuable book collections. special attention was paid to computational simulations of storage rooms. Figure 6 shows the virtual humidification and dehumidification in magazine room on the ground floor. These values. combined with air temperature. allow to estimate the extent to which virtual dehumidification/humidification would change the relative humidity. The average deviation in this aspect for all storage rooms is under 1%. Nonetheless. temporary deviations can amount to several percent. The same analysis for air temperature (cooling impact) gave average deviation of 0.9°C.

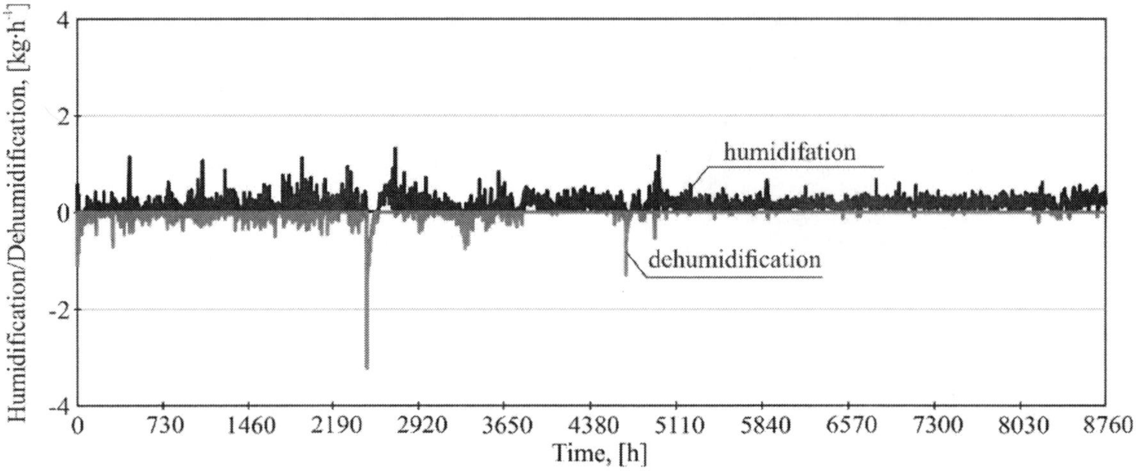

Figure 6. Virtual humidification/dehumidification of storage room on ground floor as a measure of (in)accuracy of the calculation model

Calculated heat energy. necessary to maintain measured inner air parameters in building. was 417952 kWh. The difference to real heat amount used is 10.9%. The discrepancy is due to approximate assumptions and definition of energy. Real consumption is final energy. which includes heat losses of supply pipes. Whereas energy calculated (useful energy) is the heat given off by radiators to indoor air in a convective way.

3.2 Energy optimization calculations

The only structure change proposed is improving of the thermal insulation of ceiling to unheated attic by increasing of the thickness of mineral wool from 10 to 20 cm. to mitigate influence of attics high temperature in the summer on neighboring rooms and to reduce heat loss in winter. It is an easy. low cost change. without significant intrusion into the historic character of the building. Assuming this change. three alternative climate control scenarios. to reduce energy use not compromising climatic protection level of valuable collections. were designed and tested:

I. In office rooms a minimum air temperature of 20°C is maintained. In other rooms (including book magazines) the minimum temperature is 18°C. Air humidity in all zones is free floating depending on the transient heat and moisture balance.
II. The book magazines are heated to only 7°C (protection against freezing). Other rooms heated according to scenario I. Humidity in rooms is shaped passively.
III. Heating according to scenario II. In book magazines relative humidity is maintained below 50%. In other rooms the humidity is shaped passively.

Figure 7 shows resulting temperature and RH pattern in book magazines in scenario II. Due to lower air temperature in book magazines in winter time. values of RH higher than 50% occur temporarily. This makes dehumidification from time to time necessary in this rooms (scenario III). Moisture to be removed is relatively small (see Figure 8). within drying power of portable devices used in the library. Calculation results of energy use and dehumidification. according to alternative scenarios. are comprised in Table 2.

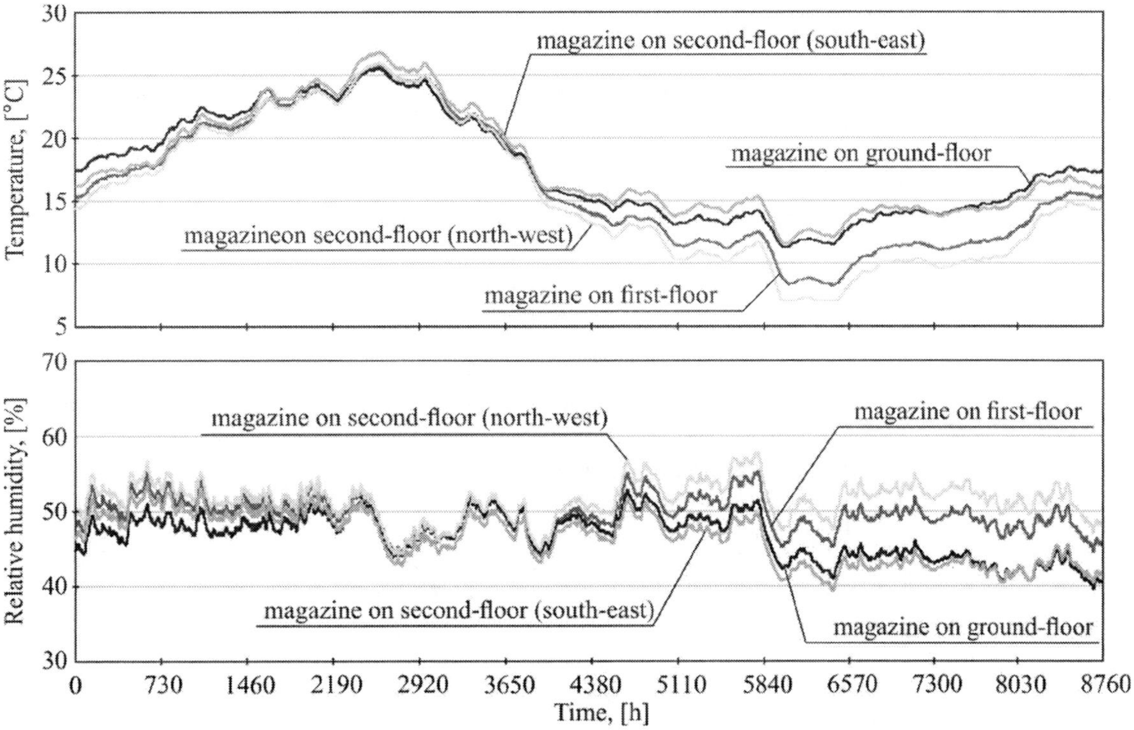

Figure 7. Inner climate pattern in magazine rooms for scenario II

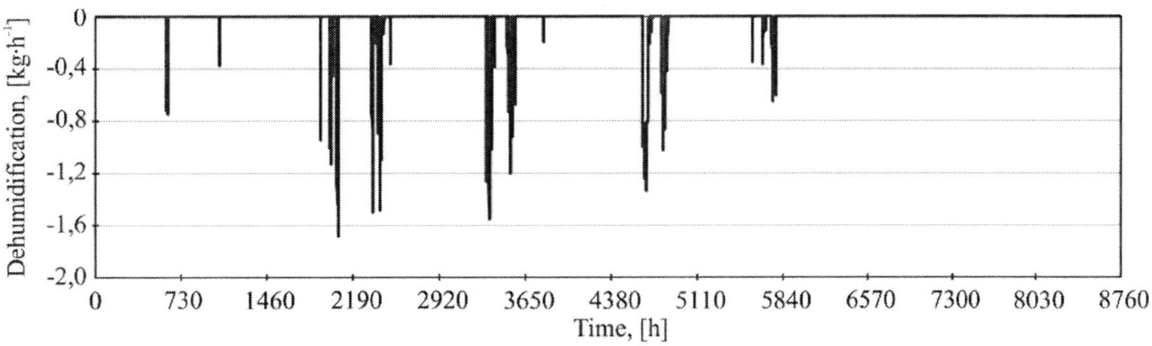

Figure 8. Dehumidification of book magazine on the ground floor. scenario III

Table 2. Energy use and dehumidification in Krasinski Palace

Specification	Energy for heating [kWh·yr^{-1}]	Dehumidification [kg·yr^{-1}]
Real energy use (measurement)	466 111	–
Energy use. model validation	417 952	–
Alternative scenario I	331 525	0
Alternative scenario II	286 427	0
Alternative scenario III	286 260	1 036

4 Summary and conclusions

Long-term measurements results of inner and outer climate parameters were used to validate calculation results in the Krasinski Palace. Uncertainty about air exchange and poor control of heating system made standard inner climate validation very difficult. Instead indirect check of calculation adequacy was applied. Using measured temperature and humidity of inner air transient heating. cooling. humidification and dehumidification loads were calculated. As only heating was available remaining loads is regarded as a measure for calculation (in)accuracy. Mean discrepancy for cooling. humidification and dehumidification was below 1%. The difference between measured and calculated energy demand for heating season 2015/2016 was about 11% for the whole building. Consequently. quantitative results of overall calculations of alternative scenarios to optimize energy use may differ within these errors. Nevertheless. the Authors believe. that following conclusions are justified considering the analysis.

Perfect heating control. 20°C in office rooms and 18°C in remaining rooms and improving the thermal insulation by double thickening of mineral wool on ceilings to the attic can save about 20% of energy for yearly heating.

As people only sporadically enter book storage rooms. they don't need to be heated for comfort. Exclusion magazine rooms from regular heating while preserving only frost protection (setpoint 7°C) as well as precise heating control in other rooms. can bring about 30% of heat savings. Lower air temperature in storage rooms will cause higher heat flow from neighboring rooms (heating trough partition walls). These rooms will acquire higher heating load power.

Due to the massive construction of the walls and ceilings of the building as well as the large mass of books. the relative humidity of the air in storage rooms is subject to much lower fluctuations than the humidity of the outdoor air. Lower air temperature in the absence of heating means higher relative humidity. The optimum RH range for book collections is 30 ÷ 50%. To maintain this range. it is necessary to dehumidify of inner air. Calculation results indicate. that instantaneous dehumidification power in largest storage room does not exceed 1.7 $kg \cdot h^{-1}$. The annual total dehumidification for all book magazines is not much above 1000 kg. Nonetheless. the energy needed for drying is much lower than energy used for regular heating.

5 Acknowledgments

Measurements and computational analysis were made as part of the "Heriverde" research project funded by The Polish National Centre for Research and Development.

6 Literature

[1] https://en.wikipedia.org/wiki/Krasińki_Palace
[2] American Society of Heating, Refrigerating and Air-Conditioning Engineers 2007 *Ashrae Handbook: Fundamentals chap. 21 Museums. libraries and archives* (Atlanta, Ga.: Ashrae)
[3] Holm A, Radon J, Künzel H M, Sedlbauer K 2004 Berechnung des hygrothermischen Verhaltens von Räumen (Calculation of the hygrothermal performance of rooms) *WTA Schriftenreihe* H **24** pp. 81-94
[4] Krus M, Rösler D, Klemm L 2013 Hygrothermal Simulations of Energy-saving Measures to Stabilize the Internal Environment of an Archive Depot *Clima* pp. 3013-3022
[5] COMSOL Multiphysics. ver. 4.3a. 2012
[6] Kupczak A, Sadłowska-Sałęga A, Krzemień L, Sobczyk J, Radoń J and Kozłowski R 2018 Impact of paper and wooden collections on humidity stability and energy consumption in museums and libraries *Energy and Buildings* vol **158** pp. 77–85
[7] Kupczak A, Kozłowski R 2017 Dyfuzja i sorpcja pary wodnej w materiałach i obiektach papierowych (Diffusion and sorption of water vapor in paper materials and objects). *18th Conference „Analiza Chemiczna w Ochronie Zabytków"(Chemical Analysis in Monuments Protection)* (Warszawa, Poland)

IOP Conf. Series: Materials Science and Engineering **415** (2018) 012030 doi:10.1088/1757-899X/415/1/012030

Low airflow measurements by means of gas tracing method

T Kisilewicz, K Nowak-Dzieszko

Cracow University of Technology, Institute of Building Materials and Structures, 31-348 Kraków, ul. Warszawska 24, Poland

Email: tkisilew@pk.edu.pl

Abstract. Tracer gas dilution decay method is a convenient tool for measurement of actual air exchange in building. The article presents the results of measurements of natural air exchange in the selected, airtight rooms. A two-point and multi-point method were used to collect data and calculate the results. The standard procedure for the selection of the optimal measurement time was also included. In tight spaces with low air exchange, the measurement results depend on the assumed duration of measurements. It was found that the duration of measurements of several hours and the multipoint method allow to obtain convergent results.

1. Introduction

Building air tightness became one the most important features of contemporary low energy buildings. Reduction of heat losses connected with air exchange is a necessary measure on the way to building low energy demand. Such a reduction is possible only when air exchange is controllable and such a control is possible only when building external envelope is air tight and expected air exchange is carried out through specially designed for this purpose openings and components of the installation. The aim of the ventilation is to maintain a proper hygienic conditions inside the space by introducing outdoor air into the room. In term of energy savings, it is also important to keep ventilation at the required rate to reduce heat losses. Air tightness of the building is necessary because it allows to avoid or minimize uncontrolled flow of air (and heat) through gaps and cracks in the building envelope. Penetration of warm, moist air across the building structure can be also a reason of interstitial condensation of water vapour, which results in decreasing of thermal insulation and durability of building envelope [1]. Uncontrolled cold air penetration to the inside decreases indoor microclimate conditions, causing local drafts and additional heat losses. This effect is even more important in low energy and passive buildings, where all kinds of heat losses should be minimized. In some cases, the achievement of the expected energy standard without a tight enclosure would be impossible, despite efficient thermal insulation and mechanical ventilation with heat recovery [2].

Airtight building demands special solutions and materials at design stage, extreme care and special experience at construction stage and experimental testing, finally.

One of the acts implementing the European Directive 2010/31/EU to Polish building law is regulation introduced by Ministry of Infrastructure. This regulation strengthens the requirements concerning building envelope insulation and reduction of primary energy demand. In accordance with this regulation, in residential, commercial and industrial buildings, all the opaque envelope elements and their joints must be designed and constructed to ensure complete tightness. Polish regulation recommends airtightness measurements and determination of n_{50} coefficient, but this kind of testing is not obligatory.

The most popular method of building tightness evaluation is currently the pressure method, according to standard EN 9972 [3]. It allows to assess tightness of the outer building envelope in the artificially created conditions of a considerable pressure difference between the interior and the building's external environment. It is a form of envelope quality diagnostics, but the assessment of actual air exchange in a building, especially with gravitational ventilation system and under real climatic conditions is still unknown. Approximate way of transition from pressure method to time averaged real conditions is included in EN 13789 [4]. In order to evaluate actual air exchange, gas tracing dilution methods have been developed and standardized EN ISO 12569 [5]. Practical aspects and problems connected with tracer gas concentration decay method will be discussed in this paper.

2. Air mixing and stratification

Standard [5] introduces three different tracer gas measurement methods:
- concentration decay method – the ventilation rate is obtained from the decaying curve of concentration observed after the end of the injection of tracer gas,
- continuous dose method – the ventilation rate is obtained from the concentration resulting from continuous injection of tracer gas,
- constant concentration method – the ventilation rate is obtained from the injection rate of tracer gas dosed for constant concentration in the space.

Selection of the method depends on building structure, ventilation system, fluctuations in ventilation rate and of course on the available testing equipment.

Tracer gas method may be used for determination of infiltration and ventilation rate, evaluation of ventilation efficiency, pollutant distribution, internal air flows between zones etc. [1,2,5,13,14,15,16].

According to the standard [5] the most commonly used tracer gases include: sulphur hexafluoride, nitrogen monoxide, carbon dioxide and helium. None of the tracer gases used in practice meets all of the ideal gas requirements. Two gases that are currently most often used for this purpose are: sulphur hexafluoride and CO_2. Due to easy access, relatively low costs of measuring equipment, security and versatility and easy mixing, the authors used in their research carbon dioxide. Measurements with the same tracer gas were described in [9,10,13,14,16]. Equipment used for concentration measurement is shown in figure 1. The Rotronic meter measures CO_2 concentration within the range of 0 to 5000 ppm with the accuracy of 30 ppm + 5% sensor indication. The temperature measurement range is -20 ÷ 60°¬C with an accuracy of up to 0.3K.

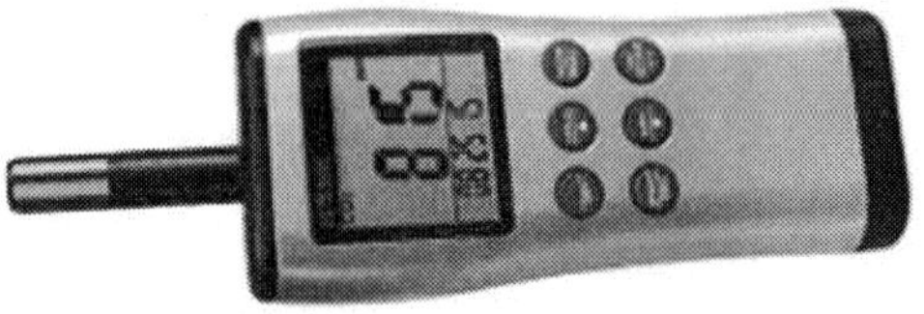

Figure 1. Carbon dioxide concentration meter Rotronic CP11 used for measurements [www.rotronic.pl].

Carbon dioxide is one of the natural compounds of atmospheric air. It is an odourless, colourless non-flammable gas, heavier than the air. The density of carbon dioxide is almost 1.5 higher than that of air. Present concentration of gas in the atmosphere is approximately 400 ppm (points per million). The use of carbon dioxide for measurements as a tracer gas requires therefore raising its initial concentration well above this value. This can be done by releasing compressed gas from the container or by exhalation by people.

The first difficulty in conducting the measurements is connected with the assumption of single zone measurement in case of multi-space building [15]. According to the European standard the uniformity of concentration within the tested space is obligatory. The concentration of tracer gas should not fluctuate more than 10% from the average value within the tested zone [5]. To meet this assumption forced mixing of air within the building seems to be necessary. In multi room zone data should be recorded separately in each room. Figure 2 presents the results of measurements conducted in office room, 3,5 x 4 m, with gravitational ventilation, by means of six gas detectors. Fresh air was supplied and exhausted by infiltration only.

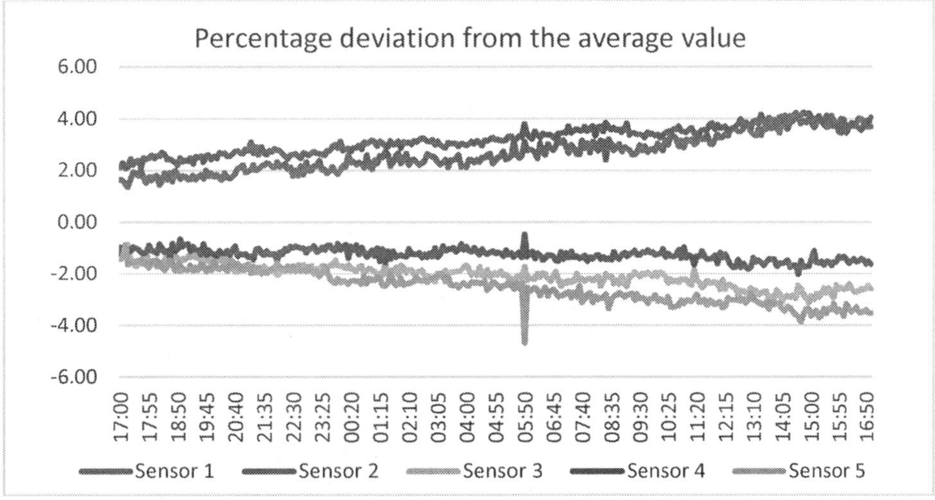

Figure 2. Percentage deviation of CO_2 concentration in tested room versus time.

CO_2 detectors were located at different height (sensor 1: 2.3 m, sensor 2: 1.7 m, sensor 3: 1.2 m, sensor 4: 0.6 m, sensor 5: 0.0 m) and in different parts of the room. Data have been gathered and stored at 10 minute intervals. Air was initially mixed. One can observe gradual increase in stratification of CO_2 in time. At the end of the recorded measurements, after 48 hours, all the fluctuations were still within the acceptable range but the deviations have doubled. A small but regular variation of CO_2 concentration in altitude was observed: above the floor (up to 1,2 m) concentration was lower than the average value and in case of the two higher located sensors recorded concentration it was higher. This tendency was getting stronger in time. It does not seem to be logical when only gas density is considered (higher density of CO_2 than of air), but the real mechanisms that influence the distribution of gases are much more complicated. More important than gravitational sorting is diffusion, convection and turbulences [6]. As a result air is quite well mixed and we have a chance to breath lighter than CO_2 oxygen even in the lowest part of atmosphere. In case of the specific building space also the other aspects of air movements within the tested space should be considered: thermal stratification, location of inlets and outlets, wind pressure distribution, ventilation shortcuts etc.

When mixing process is carried out also during the measuring period, it may affect the measured ventilation and infiltration rate, if airflow emitted from a fan directly impinges on the leakage areas in buildings [5,7,8,15].

3. Concentration 2-point decay method

In 2-point decay method the time-mean air change rate is calculated from the measurement start point to the end point. The minimum or maximum time step that should be used are not directly specified. Time-mean specific airflow rate can be calculated using the following formula [5]:

$$\bar{N} = \frac{1}{t_2 - t_1} \log_e \frac{C(t_1)}{C(t_2)} \tag{1}$$

where:

$\bar{N}$ mean time specific airflow rate, [1/h]
$C(t_1)$ room gas concentration at "t_1", [m^3/m^3]
$C(t_2)$ room gas concentration at "t_2", [m^3/m^3)]
t_1 start point of measurement, [h]

t_2 end point of measurement, [h].

According to the European standard 12569 [5] in order to achieve a good precision of measurements, the difference between initial concentration of tracer gas and at the end of test should be "sufficiently greater" than the equipment measurement error. What is a specific feature of the two-point concentration approach, the measurement method permits temporal change in ventilation rate which is in fact inevitable in real testing conditions lasting for many hours.

Presented and discussed below results of measurements have been carried out in office space, designed for four people performing light work and using desktop computers. Tested office is located on the second floor of the five-storey building. External wall of this space with two large windows is oriented north, figure 3. Doors with dimensions of 0.85 x 2.00 m are located in the southern wall of the room. Office space is ventilated by gravitational system with air supply via infiltration through windows and exhaust through ventilation duct of 18 x 18 cm, placed at a height of 2.35 m above floor.

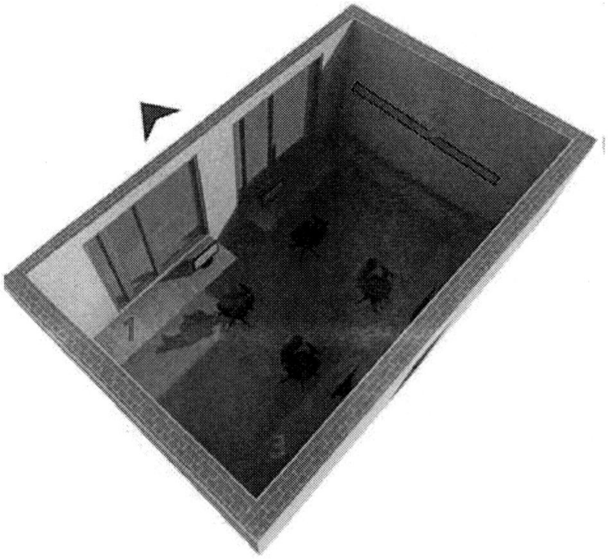

Figure 3. Tested office room and location of the 3 CO_2 sensors [9].

Three CO_2 Rotronic meters have been located as shown in figure 3, at least 2 meters away from office workers, who were the only source of carbon dioxide in this tests [10]. Two meters (1 and 3) were placed at the same height, i.e. 0.75 m above the floor, the last one (2) on the shelf, 1.75 m above the floor. Presented below results of carbon dioxide concentration were averaged for these 3 meters. Observed concentration differences always remained within the required range of 10%.

Figure 4 shows gradually increasing concentration of exhaled gas during the working day, up to 2000 ppm. Irregular changes of diagram, especially the concentration drops were caused by window opening and changing number of occupants. It should be noted that all the time concentration values significantly exceeded the requirements set in this respect for office rooms (1000 ppm) [10]. Indoor air quality in this room was very poor, gravitational ventilation practically did not work in test conditions.

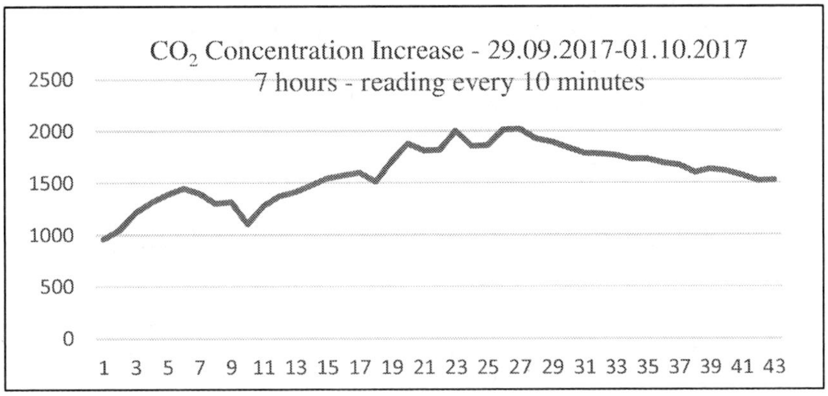

Figure 4. Average CO_2 concentration during the office hours.

Continuous carbon dioxide decay started at 4.00 pm, when all the office workers left room at the same time for business trip. Initial concentration at this moment, metabolically generated earlier by four people working in the office room, was 1472 ppm, figure 5.

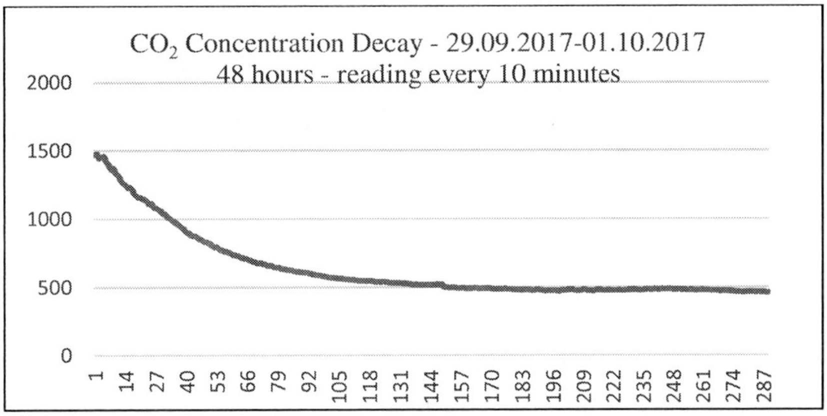

Figure 5. Average decay of CO_2 concentration in office space within 48 hours period.

The process of concentration decay was very slow, due to inefficient ventilation. Decay curve has the form of a regular exponential function. After 24-25 hours (150 readings) CO_2 concentration decay may be practically treated as linear function when internal concentration value, ca. 500 ppm, is still higher than the external (400 ppm).

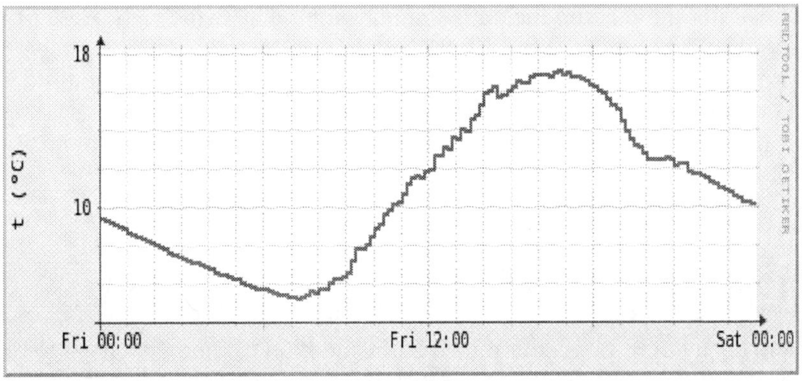

Figure 6. External air temperature on Friday, September 29-th, recorded by the automatic meteorological station in ca. 500 m distance from monitored building [http://meteo.ftj.agh.edu.pl].

Meteorological data for testing period have been measured by automatic station belonging to nearby located AGH University and are available on the university's website. Figure 6 shows the daily values of the external air temperature during the first day of measurement. Cyclic temperature changes are of a sinusoidal nature. Diurnal average air temperature was equal to 10.73 °C, minimum value 5.14 °C and maximum 17.04 °C. On the second day of testing thermal conditions were similar with average temperature 10.65 °C. Diurnal average wind speed was 1.28 m/s with prevailing direction: east-south. Based on the meteorological data review, it can be proved that the external conditions around the building were subject to the moderate and regular changes. If stack effect is considered as a most important driving force of natural ventilation it should follow the regular external air temperature fluctuation, figure 6. However, the internal air temperature was also subject to minor changes because of the weekend set back temperature. On Friday at 3.00 pm. it was equal to 20.4 °C and 48 hours later only 17.7 °C.

Figure 7 shows the values of mean specific airflow rate versus duration of the measurement period, calculated by means of two point decay method. According to the European standard, airflow rate is always calculated for the whole period, from the measurement start point to the end point.

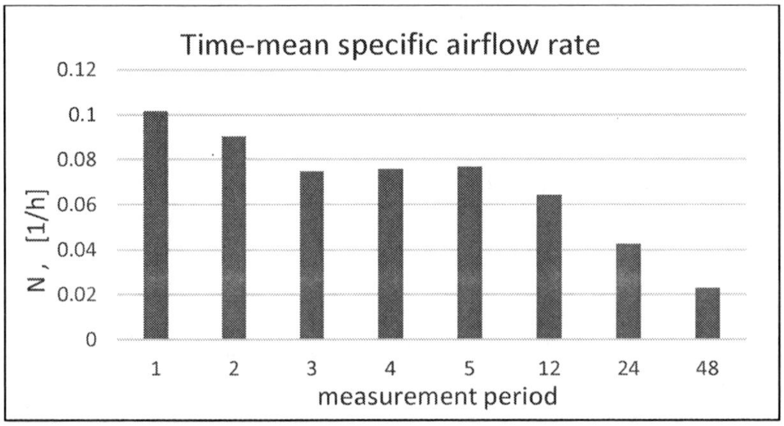

Figure 7. Calculated specific air flow rate vs. length of measurement period, two point decay method.

The results strongly depend on the length of the measurement period. Airflow rate is in the tested office space extremely low. Low airflow rate explains why during the working hours the concentration of exhaled carbon dioxide increased so intensely even though users were opening windows. Additionally it must be expected that the accuracy of measurement of such a small value will be low.

Time mean specific airflow rate changes strongly with the length of the measurement period. Observed changes do not follow fluctuation of external air temperature. The most important issue for the person conducting the measurements is the question what is the right time of measurements, and consequently which of the results shown in the graph is correct.

The standard [5] provides an iterative procedure for estimating the optimum decay time T_m which minimizes the estimation error. Gas concentration measurement variance is specific to the used equipment, so the minimum of the propagation error function has been found. For two-point decay method the following equation should be applied:

$$(N \cdot T_m)_{np=2} = 1.108\,857\,552\,88\,\ldots \tag{2}$$

where N is the initially estimated value of specific airflow.

The optimum decay time T_m is determined from equation (2) and then used to read appropriate CO_2 concentration value. Based on the selected values of gas decay a new value of airflow may be calculated [11]. If the difference between initial estimation and calculated N-value is lower than 5% of calculated value, the result can be considered acceptable. If not, the whole procedure should be iteratively repeated with the calculated N-value substituting the initially estimated one.

In case of the considered measurements it was assumed that N_{est} is equal to 0.1 [1/h], so the optimum decay time T_m is 11.09 h. Low air exchange rate means that the optimum time of measurement must be very long. Carbon dioxide concentration at 3.10 am (11 hours and 10 minutes after 16.00 pm) was 665 ppm. N_{calc} value for this period is equal to 0,066 1/h. Difference between N_{est} and N_{calc} is bigger than 5%, so the new N_{est} value is equal now to 0.066 1/h. The next T_m value calculated on this basis is equal to 16.8 hours, concentration at 8.50 is 557 ppm and N_{calc} = 0.055 1/h. Because the difference of 17% is still significantly bigger than 5%, the next iteration is necessary. The third T_m value calculated on the new basis is equal to 20.16 hours, concentration at 13.30 is 520 ppm, N_{calc} = 0.049 1/h and the difference equal to 11%. After the fourth iteration N_{calc} is 0.044 1/h and difference 10 %, after the fifth step N_{calc}= 0.041 1/h and difference 7%. Because the acceptable results were not obtained after "few steps" (required convergence was not achieved), the measurements should be taken again.

In case of the selected room, its ventilation systems and atmospheric conditions during the tests, it turned out that obtaining a reliable result is difficult or even impossible. In the conditions of light wind and a small temperature difference in the tested airtight room there was hardly any air change. During the second day of measurements, air temperature inside the building was only 1 K higher than the maximum outside temperature. With such small driving forces of air infiltration, each change in conditions translated instantly into the result of measurements. Figure 8 shows the results of measurements carried out according to a very specific scenario. It was assumed that airflow testing lasted for one hour only and was conducted for each hour separately. The results of measurements made at one-hour interval are listed next to each other, figure 8. The diagram shows how the air exchange rate based only on short 1-hour measurements, would depend on the chosen moment of measurement.

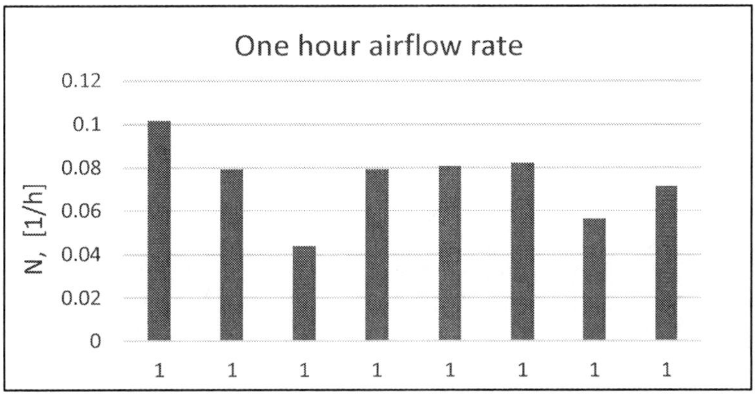

Figure 8. Calculated specific air flow rates for one hour measurement periods.

The airflow values, calculated for one hour intervals, differ significantly, although the graph of concentration decay shown in the figure 5 gives the general impression of a continuous and regular line. However, in case of extremely low air exchange in tested space and long lasting decay time, even the small momentary fluctuations of concentration may result in a noticeable change of airflow measure. The aim of above analysis was only to check the variance of short period data and to prove that such a testing approach is useless.

4. Multi-point concentration decay method

The multi-point concentration decay method can be used when the gas concentration decay process can be measured multiple times at intervals ranging from several minutes and around one hour [5,11]. Currently available measuring devices allow to record easily a big amount of data even every couple of minutes. An important assumption of this method is that the ventilation rate does not fluctuate over time. The least square method is applied when multi-point concentration decay method is used [5]. In the multi-point concentration decay method the specific airflow rate is calculated using equation (3).

$$N = \frac{\left(\sum_{j=1}^{np} t_j\right) \cdot \sum_{j=1}^{np} \log_e C(t_j) - n_p \cdot \sum_{j=1}^{np} t_j \cdot \log_e C(t_j)}{n_p \cdot \sum_{j=1}^{np} t_j^2 - \left(\sum_{j=1}^{np} t_j\right)^2} \tag{3}$$

where:

N estimated specific airflow rate,

t_j j-th elapsed time from the beginning of decay process, $t_1 = 0$,

$C(t_j)$ measured gas concentration at time (t_j),

n_p total number of measured elapsed time points (n_p is ≥ 3).

Based on the data recorded during the above presented test, the multi-point decay rate was calculated for two intervals of sampling: ½ hour and 1 hour, table 1.

Table 1. Airflow rate calculated by means of multi-point concentration decay method

decay period	interval	number of samples	N [1/h]
3 hours	½ hour	7	0.0757
3 hours	1 hour	4	0.0753
5 hours	½ hour	11	0.0736
5 hours	1 hour	6	0.0735
10 hours	1 hour	11	0.0693
11 hours	½ hour	23	0.0678
11 hours	1 hour	12	0.0672

Results of calculation do not practically depend on the selected interval of measurements, but they are slightly sensitive to the length of sampling period. The measurement time of up to 5 hours allows obtaining convergent measurement results. However, further extension of the measurement period causes that the difference between successive results increases, reaching the value of more than 11% difference in case of 11-hour measurement. This result is no longer acceptable, although 11 hour sampling period is still much shorter than the optimum time T_m (15 hours) according to the earlier mentioned standard procedure. Once again, it should be stated that low airflow rate is the cause of measurement problems and significant difficulties with convergence of the results.

5. Conclusions.

Measurement of air exchange with the trace gas method requires specific knowledge, experience and practical skills from the person carrying out the measurements.

The method of conducting the measurements must to some extent be matched to the properties of the tested object. This applies, for example, to the choice of two or multi-point method.

Particular difficulties arise when very tight building is tested. Measurement of the small values usually involves big difficulties. In the case of the low airflow rate (below 0.1 h^{-1}) it may be difficult to obtain convergence of results and final definite result of measurements, especially when using two-point decay method.

Results of measurements are sensitive to the sampling time. The optimum decay period, suggested by the standard procedure to minimize the effects of concentration measurement errors, is in case of a tight building very long and does not help to achieve expected accuracy. As it was stated in European standard EN ISO 12569, the concentration decay measurements should not last longer than several hours. These two approaches are in fact contradictory in case of a low airflow. Authors observed that testing period longer than 6 hours is not improving accuracy of the results.

It was not possible to isolate in the above presented results the impact of the changes of external conditions. It can be expected, however, that the change in the outside temperature could have resulted only in a limited change of the air infiltration to the tight space. Such fluctuations may additionally complicate testing process in a way that is difficult to predict. The above results, obtained for two and multipoint measurement method and sampling period 3-6 hours, are pretty close to each other. However, it must be stated that multipoint sampling method yielded more consistent results.

According to authors' observations, after initial forced mixing uniform concentration of carbon dioxide in air was easily achieved and sustained in the tested spaces with low specific airflow.

References

[1] Bomberg M., Kisilewicz T., Nowak K., Is there an optimum range of airtightness for a building? *Journal of Building Physics*, vol. 39, 2016, p. 395-421

[2] Nowak K., Nowak-Dzieszko K., Building airtightness testing by means of pressure method, *Technical Transactions* 2-B/2012, p. 305-313

[3] EN ISO : 2015 Thermal performance of buildings — Determination of air permeability of buildings — Fan pressurization method

[4] EN ISO 13789 : 2017 Thermal performance of buildings — Transmission and ventilation heat transfer coefficients — Calculation method

[5] EN ISO 12569 : 2017 Thermal performance of buildings and materials — Determination of specific airflow rate in buildings — Tracer gas dilution method

[6] Foucher P.Y., Chedin A., Armante R., Boone C., Crevoisier C., Bernath P., Carbon dioxide atmospheric vertical profiles retrieved from space observation using ACE-FTS solar occultation instrument, Atmospheric Chemistry and Physics, 11/6, 2011, p. 2455-2470

[7] Delmotte Ch., Proper building preparation for envelope airtightness testing, *The REHVA Eouropean HVAC Journal* 1/2013, p. 17-19

[8] Nowak K., Nowak-Dzieszko K., Influence of building preparation on results of tightness testing, Proceedings of XII-th International Research-Technical Conference on the Problem of Designing, Construction and Use of Low Energy Housing: ENERGODOM 2014, Cracow, September 10-12, 2014

[9] Marcinowski A., Determination of indoor air quality and ventilation air exchange rate in the office room using metabolically generated CO_2, Diploma Thesis, Cracow University of Technology, 2018

[10] Bulińska A., Analiza wykorzystania generowanego metabolicznie ditlenku węgla jako znacznika gazowego do określania wymiany powietrza w pomieszczeniach, Doctor Thesis, Silesian University of Technology, 2011.

[11] Okuyama H., Onishi Y., Uncertainty analysis and optimum concentration decay term for air exchange rate measurements: Estimation methods for effective volume and infiltration rate, *Building and Environment* 49 (2012), p. 182-192

[12] Sherman M., Modera M., Comparison of Measured and Predicted Infiltration Using the LBL Infiltration Model, STP No. 904 of the ASTM Symposium, Measured Air Leakage Performance of Buildings, Philadelphia, PA, April 1984, Report LBL-17001, Lawrence Berkeley Laboratory, February, 1984

[13] Zhang W., Wang L., Ji Z., Hui Y., Test on Ventilation Rates of Dormitories and Offices in University by the CO2 Tracer Gas Method, *Procedia Engineering* 121 (2015), p. 662-666

[14] Benedettelli M., Naticchia B., Carbonari A., Pascucci M.: Testing of Tracer Gas Based Measurement Procedure to Asses Air Change Rates in Buildings, ISARC. Proceedings of the International Symposium on Automation and Robotics in Construction

[15] Sherman M. H., Tracer gas techniques for measuring ventilation in a single zone, *Building and Environment*, Vol 25, No 4, 1990, p. 365-374

[16] Persily A. K. Evaluating Building IAQ and ventilation with indoor carbon dioxide. *Ashrae Transactions,* 103:1-12, 1997

IOP Conf. Series: Materials Science and Engineering **415** (2018) 012031 doi:10.1088/1757-899X/415/1/012031

Long-term monitoring of the thermal insulation properties of building envelope structures in real climatic conditions

Dušan Katunský[1*], Miloslav Bagoňa[1], Richard Baláž[1], Martin Lopušniak[2], Marián Vertaľ[2] and Marek Zozulák[2]

[1] Department of Architecture and Building Structures, Faculty of Civil Engineering, Technical University of Kosice, Slovakia
[2] Department of Building Physics, Faculty of Civil Engineering, Technical University of Kosice, Slovakia,

* Corresponding author e-mail: dusan.katunsky@tuke.sk

Abstract. The analysis and non-stationary numerical calculations of in situ temperature measurements are important details with regard to the connection between a window and a window sill. Due to its material and geometric heterogeneity, this connection is significant in terms of both surface temperature and heat flux. The structure under investigation is a panel fragment of the outer shell of an experimental chamber for in situ measurements. The aim of this paper is to compare the temperature obtained both from measurements and numerical calculation. Simulation and experimental measurement obtained in different environments, such as indoor and outdoor climates or experimental chambers, is an effective scientific tool for predicting the physical properties of building envelopes.

1. Introduction

Outdoor test cells and laboratory climate chambers are appropriate instruments for the long-term in situ monitoring of physical parameters of building envelopes and indoor climate parameters. Monitoring of the full-scale structures in non-stationary boundary conditions is a current trend in building physics research [1]. Outdoor on-site test cells are appropriate for energy and hygrothermal assessment of structures, due to their controlled indoor environment. Such cells have been applied in Dübendorf, Glasgow, Cottbus, Limelette, Almeria, Espoo, Delft etc., since 1993 [2, 3]. Outdoor test cells are a reasonable compromise between laboratory testing and full-scale building testing [3].

Results of panel research of an experimental wall structure with a window are presented in this paper [4]. Devices of the type of experimental internal and external chambers with adjustable environments (examples of which are presented in previous papers [5, 6, 7, 8]) are suitable for the comparison of experimentally obtained values with results of numerical calculations [9]. Validation of the simulation is a valuable tool for monitoring the accuracy of numerical calculations.

Thermal comfort and emission efficiency are important factors when choosing heat emission systems in low-energy buildings [10]. Outdoor test cells have also proved useful for laboratory measurements performed in Tallinn, Estonia [11].

Other institutes have full-scale outdoor laboratories, rather than simply cells, for improving research at the level of entire buildings. They mostly focus on green architecture, green roofs, wooden lightweight envelopes, etc. In Zilina, Slovakia, the full-scale monitoring of different kind of building construction, especially wooden constructions has been performed [12].

IOP Conf. Series: Materials Science and Engineering **415** (2018) 012031 doi:10.1088/1757-899X/415/1/012031

Surface temperature monitoring is not only useful for the thermal analysis of buildings and direct or indirect implementation in virtual simulations but it is equally relevant for use in the field of building physics. The measuring of surface temperatures on the building envelopes by means of commercial temperature sensors are typically applied for the field of thermal building performance aspects [13]. AdMaS (The Advanced Materials, Structures and Technologies Centre of Brno University of Technology) conducts long-term experiments under real climatic conditions. Similarly, The University Centre of Energy Efficient Building, Czech University of Technology, Prague (UCEEB) conducts long-term monitoring [14]. The main objective of the centre is the development of technologies for the reduction of energy demand and the efficient improvement of natural energy sources with regard to new constructions and reconstructions of existing buildings. This objective is achieved by a holistic approach – expert knowledge from the fields of architecture, construction, mechanical engineering, information technology, and hygiene of indoor environment and the advanced scientific equipment of the centre [14]. Another important aspect concerning building physics is summer overheating. Once again, full-scale monitoring facilities could be useful as an advanced tool for the passive cooling of green buildings [15].

The objective of this paper is to assess the accuracy of the numerical model and the justification of using the temperature field model without considering the transport of moisture. This model can be numerically modified after validation to optimise the window attachment solution.

2. Experimental chambers, monitoring cells, experimental methodology

This paper investigates the measurement and numerical calculation of the temperature values from the winter period of February 2015. The temperature at two locations and the surface temperature in the window aperture are analysed as shown in figure 1. The task was to compare the temperature results in the constructions obtained from experimental measurements with numerical calculations in order to validate the simulation model.

The assessed structure is a section of the external experimental chamber for in situ physical measurements, which is part of a laboratory of the Faculty of Civil Engineering of the Technical University in Košice. The composition of the structure is shown in table 1 and figure. 2. The structural heat transfer coefficient is U = 0.12 W/m^{2}K. The windows are made of a composite plastic material (U_f = 0.9 W/m^2·K) with a 3-fold closed glass system 4-16-4-16-4 Ar (U_g = 0.5 W/m^2·K). The values of the heat transfer coefficient are determined through calculation.

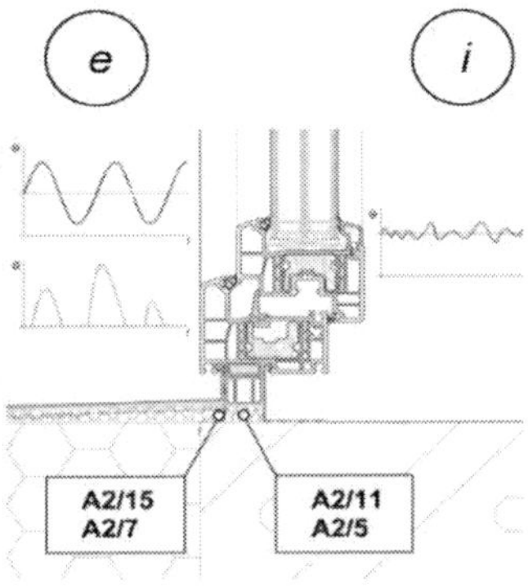

Figure 1. Southern view of monitoring chambers in the laboratory (cells → no. 3,2,1)

In this investigation, we measured values of the temperature in the window opening at the sill site. At the same time, the following parameters of the outdoor and indoor environment were monitored: air temperature, relative humidity, wind and precipitation parameters, solar radiation. The non-stationary numerical calculation of the surface temperature field was realised for part of the porous brickwork made of porous concrete. The subject of the calculation was the temperature results for the sites considered in the construction (see Fig. 2). Heat dissipation over time (for homogeneous and isotropic constructs) describes the Fourier partial differential heat conduction equation. For the numerical experiment, the Physibel simulation tool, BISTRA module, was used. The non-stationary simulation was based on the Crank-Nicolson's Finite Difference Method. This method meets the criteria of STN EN ISO 10211, Annex A for numerical computational methods.

Table 1. External wall

Name of test wall layer	c (J/kg.K)	λ (W/m.K)	ρ (kg/m3)	d(m)
1 – AAC P2 – 350	900.0	0.104	350.0	0.300
2 – adhesive PUR foam	800.0	0.040	35.0	0.010
3 – graphite extrude polystyrene	920.0	0.033	16.0	0.170
4 – adhesive mortar	900.0	0.850	1300.0	0.002
5 – primer	-	-	-	-
6 – silicone plaster	900.0	0.700	1700.0	0.002

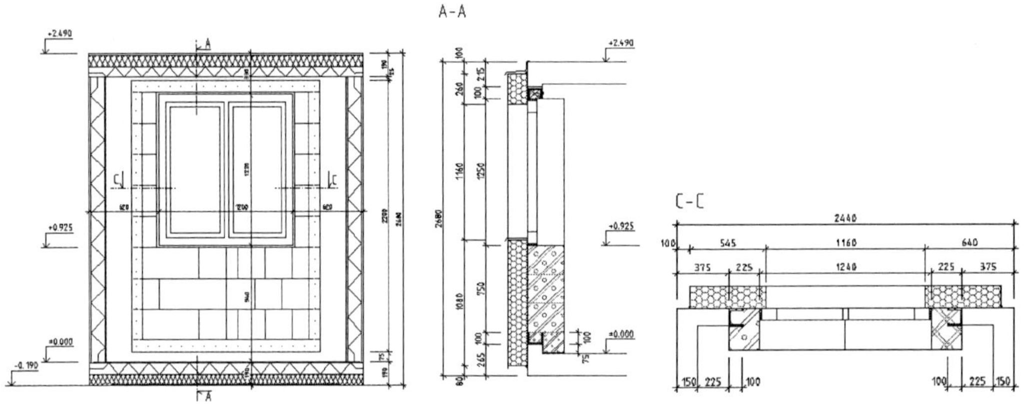

Figure 2. View and cut planes, cross section and floor plan of the contact between the wall with and window in the chamber

3. Boundary conditions of the experiment in the test cells

In the selected period, outdoor air temperatures were recorded between -6.5°C and +11°C. Solar radiation also changed from 100 W/m^2 to 500 W/m^2. In chamber 1 (south orientation) the average indoor air temperature was 22.7°C (figure. 3).

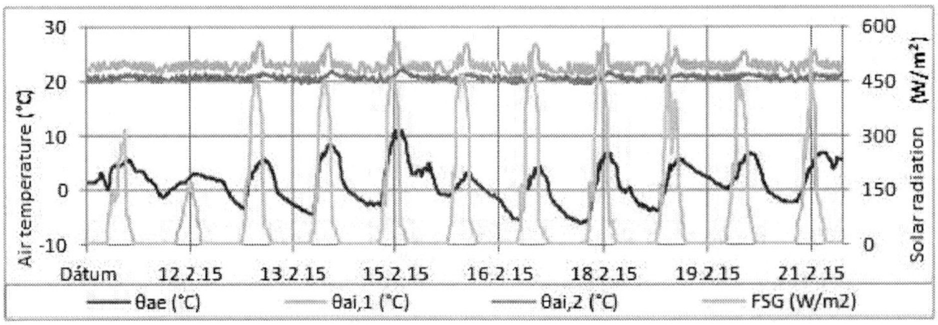

Figure 3. Measured outside and inside boundary conditions measured in February 2015

The temperature of the inside air (20°C) and the temperature of the ambient air (-13°C) are taken as the standard values for the inside and ambient temperatures. Almost identical results to chamber 1 were identified for A2/7 and A2/5 (see figure 4). Values were significantly different in the morning or when

IOP Conf. Series: Materials Science and Engineering **415** (2018) 012031 doi:10.1088/1757-899X/415/1/012031

the daily temperature reached the extreme values. Generally, the largest differences between the measured and calculated values occurred on days with a higher intensity of solar radiation.

4. Results and discussion

A ten-day period in February 2015 (11/02/2015 – 21/02/2015) was chosen for the further analysis. This winter season was selected in order to examine a fulfilment of the requirements for the minimum inside surface temperature during the heating season. Hourly measured values of ambient air temperature of Kosice in February 2015 were used. The temperatures measured in the experimental chamber during the monitored period were used as the inside temperature boundary condition. The average inside air temperature of chamber 2 (north orientation) was 20.6°C and for chamber 1 (south orientation), it was 22.7°C (Fig. 3).

The progression of the calculated and measured temperature values over a given period of time for the points considered (inner surface, A2/15 and A2/11) are shown in figures 4 and 5.

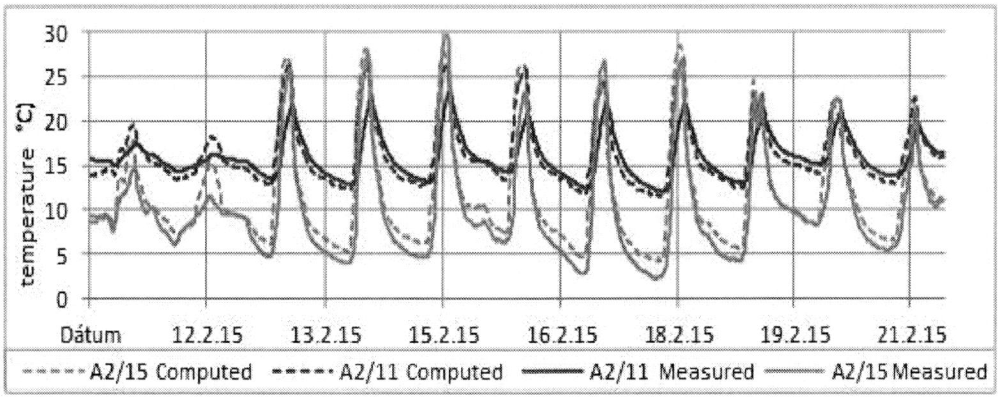

Figure 4. Measured and computed results for monitoring chamber 1 in 02/2015

Chamber 2 is orientated to the north – the effect of direct solar radiation is excluded. For A2/5, the largest difference between the measured (11.39°C) and the calculated (12.34°C) temperatures amounted to 0.95 K; this occurred on 15/02/2015 at 12:00. For A2/7, the largest difference between the measured (-1.15°C) and the calculated (2.22°C) temperatures amounted to 3.37 K; this occurred on 18/02/2015 at 00:48 (see figure 5).

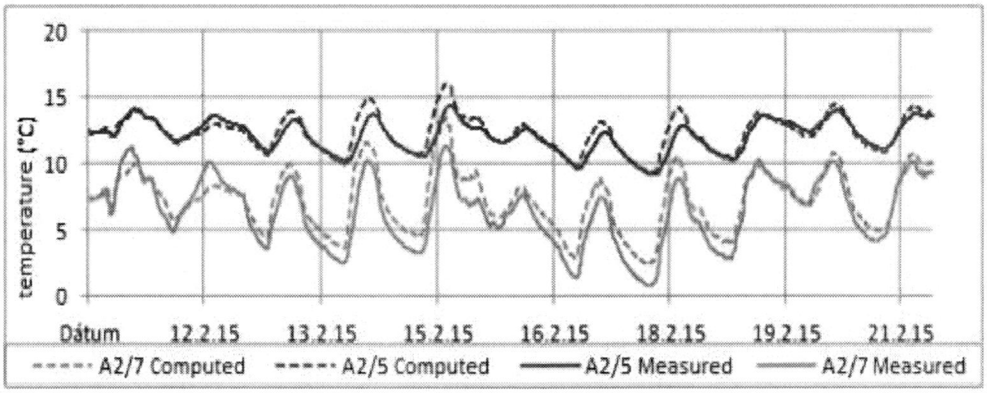

Figure 5. Measured and computed results for monitoring chamber 2 in 02/2015

The process see above note of the calculated and measured internal surface temperature shows a high consistency (figures 4 and 5). The mean internal surface temperature over eleven days is 19.33°C when obtained through calculations and is 19.31°C according to measurements. The greatest differences between the measured and calculated temperatures for A2/11 and also A2/15 are recorded at the time with the highest solar radiation. Values are significantly different in the morning and when there are daytime temperature extremes, i.e. daily minimum and maximum air temperatures (figure 4). The differences between the measured and the calculated values are caused by the neglect of several factors that affect the heat dissipation in the building structure. A simulation tool is used that analyses the temperature field without interfering with the effects of air and moisture transmission on heat dissipation, the effects of moisture content on thermal conductivity and the thermal capacity of the building material. Thus, it is shown that it is not possible to accurately describe temperature changes in the construction in this way. By using a complex simulation tool and including a correct starting humidity condition, a better match of calculated and measured values can be achieved.

The highest deviations (0.6 K) between these values are at point A2/7, unexpectedly, these values occurred in the same chamber. Variations of values in chamber 1 are from 0.9 K (A2/15) to 0.5 K (A2/11). As expected due to its orientation, we identified the highest compliance in chamber 2. The mean of median differences is 0.6 K, while in chamber 1, it is 0.75 K. Detailed analysis provides a comparison of medians and average values of measured and calculated temperature in the points (figure 6).

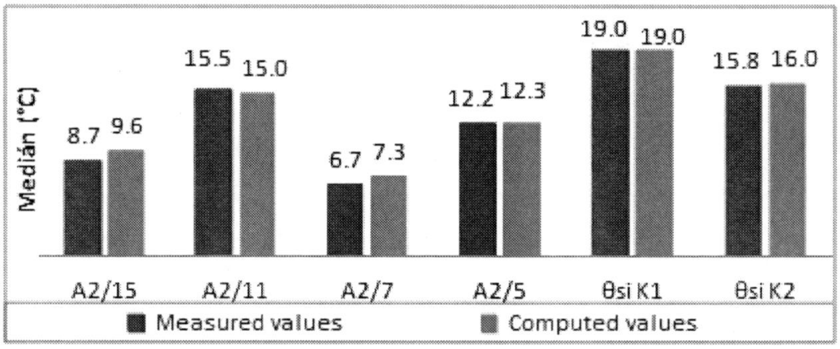

Figures 6. Measured results and comparison of medians and averages see above note of computed and measured values in the points A2/15, A2/11, A2/5 and A2/7

In the averages and medians see above note it can be seen that in both there are very similar differences in the compared values. The best match (difference of 0.1 K) of the measured and calculated values is at point A2/5 (chamber 2).

5. Evaluation of thermal instability of chosen detail

Data for hourly weather changes in Kosice was used for further analysis. This data concerns the measured outdoor air temperature and the measured intensity of global and diffused solar radiation during the month of February 2015.

The calculation was performed for three external environments:

I. constant outside temperature for Kosice, θ_e = -13.0°C

II. measured outside air temperature in Kosice without the influence of solar radiation

III. measured outside air temperature in Kosice including the influence of solar radiation

For the purposes of simplification, the internal air temperature was assumed to be 20.0°C. The temperature was constant as the experimental chamber has a quasi-stationary environment. The temperature was maintained at θ_i = 20.0°C and had a relative humidity of 50%. The boundary conditions of the analysis that were entered into the calculation of the temperature fields are shown in figure 3.

The calculation was performed using the Physibel simulation program. Three methods were selected:

1 - stationary calculation method

2 - non-stationary calculation method without the influence of the sun

3 - non-stationary calculation method with the influence of the sun

Use of the above methods allows a comparison of the results of the surface temperature values of the structure obtained by calculation for the individual states of the different conditions. The sun has a significant influence on the interior surface temperature of the structure. The largest difference between the stationary calculation and the dynamic calculation, taking into account the influence of the sun, is the value of 3.15 °C. This difference suggests the possibility of optimising the thickness of the thermal insulation system with favourable economic results. Therefore, a progressive solution for calculating the parameters of constructions is important in addition to being scientific and practical. In the paper, the detail of the window sill is solved, as it is in present form in the experimental chamber. Even on the basis of the results presented, it is obvious that if the inner window sill is installed, the surface temperature would change.

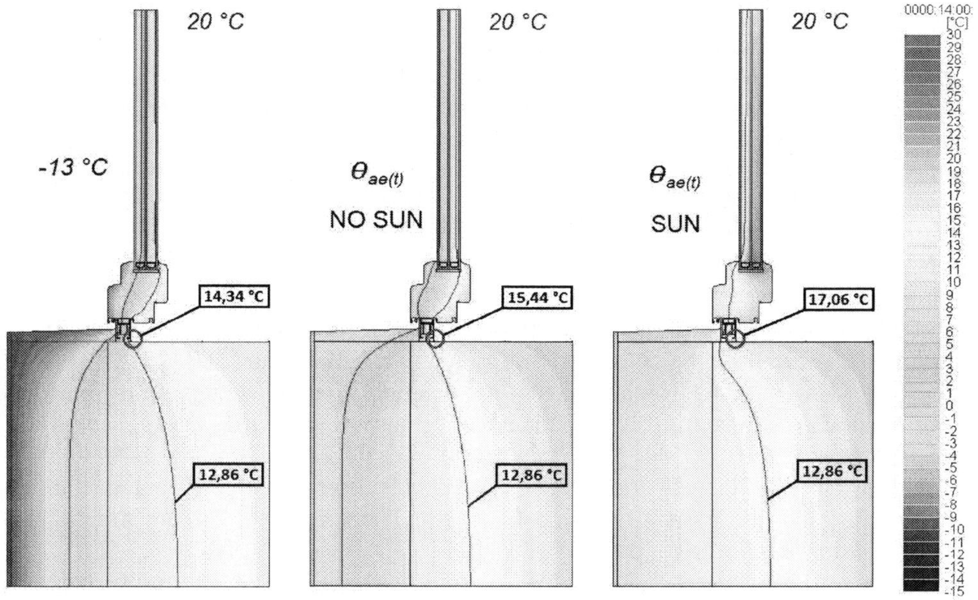

Figure 7. Differences over surface temperature values at the assessed point for individual states

The advantage of experimental chamber is the possibility of verifying the computational results and tuning the model. Validation of the model enables the obtaining of accurate data from simulation programs without the need for further verification. In this way, we can analyse the influence of the way of window profile setting, the impact of the interior sill construction, the influence of the change of the material and the height of the window sill, changes in the thickness of the external sill insulation, and changes to the surface modifications of the outer components of the structure. Thus, the verified model can calculate details of any standard buildings (historic buildings, panel systems, wooden houses) in which it has greater practical importance than it has for nearly zero energy buildings.

6. Conclusion

Validation of the simulation model is the key to solving and predicting the real physical thermal-humidity behaviour of building elements, details, building structures and their environment. Comparison of the results of calculations and measurements confirms the accuracy of the simulation model, given the consistency of the measured and calculated data. If we do not consider sunlight in such cases, we would not obtain comparable results for the south-facing wall. However, differences persist and are mainly due to the material properties, the geometric precision, the initial and boundary conditions used and the consideration of the temperature field without regard to the transport of water and air. An advanced and validated simulation model allows the solving of various types of numerical experiments with defined precision in order to optimise critical building details.

References

[1] Janssens A, Roels S, Vandaele L, Full scale test facilities for evaluation of energy and hygrothermal performances. International Workshop (2011) UG Belgium, ISBN 978-94-9069-584-2 Brussels, Belgium 162 pg.

[2] Baker P H, Van Dijk H A L , PASLINK and dynamic outdoor testing of building components, Building and Environment, **43** (2008) 143–151.

[3] Strachan P A, Vandaele L, Case studies of outdoor testing and analysis of building components, Building and Environment, **43** (2008) 129–142.

[4] Katunský D, Zozulák M, Kondáš K, Šimiček J, Numerical analysis and measurement results of a window sill, Advanced Materials Research, **899** (2014) 147-150.

[5] Zozulák M, Katunský D, Experimental temperature measurement in the window sill, Tepelná ochrana budov, **1** (2014) 32-38.

[6] Zozulák M, Vertaľ M, Katunský D. Numerical Analysis of Window Structure Seating Depth Effects on Surface Temperature and Linear Thermal Transmittance, In Advanced Materials Research Vol. **1057** (2014), pp 53-60

[7] Katunský D et al., Measuring methodology and results of heat-air-moisture performances at building envelope levels, Advanced Materials Research, **649** (2013) 85-88.

[8] Goia F, Schlemminger Ch, Gustavsen A, The ZEB Test Cell Laboratory. A facility for characterization of building envelope systems under real outdoor conditions, In Energy Procedia **132** (2017), pp. 531-536.

[9] Vertaľ M, Zozulák M, Vašková A, Korjenic A, Hygrothermal initial condition for simulation process of green building construction, In Energy and Buildings **167** (2018), 166-176.

[10] Maivel M, Kurnitski J, Radiator and floor heating operative temperature and temperature variation corrections for EN 15316-2 heat emission standard, Energy and Buildings **99** (Supplement C) (2015) 204 – 213.doi:https://doi.org/10.1016/j.enbuild.2015.04.021.

[11] Maivela M, Ferrantellia A, Kurnitski J, Experimental determination of radiator, underfloor and air heating emission losses due to stratification and operative temperature variations, In Energy and Buildings, **166** (2018), pp. 220-228

[12] Ďurica P, Juráš P, Gašpierik V, Rybárik J, Long-term Monitoring of Thermo-technical Properties of Lightweight Constructions of External Walls Being Exposed to the Real Conditions, In Procedia Engineering, **111** (2015), pp. 176-182

[13] Slávik R, Čekon M, Study of Surface Temperature Monitoring in the Field of Buildings, In Procedia Engineering, **161** (2016), pp. 1135-1143.

[14] Tywoniak J, Lupíšek A , Bureš M , Volf M Wood Based Curtain Wall for Building Retrofit – Development and Performance. In Advanced building skins 2015 Conference Proceedings. Graz: Graz University of Technology, 2015, p. 213–218. ISBN 978-3-85125-397-9.

[15] Kachkouch S, Benhamou B, Limam K Experimental Study of Roof's Passive Cooling Techniques for Energy Efficient Buildings in Marrakech climate, In International Renewable & Sustainable Energy Conference IRESC' 2017At: Tangier, Morocco, December 2017.

Acknowledgement

This contribution is a partial output of the research projects VEGA 1/0389/17 and VEGA - 1/0674/18 supported by the Scientific Grant Agency, of the Ministry of Education, Science, Research and Sport of the Slovak Republic and the Slovak Academy of Sciences.

IOP Conf. Series: Materials Science and Engineering **415** (2018) 012032 doi:10.1088/1757-899X/415/1/012032

Building of Research Centre as the experimental laboratory for civil engineering

P Juras[1], P Durica[1], D Jurasova[1], R Ponechal[1], J Rybarik[1]

[1] Department of Building Engineering and Urban Planning, Faculty of Civil Engineering, University of Zilina, Univerzitna 8215/1, 010 26 Zilina, Slovakia

peter.juras@fstav.uniza.sk

Abstract. The intelligent building of Research center was built during 2014 and 2015. Fragments of its envelope meet the requirements for passive buildings. This building is a part of the University of Zilina campus, located in Zilina, Slovakia. It consists of a high-rise part with five story and a low-rise part with laboratories. Several laboratories of departments are also in the high building. What makes the building unique are several things such as three different heating systems, use of renewable sources, soil heat accumulator, various photovoltaic panels on the roof, vacuum collectors etc. All these technologies can be compared to each other and their efficiency and suitability for different purposes can be evaluated. In this paper, a unique façade monitoring system with 36 façade meteorological stations is described. Each of these stations measures outdoor air temperature near the façade, solar radiation, wind velocity and direction in a short time interval. Recorded values can be used for a more precise simulation of energy consumption, natural air ventilation, HAM simulation, wind-driven rain calculation and also compared to other simulations.

1. Introduction

The University of Zilina has a new part since 2013, which is separated from all faculties. The idea was to create a separated workplace specialized for research and for trade of the results from research to the production only. So the story of Research Centre began. With support of the ITMS project no. 26220220183 “Research centre of the University of Zilina” funded from the Structural funds from the operational program Research and Development. The aim of the project was to create the background, which will enable the possibility to conduct state-of art research with the laboratories, equipment and material basis.

Several faculties and departments from the University: faculty of Civil Engineering, Mechanical Engineering, Electrical Engineering and faculty of Management Science & Informatics participated in the Research center project. These faculties and their departments helped with the early development stage (building design, floor plan and laboratories), technologies and the laboratory equipment.

The architectural and structural drawings were realized in cooperation with the Department of Building Engineering and Urban Planning. Ventilation, heating sources and air conditioning was designed by the other Departments. The construction was finished in November 2016 (Figure 1 and 2).

Aim of the project was not only to create the building for the staff and the laboratories, but was created the laboratory from the building itself. The indoor environment with its heating and air conditioning, technical and technological solutions, several heat sources, use of renewable energy sources etc. were aimed to create the building, which use and maintenance will be research laboratory itself. With support of the operational programe also two buildings of University Science Park (USP) were built. These buildings are very similar with its shape with the difference in the aim of the buildngs: USP building meets the ultra low energy standards and the Research centre (RC) building is an intelligent building in passive standard. Therefore another planned outcome is the direct comparison of energy performance and effectivness of the nearly zero energy building and ultra-low energy building based on the real energy consumption.

In this paper, the technological part of the building and the unique outdoor environment monitoring system which is mounted on the façade of the building is described.

Figure 1. Western view at the building of Research centre.

Figure 2. South-east view at the building.

2. Building structures design

The building (Figure 1 and 2) has five story, one basement floor and the building can be divided into two parts, the high one (main) with office and light laboratories, and the lower one with heavy laboratories including the climate chamber.

Simplified dimensions of the main building are 22.53 x 26.11 m with its height of 18.86 above the terrain. Ground level is 395.45 a.s.l.

As passive standard requires, the building envelope with all fragments need to meet the requirements of the Slovak standard [1]. The values for individual fragments are stated in the Table 1 with comparison of the Standard values and the calculated values for the building of RC and USP (low energy).

Table 1. Requirements of the Slovak standards [1, 2] on the building envelope fragments and calculated values for the buildings.

	Wall	Roof	Ground floor	Floor over the basement	Window
STN 73 0540:2012 valid since 1.1.2016 ultra-low energy	0.22	0.15	R > 2.5	0.85 – 0.30	1.00
STN 73 0540:2012 valid from 1.1.2021 nearly zero energy	0.15	0.10	R > 2.5	0.6 – 0.15	0.60
University Science Park	0.16	0.1	0.20	0.66	0.88
Research centre	0.14	0.1	0.20	0.30	0.88

Load bearing construction system is made of reinforced steel concrete. Walls are masonry from the aerated concrete bricks with th. 380 mm. These walls are insulated either with ETICS or with the aerated façade cladding. In both cases, the insulation layer is based on the mineral wool with th. 200 mm.

Roof is flat, non-walkable, covered with mPVC waterproof membrane. It is insulated with the mineral wool with total thickness of 420 mm. On the roof is made separate walkable platform from the metal mesh grating used for the maintenance of the multiple used different devices (Figure 3).

Windows are made of the 6-chambered plastic frames with a triple glazing. All windows are equipped with exterior shading blinds, which are operated automatically and with the building management system are supposed to maintain the set up indoor environment. Nonetheless, they can be operated manually in every office.

The mullion walls, which cover the staircases, lobby near the elevator are made from the aluminum and have triple glazing as well. The staircase glazing has lowered solar gains to avoid overheating of the space because of the southern orientation [3]. Entrance door are automated, with a thermal glazing and with air the layer to avoid thermal losses.

Figure 3. Rooftop with the solar equipment, vacuum tube collector, photovoltaic panel, ventilation, etc. [4]

Figure 4. Machine room in the basement. Heating cascade and pipes from the wells [4].

3. Heating, ventilation and air conditioning

Most areas of the building are equipped by the floor heating and ceiling cooling. The system of the floor heating could provide during the summer also the cooling. Combination with blinds are very effective.

Cool demand is satisfied with the two heat pump and one gas heat pump (Figure 5, 6). Primary source of cool are the wells. There are 11 wells with the average deep 150 metres. The wells are connected to the buildings distributions network in the basement machine room (Figure 4).

The cooling ceilings are located mainly in the office spaces, which have transparent opening in the envelope and are heated by the direct solar radiation. The ceilings are realized as the suspended grid ceilings combined of active (cooling parts) and common gypsum board (Figure 7, 8).

The floor heating has temperature difference 45/35 °C and the heat source is condensing gas boiler, two electricity and one gas heat pumps. Also solar system is utilized. The heating distribution network could be divided to six roots according to the actual situation or purpose.

Hot water in the storage water heater is mainly heated with the solar system, which consists of 27 vacuum tube solar collectors located on the roof (Figure 3). These collectors are managed by the solar station with regulation.

The machine room also consist of various system for automated and intelligence system for building management such as fire protection system, electricity distribution, lightning system and other energy sources such as photovoltaic panels on the roof (Figure 3) and the wells outside of the building, which are used by the heat pumps.

Another computer management system is intended to monitor the room occupancy to avoid energy losses of any kind (lighting, heating, cooling, etc.). For example, if the room is empty the lights will be automatically turned off. The governing mechanism of this is due lack of the space not incorporated in this article.

Figure 5. Machine room with the two heat pumps [4].

Figure 6. Gas heat pump located outside of the building [4].

Figure 7. Office with the combination of active and non-active parts of the ceilings.

Figure 8. Floor corridor before suspended ceiling was made. Visible cooling pipes from the divider [4]

4. Heat demand and thermal evaluation of the building

As mentioned above (Table 1), all building envelope fragments have lower *U* value as was required in the time of the building designing. Therefore also the heat demand is lower than required for this type of building. The calculated heat demand with monthly method based on the STN EN ISO 13 790 [5] (valid during the designing phase) for RC and USP are summarised in the Table 2. The calculated area is only the main building without the heavy laboratories.

The total floor are is 1244.3 m^2, building volume gross 4656.7 m^3, shape factor 0.41 1/m. For this shape factor according to the STN 73 0540:2012, the energy need for heating has to be lower than 28.9 kWh/ (m^2.a) (since 2016) and 14.5 kWh/(m^2.a) (from 2021).

In the calculation was taken into account that the intelligence building of research centre will have lower heat gains from the lights and equipment caused by the better utilizing of sources and use of devices with lower power consumption, which also means lower waste heat [3].

There was elaborated also a basic simulation in EnergyPlus during the designing phase of building, which results were for the specific energy need for heating 13.5 (with closing the windows blinds if the solar radiation will be higher than 100 W/m^2) or 12.5 (closed blinds if solar radiation > 200 W/m^2).

Table 2. Calculated energy need for heating according to the STN EN 13 790 [5].

Building	Average thermal transmittance coefficient [W/(m².K)]	Heat loss coefficient [W/K]	Ventilation heat loss coefficient [W/K]	Solar gains [kWh]	Utilization factor for the gains	Specific energy need for heating [kWh/(m².a)]
USP Ultra-low	0.34	1266	614	40919	0.66 - 0.97	39.3
RC Nearly zero	0.33	920	146	40919	0.5 - 1.0	13.2

5. General description of conducted research activities within the building

The building itself with its monitoring system enables the possibility to conduct research based on the power or energy consumption of many types, of use of energy sources (heat pump, wells, solar, photovoltaic, etc.). It can also analyse better the behaviours of the users = researches. Results of selected areas belongings to the HVAC, electricity or other science areas will be published. In this paper, the unique façade monitoring system is described.

6. Façade meteorological stations

Façade monitoring system consists of façade meteorological stations (FAMS). This system consists of 36 small devices, which are constantly monitoring the outdoor climate around the building and the influence of the building can be examined.

First FAMS (Figure 9) was fixed on the façade in August 2016. Since the end of November 2016, all 36 stations are working with some minor malfunctions and sensors change during the test period. The FAMS itself consists of metal frame, which is fixed on the façade bellow window and is powered by a cable. On the frame, the sensors measuring temperature, relative air humidity and atmospheric pressure, are fixed. The next one is the pyrgeometer for measuring of the long-wave solar radiation, pyranometer for monitoring of the global solar radiation and an ultrasonic anemometer for wind velocity and wind direction measurement. The anemometer is offset from façade 270 mm on short arm to monitor the air flow. During the autumn 2017 were on each façade changed one anemometer to the 3D ultrasonic, to monitor better the airflow.

The data collection is wireless. Each façade has its own access point located on the roof. Measured data, with a one-minute write interval, are collected and stored on a remote server. The location scheme of the stations is shown in Figure 10. There are three FAMS placed on the south façade, fourth façade, 15 on the east and 15 on the west.

Figure 9. Detailed look at façade meteorological station, equipped with following measuring sensors: air temperature, relative humidity, barometric pressure, long and short wave radiation, wind velocity and wind direction. Several FAMS are equipped with 3D anemometer, this one is with 2D.

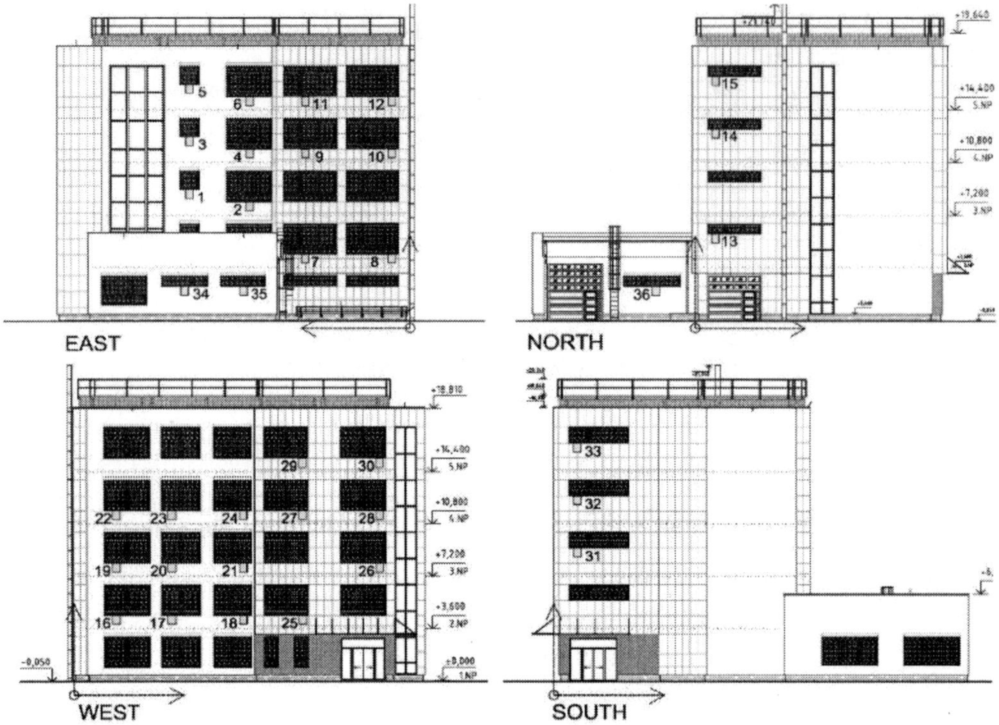

Figure 10. Scheme of the building with marked numbered façade meteorological stations (FAMS).

7. Results and Discussion

A hot sunny day (21 June) was selected to analyse summer period. Measured results are divided into the climate parameters: air temperature, global solar radiation (short wave) (Figure 11) and long wave (Figure 12) on the vertical surface. Later there is more detailed comparison on the west façade between the solar radiation, temperature and the shape of the building which blocked direct radiation on the part of façade during specific time period (Figure 13). Times stated in the graphs and in the text are in UTC, which means - 2 hours to the local summer time.

The measured air temperatures near the façade in Figure 11 are influenced by the direct solar radiation on the façade near to the FAMS position and are logically following the sun movement on the sky. During the night the temperatures are a bit lower than the air temperature measured by the weather station located on the roof of nearby building [6]. This is caused by the night sky radiation during the cloudless nights, which is cooling the surfaces. In the morning, there is for the short time direct solar radiation on the north façade (the façade is not perfectly aligned to the north). Also the east façade are receiving the solar radiation. This caused very fast heating of the façade and the air around this façade from the starting temperature 12.6 °C.

As the sun moves, the north façade came to the shade and the temperature dropped down, returning to the outdoor air temperature before 9:00 AM, which was 25.2 °C. Maximum temperature on the north façade was 29.7 °C at 7:15 AM. East façade has maximum temperature 37.7 °C at 9:35 (11:35 local time). At that time the most sunshine became the south façade which has the longest sunshine duration. The highest temperature was as expected measured in front of the west façade, exactly 39.1 °C with air temperature on the open ground 28.1 °C. After the sunset the temperature slowly returned to the air temperature. The global solar radiation graph in upper part of Figure 11 showed the south side amount lower than the east and west and as much as twice lower than the measured values by the weather station on the horizontal plane. This is caused by the vertical mounting position of the pyranometers, where the inclination of the solar beams on the surface of pyranometers is crucial. During the winter time period, the courses are opposite - radiation on the horizontal plane is lower than on the vertical plane.

IOP Conf. Series: Materials Science and Engineering **415** (2018) 012032 doi:10.1088/1757-899X/415/1/012032

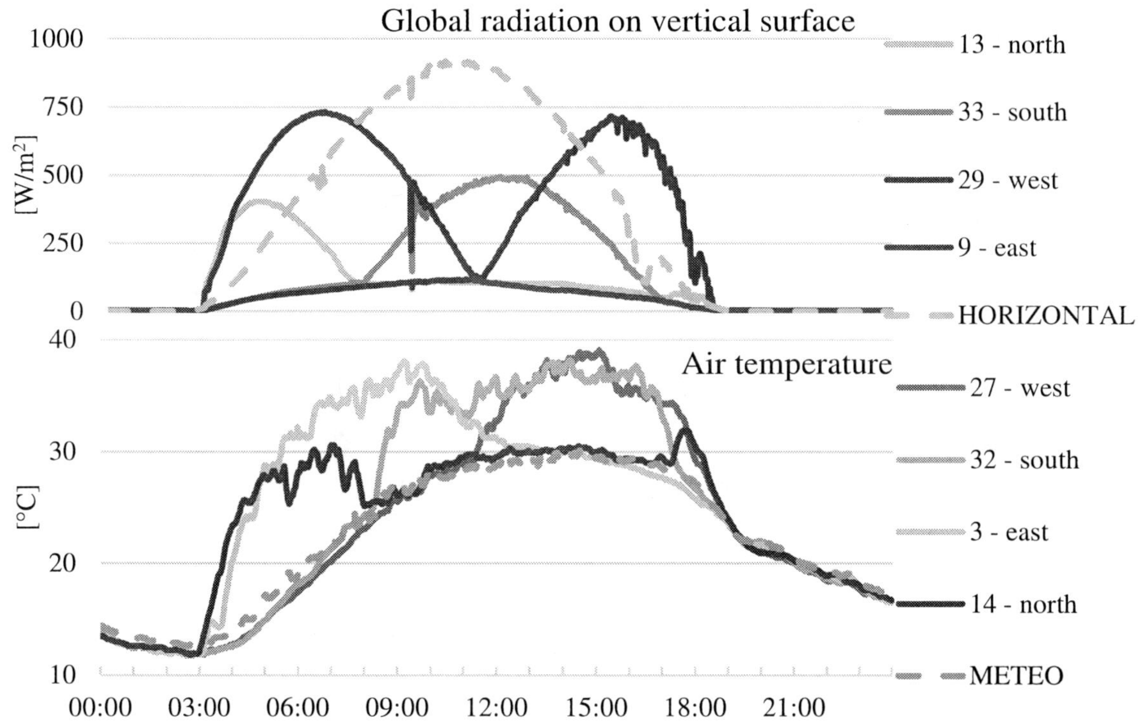

Figure 11. Courses for the 21 June 2017, global solar radiation and the air temperature.

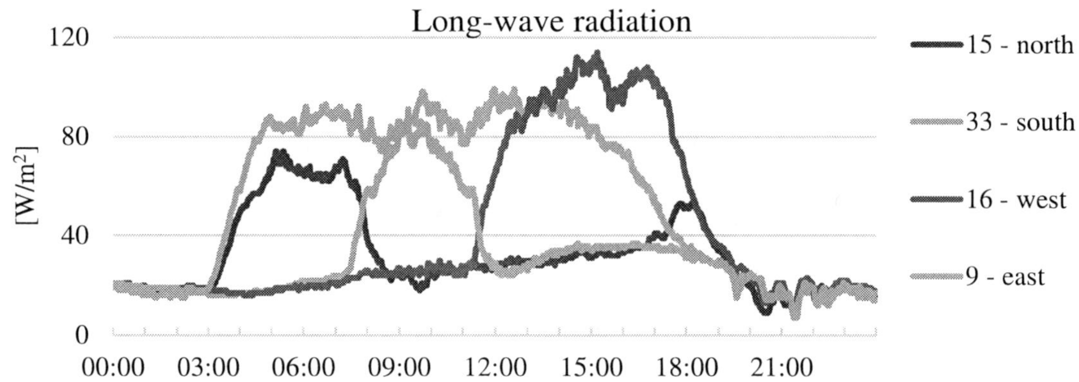

Figure 12. Long wave radiation courses for 21 June 2017

In Figure 12 are shown daily courses for long wave radiation measured by the pyrgeometers, one value for each side. Unfortunately, during that day the pygeometer on the weather station did not work so the comparison was not possible. The courses are also following the sun movement on the sky, firstly the north and east FAMS´ temperatures rises, later the south and lastly the west one, which had also the highest amount. Specially, the pyrgeometers should be influenced by the reflectance of the ground, the analysis of FAMS with different heights could be interesting.

The shape of the façade has small overhang on the right side (visible in Figure 10), which caused that the third row (from left) of FAMS are shaded as the sun moves from the south. This is represented as the jump (Figure 14 - FAMS no. 24) when the FAMS is no longer in the shade. Also the temperature rises quickly after that and in this specific case became the highest (Figure 13).

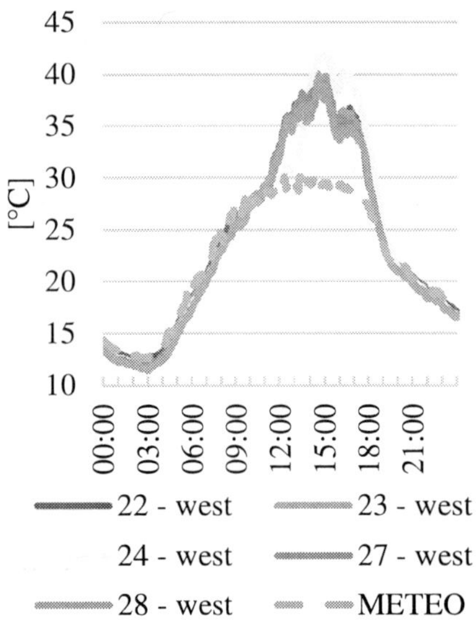

Figure 13. Differences between the temperatures measured on the western façade.

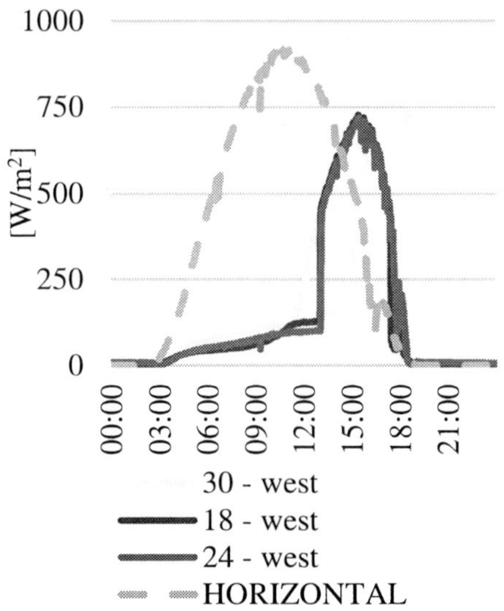

Figure 14. Differences between the solar radiation measured on the western façade.

8. Conclusion

In this paper was briefly summarized the idea and the realization of the Research centre building, which serve as the basis for the research conducted not only in the laboratories but in-situ, in the building with the researches in it. More articles about the indoor environment and results of use of renewable energy sources will be published by the other departments.

In this paper was described not only the building but also the façade monitoring system. System was described and summer day courses were analysed.

Future plan is to use the results in natural ventilation simulation (different outdoor air temperature around the façade through the day), impact on the energy consumption, analysis of the air flow around the building etc.

Acknowledgments

The research was done with the support of grant project VEGA no. 1/0945/16.

References

[1] STN 73 0540:2012: Thermal protection of buidings. Thermal performance of buildings and components. (In Slovak)

[2] STN 73 0540:2012 - 2. Z1: Part 2. Functional requirements, Change 1

[3] Ponechal R et al. 2017 Mullion wall in building of Research Centre *Proc. Int. Conf. Thermal Protection of Buildings* (Bratislava BB Print) pp. 145-150

[4] Durica P et al. 2017 *Energetické a environmentálne navrhovanie a hodnotenie budov* (Zilina: EDIS) vol. 1 403p

[5] STN EN 13 790:2008 Energy performance of buildings. Calculation of energy use for space heating and cooling

[6] Staffenova D et al. 2014 Climate Data Processing for Needs of Energy Analysis. *J. Advanced Materials Research: enviBUILD* vol. 1041 pp. 129-134

Thermal insulation versus overheating - monitoring results

T Kisilewicz

Cracow University of Technology, Institute of Building Materials and Structures, 31-348 Kraków, ul. Warszawska 24, Poland

Email: tkisilew@pk.edu.pl

Abstract. Initial design decisions regarding building external envelope may have important consequences for thermal comfort and total energy demand not only for heating but also for cooling. Before reaching for more energy-intensive solutions, effective use of passive measures to avoid overheating is necessary. In the article, special attention was paid to the influence of external wall structure on space overheating. On the basis of monitoring results it was found that the standard thermal insulation prevented excessive energy conduction into the space on very hot days and it should not be considered as a barrier for heat dissipation. The standard opaque wall was also compared with the uncovered transparent insulation, which is in summer an intensive source of energy gains. The basic dynamic characteristics of the monitored wall were analytically analysed by means of the periodic waves method.

1. Introduction

Low energy buildings require new experience, systems and tools for the design, which for obvious reasons are not commonly available. While conventional buildings may be characterized by a considerable tolerance to design errors in the field of building physics, the new ones are extremely vulnerable and demand new approach. The article deals with the influence of external wall properties on the comfort of office space use. The specific difficulty of the use of office buildings is associated with relatively high internal heat gains [1]. While the issues regarding heating demand are already quite well recognized, the protection of the building against overheating is usually left to the designers of the installation rather than the architects or constructors of the building. These initial design decisions can have important consequences for thermal comfort and total energy consumption. In the climate of Central Europe, before reaching for more energy-intensive solutions, effective use of passive solutions to protect against overheating is necessary.

Passive ways to protect the building from overheating are characterized by great simplicity, usually do not require significant investment expenditures, do not consume non-renewable energy, do not require costly system maintenance. To a large extent, these are solutions well known since a long time, such as appropriately selected size and orientation of the glazing, large heat capacity of the interior, night cooling etc. [2,3,4,5]. For a long time these were the only means to control the conditions in buildings. Nowadays it's necessary to specify again the conditions of their application and the rules of rational design [6].

The article focuses on the specific influence of external wall structure on indoor conditions in the warmest period of the year. Although no one disputes nowadays the sense of thermal insulation use, very thick insulation layers, as in passive buildings, are considered as the cause of space overheating [3,6,7]. According to the widespread opinion efficient thermal insulation of external building shell is a barrier for energy discharge to external environment during the hot period of the year.

2. Dynamic thermal characteristics of a building component

Ambiguities and myths associated with dynamic heat exchange through building components result to a large extent from computational difficulties related to dynamic conditions. Analytical calculations are very complicated and the numerical modelling is usually not used in the everyday design process. Designing process is therefore necessarily based on observations related to stationary heat exchange conditions. Meanwhile, the processes of heat storage and thermal inertia of the partitions have a great influence on the heat exchange and the actual conditions inside the buildings [3,7,8].

One of the simplest and best known heat transfer dynamic models is a theory of harmonic heat waves, based on Fourier analysis of the cyclic functions. Due to a quasi-periodic course of ambient temperature and Fourier analysis tool, this analytical approach allows accurate modelling of real conditions [5]. In International Standard EN ISO 13786 – 2017 [9] only sinusoidal boundary conditions are considered: building boundaries are submitted to sinusoidal variations of temperature or heat flow rate. Two basic dynamic features of building component are usually considered. Thermal admittance relates heat flow rate to temperature variations on the same side of the component. Transmittance is a complex quantity defined as a complex amplitude of the density of heat flow rate through the surface of the component adjacent to zone m, divided by the complex amplitude of the temperature in zone n when the temperature in zone m is held constant.

Combined thermal characteristics of the building component under the steady and periodic conditions may be expressed as a decrement factor i.e. ratio of the modulus of the periodic thermal transmittance to the value of the steady-state thermal transmittance U.

In this work thermal transmittance Y and decrement factor f were used as the dynamic wall characteristic. *D'Thermal* software was used to run the computations needed for this paper.

The structure and material data of the tested external envelope have been shown in table 1.

Table 1. Wall structure and material data.

	thickness d [m]	density ρ [kg/m³]	thermal conductivity λ [W/mK]	specific heat c [J/kgK]
internal plaster	0.02	1700	0.8	840
ceramic blocs	0.188	800	0.21	880
expanded polystyrene	0.1	10	0.045	1460
external plaster	0.01	2000	1.0	840

The standard dynamic characteristics, according to EN ISO 13786 (2017) and for the period of 24 h, have been calculated and the results of calculation have been compiled in table 2.

Table 2. Dynamic characteristics of the tested wall – 24 hour period (surface thermal resistances excluded).

	complex number	modulus [W/(m²K)]	time lead/shift [h]
external admittance	0.416+1.292i	1.357	4.811
internal admittance	2.513+4.003i	4.726	3.859
transmittance	-0.051-0.082i	0.096	-8.136

Calculated internal areal heat capacity of the wall (without surface resistances) is equal to 66.316 kJ/m^{2}K, while external areal heat capacity is only 19.951 kJ/m^{2}K. Decrement factor is equal to 0.220.

Periodic thermal transmittance is responsible for propagation of thermal waves across the building component. External environment variations are the result of ambient temperature changes but also of solar radiation absorbed upon external surface of a building shell. Low transmittance on one hand may prevent heat dissipation but on the other hand it protects space against conduction of absorbed by external face solar energy [8]. Large phase shift is an important aspect of heat transfer, that allows to separate in time solar gains through the windows and those due to the conducted heat wave [5]. "Delayed' energy may be efficiently removed by intensive night ventilation [10].

In order to exemplify the significance of a massive material for non-stationary heat transfer, the dynamic characteristics of the wall without the ceramic blocs have been calculated. Thermal resistance of the removed ceramic layer was compensated by increased up to 14 cm expanded polystyrene thickness. Major reduction of thermal capacity would result in the significantly increased value of the transmittance modulus: 0.317 W/(m^{2}K) and heavily reduced phase shift: -0.423 h. In spite of the same thermal resistance in the stationary conditions, the lightweight wall would transmit in real dynamic conditions much more energy than the heavyweight one.

A specific feature of non-stationary conditions is the sensitivity of building elements to the frequency of changes. As a result the same element has different properties depending on the environmental variation conditions. All the above dynamic features have been calculated for 24 hour period of the boundary condition changes. The daily changes of the boundary conditions are, of course, of a great practical importance due to the standard variability of the outside temperature, solar radiation and interior use. However, it is also important to consider the changes taking place in a much longer period of time, e.g. a few days. Long heat waves occur many times every year, regardless of global warming.

Calculated dynamic features of the tested wall for 72 hour period are presented below in table 3.

Table 3. Dynamic characteristics of the tested wall – 72 hour period (surface thermal resistances excluded).

	complex number	modulus [W/(m^{2}K)]	time lead/shift [h]
external admittance	0.366+0.471i	0.596	10.430
internal admittance	1.299+2.150i	2.512	11.772
transmittance	0.111-0.199i	0.228	-12.183

Decrement factor for 72 hour period is 0.621, it is almost 3 times bigger than for 24 h period, internal heat capacity due to the bigger than before penetration depth is now 108.602 kJ/m^{2}K and external capacity is 29.572 kJ/m^{2}K. Much bigger than before values of transmittance modulus and decrement factor mean that the longer heat wave will easily propagate across external building shell and despite low amplitude would significantly affect internal thermal comfort.

3. Field testing of the two-layer external wall

Figure 1. South oriented wall of the tested space with two windows and the transparent insulation panel in between

The author had an opportunity to monitor long time thermal performance of the south oriented office building wall, figure 1. The central part of this wall was covered with the transparent capillary insulation.

Results related to transparent insulation have been in detail discussed in the former papers, in this paper mostly data connected with the standard opaque insulation will be presented and discussed.

The structure of the standard south oriented double layered wall was the same as in Table 1: internal lime plaster of 0.02 m, ceramic blocks of 0,188 m, **E**xternal **T**hermal **I**nsulation **C**ompound **S**ystem (ETICS) with 0.10 m of expanded polystyrene and mineral plaster. Calculated value of the thermal (stationary) transmittance coefficient was 0.301 W/(m^2·K).

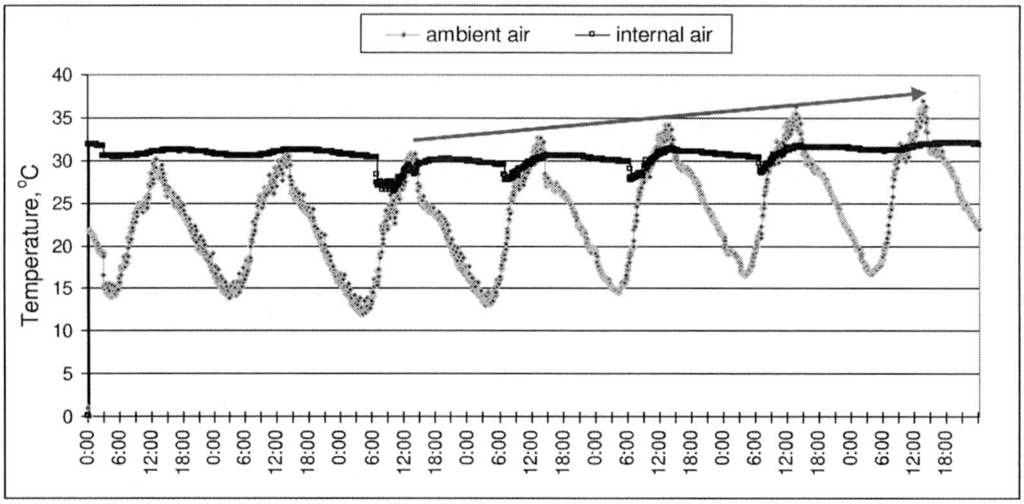

Figure 2. Hot summer week in July, internal and external air temperature

The following sensors have been applied to monitor tested elements:

- Pt-100 resistors, class A, to measure temperature of external and internal air and temperature of internal surface of the tested wall, accuracy ±(0.15 +0.002[t]),
- heat flow transducer placed on internal surface of the wall, accuracy 5%,
- vertical pyranometer, accuracy 5%.

All the sensor data have been gathered, digitalized and stored in the 16-channel scanner MPI-LAB with five minute intervals.

In figure 2 data of internal and ambient air temperature, recorded during one hot summer week, have been presented. Due to the large glazed area and transparent insulation panel, the tested space was heavily overheated. Internal air temperature was very high during the whole period with merely slight fluctuations in diurnal cycle. Only a small drop of internal air temperature during night may be partially explained by intensive heat flux released to the space in this time from massive accumulator in transparent system. During the whole analyzed period external air temperature was significantly fluctuating, falling down in night even below +15 °C, thus making possible night cooling of intensely overheated space [7,8,10,11]. Unfortunately, this opportunity was practically not used, with the exception of short morning periods, when the windows were opened by the cleaning staff. It may be easily observed on the third day of analyzed period, figure 2, as a sudden drop of internal air temperature at 6.00 a.m.

As it may be observed in figure 2 (red arrow), mean ambient air temperature was since the third day slowly but constantly rising reaching its maximum of 37°C on the 7-th day. So in fact the tested wall was exposed not only to 24 h heat waves but also to much longer wave with low amplitude. Accumulation capacities of massive buildings, with a thickness of heavyweight structural layers up to 25 cm, are sufficient to effectively dampen the daily fluctuation of boundary conditions. In the case of longer waves, with a few-day period of changes, the wave attenuation is very limited, as shown in Table 3, and mean internal air temperature is gradually rising.

Solar irradiance within the analyzed summer week was very intensive and followed nearly the same pattern each day, figure 3. Due to the building shape, pyranometer located close to the corner was partially shaded against direct radiation in the afternoon, what resulted in abrupt irradiance drop after 14.00 (right side of each peak).

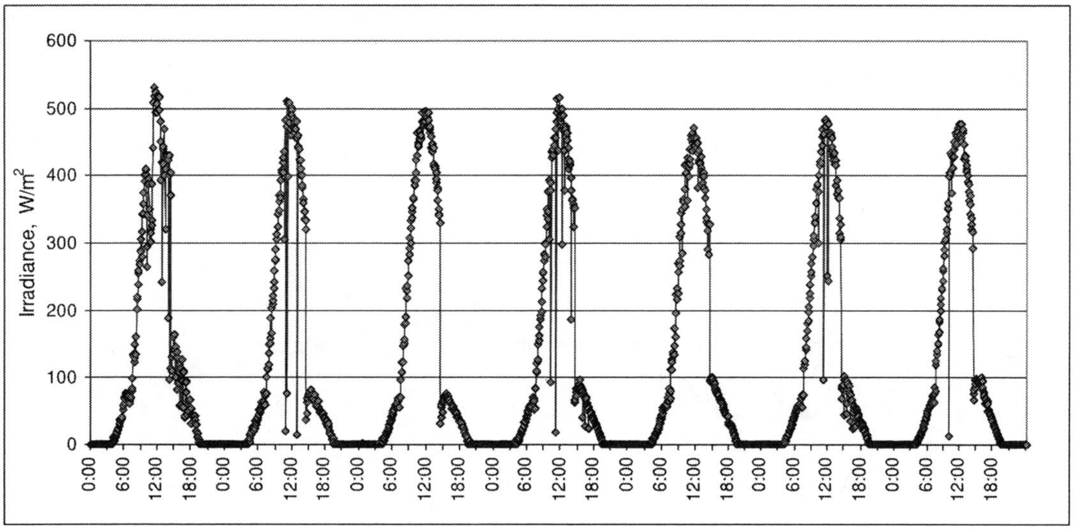

Figure 3. Solar irradiance measured at south oriented wall

Heat flux intensity, measured on internal surface of the discussed double-layer wall has been shown in figure 4. Positive number means heat flow into the wall while negative number heat delivered to the interior.

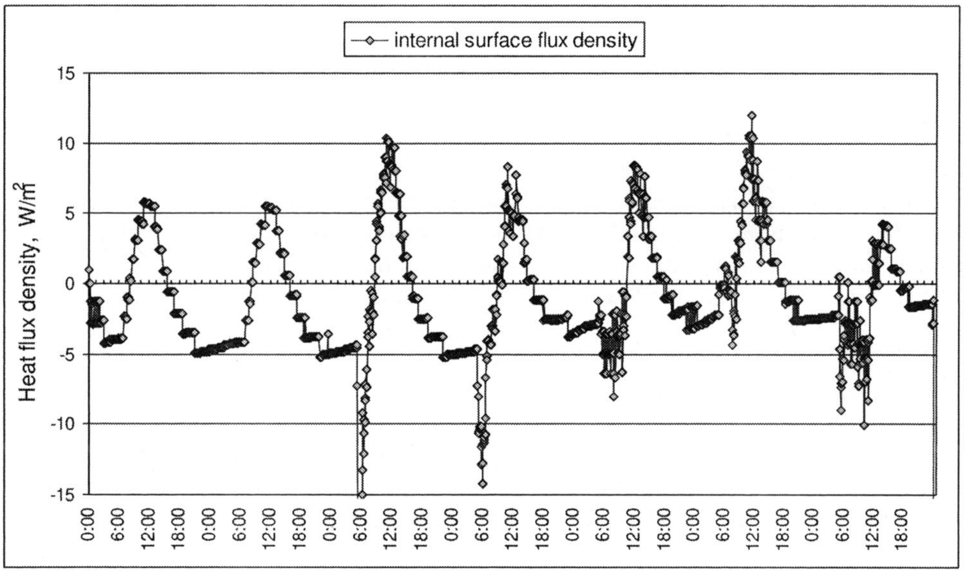

Figure 4. Hot summer period in July, heat flux intensity on internal face of the south oriented wall

In spite of the momentary temperature difference between internal and external environment, direction of heat transfer across internal surface of the wall was changing in a regular way: into the wall during daytime and out of the wall during night. The relation between external and internal air temperature is definitely different. Within the observed period heat flow was not induced by the instantaneous temperature difference between the two environments but by the local conditions at the internal surface of the massive structural part of this wall. Discussed earlier sudden drop of the internal temperature at 6.00 a.m. resulted immediately in intensive heat flux change and maybe easily observed in figure 4 in form of regular variations. Weekly heat flow balance at internal face of the tested wall is negative and equal to -193.5 Wh/m^2, while the sum of the absolute values of heat flow is three times higher: 586.4 Wh/m^2. Sum of the absolute values was intended here as a measure of total heat exchange at internal face. During the whole week more energy was delivered to the space than to external environment. In spite of the recorded big environmental temperature differences, there was no effect of energy discharge to external environment.

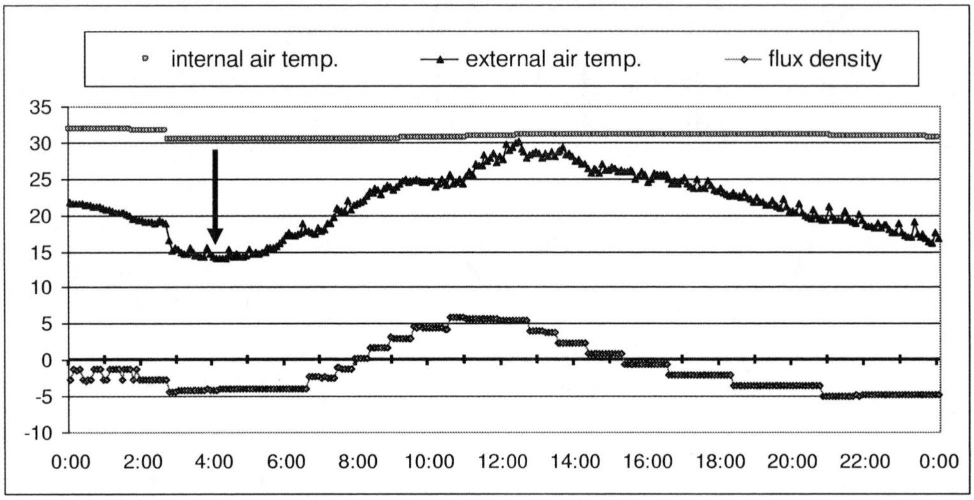

Figure 5. First day of the hot summer week in July, internal and external temperature (upper curves) and internal face heat flux density (bottom curve),

On the first day of monitored period, figure 5, despite the significant temperature difference, especially in night, heat stored in the massive wall was still delivered to the space (negative heat flux).

According to the simplistic stationary understanding of heat transfer across the external building wall, heat flux direction and its intensity would depend on the instantaneous environmental conditions with no space for history of temperature fluctuations and energy storage. In reality, in case of the massive and well insulated wall, momentary night conditions in both environments do not result in the immediate heat conduction to external air and discharge of the accumulated gains.

The reversed heat flow, that started at ca. 8 o'clock a.m., is partially an effect of the transmitted external temperature wave [8], but also a result of the morning solar gains delivered through the big glazed area, figure 5. Daily heat flux balance for internal surface is also in this case negative and equal to -26.7 Wh/m^2, while the total energy exchange within 24 h period was – as before - three times higher and equal to 78.0 Wh/m^2.

In figure 6 heat flux density values recorded on the internal faces of the standard double layered wall and the wall covered with transparent insulation have been compared. Although both diagrams look apparently similar, it should be noted that in case of the standard wall heat flux values were positive only during day hours i.e. energy was stored in the wall. While in case of the wall covered with transparent insulation intensive and continuous heat delivery to the tested space may be observed. Heat flux intensity is strongly fluctuating but its direction is uniform during the whole analyzed period.

The results of the measurements presented above concerned the hottest short period of the year, design-relevant results should be related to the longer periods. Taking into account the whole summer period (June – September, the total measured balance of heat flux was positive. More energy collected in this space was discharged by conduction to external environment than transmitted inside. Such a result was, of course, easy to predict.

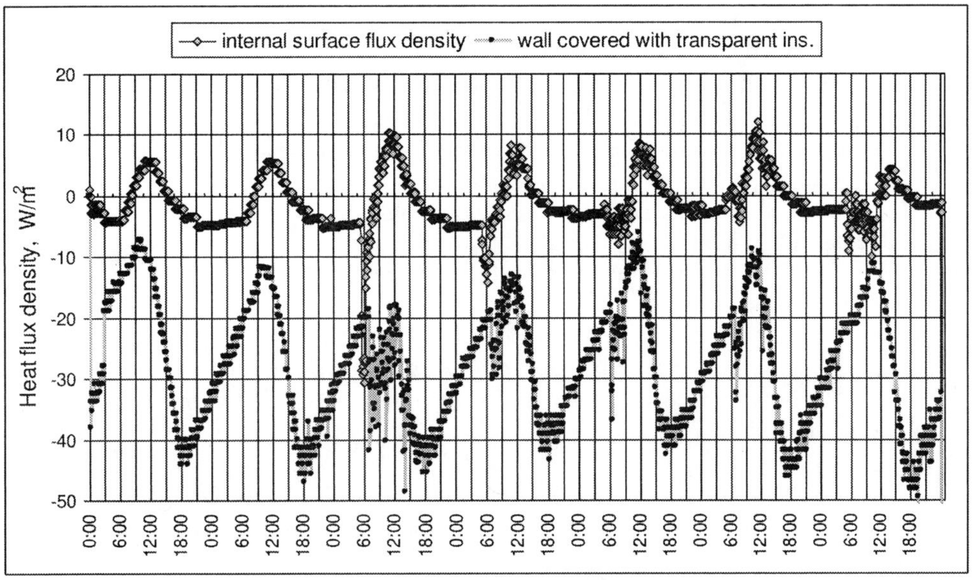

Figure 6. Hot summer period in July, heat flux intensities on internal surface of the wall with standard and transparent insulation

4. Conclusions

Rational design of low energy buildings requires linking basic architectural and structural decisions with their effects. Such a feedback should be observed already at the stage of preliminary design concept, even when a basic concept regarding wall structure and insulation thickness is considered. Improper design decisions often result in increased energy consumption and discomfort. For the future user, these aspects may be particularly troublesome and difficult to accept.

Presented above results of building monitoring and related to them results of calculations authorize the following conclusions:

1. It is not commonly understood that the massive layers of external building wall reduce significantly its dynamic thermal transmittance. Thermal transmittance modulus of the lightweight variant of the wall with the same resistance as the massive one, would be 3.3 times bigger and its time shift even 19 times shorter than of the analyzed and tested massive wall. It means that the external heat wave would propagate much easier across the lightweight wall, thus increasing thermal load and overheating intensity. Big time shift of a massive wall allows to separate heat gains in time and reduce maximum heat load during a day.
2. Dynamic heat transfer is not only a function of a wall structure but also of a heat wave length (or frequency). In case of the analyzed massive structure, decrement factor for 72 hour heat wave is almost 3 times bigger than for 24 h period. The damping effect of this wall is significantly reduced in such circumstances and slow changes of external conditions will be easily transferred inside the building.
3. Collected results of measurements confirm that night ventilation could be a reasonable way to reduce overheating. Unfortunately in the tested building this passive method was used only to a negligible extent, despite encouraging external conditions (low temperature at night).
4. During a hot and sunny week more energy was delivered to the space than transferred into the wall. In spite of the recorded big environmental temperature differences, there was no evident effect of surplus energy discharge to external environment during this week.
5. Unshaded transparent insulation is a potential source of intensive space overheating. Even though it had a horizontal capillary structure, that was supposed to block efficiently radiation coming from the sun high above the horizon, transparent insulation was not "switched off". An efficient shading system is necessary.

References

[1] Pfafferott JŰ, Herkel S, Kaltz DE and Zueschner A, 2007, Comparison of low-energy office buildings in summmer thermal comfort using different criteria. *Energy and Buildings* 39: 750-757.

[2] D'Ambrosio AFR, Olesen BW, Palella BI, Riccio G, 2014, Thermal comfort: Design and assessment for energy saving. *Energy and Buildings* 81: 326–336.

[3] Stazi F, 2017, *Thermal inertia in energy efficient building envelopes*, Oxford: Butterworth-Heinemann.

[4] Kisilewicz T, 2015, Passive control of indoor climate conditions in low energy buildings. *Energy Procedia* 78: p. 49-54.

[5] D'Orazio M, Di Perna C and Di Giuseppe E, 2014, A field study of thermal inertia of roofs and its influence on indoor comfort. *Journal of Building Physics* 38(1): 50-65.

[6] Evola G, Marlettaa L, Costanzoa V and Carusoa G, 2015, Different Strategies for Improving Summer Thermal Comfort in Heavyweight Traditional Buildings. *Energy Procedia* 78: 3228 – 3233.

[7] Hooff T, Blocken B, Hensen JLM and Timmermans HJP, 2014, On the predicted effectiveness of climate adaptation measures for residential buildings. *Building & Environment* 82: 300-316.

[8] Hudobivnik B, Pajek L, Kunic R and Košir M (2016) FEM thermal performance analysis of multi-layer external walls during typical summer conditions considering high intensity passive cooling. *Applied Energy* 178: 363–375.

[9] EN ISO 13786 - 2017 Thermal performance of building components - Dynamic thermal characteristics - Calculation methods.

[10] Serghides DK, Georgakis ChG, 2012, The building envelope of Mediterranean houses: optimization of mass and insulation. *Journal of Building Physics* 36(1): 83-98.

[11] Ibrahim A, Pelsmakers SLJ, 2018, Low-energy housing retrofit in North England: Overheating risks and possible mitigation strategies, *Building Services. Engineering Research and Technology* 39(2): 161–172.

IOP Conf. Series: Materials Science and Engineering **415** (2018) 012034 doi:10.1088/1757-899X/415/1/012034

Towards the definition of a reference ideal radiator for the assessment of heat emission efficiency in buildings

A Ferrantelli[1], K-V Võsa[1] and J Kurnitski[1,2]

[1] Tallinn University of Technology, Department of Civil Engineering and Architecture, Tallinn, Estonia
[2] Aalto University, Department of Civil Engineering, P.O.Box 12100, FI-00076 Aalto, Finland

E-mail: andrea.ferrantelli@ttu.ee

Abstract.

Radiators, as the primary devices used in space heating, cover a fundamental role in the energy efficient use of buildings. In the search of an optimal configuration, from both energy consumption and thermal comfort viewpoints, comparison of different types and emitter size is crucial. However, an ideal heater with the highest maximum heat emission efficiency that can be used as a baseline still needs to be defined.

This study contributes to the development of the heat emission standard EN 15316-2-1:2007. We perform an analysis of heat transfer in enclosures with a 21- and 10-type panel radiators, using an ideal convector as a reference heater. Accounting for room side effects of thermal radiation, we compute the induced operative temperature both analytically and numerically. The numerical simulations are carried out for a European reference room specified in the CEN TC130 standardisation committee, introducing a number of assumptions for defining the benchmark radiator. For instance, back wall losses are neglected for ideally isolated emitters.

First we find that, contrary to what might be expected, the convector is the worst performing heater. On the other hand, studying analytically the operative temperature as a function of radiator type and size, we identify the most performing heater in the 10-type.

Furthermore, our analytical solution highlights a simple predictive method for computing the operative temperature for any panel size, in complete generality. As an example, we provide a series of analytical formulas for calculating the operative temperature of the radiators addressed in this study. The results of this work therefore provide a useful background for a precise and rigorous definition of an ideal radiator.

1. Introduction

Space heating constitutes the primary source of energy consumption in the building sector [1, 2, 3]. Heating can be provided by several different types of emitters, such as panel radiators, convectors, underfloor, ceiling and air heating. It is generally understood that each type of system can impact the energy demand very differently [4, 5, 6, 7, 8].

In particular, panel radiators are a common solution to heating demands in residential buildings. Assessment of their performance on both the view points of energy consumption and thermal comfort has produced a rich literature, mostly concentrated on experimental or numerical methods [9, 10, 11, 12, 13].

Such investigations are necessarily involved, as a large number of factors determine energy consumption and thermal comfort. These range from heater design to their specific type and location in the room (e.g. close to a window or slightly detached from a wall) [4, 13, 14, 15]. For example, laboratory measurements [16] favour low temperature panel radiators instead of conventional heaters with a higher heating curve. Also, serial and parallel connected radiators provide different results; for instance, recent experimental tests found a better performance of serial-connected 11-type and 22-type [17].

In this paper we address the issue of defining an ideal radiator with the highest maximum heat emission efficiency, which should then be used as a reference baseline. To our knowledge, such a benchmark heater is not yet present in the literature [18]. Specifically, we examine the performance of 10- and 21- type panel radiators with heat output $\sim$ 134W, in an average enclosure provided by the Standard CEN TC130. Keeping the heat output constant, we study how different panel sizes affect the thermal comfort in the room; this is a non-trivial effect, as the surface temperature decreases with size. This is accomplished by means of the operative temperature t_{op}, defined as the uniform temperature of an enclosure in which an occupant would exchange the same amount of heat by radiation and convection as in the existing non-uniform environment [19]. Such quantity is here computed directly from the mean radiant temperature [15, 19, 20] in two ways: numerically, by performing simulations in steady state with the software IDA ICE [21], and analytically, following the standard ISO 7726 [19].

With these two methods, we obtain operative temperature curves in function of the panel size for both the 21- and 10-type. Comparison of our t_{op} values with the one of an ideal convector (100% convection) with the same heat output shows that, contrary to what might be expect, the ideal heater cannot be identified with the convector, which is outclassed by both the 21- and 10-type panel radiators. Rather, we find the 10-type to be the most performing. This restricts the range of candidates and sets some ground for future developments.

Additionally, our analytical study introduces a simple, yet general and predictive method for computing the operative temperature for *any* panel size and radiator type. For the case at hand, we list a collection of analytical formulas for calculating t_{op} for radiators of 10- or 21- type, with panel size in the range considered and no back wall losses.

The paper is structured as follows: in Section 2 we describe our setup and how IDA ICE and the analytical method compute the operative temperature, then in Section 3 we provide plots and performance curves for each radiator, identifying the most performing emitter and describing the predictive method for t_{op}. Our results are finally summarized in Section 4.

2. Method

In this study we consider a room with a single external wall and adiabatic internal walls, floor and ceiling under steady-state conditions (Table 1). Thermal layer properties and room dimensions are chosen according to CEN TC130 European Committee specification, namely the U-value of external wall is 0.25 W/m2K. A 21- and 10-type panel radiators with the same heat output $\sim$ 134W are then used in IDA ICE simulations to offset the heat loss through the external wall.

We consider panel radiators of varying size (figure 2) and type (10 and 21-type radiators) installed in the middle of the external (cold) wall, at 15cm from the floor. Catalogue values of a well-known panel radiator manufacturer are used as input values for the characteristic equation in the IDA ICE radiator model, namely the nominal heat output ϕ_n and power law exponent n. A supply temperature of 40°C is used for both types of the radiator.

The wall portion directly behind the radiator is modelled as an ideal insulator to eliminate back wall losses. Accordingly, the U-value of the external wall is modified as the size of the radiator changes, so that the overall UA-value of the wall and according heat loss stay constant.

The software calculates numerically, with the finite difference method, the surface temperatures of radiator and enclosure walls, together with the operative temperature for a person sitting in the centre of the room (figure 1). We locate the calculation point at 0.6m above the floor in the room centre (i.e. at 2m distance from each wall) [19]. The most performing heater configuration will then be the one for which the operative temperature is maximal, with the same heater nominal output.

The surface temperatures obtained with IDA ICE are then implemented into an analytical computation of the operative temperature t_{op}, which follows the ISO 7726 standard. For a fixed radiator height h, we obtain an expression for $t_{op} = t_{op}(w)$ that is a function of the width w, then study its behaviour in the

Table 1. Boundary conditions for steady-state simulation.

Parameter	Value
Indoor air temperature	20 °C
External air temperature	-15°C
External relative humidity	85%
Direct normal irradiance	0 W/m^2
Diffuse horizontal irradiance	0 W/m^2

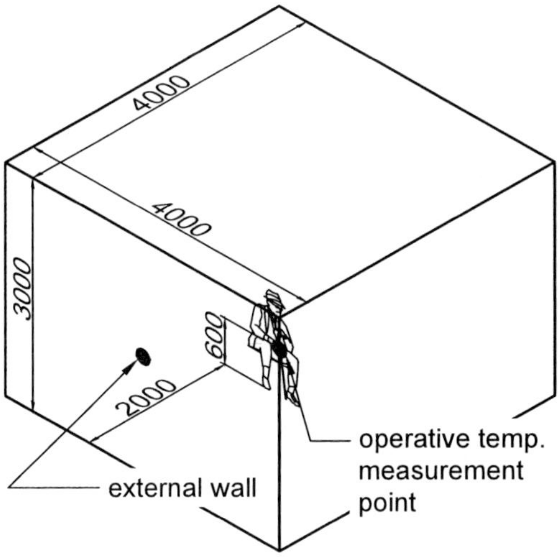

Figure 1. Room layout.

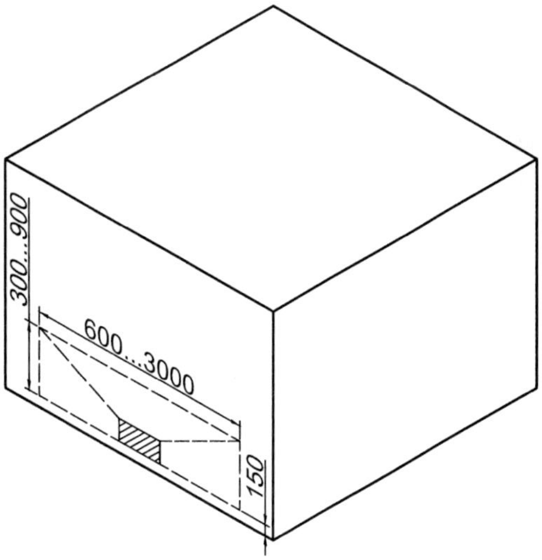

Figure 2. Panel radiator setup on the external wall.

range considered by the simulations.

The numerical and analytical methods compute the operative temperature differently. IDA ICE evaluates it as the arithmetic average of air temperature t_{air} and mean radiant temperature $\bar{t}_r$ [21],

$$t_{op} = \frac{t_{air} + \bar{t}_r}{2}, \tag{1}$$

(throughout this paper, $[t_i]$=[°C] and $[T_i]$=[K]). This differs from the exact definition given in the ISO 7726 [19],

$$t_{op} = \frac{h_c t_{air} + h_r \bar{t}_r}{h_c + h_r} \equiv A t_{air} + (1 - A)\bar{t}_r \,, \tag{2}$$

where the average is weighted by the radiation and convection heat transfer coefficients h_r and h_c at the calculation point, as explained in e.g. [15].

Another difference between the analytical solution (that follows the ISO standard) and IDA ICE consists of a unique way of the latter for computing the mean radiant temperature $\bar{t}_r$. The numerical software considers the radiator heat flux only on the surfaces that are *parallel* to the principal calculation surface; $\bar{t}_r$ is then obtained as the average of mean radiant temperatures from the six principal directions, weighted by the respective view factors.

In contrast, the ISO 7726 prescribes that for each direction one considers the radiator heat flux on

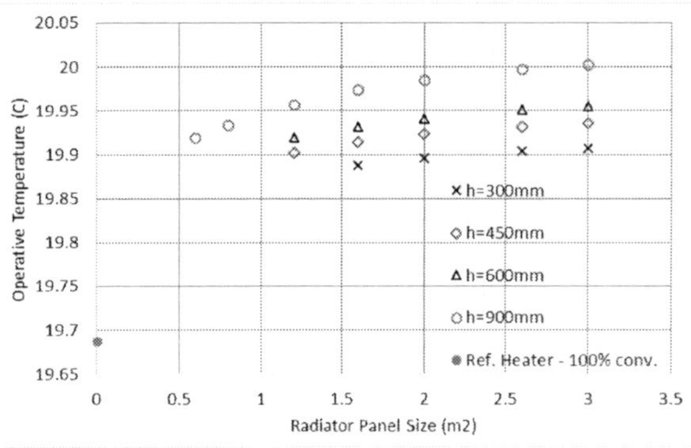

Figure 3. IDA ICE operative temperatures for a 10-type radiator and convector in function of the panel area.

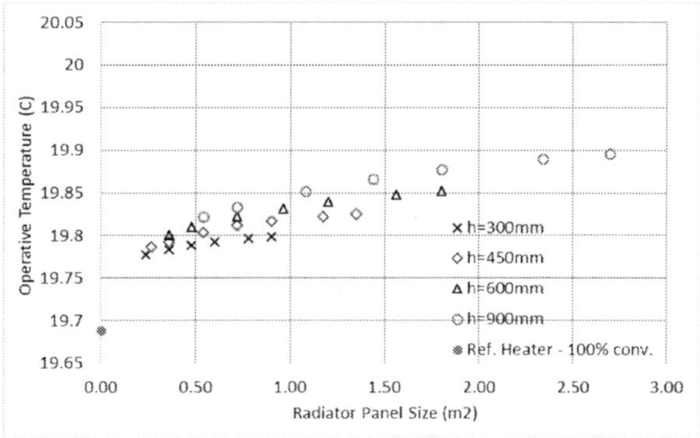

Figure 4. IDA ICE operative temperatures for a 21-type radiator and convector in function of the panel area.

both parallel *and* perpendicular surfaces, calculating the mean radiant temperature [19],

$$\bar{t}_r = \sqrt[4]{\frac{\sum_{i=1}^{6} \beta_i \overline{T}_{pr}^{(i)}}{\sum_{i=1}^{6} \beta_i}} - 273.15, \tag{3}$$

from the plane radiant temperatures $\overline{T}_{pr}^{(i)}$, which are computed from the six surface temperatures $T_{sj}^{(i)}$ and the view factors in all the six directions. $\bar{t}_r$ is therefore a weighted average over the projected area factors of a person $\beta_i = 0.18, 0.22, 0.30$ in the vertical, lateral and frontal directions respectively [19].

Throughout this paper, all the analytical results are calculated with Eqs.(2) and (3), and follow the ISO standard completely. We will show that t_{op} values computed with both methods do not differ remarkably; however, our conclusions are drawn mostly from the analytical results, to fully comply with ISO 7726 and to account for the full physical processes. The only point in common with IDA ICE consists of the surface temperatures $T_{sj}^{(i)}$, which are written as polynomial interpolations from the data provided by the software. Since both the definitions of $\bar{t}_r$ and t_{op} are different, the analytical operative temperatures are independent of the numerical ones: the interpolations are implemented for having a more general form of the operative temperature than by using the raw data for the surface temperatures. This way, instead of calculating t_{op} for each point, we can write $t_{op} = t_{op}(x)$ and obtain more general and predictive results, which are described in the next section.

3. Results

We first notice, as illustrated in figures 3 and 4, that it is not advisable to assess the performance in relation to the area only. The t_{op} values show that the efficiency varies with height, given the same area.

More importantly, a reference ideal convector (100% convection and 0% radiation) returns the lowest operative temperature out of the three. It therefore cannot be identified with a hypothetical "ideal" heater.

Operative temperatures for fixed heights are plotted in figures 5 and 6 for a 10-type and a 21-type panel radiator, respectively (some values for the 10-type are missing, as it could not reach the required power output of about 134W). Overall, the numerical and analytical solution do not differ appreciably. This is mostly evident for the 21-type radiator, while for the 10-type we see a slightly larger difference. Remarkably, the 10-type is the most performing heater: t_{op} values are constantly larger by ~0.1°C, approaching for $h = 0.9$m the air temperature 20°C. Namely, for the study at hand, the 10-type can be regarded as our "ideal" radiator.

Furthermore, it comes clearly that for a given width, the t_{op} values are linearly distributed along

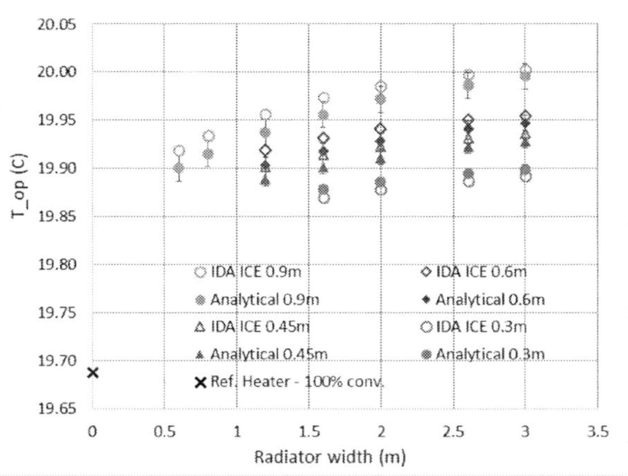

Figure 5. Operative temperatures for a 10-type panel radiator, with $h = 0.3, 0.45, 0.6, 0.9$m.

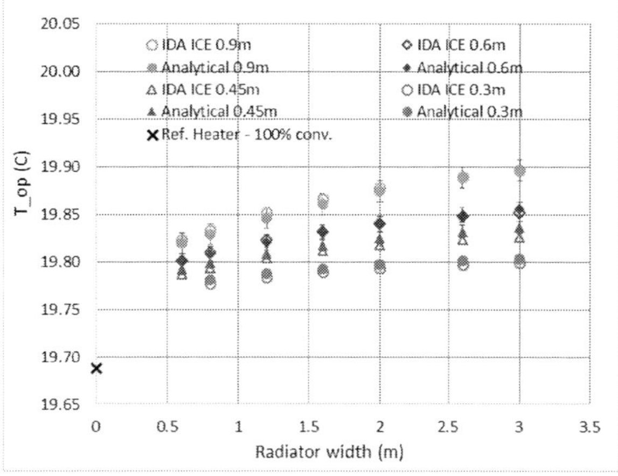

Figure 6. Operative temperatures for a 21-type panel radiator, with $h = 0.3, 0.45, 0.6, 0.9$m.

different heights h. This can be proven by using an analytical method first introduced in [22]: interpolating by least squares the operative temperature versus the height for a fixed width, we obtain the curves listed in Table 2 for both radiators. These can be formally written as

$$t_{op}(h, w) = A(w)h + B(w) , \tag{4}$$

which returns the operative temperature for *any* desired height in the range $0.3m \leq h \leq 0.9m$.

On the other hand, for a given height the t_{op} values are *not* linearly distributed along different widths w: they actually take the quadratic form

$$t_{op}(h, w) = A(h)w^2 + B(h)w + C(h) . \tag{5}$$

Specifically, as it is shown in Table 3, $A(h) < 0$ for any height. Since instead t_{op} grows linearly as the width increases ($A(w) > 0$), this proves rigorously that the operative temperature is more dependent on the height than on the width, as expected for this situation on physical grounds.

Finally, one can prove that Eqs.(4) and (5) are equivalent, namely by substituting one value for h and w they return the same t_{op} (discrepancy of order ~ 0.001, around 0.02%).

Further conclusions can be made by means of the analytical solution. As an example, consider the 21-type radiator. In general, the explicit form of the operative temperature is going to be highly non linear, but plotting the first derivative $Dt_{op} \equiv dt_{op}/dw$ with respect to the width provides with additional useful information. Figure 7 shows that while the increase of t_{op} with w is generally dropping, this happens first linearly, until a "plateau" starts at $w \sim$1m, ending at about 2m. Here the decrease with w is less steep, providing a range Δw where t_{op} is optimised with respect to width increase.

In other words, there exist widths which are most advantageous. Investigating the second and third derivatives $D^2 t_{op} \equiv d^2 t_{op}/dw^2$ and $D^3 t_{op} \equiv d^3 t_{op}/dw^3$, we locate the beginning and ending of the plateau precisely at $w = 0.87$m and $w = 1.86$m, respectively. The latter point corresponds to a minimum of $D^3 t_{op}$, which identifies a change of concavity in $D^2 t_{op}$. Moreover, the second derivative also gives the exact point for which the increase is minimal, at $w = 2.736$m. Notice how a similar value holds for $h = 0.9$m, figure 8.

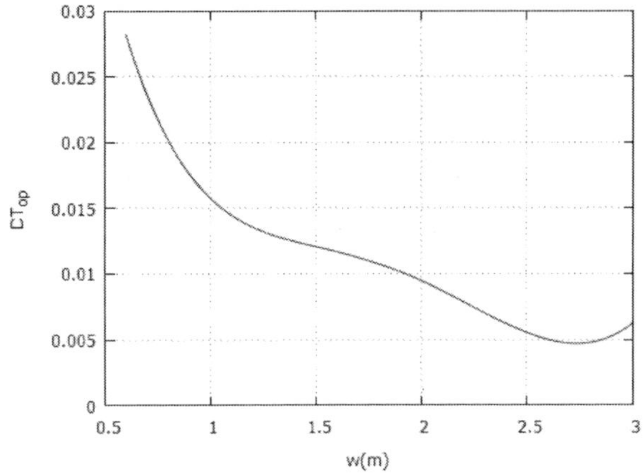

Figure 7. First derivative, analytical t_{op} for a 21-type radiator, $h = 0.3$m.

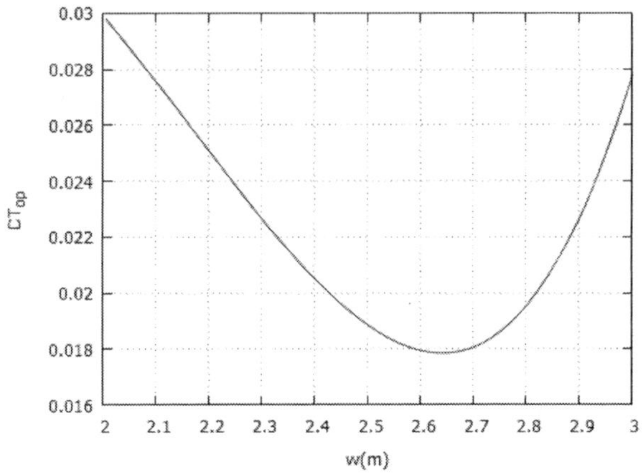

Figure 8. First derivative, analytical t_{op} for a 21-type radiator, $h = 0.9$m.

Table 2. Operative temperature t_{op}(°C) in function of panel width and height.

		10-type	21-type
Height (m)	0.30	$-0.0046w^2 + 0.0357w + 19.833$	$-0.0031w^2 + 0.0216w + 19.766$
	0.45	$-0.0066w^2 + 0.0494w + 19.839$	$-0.0049w^2 + 0.0355w + 19.773$
	0.60	$-0.0069w^2 + 0.053w + 19.851$	$-0.0054w^2 + 0.042w + 19.778$
	0.90	$-0.0111w^2 + 0.0786w + 19.859$	$-0.007w^2 + 0.0573w + 19.788$
Width (m)	0.60	-	$0.0593h + 19.765$
	0.80	-	$0.0773h + 19.761$
	1.20	$0.1079h + 19.84$	$0.0938h + 19.763$
	1.60	$0.1268h + 19.842$	$0.1101h + 19.764$
	2.00	$0.1404h + 19.846$	$0.125h + 19.764$
	2.60	$0.1489h + 19.853$	$0.1408h + 19.764$
	3.00	$0.1403h + 19.86$	$0.1509h + 19.763$

Table 3. Formulas for the coefficients entering Eqs.(4) and (5).

	10-type	21-type
$A(h)$	$-0.1074h^3 + 0.1828h^2 - 0.1045h + 0.0132$	$-0.0061h - 0.0017$
$B(h)$	$0.6012h^3 - 1.0361h^2 + 0.6114h - 0.0707$	$0.3025h^3 - 0.5728h^2 + 0.3929h - 0.0529$
$C(h)$	19.846	$0.0741h^3 - 0.1444h^2 + 0.1233h + 19.74$
$A(w)$	$-0.0234w^2 + 0.1174w + 3 \times 10^{-5}$	$-0.0084w^2 + 0.0667w + 0.0254$
$B(w)$	$0.0039w^2 - 0.0054w + 19.841$	19.763

4. Conclusions

In the present study we addressed the performance of 10- and 21-type panel radiators using both an analytical and a numerical approach, in the search for identification of an ideal radiator. We performed a rigorous investigation on the operative temperature sensed by a user sitting in the middle of a test room of average size, in compliance with the standard CEN TC130. The overall behaviour of operative temperature t_{op} as a function of radiator size and type was determined quantitatively with two different methods: the numerical simulation software IDA ICE and an analytical computation based on the standard ISO 7726.

Comparison of the two radiators with an ideal convector providing the same output $\sim$ 134W shows that the most performing emitter is the 10-type. For instance, a common panel size with $h = 0.6$m, $w = 1.6$m returns $t_{op} = 19.83$°C and $t_{op} = 19.92$°C resp. for 21- and 10-type, while $t_{op} = 19.69$°C for the convector.

For larger sizes, the 10-type radiator even reaches 20°C operative temperature, which is equal to the air temperature. In other words, it constitutes our "ideal" radiator for the setup considered.

Overall, we found that larger panels provide a higher t_{op}, as naturally expected; nevertheless, since the surface temperature of the panel, with same power output, decreases with size, the operative temperature grows linearly only with height h. The increment along the width w is instead less pronounced, due to a negative quadratic term $\propto -w^2$ in $t_{op}(h, w)$. In other words, the performance of radiators is more sensitive to the height than to the width. The formulas listed in Tables 2 and 3 allow to quantify this behaviour precisely.

Furthermore, the analytical study highlights various features. The performance increment with larger panels is evident mostly with 10-type panels, where it approaches a linear behaviour. For the 21-type radiator we found instead that for small $h < 0.6m$ there is no substantial increase in the operative temperature for $w \gtrsim$ 2m. Accordingly, an optimised 21-type panel radiator should lie within this range. We have thus shown that for smaller h there exists a width interval for which adopting larger panels is most profitable.

More generally, through our analytical calculation we could highlight a simple predictive method for computing the operative temperature for *any* panel size and room configuration, in complete generality. In particular, this paper provides precise analytical formulas for calculating t_{op} for radiators of 10- or 21- type with dimensions $0.3m \leq h \leq 0.9m$ and $1.2m \leq w \leq 3m$, assuming no back wall losses, in a room corresponding to the Standard CEN TC130. It would then be very interesting to extend the above results and method to a more standard situation, e.g. adding back wall losses to assess their impact on our quantitative results.

Finally, we should remark that this study constitutes only an early attempt in formalizing the heating performance studies of radiators; it can and should be extended to other types of emitters and to parametric studies on the more general impact of view factors, room and heater geometry. These should eventually aim to identify a more general pattern, for practical utility and towards a rigorous definition of an ideal or benchmark emitter for each relevant case.

Acknowledgments

The research was supported by the Estonian Research Council with Institutional research funding grant IUT1-15. The authors are also grateful to the Estonian Centre of Excellence in Zero Energy and Resource Efficient Smart Buildings and Districts, ZEBE, grant 2014-2020.4.01.15-0016 funded by the European Regional Development Fund.

References

[1] Serrano S, Ürge Vorsatz D, Barreneche C, Palacios A and Cabeza L F 2017 *Energy* **119** 425 – 434 ISSN 0360-5442
[2] D'Agostino D, Cuniberti B and Bertoldi P 2017 *Data in Brief* **14** 759 – 762 ISSN 2352-3409
[3] Yao R and Steemers K 2005 *Energy and Buildings* **37** 663 – 671 ISSN 0378-7788
[4] Olesen B W, Mortensen E, Thorshauge J and Berg-Munch B 1980 *ASHRAE transactions* **86** 34–48
[5] Inard C, Meslem A and Depecker P 1998 *Building and Environment* **33** 279 – 291 ISSN 0360-1323
[6] Olesen B W and de Carli M 2011 *Energy and Buildings* **43** 1040 – 1050 ISSN 0378-7788 tackling building energy consumption challenges - Special Issue of ISHVAC 2009, Nanjing, China
[7] Léger J, Rousse D R, Borgne K L and Lassue S 2018 *Building and Environment* **128** 161 – 169 ISSN 0360-1323

[8] Maivel M, Ferrantelli A and Kurnitski J 2018 *Energy and Buildings* **166** 220 – 228 ISSN 0378-7788
[9] Hasan A, Kurnitski J and Jokiranta K 2009 *Energy and Buildings* **41** 470 – 479 ISSN 0378-7788
[10] Shati A, Blakey S and Beck S 2011 *Energy and Buildings* **43** 400 – 406 ISSN 0378-7788
[11] Munaretto F, Recht T, Schalbart P and Peuportier B 2017 *Journal of Building Performance Simulation* **0** 1–22
[12] Sevilgen G and Kilic M 2011 *Energy and Buildings* **43** 137 – 146 ISSN 0378-7788
[13] Jahanbin A and Zanchini E 2016 *Applied Thermal Engineering* **105** 467 – 473 ISSN 1359-4311
[14] Ali A H H and Gaber Morsy M 2010 *Energy Efficiency* **3** 283–301 ISSN 1570-6478
[15] Kalmár F and Kalmár T 2012 *Energy and Buildings* **55** 414 – 421 ISSN 0378-7788
[16] Maivel M and Kurnitski J 2014 *Energy and Buildings* **69** 224 – 236 ISSN 0378-7788
[17] Võsa K V, Ferrantelli A, Kull T M and Kurnitski J 2018 *Applied Thermal Engineering* **132** 531 – 544 ISSN 1359-4311
[18] EN 442-1:2014 Radiators and convectors. Technical specifications and requirements. Standard. CEN. Bruxelles, BE.
[19] ISO 7726:1998 Ergonomics of the thermal environment - Instruments for measuring physical quantities. Standard. International Organization for Standardization. Geneva, CH.
[20] d'Ambrosio Alfano F R, Dell'Isola M, Palella B I, Riccio G and Russi A 2013 *Building and Environment* **63** 79 – 88 ISSN 0360-1323
[21] IDA ICE - Indoor Climate and Energy. Tech. rep. EQUA Stockholm, Sweden, 2013.
[22] Ferrantelli A, Ahmed K, Pylsy P and Kurnitski J 2017 *Energy and Buildings* **143** 53 – 60 ISSN 0378-7788

IOP Conf. Series: Materials Science and Engineering **415** (2018) 012035 doi:10.1088/1757-899X/415/1/012035

Hygienic Water Audit Scheme as a Tool in the Fight against Legionella - problems and goals of the research project

D Kaposztasova* and Z Vranayova

Institute of Architectural Engineering, Department of Building Services, Faculty of Civil Engineering, Technical University of Kosice, 04200, Slovak Republic

*Email: daniela.kaposztasova@tuke.sk

Abstract. Water and air they mean life, but their quality represents our health. The main aim of the article is to introduce our research project „*Hygienic Water audit scheme as a tool in the fight against Legionella"* in terms of specific requirements from companies to minimize the energy efficiency for elimination of microbiological contamination of water supply and air conditioning systems (replace thermal disinfection by another effective method). The issue of water contamination, resp. air of *Legionella pneumophila*, its appearance and impact on the design and operation of water supply and air conditioning systems is relatively young. Little known problem in Slovakia has in the context of the European Union a serious position as it causes high mortality. The aim of the interdisciplinary research is to analyze the current state of hospital buildings, of the individual factors on a working materials and working environment in terms of their potential for microbiological contamination.

1. Introduction

Microbiological contamination of water and air and health risks caused by germs that colonize technical systems causes various problems and illnesses. Among these conditions is Legionnaires' disease - Legionellosis. Caused by ubiquitous water bacteria *Legionella pneumophila*, having a diameter of 0.2 to 0.7 microns which is dangerous in the form of contaminated aerosol, which may be formed, for example during showering by inhalation. Are we taking Legionnaires' disease seriously enough? The reported cases not only in hospitals are listed in [1]. Forty-Six Legionella infections have been diagnosed among patients of a hospital in the Lisbon area of Portugal since October 31, 2017. Four of the patients have died. Portuguese health officials suspect the source of the outbreak was the hospital's potable plumbing system [1].

Our measurements showed that Legionella is present in the distribution of water [2] and air [3] in Slovakia. For the water supplier and the operator of the building is an obligation in the water system to ensure the prescribed quality and temperature at each sampling site, as well as providing air ducts from a hygiene point of view. Health flawless drinking water is provided for in the Act of NR- No. 247/2017 Coll. in Slovak Republic on protection, support and development of public health and on amendments to certain laws, which is valid from October 2017 [4].

Health risk is evaluated according to the quality indicators and hygienic limits. The quality of supplied drinking water to the public water supply and sanitation are responsible in our water companies. The quality of water supplied to consumers is monitored by regional public health within the meaning of Slovak Government Regulation no. 354/2006 Coll. [5]. Sample sites should be selected in areas or buildings where water flows from outlets that are normally used for human consumption. For domestic hot water use standard STN 83 0616 Hot water quality [6] is valid for that bacteriological and biological indicators. Hot water quality must meet the criteria for drinking water. Designers should respect the STN EN 806-2 Specifications for installations inside building conveying water for human consumption - Criteria for internal water pipes carrying potable water, Part 2 - Design [7] and as well as technical management TC 164 WI 164353 - Recommendations for prevention of Legionella growth for installations inside buildings conveying water for human consumption which specifies a standard in view of the presence of *Legionella pneumophila* [8]. These fundamental rules are valid in all member states of the European Union. In the Czech Republic they apply the decree of the Ministry of Health. 252/2004 Coll. showing the desired parameters of the hot water, which should be supplied to the distribution points of over 50 ° C, preferably above 55 ° C. United Kingdom adopted as a national standard method for assessing the impact of materials in contact with drinking water test MDOD (The Mean Dissolved Oxygen Difference) and is governed by regulations BS 6920.

The problem of safe ventilation equipment lies in the fact that often the focus has been only on their proper functioning and good condition, while control of hygiene and disinfection required is neglected. Directive Association of German Engineers VDI 6022 "Hygienic requirements for ventilation equipment - office and assembly areas" describes the necessary procedures for planning, the actual implementation, operation and maintenance of HVAC equipment in order to achieve the desired quality and hygiene of the indoor air [9]. While ANSI / ASHRAE 188-2015 provides basic requirements for these systems, but the overall execution and risk identification is already the owner of the building. [10] Among the possibilities for water protection against legionella we include for example: the ensuring sanitary water with chlorine dioxide, ozone, UV radiation, thermal disinfection, physical methods such as the use of filters and much more. It is necessary to examine methods to eliminate bacteria in addition to freight thermal disinfection, also is in the world a general need to define both efficient and safe technical measures in domestic hot water systems and air conditioning. Criteria for selection of a method, the use of combinations of methods, parameters for installation, pitfalls in implementation and a plan for subsequent environmental surveillance, their energy consumption will be discussed in detail. Improvement of the energy efficiency of disinfection methods is met as the selection of the most suitable, cost effective and risk elimination method for each case. [11].

2. Project objectives, methods and their prospects of success

Theoretical and experimental analysis of systems technology environment in connection with their possible microbiological contamination assess the possibility of eliminating health risks with regard to their energy consumption and evaluation in hygienic audit.

Our team is divided into experts in the field of water and air in buildings, in cooperation with environmental and material chemist experts. This team is composed of knowledgeable professionals, respectively, young scientists who already have experience of dealing with projects at national, respectively international level. They can guarantee the fulfillment of the objectives.

The goals were set as follows:

1. State of art regarding to the quality of supplied hot water to heat sources focused on hospitals in Slovak conditions.
2. Examination of the state of art regarding to the quality of supplied air conditioning equipment and indoor quality for user focus on hospitals.
3. Decomposition of the factors influencing the potential for microbiological contamination of the water supply and air.
4. Analysis of the impact of material base distribution systems technology environment for preserving the quality of the working material.
5. Analysis of the impact of age, biofilm and sediment in the pipes for preserving the quality of the working fluid.

6. Analysis of the possibilities of technical measures to eliminate potential microbiological contamination.
7. Analysis of the effects of selected methods to minimize the health risk in terms of their impact on the pipe material and energy consumption (cost-benefit analysis will be reported).
8. Draft set of decision criteria that reflect the effects of acceleration increase the bacterium *Legionella pneumophila* and selection of appropriate methods to eliminate the risk of contamination.
9. The draft methodology of determining the degree of importance of the criteria taking into account the effects of accelerating increase of microbiological contamination in the systems of from supplier to end user.
10. Processing model of risk analysis, defining boundary conditions and principles to minimize the microbiological contamination of water and air.
11. Implementation of the proposed methods for reducing the rate of bacteriological pollution in real systems technology environments by evaluating their effectiveness and energy efficiency.
12. Development of technical report and sheme of "Hygienic Audit" for safe design, operation and maintenance of hospital systems.

2.1. Project stages

The proposed methodology solutions, justification of selection and measure of the ability to meet the declared objectives. The project will benefit by the following characteristics:
• *Classification* - sorting the systematic factors related to *Legionella pneumophila* bacteria colonization in the water and air pipes, and the risks identification,
• *Methodology* - draft set of decision criteria, taking into account the effects of accelerating the proliferation bacteria causing legionellosis, including the determination of levels of importance criteria
• *Model* - Risk Analysis techniques for system protection, link with virtual microbiological contamination of the water supply, respectively air,
• *Know-how* - technological instructions as a set of experiences, knowledge and key information regarding the application of appropriate methods and technologies for the management of risks arising from the process of ensuring the supply of hygienic water users, respectively air. Variability solution is built directly into a sequence of steps to achieve project goals. Model of risk analysis preceded by analysing individual factors related to microbiological contamination of working materials (water and air) necessary for human life.

At different stages of the project we will use a variety of qualitative and quantitative methods: starting by study and critical analysis of available publications and carried out research on a subject which will be continuously updated throughout the study was performed.

2.2. Project benefits to innovation

The anticipated benefits of the project are planned on several levels:

- *State of art research*: which shows the rate of microbiological contamination of the water supply and air condition for hospitals in Slovakia
- *Theoretical level*: in terms of set goals, assess their analysis of the factors accelerating the reproduction of the bacteria *Legionella pneumophila*, in terms of investor, operator and user of vulnerable systems technology environment
- *Methodology level*: will be processed by methodological model of risk analysis associated with the solution of safe design, operation and maintenance of systems of Environmental Engineering
- *Practical level* of benefits: it is planned to verify the model of risk analysis, as well as the proposed methods of reducing the level of bacteriological contamination in real systems technology environments by evaluating their effectiveness, with special regard to energy efficiency solutions.

3. Current state of art

The main goal of our project described in this paper is to find out the microbiological contamination level in selected hospitals. The selection was directly aimed at risky departments in hospitals (geriatrics, gynecology clinic, clinic of pneumoniae etc.) where the factors that encourage bacteria growth can occur and be dangerous for patients with weakened immunity. We started with 20 samples from 4 hospitals in first stage (Table 1). Nowadays the sampling continues and more than 20 samples are prepared for the examination.

Table 1. Hospital survey – stage 1 (source shower heads)

Hospital 1	Hospital 2	Hospital 3	Hospital 4
Water samples – 5	**Water samples – 5**	**Water samples – 5**	**Water samples – 5**
•Department of long-term illness	•Department of Gynecology	•Department of Gynecology	•Geriatrics and Gerontology
•Department of Internal Medicine	•Department of Neurology	•Department of Hematology	•Geriatrics and Gerontology
•Department of Gynecology	•Clinic of Infectious Diseases	•Department of Balneology etc.	•Department of Long-term illness
•Department of Gynecology	•Clinic of Pneumonia	•Department of Psychiatry	•Department of Long-term illness
•Centre for Burn Injuries	•Department of Internal Medicine	•Department of Internal Medicine	•Institute of Preventive Medicine

As it is set in STN ISO 11731- part 2 [12] for Legionella presence all water samples were analysed. Hospital 1, 2, 4 were in all cases negative. In number 3 all 5 samples were positive. We can see the results in table 2. Now we continue with sample collection in other hospitals. In case of positive findings we are planning to apply the hygienic water audit scheme to proof it in practice.

Table 2. Hospital survey – positive samples

	Positives samples			**Legionella species**	
	Water samples		**source**	**CFU/100ml**	**Serotype**
Hospital 3	1	Department of Gynecology	shower	**3400**	3
	2	Department of Hematology	shower	**1200**	1
	3	Department of Balneology etc.	shared shower	**1600**	3
	4	Department of Psychiatry	shower	**4600**	3
	5	Department of Internal Medicine	shared shower	**2800**	3

Positive samples only confirmed the need of such research and prevention, to protect public health.

4. Conclusion and Discussion

The presented project is gradually taking the shape depending on the level of science and technology, health and other fields of science. The issue of water contamination, respectively air *Legionella pneumophila*, is relatively young and its appearance and also their integration and impact on the design and operation of systems technology environment - especially water supply and air conditioning equipment. In this project we use know-how of Slovak and foreign companies focused on technical measures to eliminate the health risks of supplied water and air. The research team aims to develop a custom solution that will build on the strengths of known approaches and to eliminate weaknesses detected in real operation.

Outputs and benefits of the project will be a contribution to knowledge, classification and propose suitable measures as a platform of so called Hygienic audit using experimental measurements and parameters of cold, hot water and air quality. In terms of specific requirements to minimize the energy efficiency of thermal plants will be processed by model of risk analysis and technological process, providing hygienic harmlessness supply water and air. To improve energy efficiency targets and reduce water usage should be assed at the design stage if possible and all energy saving measures should be risk-assessed for their impact on Legionella control. In previous research, positive legionella bacteria in up to 8% of water samples taken in the exchange stations were identified. Since thermal disinfection, referred as a measure in the first place, is unsystematic and energy-intensive solution, verification of other progressive options that are less energy-consuming for suppliers, but also for users with no risk were done. The scientific report upon request from main heat suppliers in recent years just confirmed that the European Union preferred method of eliminating bacteria by periodic thermal disinfection increased financial and time requirements of the system and does not correspond with the concept of reducing energy consumption being introduced throughout the European Union. It is necessary to examine other possible methods for bacteria elimination in addition to unsystematic thermal disinfection, also is in the world a general need to define technical measures in hot water systems and air conditioning. Heat distribution systems of water and air in our larger and older buildings have many common health deficiencies as age of material, composition, lack of control, neglected maintenance, etc. Each disinfection technology has strengths and weaknesses, limits and different costs. Slovakia has not implemented across the board targeted collection and evaluation of the supply air on the content of *Legionella pneumophila*, which is why our aim of mapping out the real situation and proposed solutions in buildings caused to health.

Principal investigator and the team has been dealing with this issue of contamination of working materials in the building for years. Selection of the appropriate technology depends on the individual needs of the operator, water quality and requirements for operation and maintenance of equipment, but it is feasible to bring into practice the general rules that we have the ambition to put into practice than the recommended / required "hygiene audit water and air", with similar output structure as in energy certification - classification of buildings according to the degree of health risk and the necessary technical measures.

Potable hot and cold water quality and air quality must by ensured by using a regular monitoring scheme as Hygienic water audit to ensure public health safety. This tool should help the hospitals to control the legionella risk and set the best solution for their case if needed.

5. Acknowledgement

This work was supported by the Slovak Research and Development Agency under the contract No. VEGA 1/0697/17 Hygienic water audit platform as transition tool to Legionella free water and HVAC systems in hospitals.

References

[1] HC info recent outbreaks. Available online: https://hcinfo.com/about/outbreaks/recent/ (accessed on 25/02/2018)

[2] Vranyova Z, Kosicanova D and Očipova D 2011 *Hot water supply systems and their microbiological contamination* (Kosice: Technical University of Kosice Press) p 105

[3] Vranyova Z, Kosicanova D and Knizova K 2011 *Air supply systems and their microbial contamination* (Kosice: Technical University of Kosice Press) p 125

[4] Act of NR SR No. 247/2017 Coll Laying down details of drinking water quality, control drinking water quality, monitoring program and risk management at drinking water supply

[5] No. 354/2006 Coll Government Order of the Slovak Republic laying down requirements for water intended for human consumption and quality control of water intended for human consumption (ended 14.10.2017)

[6] Slovak standard STN 83 0616 Hot water quality

[7] STN EN 806-2 Specification for installations inside buildings conveying water for human consumption - Part 2: Design.

[8] Technical management TC 164 WI 164353 - Recommendations for prevention of Legionella growth for installations inside buildings conveying water for human consumtion which specifies
[9] Directive Association of German Engineers VDI 6022 "Hygienic requirements for ventilation equipment - office and assembly areas"
[10] ASHRAE 188-2015 Standard 188 establishes minimum legionellosis risk management requirements for building water systems
[11] Sincak P, Ondo J, Kaposztasova D, Vircikova M and Vranayova Z 2014 *I. J. of Envi. Research and Public Health,* DOI: 10.3390/ijerph110808597, MDPI
[12] STN ISO 11731 - part 2 Water quality. Detection and enumeration of Legionella. Part 2: Direct membrane filtration method for waters with low bacterial counts (ISO 11731-2:2004)

Impact of heating concept on the energy demand of family house

P Durica[1], J Jandacka[2], P Juras[1]

[1] Department of Building Engineering and Urban Planning, Faculty of Civil Engineering, University of Zilina, Univerzitna 8215/1, 010 26 Zilina, Slovakia
[2] Department of Power Engineering, Faculty of Mechanical Engineering, University of Zilina, Univerzitna 8215/1, 010 26 Zilina, Slovakia

peter.juras@fstav.uniza.sk

Abstract. This paper presents results of a parametric study, which analyses the energy demand of a family house. Assessed house has various shape factors and the thermal-technical properties of the building envelope. Also the recuperation unit and the heat source is taken into account. The house in the first case is a one story bungalow, the next version is a partly two story house and in the third case it is a complete two story house. Each of these geometric variants is calculated for four various building envelopes, which are considered for different energy performance standards including low energy and passive standard. Air exchange rate is considered according to the Slovak standard as natural or mechanical with use of recuperation. As heat sources are considered gas, heat pump, pellets, electricity and wood. Also two combined sources are analyzed: gas and pellets with solar collectors for preparing the hot water. The results are analyzed according to the Slovak standard and directives, which are in terms of delivered energy, primary energy and CO_2 emissions.

1. Introduction

According to the European Directive 2010/31/EU [1], the building sector is the largest single energy consumer in Europe, which is absorbing 40 % of final energy. About 75 % of buildings are energy inefficient [2]. The outcome of the directive is to reduce 20 % the Union's energy consumption by 2020.

The Energy Performance of Buildings Directive requires all new buildings to be nearly zero-energy by the end of 2020. All new public buildings must be nearly zero-energy by 2018. This should be done individually by the EU member countries according to the National plans [3]. In Slovakia, this was done by the new standard dealing with the thermal protection of buildings, STN 73 0540:2012 [4]. Last year it was updated to Z1 version [5]. This standard creates three steps to the nearly zero energy buildings. These were mandatory since 2013 to 2016, since 2016 and finally will be from 2021. The act No. 555/2002 [6] and 364/2012 [7] made this Standard and its requirement mandatory and also created classification depending on the primary energy need, which depends mostly on the heating source.

In this article a family house is analyzed in terms of meeting the Standard, which is valid now (ultralow-energy) and the future for the nearly zero building. This family house was analyzed from the point of view of building envelope, which consists of fragments such as wall, roof, floor etc., shape of the building and ratio between building envelope and the enclosed volume and also various heating sources were taken into account.

2. Slovak standard requirements

Slovak standard describes five requirements, which new or refurbished building have to comply. In this paper, only the energy need for heating, which is the one most complicated to reach, was analyzed. The others is the *U*-value for the individual fragments creating the building envelope, minimum surface temperature, natural or mechanical ventilation. The building itself usually do not meet the energy for heating requirements even though all the fragments meet the *U*-value requirements.

Required energy for heating according STN 73 0540 Z1 [5] is stated in the Table 1. The values are stated as a specific energy need for heating per square meter (m^2) or cubic meter (m^3), divided into four classes depending on the year of designing of the building. Last parameter is the shape factor, which is ratio between the sums of building envelope fragments (m^2) to enclosed heated volume (m^3). Nowadays the Recommended values, which differ from 25 to 50 (kWh / (m^2.a)) depending on the shape factor, are valid.

Table 1. Requirements of the Slovak standards for Energy need for heating [5].

Building Shape factor (1/m)	Energy need for heating from STN 73 0540:2012 Z1[5]							
	Maximal value $Q_{H,nd,max}$		Standard value $Q_{H,nd,N}$ valid since 1.1.2013		Recommended (ultra-low building $Q_{H,nd,r1}$ valid since 1.1.2016		Final value (nearly zero building) $Q_{H,nd,r2}$ valid from 1.1.2021	
	$Q_{H,nd,max1}$ (kWh/(m^2.a))	$Q_{H,nd,max2}$ (kWh/(m^3.a))	$Q_{H,nd,N1}$ (kWh/(m^2.a))	$Q_{H,nd,N2}$ (kWh/(m^3.a))	$Q_{H,nd,r1,1}$ (kWh/(m^2.a))	$Q_{H,nd,r1,2}$ (kWh/(m^3.a))	$Q_{H,nd,r2,1}$ (kWh/(m^2.a))	$Q_{H,nd,r2,2}$ (kWh/(m^3.a))
≤0.3	70.00	25.00	50.00	17.90	25.00	8.93	12.50	4.47
0.4	78.60	28.10	57.10	20.40	28.55	10.20	14.28	5.10
0.5	87.10	31.10	64.30	23.00	32.15	11.49	16.08	5.75
0.6	95.70	34.20	71.40	25.50	35.70	12.75	17.85	6.38
0.7	104.30	37.50	78.60	28.10	39.30	14.04	19.65	7.02
0.8	112.90	40.30	85.70	30.60	42.85	15.31	21.43	7.66
0.9	121.4	43.40	92.90	33.20	46.45	16.60	23.23	8.30
≥ 1.0	130.0	46.50	100.0	35.70	50.00	17.86	25.00	8.93

3. Analyzed house description

The analyzed family house is based on the single floored house (bungalow). This type is in the last decade very popular in Slovakia. But nowadays it has a problem to meet the standard requirements, because of a not suitable shape factor. Many types have problem with energy certification even though the fragments have much lower *U*-values for fragments than required. That's why a partly two floored house and complete two story house with flat roof was created from the single floor house. The 3D models of these three variants are shown in Figure 1 also with its shape factors.

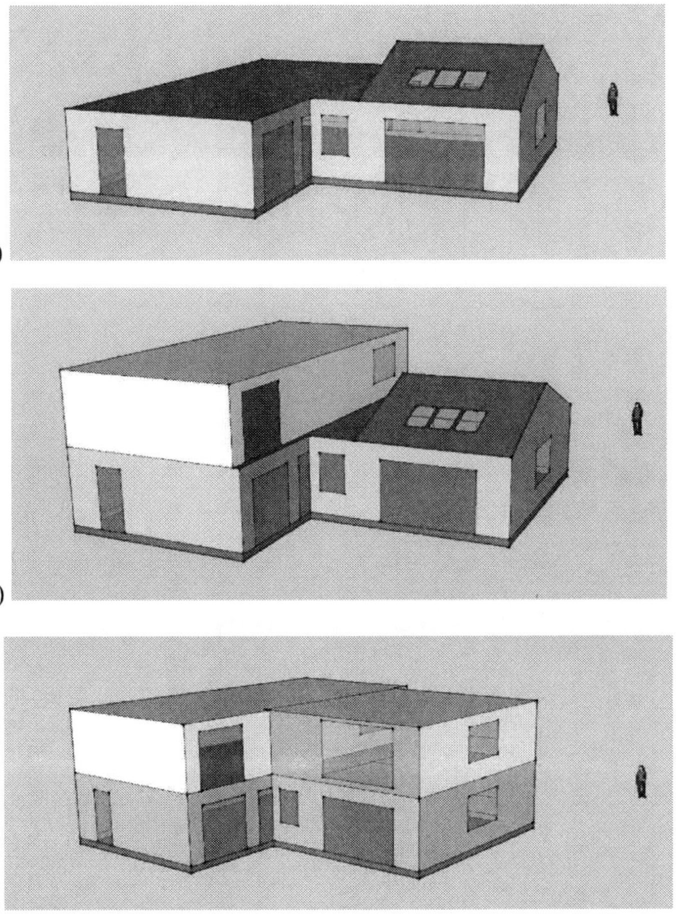

a)

b)

c)

Figure 1. Three variants of analysed house: building shape factor a) 0.973, b) 0.8, c) 0.672.

U-values of the fragments and calculated energy need for heating are summarized in Table 2. The basic variant is from the project, which has several fragments, more than usual based on the fact, that the living room has sloped roof without ceiling, visible from Figure 1 a), this feature is kept also in the variant b).

The not suitable shape factor (lower number – better use of enclosed volume) and well insulated walls cannot meet the required objectives. For the highly insulated buildings and nearly zero energy building, the mechanical ventilation is needed, because the heat loss from ventilation in order to meet the minimal air change is crucial.

IOP Conf. Series: Materials Science and Engineering **415** (2018) 012036 doi:10.1088/1757-899X/415/1/012036

Table 2. Calculated energy for heating for three variants of the house and four different types based on the *U*-values.

													Q calculated		
A	bungalow														
	Shape factor	0.973	Wall	Wall th. 250	Wall th.115	Wall plinth	Insulated ceiling	Shaped roof	Floor on the terrain	Window	Roof window	Avg $U_{e,m}$	n = 0.50	n = 0.21	n = 0.17
	Recommended value	49													
	Final value	24.1													
1	Design values from project		0.09	0.13	0.14	0.11	0.09	0.11	0.18	0.76	1.10	0.201	53.34	32.61	29.75
2	$U_{r,1}$ - standard	1.1.2016	0.22	0.22	0.22	0.22	0.15	0.15	0.18	1.00	1.40	0.275	73	52.27	49.41
3	$U_{r,2}$ - standard	1.1.2021	0.15	0.15	0.15	0.15	0.10	0.10	0.18	0.60	1.00	0.206	54.78	34.05	31.19
4	U-value to meet the standard	1.1.2021	0.09	0.09	0.09	0.09	0.09	0.09	0.14	0.60	1.00	0.172	45.88	25.15	22.29
	partly two story house														
B	Shape factor	0.8	Wall	Wall th. 250	Wall th.115	Wall plinth	Insulated ceiling	Shaped roof	Floor on the terrain	Window	Roof window	Avg $U_{e,m}$	Q calculated n = 0.50	n = 0.21	n = 0.17
	Recommended value	42.85													
	Final value	21.43													
1	Design values from project		0.09	0.13	0.14	0.11	0.09	0.11	0.18	0.76	1.10	0.199	40.9	21.55	18.88
2	$U_{r,1}$ - standard	1.1.2016	0.22	0.22	0.22	0.22	0.15	0.15	0.18	1.00	1.40	0.286	58.48	39.13	36.46
3	$U_{r,2}$ - standard	1.1.2021	0.15	0.15	0.15	0.15	0.10	0.10	0.18	0.60	1.00	0.209	42.99	23.64	20.97
4	U-value to meet the standard	1.1.2021	0.09	0.12	0.12	0.10	0.09	0.09	0.16	0.60	1.00	0.178	36.63	17.29	14.62
C	two story house														
	Shape factor	0.672	Wall	Wall th. 250	Wall th.115	Wall plinth	Insulated ceiling	Shaped roof	Floor on the terrain	Window	Roof window	Avg $U_{e,m}$	Q calculated n = 0.50	n = 0.21	n = 0.17
	Recommended value	38.5													
	Final value	19.2													
1	Design values from project		0.09			0.11	0.09		0.18	0.76		0.196	34.97	16.11	13.51
2	$U_{r,1}$ - standard	1.1.2016	0.22			0.22	0.15		0.18	1.00		0.286	49.91	31.05	28.45
3	Ur,2 - standard	1.1.2021	0.15			0.15	0.10		0.18	0.60		0.208	36.88	18.02	15.42
4	U-value to meet the standard	1.1.2021	0.10			0.15	0.10		0.18	0.60		0.188	33.69	14.83	12.22

variant does not meet the standard | variant meets the required values since 1.1.2016 | variant meets the required values from 1.1.2021

Totally 36 combinations were calculated (Table 2), which differ from each other by the shape and *U*-value of the individual fragments (every time for all fragments). The four types were calculated in each variant:

- based on the project - design values,
- individual *U*-values meet the standard required values (since 2016) and accordingly the energy need for heating is stated and compared to the standard - ultra low energy level,
- individual *U*-values meet the standard final values and accordingly energy need for heating is stated and compared to the standard – nearly-zero energy level,
- *U*-values are calculated so the type comply the standard requirements.

4. Results and discussion

The energy need with different air change rate is also calculated for each type. As can be seen, the A-variant, without mechanical ventilation does not comply the required value, so, the project should not get the planning permission.

Required values for energy need for heating based on Table 1 for calculated shape factor of three variants are shown in Figure 2. Figure 3 shows the dependency of calculated energy need on the U-value for the wall for different values from the Standard. The decreased energy need for the better shape factor and air change ratio are showed in Figure 4.

With decreasing the shape factor, not so low *U*-values are needed to comply the standard. The variant C complies the nowadays required value without mechanical ventilation while the variant A does not. Even bigger problem has the variant A with the final values for nearly zero energy building. Also with mechanical ventilation with high efficiency, highly insulated walls and roof does not comply the final values.

Comparing the three variants (Figure 5 - 7) with same *U*-values (type 2 or 3), the energy need is reduced from 73 to 58 (variant B) to 50. This is based on the floor area and caused by the better utilizing shape.

As stated before, the heat loss created by the ventilation is crucial by the low energy houses. By the mechanical ventilation, the energy need can be reduced to half, depending on the effectiveness of the mechanical ventilation and energy recuperation (Figure 6 and 7).

In terms of primary energy, where is taken into account the heat source and its influence on the environment by using the national conversion factors. Also heating of the hot water is added. All variants and all four types for each variant with air change ratio 0.5 are summarized in Figure 8.

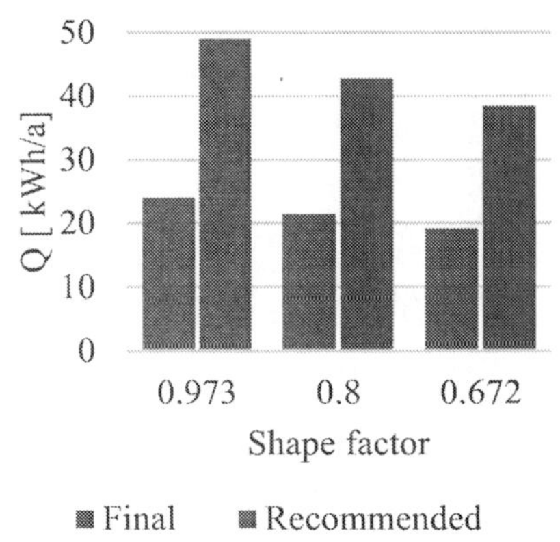

Figure 2. Energy need for heating values based on building shape factor from the standard STN 73 0540.

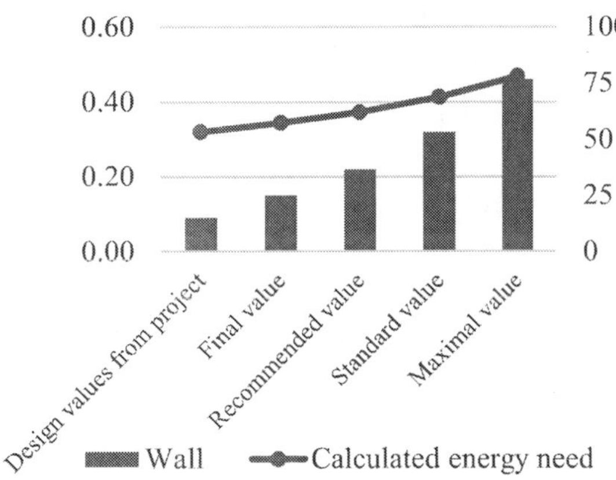

Figure 3. Energy need for heating dependency based on the *U*-value for wall.

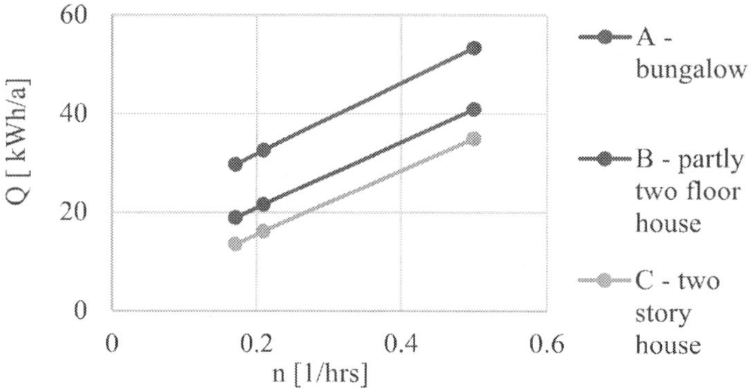

Figure 4. Energy need for heating for three different variants with different air change rates.

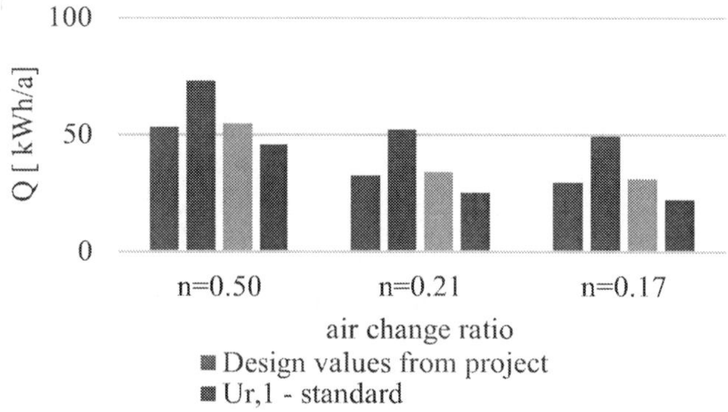

Figure 5. Energy need for heating depend on the *U*-value meeting the required values for individual fragments and different air change rates.

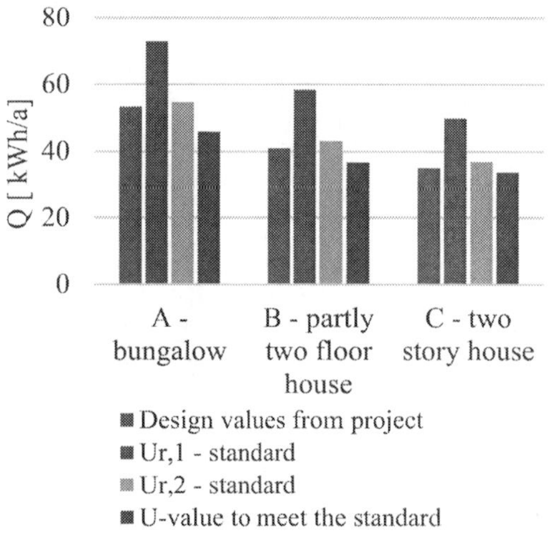

Figure 6. Energy need for heating depend on the *U*-value meeting the required values for individual fragments and air change ratio 0.5.

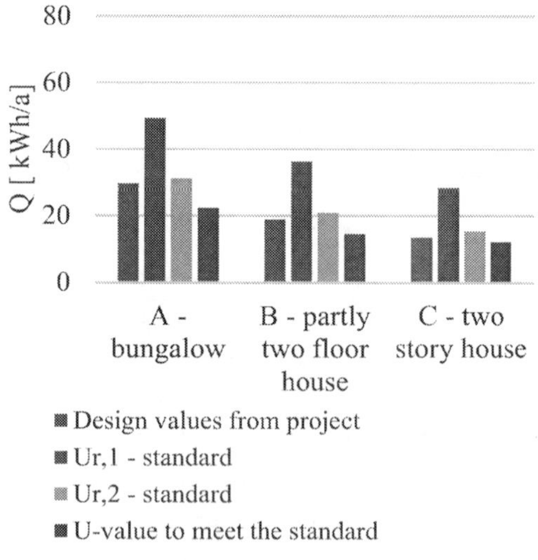

Figure 7. Energy need for heating depend on the *U*-value meeting the required values for individual fragments and air change ratio 0.17.

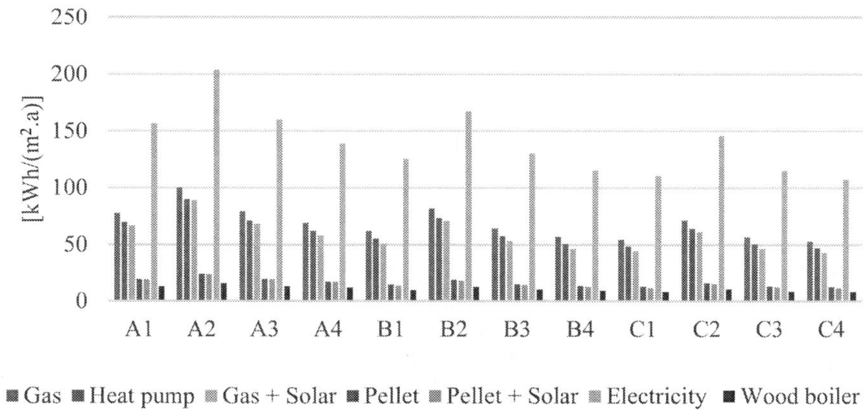

Figure 8. Primary energy for all variants dependent on the heat source. Air change rate 0.5.

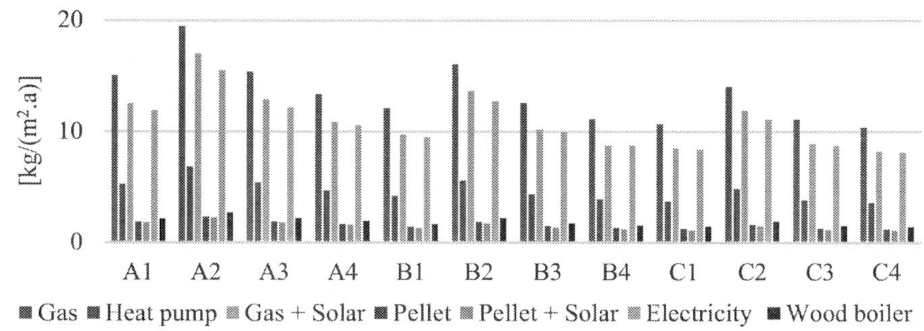

Figure 9. CO_2 emissions for all variants dependent on the heat source. Air change rate 0.5.

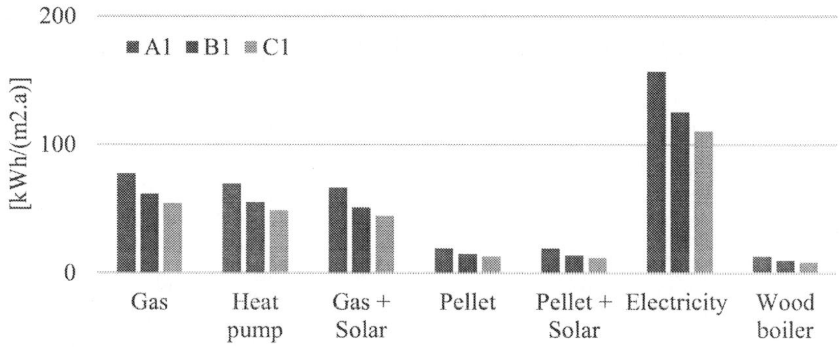

Figure 10. Primary energy for the first type of three geometric variants based on the heat source. Air change rate 0.5.

CO_2 emissions for all variants also with air change rate 0.5 are summarized in Figure 9. Presented results showed that the national conversion factor for each energy source is the key factor. This is shown in the Figure 10, where the electricity boiler has the worst results (conversion factor 2.2), gas boiler is in the middle (conversion factor 1.1) and the lowest primary energy has the wood and pellet heating (conversion factor 0.2). In the future there is accepted to lower the conversion factor for gas.

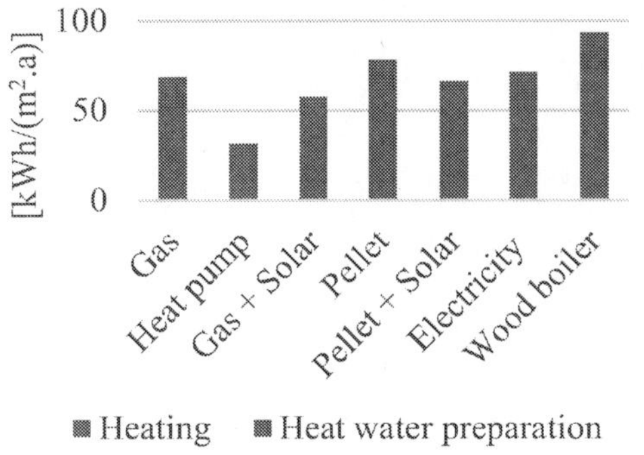

Figure 11. Delivered energy for A1 variant and air change ratio 0.5.

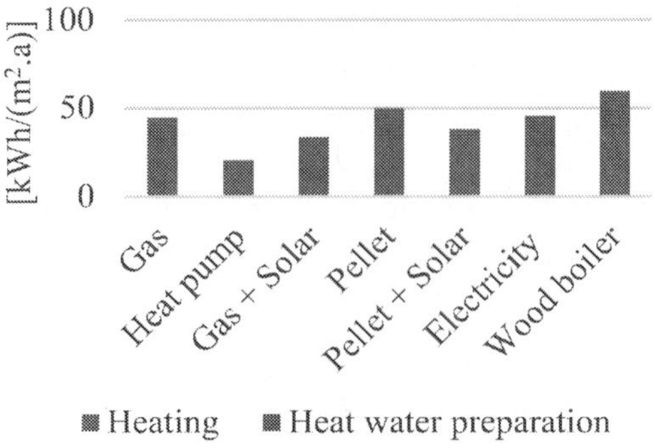

Figure 12. Delivered energy for A1 variant and air change ratio 0.17.

Finally, the delivered energy, which is sum for energy for heating and for hot water preparation and influence of air change ratio are in Figure 11 and 12. In real project, there will be also added energy for cooling and energy for lighting. This total sum is multiplied by the conversion factor and that's why for example the wood boiler changed from the worst alternative by the delivered energy to the best in primary energy. Every building had to be put in the building energy performance classes. Nowadays, the building need to get in the class A1 or A0 in global indicator - primary energy, table F [7].

5. Conclusion

This presented heating study shows the influence of building shape factor, *U*-values for the building envelope fragments and air change ratio based on the natural or mechanical ventilation. Nowadays, with implementing the mandatory Energy performance of building directive is hardly possible to meet the designing assessment without mechanical ventilation for ultra-low energy buildings. The criteria, which will be valid from 2021 was possible to comply only with variants with better building shape factor and mechanical ventilation.

In term of global indicator - primary energy, the results is based on the source of energy, best sources are today in Slovakia representing by the conversion factors 0.2 for renewable energy sources, such as wood or pellets. The worst is electricity. The gas is still a good alternative. The use of additional solar system can reduce the energy demand for water heating. In the future, more detailed comparison will be conducted.

Acknowledgments
The research was done with the support of grant project VEGA no. 1/0945/16.

References

[1] Directive 2010/31/EU Energy performance of buildings
[2] 2016/1318 Commision recommendation
[3] National plan for increasing the number of nearly zero-energy buildings - Slovakia
[4] STN 73 0540:2012: Thermal protection of buidings. Thermal performance of buildings nad components. (In Slovak).
[5] STN 73 0540:2012 - 2. Z1: Part 2. Functional requirements, Change 1.
[6] Act No. 555/2005 (In Slovak)
[7] Act No. 364/2012 (In Slovak)

IOP Conf. Series: Materials Science and Engineering **415** (2018) 012037 doi:10.1088/1757-899X/415/1/012037

Reducing energy demand in existing buildings as a tool to reduce air pollution

P Lis

1Czestochowa University of Technology, Faculty of Infrastructure and Environment, J.H. Dąbrowskiego 73, 42-201 Częstochowa, Poland

Corresponding author: piolis@is.pcz.czest.pl

Abstract. Buildings, on average, account for approximately 41% of total energy consumption in the European Union. This energy consumption also translates into hundreds of millions of tonnes of emitted CO2 and other air pollutants. Due to a large share in the total energy consumption, the buildings sector has a significant potential to reduce the energy intensity of building operation, increase the share of renewable energy sources (RES), and thus significantly reduce emission of air pollutants. Existing buildings in Poland, both residential and public, are characterized by high potential in terms of energy and environment. One way to improve the energy efficiency of existing buildings and reduce emissions (in particular greenhouse gases) is through thermos modernization which adapts existing buildings to current and future energy demand requirements. The paper presents the expected environmental effects of measures which adjust the existing residential buildings to the requirements in force in Poland since 2021. It has been assumed that the energy demand for heating buildings will be limited to 55-60 (kWh/(m^2year)). The calculations show that such measures will result in a reduction of total air pollutant emissions from households due to the reduction of energy demand for heating of dwellings from about 30% to about 67%, depending on the type of pollution.

1. Introduction

Given the high dependence of the modern world on various forms of energy, the depletion of conventional energy resources and the environmental impact of their processing, the efficient use of energy should be the subject of special measures.

In the European Union, legislative foundations have already been created to take appropriate actions in the form of binding provisions on the need to achieve, by 2020, all EU countries, objectives that are conventionally 3 x 20.

These goals were agreed at the European Council summit in March 2007 and they are:

- reducing CO_2 emissions by 20%,
- increasing the share of energy from renewable sources to 20%,
- increasing energy efficiency by 20%.

Recent years have seen two dominant and interdependent trends in the discussion of economic development problems:

- increase in demand for energy resulting from development with simultaneous depletion of non-renewable fuels and searching for sources to meet the growing needs,
- emission of pollutants and their impact on climate change.

The presence of these thematic currents in a broad debate led the author to carry out considerations and author's simplified quantitative analyses, limited to residential buildings and households. The quantitative analysis was made on the basis of GUS statistical data referring to 2011, when the last National Census was conducted, which were subject to transformations to the form desired by the author. Their partial results in terms of reducing the emission of pollutants into the air as a result of a possible reduction of energy consumption for heating buildings are presented in this paper.

2. Emission of pollutants into the air

Another important aspect of the production of energy from non-renewable fuels using conventional energy technologies is also related to the subject of improving energy efficiency. This aspect is the emission of pollutants into the air. The issue sometimes even seems to dominate over energy efficiency, which is not fully justified in the context of a very broad interdependency.

In the EU, ambitious rules have been set to reduce pollutant emissions. This situation, due to the monoculture of coal in the Polish economy, sets additional difficult goals and tasks for our country. According to the author, these targets are too high in terms of the time of implementation due to the economic and technical tasks necessary to reduce CO_2 emissions into the air which is maintained in Poland at the level of 300 million tones / year (GUS. Information Portal. State and protection of the environment 2016) [1].

There is no doubt that efforts should be made in this direction. However, it is important to quantitative and qualitative manner determine the scope of activities undertaken this route. According to the author, the set emission limits will probably not be met by the use of eco-efficient energy technologies alone, and some opportunities in this respect arise while saving energy and reducing energy consumption (fig. 1 and 2).

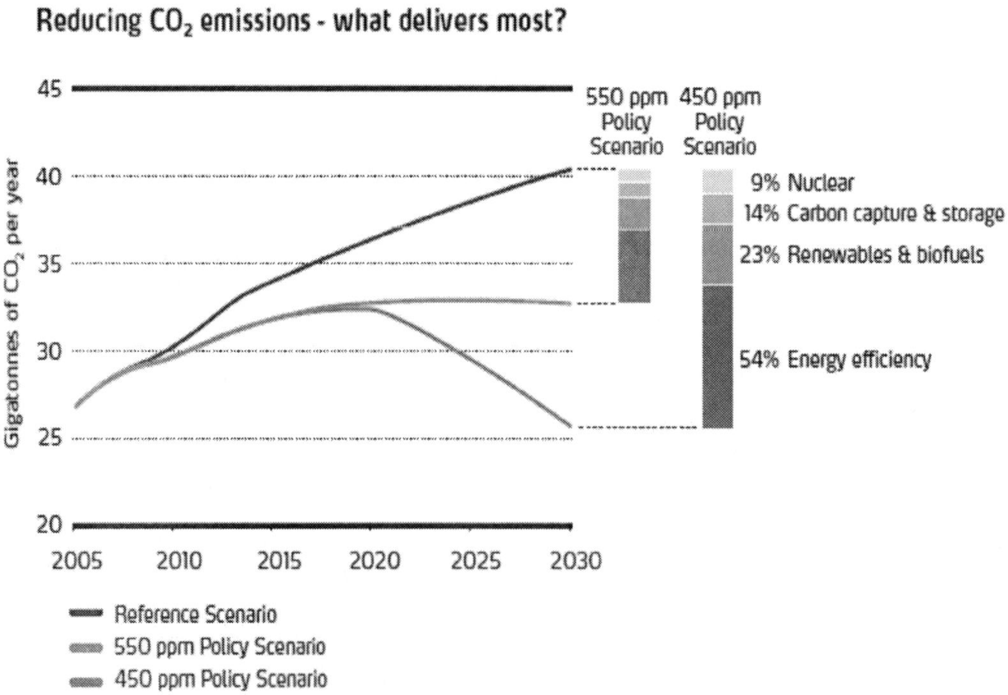

Figure 1. Carbon dioxide CO_2 reduction potentials [2]

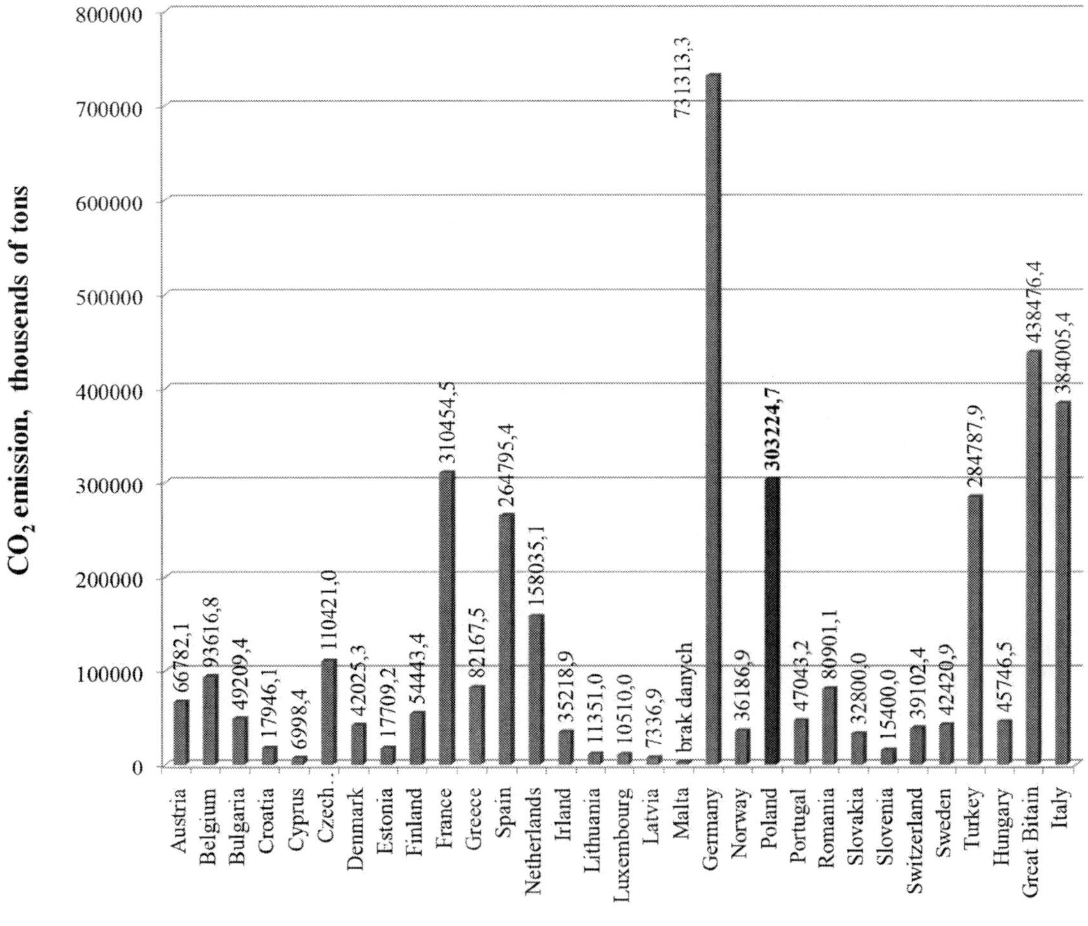

Figure 2. Carbon dioxide CO_2 emissions from fuel combustion [1]

3. Energy consumptions in buildings

The search for the largest opportunities in the use of energy efficient and effective should focus where there is the highest energy consumption. The subsector of buildings with the majority share of residential buildings plays the dominant role here. It is one of the main consumers of energy in the modern economies of developed countries, and above all in the operation phase of these facilities. On average, buildings account for approximately 41% of the total energy consumption in the European Union, which also translates into the emission of hundreds of millions of tons of CO2 [3,4].

Due to a large share in the total energy consumption, the buildings sub-sector has a significant potential to reduce the energy intensity of building operation, increase the share of RES, and thus significantly reduce emission of, inter alia, CO2 (Fig. 3).

The quoted data only strengthen the thesis that reducing energy consumption in buildings is very important for rational energy management and reducing emissions of pollutants into the air. Residential buildings (the largest functional group of buildings) play a dominant role in this process and have therefore been the subject of particular attention by the author of this publication. Considerations and analyses carried out later in this paper refer to the year 2011 due to the fact that in that year the National Census (abbreviated to NSP 2011) was conducted.

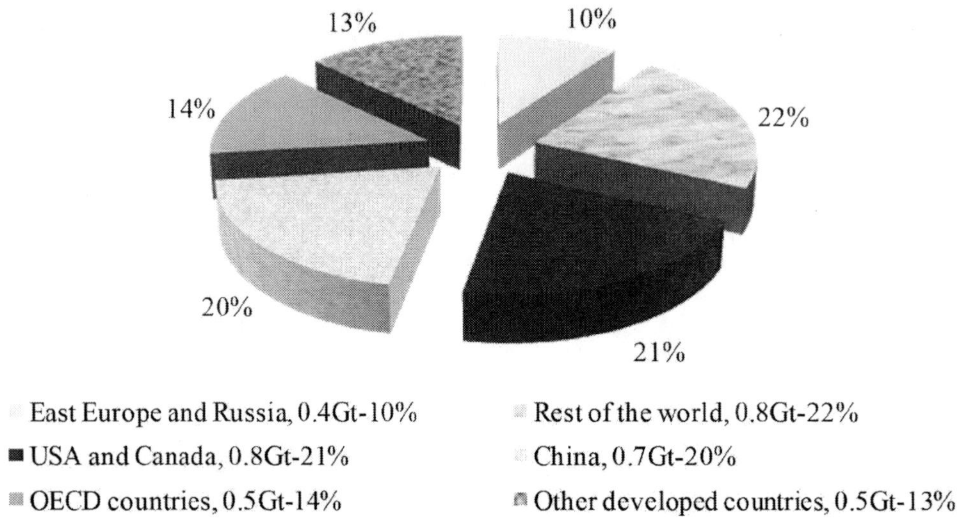

Figure 3. Potential possibilities to reduce CO_2 emissions resulting from the exploitation of buildings [5]

4. Energy consumption in households

In 2011, there were 12,060,000 dwellings in occupied residential buildings (Table 1) which were used by 13,568,000 households [6,7].

Energy consumption in households (including passenger cars in operation) reached the level of 1117 PJ [8] and constituted about 27% of the national energy supply with the dominant share of heating amounting to 68.8% (Fig. 4) [8].

"In terms of volume, solid coal fuels and district heating, which is also produced from these fuels in approx. 75%, played a leading role in space heating" [8]. It should be added that the production of network heat is based in about 75% also on solid coal fuels [8]. Natural gas was the third most used energy carrier. However, only 9.2% of households used this more environmentally friendly fuel than coal as their primary carrier [8].

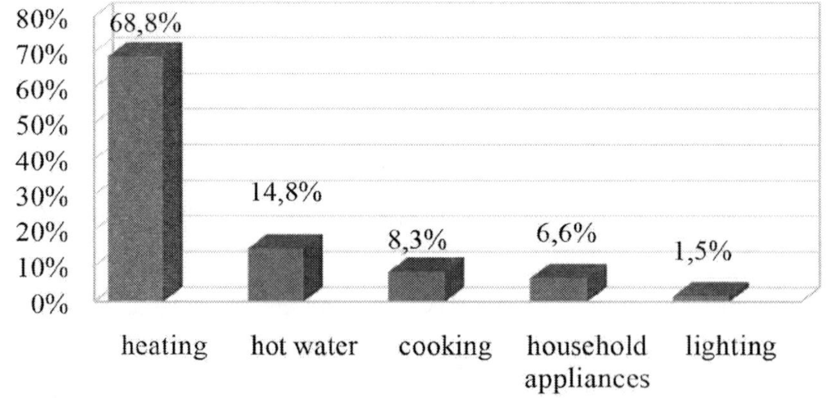

Figure 4. Structure of household energy consumption by purpose [8]

5. Energy efficiency of residential buildings and reduction of air pollutant emissions

The analysis of the possibilities to reduce the energy consumption for heating residential buildings and pollutant emissions inherent thereto in relation to the energy use objectives of buildings clearly shows the greatest potential for measures taken in the scope of space heating.

Table 1. Reduction of energy consumption for heating buildings and flats built in Poland at various times as a result of reduced unit energy demand for heating up to: 60 kWh/(m^{2}year) for single family buildings, 55 kWh/(m^{2}year) for multi-family buildings (own elaboration based on [6,7,8,10]).

The construction period	Energy demand for heating (kWh/(m^{2}year))		Energy demand for heating after reduction (kWh/(m^{2}year))		Reduction of energy demand for heating (kWh/(m^{2}year))	
	in single-family buildings	in multi-family buildings	in single-family buildings	in multi-family buildings	in single-family buildings	in multi-family buildings
before 1918	367.66	264.31	60	55	307.66	209.31
1918-1944	306.10	191.31	60	55	246.10	136.31
1945-1970	265.22	172.74	60	55	205.22	117.74
1971-1988	230.25	156.52	60	55	170.25	101.52
1989-2010	183.36	125.67	60	55	123.36	70.67

Such measures in existing housing may be carried out:

- in terms of ("quantitative") reduction of energy demand for space heating by adjusting it to future heat and energy requirements,
- in the scope of ("qualitative") reduction or elimination of emissivity of sources of power generation for heating. This can be achieved by using appropriately cleaner technologies and fuels or by introducing on a large scale renewable energy sources which can be used in our climatic conditions to heat buildings. Such activities contribute to the reduction of the emission of pollutants into the air while limiting the consumption of primary energy for heating. However, they do not affect the final energy consumption
- by optimally combining the activities mentioned above in an optimal way.

In the following, the results of the quantitative analysis of the potential to reduce the energy demand for heating residential buildings are presented by their adaptation to the level 60 kWh/(m^{2}year) and 55kWh/m^{2}year), respectively for multi-family and single-family buildings used in Poland in 2011 (Table 1).

In the next step, a percentage reduction of the total energy demand for heating of the buildings under consideration was assumed at the calculated average level of 72.29% (Table 2). This value allowed for a simplified calculation of the reduction of emissions of selected pollutants to the air as a result of the reduction of energy demand for heating buildings by raising their energy standard. For the calculations, a simplified assumption was made that the percentage reductions in energy demand and pollutant emissions were equal. The calculations were carried out starting from the elaboration of the obtained data concerning the emission of selected pollutants to the air from heating and total in households, and ending with their percentage reduction resulting from the previously calculated reduction of energy demand for heating. Due to editorial limitations, Table 3 and 4 presents only the results of the last stage of calculations for households in 2011 in the form of reduction of the types of emissions to the air considered in the work as a result of reduction of energy demand for heating.

Table 2. Reduction of energy consumption for heating buildings and flats built in Poland at various times as a result of reduced unit energy demand for heating up to: 60 kWh/(m^2year) for single family buildings, 55 kWh/(m^2year) for multi-family buildings (own elaboration based on [6,7,8,10]).

The construction period	Reduction of total energy consumption for heating (TWh/year)		Reduction of total energy consumption for heating (TWh/year)	Reduction of total energy consumption for heating (% /year)		Reduction of total energy consumption for heating (% /year)
	in single-family buildings	in multi-family buildings		in single-family buildings	in multi-family buildings	
before 1918	11.72	6.34	18.05	83.7	79.2	82.05
1918-1944	12.86	5.70	18.56	80.4	71.3	77.35
1945-1970	20.89	9.54	30.43	77.4	68.2	74.23
1971-1988	27.36	12.97	40.33	73.9	64.9	70.76
1989-2011	16.15	7.31	23.46	67.3	56.2	63.40
TOTAL (AVERAGE)	88,98	41.86	130.84	(75.4)	(66.4)	(72.29)

Table 3. Annual air pollutant emissions from the heating and the total in households (own elaboration based on [11,12]

Specification of emission	Air pollutant emissions from household heating (thousand tons)	Air pollutant emissions from household (thousand tons)	Total air pollutant emissions in Poland (thousand tons)
AIR EMISSIONS FOR GREENHOUSE GAS EXPRESSED CARBON DIOXIDE EQUIVALENT			
Emission of carbon dioxide CO_2	49440.568	62566.758	330309
Emissions of nitrous oxide	280.575	435.758	27241
Emissions of methane	2538.768	2572.661	35538
AIR EMISSION OF SELECTED POLLUTANTS			
Emissions of nitrogen oxides NO_x	67.506	117.448	851
Emission of sulfur oxides SO_x	219.180	219.180	910
Emissions of ammonia NH_3	0.510	0.819	270
Emission of non-methane volatile organic compounds	102.769	229.456	652
Emission of carbon monoxide CO	1622.308	1873.686	2916
Dust emission PM_{10}	103.756	109.877	245
Dust emission $PM_{2,5}$	61.348	66.813	151
Emission of total suspended dust	141.360	x	414
Emission of dioxins and furans, i-TEQ	134.9	x	269
Emissions of hexachlorobenzene (HCB)	2.8	x	14
Emission of polychlorinated biphenyls (PCB)	0.0004353	x	0.0007246
Emission of polycyclic aromatic hydrocarbons	0.1230266	x	0.1437719

Table 4. Annual reduction of air pollutant emissions from the heating and the total in households (own elaboration based on [11,12]

Specification of emission	Reduction of air pollutant emissions from household due to reduction of energy demand for heating in households (thousand tons)	Reduction of air pollutant emis. from household due to reduction of energy demand for heating in relation to air pollutant emissions from household (%)	Reduction of air pollutant emis. from household due to reduce of energy demand for heating in relation to total air pollutant emissions in Poland (%)
AIR EMISSIONS FOR GREENHOUSE GAS EXPRESSED CARBON DIOXIDE EQUIVALENT			
Emission of carbon dioxide CO_2	35738.35	53.33	10.82
Emissions of nitrous oxide	202.81	43,45	0.74
Emissions of methane	1835.16	66.60	5.16
TOTAL	37776.33	x	x
AIR EMISSION OF SELECTED POLLUTANTS			
Emissions of nitrogen oxides NO_x	45.56	41.55	5.74
Emission of sulfur oxides SO_x	147.92	72.29	17.41
Emissions of ammonia NH_3	0.34	45.03	0.14
Emission of non-methane volatile organic compounds	69.36	32.38	11.39
Emission of carbon monoxide CO	1094.89	62.59	40.22
Dust emission PM_{10}	70.02	68.26	30.58
Dust emission $PM_{2,5}$	41.40	66.37	29.29
Emission of total suspended dust	95.40	x	24.69
Emission of dioxins and furans, i-TEQ	91.04	x	36.25
Emis. of hexachlorobenzene (HCB)	1.89	x	14.88
Emis. of polychlorinated biphenyls (PCB)	0.000315	x	43.43
Emission of polycyclic aromatic hydrocarbons	0.08	x	61.86
TOTAL	1775.74	x	x

6. Summary

The potential for energy savings in buildings (mainly residential) is estimated at: 33-60% for improving thermal insulation of walls, 16-21% for modernizing ventilation system, 14-20% for improving thermal insulation of transparent partitions, 10-12% for regular inspection and repairs of central heating boilers, 50-80% for modernizing hot water production system with the use of RES [13]. The calculations show that as a result of adjusting the energy demand for heating in residential buildings to the level of 50-60 kWh (m^2/(a year)) (based on the requirements in force in Poland since 2021), the demand in existing residential buildings may be reduced by an average of approx. 67% as compared to the situation in 2011.

At the same time, such measures will reduce the overall emissions of air pollutants from households due to the reduction of energy demand for heating of dwellings from about 30% to about 67% (Table 3 and 4), depending on the type of pollutant. In the scale of the whole country, this will reduce the air pollutant emissions considered in this paper from about 0.7% to about 62% (Table 3 and 4) in comparison to the national emissions of these pollutants in 2011. It should also be noted that the low percentage reduction for domestic emissions is mainly due to the low share of these household emissions in the total Polish air emissions account.

Acknowledgements: The material was prepared as part of the statutory research of the Czestochowa University of Technology BS / PB-407-302 / 11 task 6.

References

[1] Główny Urząd Statystyczny 2016 *Stan i ochrona środowiska 2016. Emisja CO_2 ze spalania paliw* (http://stat.gov.pl/statystyka-miedzynarodowa/porownania-miedzynarodowe/tablice-o-krajach-wedlug-tematow/stan-i-ochrona-srodowiska/ 2.1_emis_co2_przy_spal_paliw_ii_2017_r.xls) (12.04.2017)

[2] Rockwool - Climate & Environment 2009 *Climate & Energy. The fight against climate change starts at Home*

[3] Lis P. 2013 *Cechy budynków edukacyjnych a zużycie ciepła do ogrzewania* (*Seria Monografie*) vol 263 (Częstochowa: Wydawnictwo Politechniki Częstochowskiej) p 361

[4] Lis. P., Sekret R. 2016 Efektywność energetyczna budynków - wybrane zagadnienia problemowe *Rynek energii* **6/2016** pp 29-35

[5] Vattenfall, Publisher: Mogren A., Co-publisher: Pettersson L., Editor: Fager Ch., Assisting editor: Bratt J., Design: Linder D. 2007 *Vattenfall's climate map 2030* (Printing: Trydells) (http://www.eiif.org/awm/downloads/Vattenfalls_climateMAP.pdf) (12.04.2017)

[6] Matulska- Bachura A. - kierownik, Andrusiuk K., Dec A., Knyszewska E., Przybylska M., Różańska B., Sobczyk M. 2013 *Mieszkania. Narodowy Spis Powszechny Ludności i Mieszkań 2011.* (Warszawa: GUS Zakład Wydawnictw Statystycznych)

[7] Matulska- Bachura A. - kierownik, Andrusiuk K., Dec A., Knyszewska E., Przybylska M., Różańska B., Sobczyk M. 2013 *Zamieszkane budynki. Narodowy Spis Powszechny Ludności i Mieszkań 2011.* (Warszawa: GUS Zakład Wydawnictw Statystycznych)

[8] Matulska- Bachura A. - kierownik, Andrusiuk K., Knyszewska E., Przybylska M., Różańska B., GUS Departament Handlu i Usług 2014 *Warunki mieszkaniowe gospodarstw domowych i rodzin. Narodowy Spis Powszechny Ludności i Mieszkań 2011.* (Warszawa: GUS Zakład Wydawnictw Statystycznych)

[9] Tkaczyk W., Kozieł A., Mikołajuk H., GUS - Departament Produkcji, Ministerstwo Energii, Agencja Rynku Energii S.A. (ARE S.A.) 2015 *Zużycie energii w gospodarstwach domowych w 2015 r.* (Warszawa: GUS Zakład Wydawnictw Statystycznych)

[10] Ürge-Vorsatz D. - pod kier., Wójcik-Gront E., Tirado Herrero S., Labzina E., Foley P. 2012 *Wpływ na rynek pracy programu głębokiej modernizacji energetycznej budynków w Polsce.* (Raport końcowy 17.01.2012) (Budapeszt: The Center for Climate Change and Sustainable Energy Policy (3CSEP). European Climate Foundation. Central European University)

[11] Bochenek D. - kierownik, Bancarzewska M., Górska A., Karczewska K., Lizuraj M., Pawłowska T., Reńska A., Sulik J., Wojciechowska M., Wrzosek A., Zagórska K., Zieleźny K., 2013 *Ochrona środowiska* 2013 (Warszawa: GUS Zakład Wydawnictw Statystycznych)

[12] Departament Badań Regionalnych i Środowiska GUS 2016 *Rachunki ekonomiczne środowiska. GUS. Notatka informacyjna. Załącznik nr 1 Rachunki emisji do powietrza.* (Plik arkusza kalkulacyjnego Excel). (http://stat.gov.pl/obszary-tematyczne/srodowisko-energia/srodowisko/rachunki-ekonomiczne-srodowiska,7,1.html (12.04.2017)

[13] Balaras C.A., Gaglia A.G., Georgopoulou E., Mirasgedis S., Sarafidis Y., Lalas D.P. 2007 European residential buildings and empirical assessment of the Hellenic building stock, energy consumption, emissions and potential energy savings *J. Building and Environment 2007* Vol **42**, pp 1298–1314

[14] GUS Departament Produkcji. Wydział Bilansów Paliw, Surowców i Materiałów2016 *Energia* (Warszawa: GUS Zakład Wydawnictw Statystycznych) (https://stat.gov.pl/files/gfx/portalinformacyjny/pl/defaultaktualnosci/5485/1/4/1/energia2016.pdf)

Ice rink – design of size of water storage tanks with small heat source for ice resurfacing machine

M Kovac, K Kovacova

Technical University of Kosice, Faculty of Civil Engineering, Vysokoskolska 4, 040 01 Kosice, Slovakia

E-mail: martin.kovac@tuke.sk

Abstract. The paper describes a method of design of storage tanks volume for an ice resurfacing machine in ice rink by using heat source with small thermal power (idea of downsizing). The main aim is to minimize investment costs for boiler room (the heat pump water-water). The achieved results show saving of investment costs about 35%. The request is the minimum water temperature at the value of 45°C before each filling process of resurfacing machine. Therefore it is very important to design the storage tanks volume where the water temperature will not be below the 45°C before the each filling an ice resurfacing machine during the hockey match or training. The required volume of storage tanks was calculated at the value of about 2000 litres if there is used the heat source with thermal power 45 kW. The number of storage tanks is given by boiler room dimensions too. In this case 3 storage tanks were designed.

1. Introduction

Most rinks use heated water to resurface the ice. Hot water flows more evenly onto the ice and bonds more strongly to the old ice than does cold water. The reason for this is that hot water melts some of the old ice, smoothing out small cracks and flaws in the ice [1]. From the surveys mapping the energy consumption of ice rink we can see that the energy for ice resurfacing is about 6 % - 10 % of total energy use in modern ice arenas [2, 3, 4].

The paper describes the technical design of storage tanks volume with small heat source for ice resurfacing machine in the ice rink. The design of heat source is based on the idea of „downsizing“, where the heat source with low thermal power is used. This heat source is combined with so volume of water in storage tanks in which the temperature does not drop below 45°C before each filling process of resurfacing machine.

The consumption of water for resurfacing machine is from 400 to 700 litres per one operation [5]. In this case it is needed the storage tank with minimum volume 700 liters. So the heat source with small thermal power (e.g. 45 kW) will have a problem with warming of water in storage tank (700 liters) to the required value of temperature during the short time interval (length of one hockey period is 20 minutes + (plus) resurfacing of ice ~~is~~ during the break – (minus) time for filling a resurfacing machine). By using of heat source with higher thermal power (e.g. 90 kW) we will achieve the required temperature of water in storage tanks in short time but the investments cost of that heat source will be higher. Using a heat source with small thermal power in combination with adequate volume of storage tanks could by a good way (idea of downsizing). The maximum attention is needed in the process of calculation of required volume of storage tanks in the combination with heat source with small thermal power.

2. General requirements for resurfacing of ice

Resurfacing of ice surface is made by using of resurfacing machine. A blade shaves a thin layer from the surface of the ice. After a horizontal screw (auger) gathers the shavings, a vertical screw propels the into the snow tank. Water is fed from a wash-water tank to the „conditioner", which rinses the ice. Dirty water collected in front of a squeegee is vacuumed, filtered, and returned to the tank. Clean water from the ice making tank is delivered to the ice through a pipe and spread evenly by a towel pulled across the ice behind the „conditioner"[6].

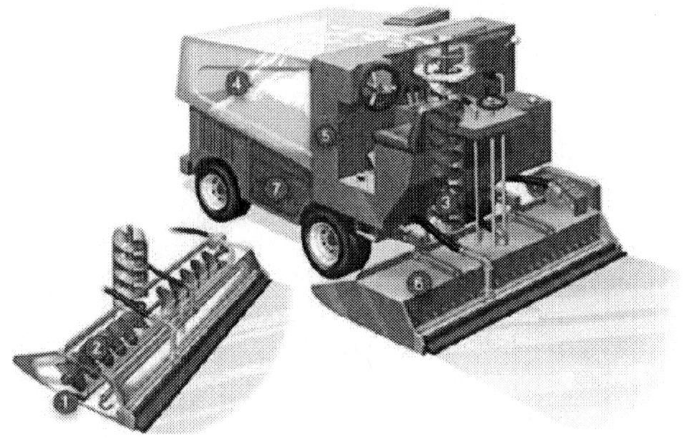

Legend:

1 - blade
2 - horizontal screw
3 - vertical screw
4 - snow tank
5 - wash-water tank
6 – „conditioner"
7 – ice making tank

Figure 1. Typical resurfacing machine [6].

The consumption of water for resurfacing machine is from 400 to 700 litres per one operation. The used water has to meet the required chemical properties, with neutral pH value and free from mechanical particles. The temperature of water is from 30 to 50 °C. The average water temperature for resurfacing machine is 45°C. The filling time of resurfacing machine is from 3 to 11 minutes. The ice is usually resurfaced 5 – 10 times per day [7].

3. Design of storage tanks volume

In this case will be the gas boiler used as a heat source. But the owner of ice rink wants to use the electric heat pump (water to water) instead of gas boiler in near future. Logically, the request is to minimize the investment costs for heat pump. The energy source for heat pump will be waste heat from ice cooling technology. In analyze is considered with thermal power of heat source (gas boiler, electrical heat pump) at the value of 45 kW. The request is the minimum water temperature at the value of 45°C before each filling process of resurfacing machine. During a hockey game we have three period with standard length of 20 minutes and two brakes with 15 minutes. In real the each period is longer than 20 minutes. The resurfacing of ice surface is made before the start of game and then after first and second period of a game. When the resurfacing machine is filled by the water from the storage tanks the water temperature drops in storage tanks (filling with cold water from a water supply). The used heat source (gas boiler or heat pump) heats the water in storage tanks continuously.

If we know the thermal power of heat source, operation time of heat source, the amount of cold water and its temperature we can calculate the water temperature after heating by using the following equation (1). This equation describes the thermal energy in water. For our case we have modified the equation (1) into the equation (2).

$$Q = mc\Delta T \tag{1}$$

where Q is thermal energy in liquid; m is mass of liquid; c is specific heat capacity of liquid and ΔT is temperature diference of liquid.

$$\begin{aligned} P_{\mathrm{gen}}\eta_{\mathrm{gen}}t_{\mathrm{gen}} &= m_{\mathrm{w}}c_{\mathrm{w}}\Delta T_{\mathrm{w}} \\ P_{\mathrm{gen}}\eta_{\mathrm{gen}}t_{\mathrm{gen}} &= \frac{V_{\mathrm{w}}\rho_{\mathrm{w}}}{3600}c_{\mathrm{w}}\left(\theta_{\mathrm{w,warm}} - \theta_{\mathrm{w,cold}}\right) \end{aligned} \tag{2}$$

where P_{gen} is thermal power of heat source; η_{gen} is efficiency of heat source; t_{gen} is operating time of heat source; m_{w} is mass of water; V_{w} is volume of water; c_{w} – is specific heat capacity of water; ρ_{w} is density of water; ΔT_{w} is temperature diference of water; $\theta_{\mathrm{w,warm}}$ is temperature of hot water and $\theta_{\mathrm{w,cold}}$ is temperature of cold water.

By using equation (2) we can calculate the temperature of water which was supplemented into the storage tanks after filling of resurfacing machine. This volume of supplemented warm water with calculated water temperature is mixed with water in storage tanks. So we need to calculate the final water temperature in storage tanks by using the following equation (3).

$$V_1\theta_1 + V_2\theta_2 = V_3\theta_3 \tag{3}$$

where V_1 is volume of water remaining in storage tanks after filling a resurfacing machine; θ_1 is temperature of water remaining in storage tanks after filling a resurfacing machine; V_2 is volume of water added from a water supply into the storage tanks after filling a resurfacing machine; θ_2 is temperature of cold water added into the storage tanks from a water supply after filling a resurfacing machine; V_3 is volume of water in all storage tanks and θ_3 is temperature of water in all storage tanks after warming by using of heat source.

The next figure (Figure 2) shows the water temperature in storage tanks during a hockey game. The temperature of water was calculated by using equation (2) and (3). The initial conditions which were used in calculation of water temperature in storage tanks are shown in the following table (Table 1).

Table 1. Initial conditions.

Parameters	Values
Thermal power of heat source	45 [kW]
Number of heat sources	1 [-]
Thermal power of all heat sources	45 [kW]
Time for filling a resurfacing machine	6 [minutes]
Temperature of cold water added into the storage tanks	7 [°C]
Temperature of hot water in storage tanks before a hockey game	60 [°C]
Heat capacity of water	4.186 [kJ/(kgK)]
Density of water	999.7 [kg/m^3]
Real volume of storage tanks	2289 [l]
Maximum consumption of water for resurfacing machine	700 [l]

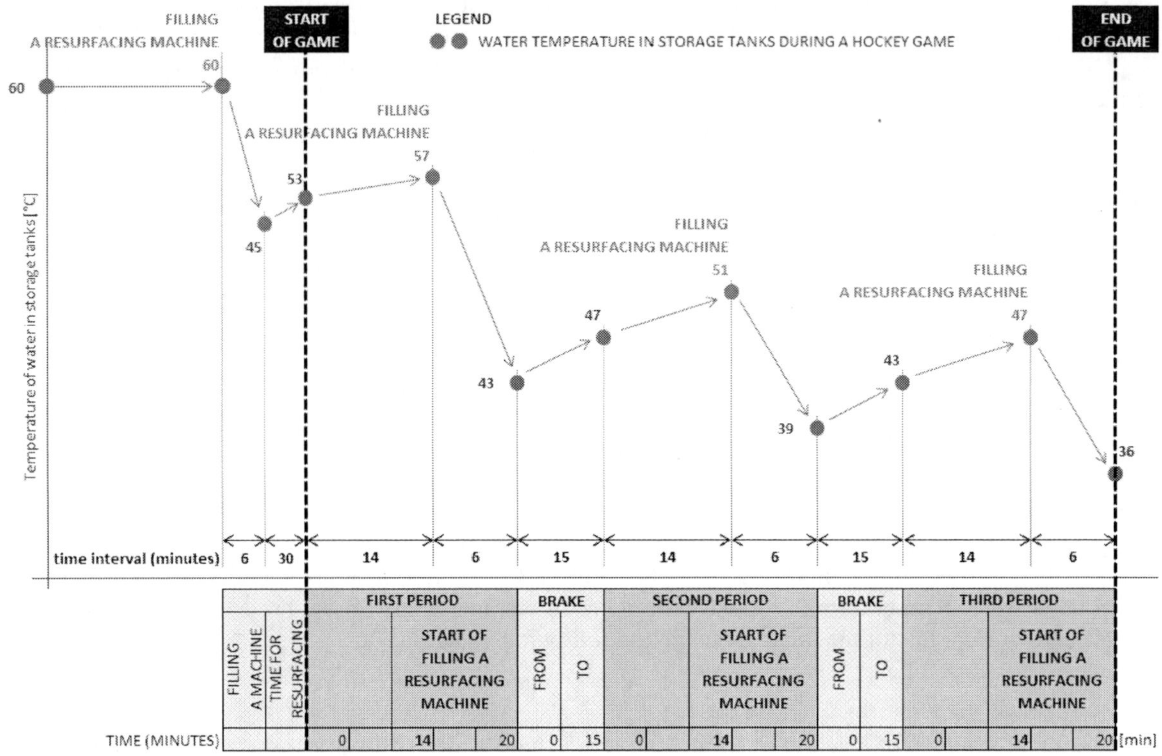

Figure 2. Water temperature in storage tanks during a hockey game by using a heat source with small thermal power (45 kW).

4. Functional scheme of boiler room

The next figure (Figure 3) shows the functional scheme of suggested heat source with small thermal power (gas boiler or planned installation of heat pump in near future) and with water storage tanks. In order to achieve the maximum heat transfer from the heat source (gas boiler or planned installation of heat pump in near future) into the storage tanks we use the tanks without twisted heat exchangers. The maximum effective heat transfer from the heat source (gas boiler or planned installation of heat pump in near future) into the water in storage tanks is designed through the flat plate heat exchanger. The storage tanks are connected with pipes (cold water into the tanks, warm water from tanks for resurfacing machine, heating pipes) through the Tichelmann solution.

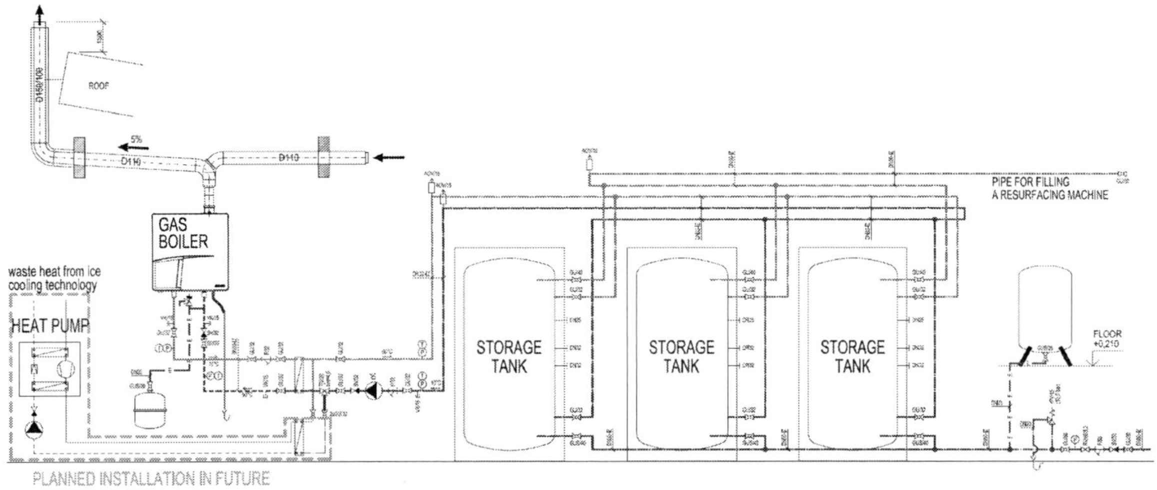

Figure 3. Functional scheme of gas boiler installation with storage tanks of water.

5. Conclusion

This paper describes a design of storage tanks volume for resurfacing machine by using heat source with small thermal power (idea of downsizing). The aim of paper was to show the possibility of using the heat source with small thermal power in order to supply hot water for ice resurfacing machine. In this case it is need to pay high attention to design of required volume of storage tanks where the water temperature will not be below the value 45°C before the filling a resurfacing machine. There was done the financial analyze of investment costs for electric heat pump because the owner of ice rink wants to buy them in near future. There were compared two variants. In first variant there is used the heat pump with small thermal power (45 kW) in combination with 3 storage tanks. Their water volume is about 2000 liters. In second variant there is used the heat pump with higher thermal power (90 kW) in combination only with 1 storage tank. The water volume of storage tank is about 700 liters in this case. The investment costs of first variant were calculated at the value of about 21 000 Euro. For the second variant the investment costs are about 33 000 Euro. If we compare these two variants we can say that using the idea of "downsizing" will bring savings in investment costs about 35%. If we compared gas boilers as heat source, the difference in investments cost would be minimal.

Acknowledgments

This paper is the outcome of the Projects implementation: KEGA 052 TUKE-4/2013 - The implementation of a virtual laboratory for designing energy-efficient buildings, and University Science Park TECHNICOM for Innovation Applications Supported by Knowledge Technology, ITMS: 26220220182, supported by the Research & Development Operational Programme funded by the ERDF.

References

[1] Dietrich B K and McAvoy T J 1980 *Energy Conservation in Ice Skating Rink* (Tennessee: Technical Information Center Oak Ridge)

[2] Pachai A C 2009 *Energy consumption and heat recovery of refrigeration system in modern Arena* (Hoejbjerg: Johnson Controls Building Efficiency)

[3] Linhartova V and Jelinek V 2016 Heat Recovery of the Refrigeration System in the Ice Arena *CLIMA 2016 - proceedings of the 12th REHVA World Congress* vol 9 ed P K Heiselberg (Aalborg: Aalborg University, Department of Civil Engineering)

[4] Kolasniewski Y 2017 *Evaluation and modelling of ice ring energy usage* (Stockoholm: INSA

Strasbourg)
[5] ASHRAE 2014 *Refrigeration Handbook* (Atlanta: American Society of Heating, Refrigeration and Air-Conditioning Engineers) chapter 44
[6] Kaya R 2017 *Energy Usage in Ice Rink Resurfacing* (Stockholm, KTH School of Industrial Engineering and Management)
[7] IIHF 2016 *Ice Ring Guide* (Zurich, Internationa Ice Hockey Federation)

Simulation of visual comfort in selected industrial hall

E Dolnikova[1] and D Katunsky[2]

[1] Department of Building Physics, Institute of Architectural Engineering, Faculty of Civil Engineering, Technical University of Kosice, Slovak Republic
[2] Department of Architecture and Building Structures, Institute of Architectural Engineering, Faculty of Civil Engineering, Technical University of Kosice, Slovak Republic

Correspondence: erika.dolnikova@tuke.sk; Tel.: +421-055-602-4162

Abstract. Industrial halls are usually subjects to high energy demand. It is due to the many manufacturing processes, lighting, and the corresponding amount spent on space conditioning. The industrial buildings is one of the heaviest consumers of energy. Natural daylight is a vital element in our daily life. Providing natural daylight into the working environment is of fundamental importance for the comfort, efficiency and safety for the people in that environment. This paper deals with visual comfort evaluation in selected hall. Its calculated by the radiance simulation program.

1. Introduction

Through vision, one gets 75% to 90% of information from the surrounding environment. Visual comfort is the psychological state in which the entire visual system performs its function optimal. Good daylighting is an important factor. Convenience affects the type of light sources, the type and layout of the luminaires, the level of luminance and their uniformity in different planes and also the distribution of brightness in the space. The geometric parameters of the space, surface properties, including colour, also play an important role in the assessment of suitable working conditions. What matters is whether or not there is glare. Lighting should be sufficient and even, without excessive brightness and contrast. When the light is properly distributed in space, the objects appear plastic. Under well-designed lighting conditions, a person can feel as if they are clearly seated with a longer stay [8]. Many authors deal with visual comfort and discomfort.

Drahos et al. studied the problems of measurement and assessment of lighting in the working environment of industrial plants have been occupied by a relatively large number of households in the rest of the decade.

Turekova et al. studied several of them point to the fact that in the comprehensive assessment of lighting in the work environment, the individual qualitative and quantitative lighting parameters should be assessed not only on the basis of objectivity - measurement, but also the subjective response of employees. The subjective assessment gives a closer picture of the working environment and presents the satisfaction and well-being of the workforce in the workplace.

Reinhold et al. investigated relationship between office lighting conditions and occupational health. Measuring health aspects in combination with measuring lighting requires knowledge of both fields. Bellia et al. investigated the recent discoveries in photobiology to those interested in lighting design. Tzempelikos A investigated present new developments on daylighting research, visual comfort and non-visual effects related to health and well-being. Present new information on daylight prediction models and simulation techniques, glare evaluation methods.

Konis K studies daylighting performance over daily and seasonal changes in sun and sky conditions in core and perimeter zones of the building. Results show a high frequency of visual discomfort responses at both perimeter and core workspaces and observations reveal a large percentage of façade glazing covered by interior shading devices.

Tabadkani et al studies an improvement in the performance of dynamic facades in comparison with the static systems.

Konis K studies 15 single-variable probabilistic visual discomfort models implemented in software simulation workflows to place quantitative data in context with subjective outcomes. Results show a study conducted in the core zones of a side-lit office building.

The simulations presented in this contribution were performed using the original UNIX-based Radiance Lighting Simulation System [5].
There are different types of glare index calculation:

- DGI –Daylight Glare Index;
- VCP-Visual Comfort Probability;
- DGP-Daylight Glare Probability;
- UGR-Unified Glare Rating;
- CGI-CIE Glare INDEX.

In the article, visual comfort is evaluated through the glare index Visual Comfort Probability (VCP) also known as Guth Visual Comfort Probability. It is defined as the percentage of people that will find a certain scene (viewpoint and direction) comfortable with regard to visual glare. It was defined by Sylvester K. Guth in 1963 [5].

Interestingly, increasing background luminances decreases discomfort glare. The Visual Comfort Probability, or VCP, of a lighting system is a rating that indicates the percentage of people that will find a given discomfort glare acceptable. The VCP rating is applicable direct lighting fixtures with the viewer in a specific location and looking in a particular direction. A VCP rating of 70% is usually considered acceptable, but 80% is required for areas with computer monitors. VCP differences between fixtures of 5 or less are not significant [5].

2. Daylight measurements and calculations

Daylight measurements was realized in the hall in Košice. The hall is located in the Slovak Republic. It is a one-storey building with reinforced concrete pillars and connected windows and toplighting by skylight (Figure 1). The ceiling structure is made up of a space barrier beam located on the pillars. The hall is designed for work on the metal working and forming, production of sheet metal blank. Interior dimensions of hall are 15 m x 60 m x 8,5 m (Figure 2). Sidelighting is created by the windows with dimensions 5 m x 1.8 m and 3 m x 1.8 m, and toplighting is created by one saddle skylight with dimensions 2.4 m x 48 m x 1.1 m (Figure 3). The object is devided into three longitudinal tracts. There are two smaller tracts on the sides, where are offices, laboratories and classrooms.

Daylight measurements were performed on two different days. They were performed on the first day twice and the other days three times. The measurements were made in December when the sky was cloudy. The instruments were two "Hagner EC1 type" lux meters, a "Hagner Universal Photometer type S2 with an accuracy of 5%, and the "Hagner Reference calibration sample" with factor value and a reflection of 0.966. On the selected days, the value of the outside light in a cloudy sky ranged from 9500 lx-10,500 lx on the first day and 10,500 lx-12,000 lx on the second day.

The boundary conditions for calculations have been observed in the measurement of daylighting.
The fenestration systems are created by single wired glass. In the calculation the following coefficients were considered: light loss coefficients 0.8, maintenance factor of glazing on exterior surface 0.9, maintenance factor of glazing on interior surface 0.85, reflectance factor of external terrane 0.15 - dark ground). Table 1 present the light coefficients of reflections. The neighbouring objects are in distance, which does not shade the hall [1], [2].

Table 1. Reflectance factor of surfaces in the hall.

Surface	Reflectance
Wardrobe	0.3
Machine 1	0.2
Machine 2	0.5
Ceiling	0.7
Wall	0.4
Floor	0.2

The hall is orientated by longitudinal dimensions in the east-west direction. Figure 1 shows the view of the building from the outside. Figure 2 shows the view of the building from indoor and the internal view as a result of daylight simulation with the Radiance simulation program. External view of skylight is illustrated in Figure 3.

The following pictures documented the results of simulation luminances ($cd.m^{-2}$) for interior environment of the hall. For saddle skylight simulation for single wired glass was realized. The calculations of simulation from three views from working place of interior hall was realized.
Some points of luminance ($cd.m^{-2}$) and values of glare sources ($cd.m^{-2}$) are shown in Figure 4,5,6,7. The evaluation of visual comfort was realized on based of calculation glare index (Figure 8,9). The single wired glass and diffuse glass were considered.

Figure 1. Exterior view of the hall.

IOP Conf. Series: Materials Science and Engineering **415** (2018) 012039 doi:10.1088/1757-899X/415/1/012039

Figure 2. Internal radiance image of hall-view on the ceiling structure.

Figure 3. External radiance image-view on the saddle skylight .

Luminance within the visual field should be limited, for the purpose of avoiding glare and also excessive luminance ratios. For visual comfort to be achieved luminance ratios should also not exceed prescribed values.

Specific recommendations assume a ratio of 1:3 between near surfaces and the visual task in the field. A ratio of 1:2 for the distant surfaces in the visual field. The maximum permissible luminance ratio between the task and surface is 1:40 in the field of view. A deviation for a ratio of 1:50 has been observed when the luminance from the window filled a small portion of the visual field.

Figure 4. Internal radiance images of hall-luminance; potencial glare sources (cd.m^{-2}).

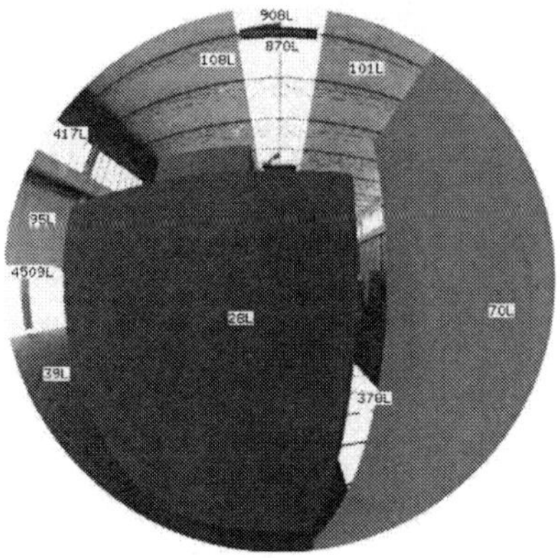

Figure 5. Internal radiance images-east view; luminance values (cd.m^{-2}).

Figure 6. Internal radiance images-north view; luminance values (cd.m^{-2}).

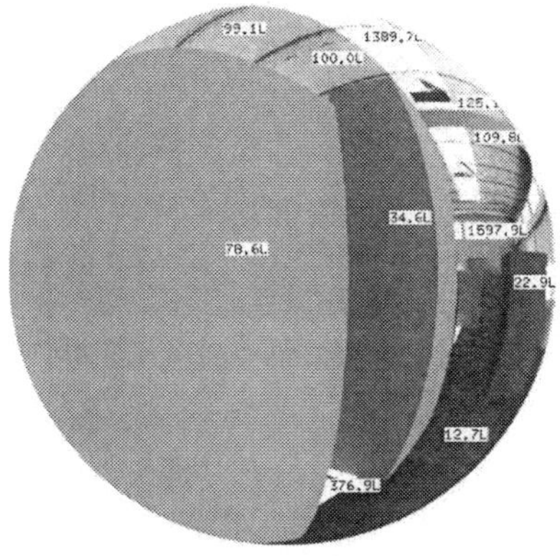

Figure 7. Internal radiance images-south-west view; luminance values (cd.m^{-2}).

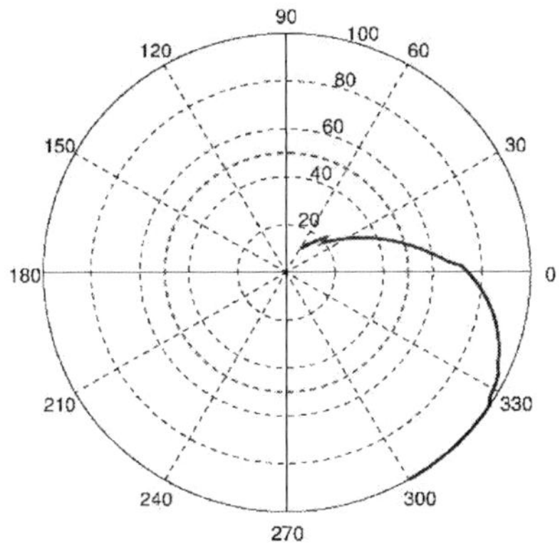

Figure 8. Evaluation of visual comfort by Guth VCP – single wired glass.

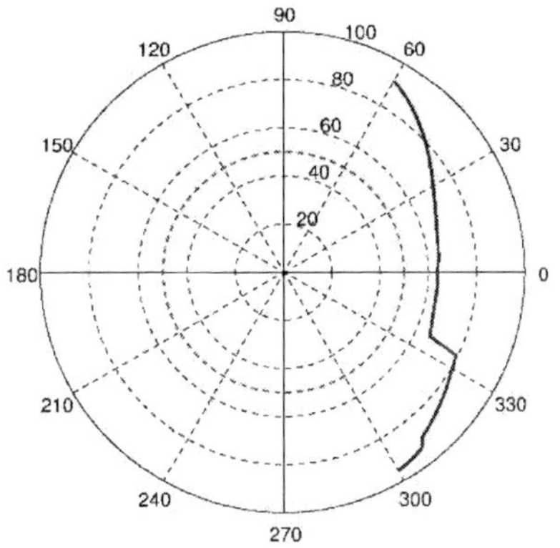

Figure 9. Evaluation of visual comfort by Guth VCP – diffuse glass.

3. Results

Guth Visual Comfort Probability (Guth VCP) is a metric used to rate lighting scenes. The Guth VCP provide an indication of the percentage of people in a given space that would find the glare from a fixture to be acceptable. A minimum VCP of 70 % is recommended. The most important factor with respect to lighting quality is glare. Glare is a sensation caused by luminances in the visual field that are too bright. Discomfort, annoyance or reduced productivity can result [5].

The visual comfort is affected for relation of illuminances 1:10 and glare for relation 1:100 is occurred [18], [19]. Potencial glare sources (cd.m^{-2}) are identified as a red circle with the intensity of the source at the centre (Figure 4).

For evaluation of visual comfort for combined daylighting system and single wired glass is showed that in case single wired glass. In case diffuse glass visual comfort is affected. In the pictures (Figure 5,6,7) are represented the luminance values (cd.m^{-2}).

The graphs (Figure 8, 9) shows percentage of people who would be satisfied. In case single wired would be satisfied with visual comfort 10-90 % of people (Figure 8). In case diffuse glass would be satisfied 60-90 % of people (Figure 9). Visual comfort is strongly influenced by view direction.

4. Conclusions

This paper provides new insight on glare evaluation. Our task was to evaluate the visual comfort in selected areas of the factory with production of sheet metal blank that is located in the Slovakia. The visual comfort is evaluated by Guth Visual Comfort Probability. Simulation calculations were made from different views of the interior in clear sky. According to standard STN EN 12464 - 1 users in interior spaces must be protected against glare. The workers must feel a connection with the exterior environment, if it is not, their productivities decrease. The contribution shows that using diffuse glazing improved visual comfort - so satisfaction was higher by 40-100% depending on the viewing angle.

5. References

[1] Dolníková E 2001 Vplyv horného osvetľovacieho systému na svetelnú pohodu v priemyselných halách, kandidátska dizertačná práca, Svf TU, Košice, 148p.

[2] Dolníková E and Katunský D 2017 Hodnotenie svetelnej pohody pri kombinovanom osvetľovacom systéme v priemyselných halách; TUKE Košice: Košice, Slovakia, 151p, ISBN 978-80-553-2809-6.

[3] Drahos R and Drahos M 2011 Zdravotné aspekty denného osvetlenia pri práci. Fyzikálne faktory prostredia: Ochrana pracovného prostredia pred hlukom a vibráciami : Hodnotenie kvality prostredia, **1,** 81–87.

[4] Hnilica R and Dado M 2008 Hodnotenie osvetlenia v prevádzke drevospracujúceho priemyslu. Rizikové faktory pracovného prostredia v drevospracujúcom priemysle: kolokvium ku grantovej úlohe VEGA 1/3445/06, 28–33

[5] Ward G. RADIANCE Tutorial [online]. c1993, last revision 1999. http://radsite.lbl.gov/radiance/refer/tutorial.html

[6] Bellia L, Pedace A and Fragliasso F 2015 The impact of the software's choice on dynamic daylight simulations' results: A comparison between daysim and 3ds max design®. Sol. Energy, **122**, 249–263.

[7] Tureková I and Rusko M 2012 Objektivizácia umelého osvetlenia v priemyselných logistických halách. Fyzikálne faktory prostredia : Mimoriadne číslo časopisu o problematike fyzikálnych faktorov prostredia, **2**, 131–142.

[8] Yao J, Chow D H C and Chi Y 2016 Impact of Manually Controlled Solar Shades on Indoor Visual Comfort. Sustainability, **8**, 727, doi:10.3390/su8080727.

[9] Lee J H, Moon J W and Kim S 2014 Analysis of occupants' visual perception to refine indoor lighting environment for office tasks. Energies, **7**, 4116–4139.

[10] Lyons L S 2013 Handbook of Industrial Lighting; Elsevier: Amsterdam, The Netherlands, 224p, ISBN 0-408-00525-4.

[11] Gonzalez J and Fiorito F 2015 Daylight design of office buildings: Optimisation of external solar shadings by using combined simulation methods. Buildings, **5**, 560–580.

[12] Costanzo V and Donn M 2017 Thermal and visual comfort assessment of natural ventilated office buildings in Europe and North America. Energy Build., **140**, 210–223.

[13] Tzempelikos A 2017 Advances on daylighting and visual comfort research. Building and Environment, **113**, 1–4.

[14] Konis K 2013 Evaluating daylighting effectiveness and occupant visual comfort in a side-lit open-plan office building in San Francisco, California. Build. Environ.,**59**, 662–677.

[15] Konis K 2014 Predicting visual comfort in side-lit open-plan core zones: results of a field study pairing high dynamic range images with subjective responses. Energy and buildings., **77**, 67–79.

[16] Nasrollahi N and Shokri E 2016 Daylight illuminance in urban environments for visual comfort and energy performance. Renew. Sustain. Energy Rev., **66**, 861–874.

[17] Reinhold K and Pint P 2009 Lighting of workplaces and health risks. Electronics and electrical engineering., **90**, 11–14.

[18] Tabadkani A, Banihasmeni S and Hosseini R 2018 Daylighting and visual comfort of oriental sun responsive skins: A parametric analysis. Building simulation., **11**, 1 –14.

[19] Standards STN 730580. Daylighting in Buildings, Part –1 Basic Requirements, 1986 Part–2; Daylighting of Residential Buildings; Slovak Republic Office of Standards, Metrology and Testing: Bratislava, Slovakia, 2000.
[20] Standards EN 12464-1:2012. Light and Lighting; Lighting of Work Places-Part. 1: Indoor Work Places; Slovak Republic Office of Standards, Metrology and Testing: Bratislava, Slovakia, 2012.As part of the production system for *Journal of Physics: Conference Series*, online versions of all reference lists will, wherever possible, be linked electronically using CrossRef.

6. Acknowledgments

This paper was elaborated with the financial support of the research project VEGA 1/0674/18 of the Scientific Grant Agency, the Ministry of Education, Science, Research, and Sport of the Slovak Republic and the Slovak Academy of Sciences.

IOP Conf. Series: Materials Science and Engineering **415** (2018) 012040 doi:10.1088/1757-899X/415/1/012040

Determination of daylight conditions in office room using digital images as a light source.

D Knera[1] and D Heim

Department of Environmental Engineering, Lodz University of Technology, Poland

[1]dominika.knera@p.lodz.pl

Abstract. The paper presents comparative analysis of daylight conditions in office room simulated with two techniques of sky and surrounding modelling. The luminance distribution in the surrounding environment was determined using CIE standard sky model and image based lighting (IBL) technique. In IBL method the sky and surrounding environment was defined by high dynamic range (HDR) image. Used HDR images were captured by camera with fisheye lens with 180° viewing angle which enables the projection of the entire hemisphere of the sky and surrounding environment. Determination of daylight conditions were performed for two days characterised by overcast sky conditions. Simulations were conducted using Radiance Lighting Simulation and Visualization software. Obtained results show differences in luminance and illuminance distribution in analysed room for both techniques of sky and surrounding modelling. Reason of the observed divergences is uneven cloud layer distribution at HDR images and influence of surrounding environment visible at HDR images. IBL technique allow more precise modelling of daylight distribution, especially in urban environment where surrounding buildings and vegetation significantly effect on the light availability.

1. Introduction

Providing the appropriate lighting distribution in building is significant to ensure comfortable environment for occupants. Visual comfort effects on occupant work capability and efficiency as well as their well-being and health. Sufficient lighting conditions are achieved by ensuring required lighting level, uniformity of illuminance values, suitable contrast conditions and avoiding the glare effect. Moreover, access to daylight is also very important in the human circadian system's regulation. Therefore, detail analysis of daylight distribution in indoor environment is very important for building assessment and design process. Daylight conditions in indoor environment can be assessed by different calculation methods: simple mathematical models, scale models or complex numerical modelling by computer tools [1]. All of them consider many parameters connected with solar radiation (direct, diffuse, reflected radiation) incident on analysed surfaces, sun position, building geographical location or properties of material and glazing elements. In detail analysis all mentioned parameters should be precisely defined.

One of the most frequently used computer simulation tool is Radiance Lighting Simulation and Visualization software – professional and complex tool kit for visualizing and simulation of lighting in virtual environments [2]. Radiance enables the lighting assessment in specific point at a time but also annual analysis. Moreover, it has many possibilities for modelling of sky: using generic CIE, Perez weather or image based sky models [3]. CIE sky models represent the sky brightness based on location, date, time, and sky type ranging from cloudless skies to overcast skies and they are commonly used in simulation analysis. However, detailed experimental analysis shown that CIE sky models insufficiently map the real sky conditions and the more advanced sky models should be considered for detail simulations [4, 5]. One of the methods which allow modelling the real parameters of sky and surrounding for analyses location and time step is Image Based Lighting (IBL) technique [6, 7]. This method allows illuminating the analyzed scene with light from image of the real world. In IBL were used High Dynamic Range (HDR) images which represent real-world luminance through capturing of scene images with extended dynamic range [8, 9].

The main aim of this paper is analysis of the daylight conditions in indoor space using two techniques of sky and surrounding modelling: CIE and IBL methods. Simulations were performed by Radiance computational tools for overcast sky conditions. Obtained results allow evaluating the daylight distribution in office room with comparison of the impact of sky and surrounding models.

2. Case study

HDR images used in analysis were performed according to the HDR image technique [9] by the digital camera Nikon D80 equipped with wide-angle lens Sigma Fish-eye. Lens with 180° viewing angle enables the projection of the whole hemisphere including sky and surrounding environment noticeable by the window at a single digital image. In every time step were made set of five digital images with the same aperture and different shutter speeds. Each image was made in the same position capturing entire hemisphere of the sky and surrounding environment visible by the window of analysed room. Next, obtained digital images were merged to high dynamic range image using the computer program Photosphere with consideration of the previously determined camera response curve and vignetting correction. In table 1 were presented performed HDR images used in daylight analysis.

Table 1. HDR images used in analysis.

	9:00	12:00	15:00
III			
IX			

Daylighting analysis was performed for modeled office cell representing the existing experimental office room. Dimensions of indoor space are respectively width, length and height: 2.6 m, 4.4 m and 2.6 m. Window is situated in a center of the exterior wall at the height of 0.7 m and dimensions of the glazed unit are 1.1 m of width and 1.1 m of height. Exterior wall is oriented on the west. The RGB values of the interior surfaces of walls, floor and ceiling were determined based on the measurements using spectrophotometer Konica Minolta CM-2500d. Furthermore, analyzed office rooms were located at the level of 4th floor which corresponds with real location of the experimental room.

Analysis was performed for the location of the office room in the Central Europe (51°46'N 19°27'E) under the temperate climatic conditions. Simulations of the daylight were made for two selected days close to the equinox day in March and September characterised by overcast sky conditions (hereinafter named III and IX respectively). Furthermore, three selected hours where considered: 9:00, 12:00 and 15:00. For each time step were defined two models of the luminance distribution in the sky and surrounding environment using: CIE standard overcast sky (hereinafter named CIE) and image based lighting technique with previously prepared HDR images (hereinafter named IBL). Simulation of the daylight distribution in indoor space using Radiance tool were performed for each mentioned time step.

3. Results

3.1. Luminance

Results of the daylight simulation obtained from simulations made using Radiance tool were divided into two parts. In the first part, daylight luminance is presented in the form of false color images of the analyzed office room for each time step and luminance of the sky and surrounding model (table 2). Images present view of the room from the position of internal wall situated opposite to the window represented by square surface in the middle of the image. It can be noticed that luminance in room is rather low, only for IBL model image in IX at 12:00 luminance exceed 275 cd/m^2. Luminance distribution for all cases with CIE model is similar regardless the hour and day because position of the sun and hence the radiation intensity of overcast sky for both days are similar. On the other hand, luminance distribution for cases with IBL method differs slightly. In III (9:00 and 15:00) and IX (9:00) luminance distribution for cases with IBL technique is very similar to cases with CIE model, only little higher on the floor surface. Remaining cases with IBL method are characterized by significantly higher luminance. The reason of the noticed differenced is uneven cloud layer at HDR images defining light sources. Sky captured at HDR images in III (12:00) and IX (12:00 and 15:00) are not uniformly overcast, the cloud layer is lower, therefore solar radiation intensity is higher than for CIE standard overcast sky model.

3.2. Illuminance

Second part of the analysis concern on illumiance in indoor environment determined for analyzed cases. Results obtained for illuminance are presented in the form of the images with contour lines shown in table 2. It can be noticed that similarly to the luminance analysis, illuminance in III (9:00 and 15:00) and IX (9:00) for both sky and surrounding models is similar when in remained cases illuminance for IBL method is significantly higher. Described divergence is also caused by the differences between cloud layer in the IBL and CIE models.

However, in illuminance maps for III (9:00 and 15:00) and IX (9:00) can be observed additional small differences between cases with CIE and IBL models. Illuminance in indoor space for IBL technique is lower and not symmetric - higher values are at the left side of the room. Noticed differences are effects of surrounding urban environment, visible through the window, which influence on the uneven illuminance distribution in the room.

IOP Conf. Series: Materials Science and Engineering **415** (2018) 012040 doi:10.1088/1757-899X/415/1/012040

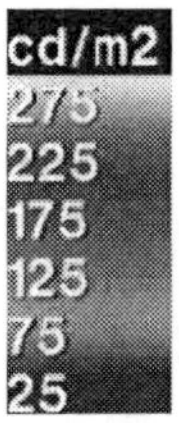

Table 2. False color images of luminance distribution in the analysed office room.

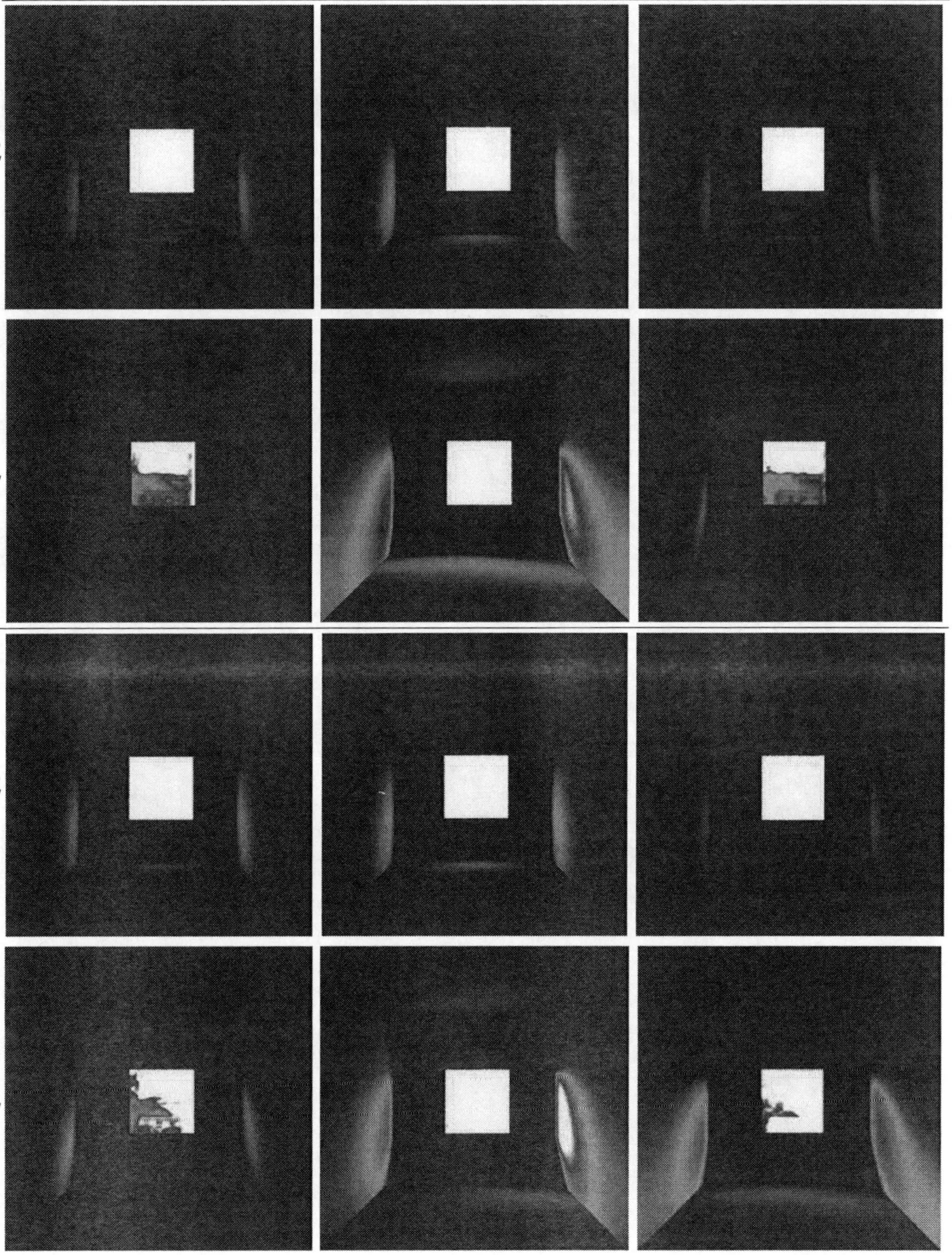

IOP Conf. Series: Materials Science and Engineering **415** (2018) 012040 doi:10.1088/1757-899X/415/1/012040

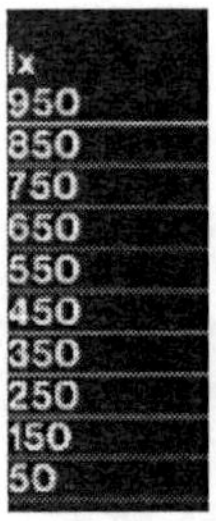

Table 3. Contour line images of illuminance distribution in analysed office room.

		9:00	12:00	15:00
III	CIE			
	IBL			
IX	CIE			
	IBL		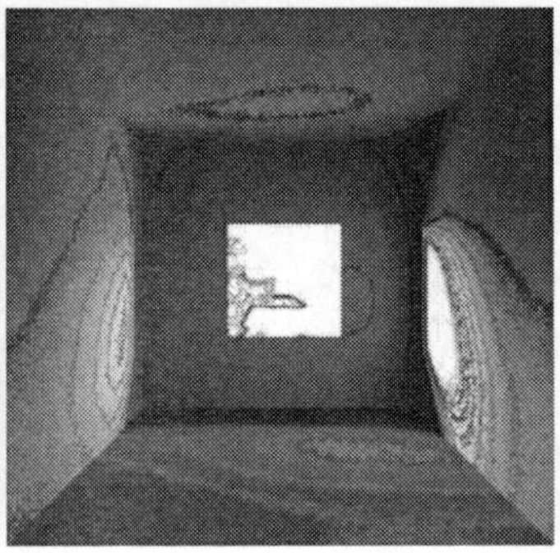	

4. Conclusions

In presented paper the numerical analysis of the daylight in office room under overcast sky conditions is performed. Two techniques of sky and surrounding modelling were considered. The luminance distribution in the sky and surrounding environment was determined using CIE standard sky model and image based lighting (IBL) technique. Results obtained for both modelling techniques show differences in luminance and illuminance distribution in indoor space. Noticed variances are mainly results of uneven cloud layer at HDR images comparison to uniformly overcast CIE sky model. Moreover, of radiation reflected from surrounding environment captured at HDR images can also influence on the luminance and illuminance distribution in analysed indoor space. CIE standard sky models are commonly used for general analysis of light distribution in indoor environment. However, for analysis requiring detail modelling in the real location, especially urban environment, more comprehensive model of sky and surrounding should be introduced, like image based lighting models.

References

[1] Wong I L 2017 A review of daylighting design and implementation in buildings *Renewable Sustainable Energy Rev.* **74** 959–68

[2] Ward G Shakespeare R 1997 *Rendering with Radiance* (San Francisco: Morgan Kaufmann Publishers)

[3] Inanici M Hashemloo A 2017 An investigation of the daylighting simulation techniques and sky modeling practices for occupant centric evaluations *Build Environ* **113** 220-31

[4] Heim D Szczepanska E 2011 Daylight Distribution in a Building Space: a Comparison of Real Conditions and Theoretical Sky Models *Proc. 12th Conf. of Int. Building Performance Simulation Association* Sydney 14-16 November pp 14–6

[5] Szczepanska-Rosiak E Heim D Gorko M 2013 Visual comfort under real and theoretical, overcast and clear sky conditions *Proc. of 13th Conf. of Int. Building Performance Simulation Association* pp 2765–73

[6] Debevec P 2002 Image-based lighting *IEEE Comput. Graph. Appl.* **22 (2)** 26-34

[7] Inanici M 2010 Evaluation of high dynamic range image-based sky models in lighting simulation *Luekos J. Illum. Eng. Soc.* **7 (2)** 69-84

[8] Jacobs A 2007 High dynamic range imaging and its application in building research *Adv. Build. Energy Res.* **vol 1** 177-202

[9] Reinhard E Heidrich W Debevec P Pattanaik S Ward G Myszkowski K 2010 *High Dynamic Range Imaging Acquisition, Display, and Image-Based Lighting* ed 2nd (Burlington: Morgan Kaufmann Publishers)

Determination of total solar and visual radiation transmitted through triple glazing component with PCM layer

D Heim[1], D Knera[1], M Krempski-Smejda[1] and A Wieprzkowicz[1]

[1]Department of Environmental Engineering, Lodz University of Technology Wolczanska 213, 90-924 Lodz, Poland

E-mail: dariusz.heim@p.lodz.pl

Abstract. Modification of a glazing unit by application of PCM layer can lead to the increase of the heat capacity of the component and effective control of solar heat gains in buildings. The purpose of this study was to investigate the effect of the PCM layer on total solar and visual transmittance in solid, mushy (transition) and liquid state of PCM. The experiment was conducted in laboratory scale using triple glazed window unit sample and artificial sun as a source of solar radiation. The transmittance (total and visual) was calculated based on the results from pyranometers and illuminance meters taking into account the intensity of solar radiation at the level of 1000 W/m^2. Measurements revealed that the component is almost blind in the solid state why in the liquid the transmittance is lower than for empty - traditional triple glazing. The transmitted radiation (total and visible) depends on the location of PCM layer (internal or external window chamber). The registered total solar radiation obtained in solid state was even 0 W/m^2 (when material was on the inner side), while the maximum values for liquid was 142 W/m^2 (material is located on the outer side). The light transmittance for more transparent case was around 1.6 klx and ten times higher in liquid than solid state.

1. Introduction

The application of PCM in transparent constructions became common research challenge during last five years since the following monography became published by Heim [1]. In 2012 Goia et al. [2] determined the optical characteristics of advanced glazing system with PCM. The glazing was developed by the incorporation of a paraffin-based Phase Change Material (PCM) into a transparent component. The thickness of paraffin was e.g. 15 mm. The spectral transmissivity, reflectivity and absorptivity of the PCM glazing system were measured between 400 and 2000 nm. Based on the obtained results for the whole component the integral values of total solar transmittance were 0.75 for liquid and 0.46 for solid state respectively. These values, especially for solid phase seem to be very high in comparison with other research studies done by Gowreesunker et al. [3], where measured transmittance in solid phase was 0.20. The case studied in [3] was a double glazed unit which consisted of 4 mm glass pane and 16 mm of PCM. Moreover, the experiments done by Bianco et al. [4] for 10 mm thick, polycarbonate crystal panels gave the results of solar transmittance 0.32. Additionally, in paper [5] authors provided the transmission coefficient of glazed unit (4/15/4 mm) at the level of 0.03 in solid state. It was underlined that the transmission and reflection coefficients measured by the commercial spectrophotometer are low in comparison with basic observations of the component in practice and from previous experimental analyses [2]. With this set-up, the absorptance appears to be the dominant feature in both the visible and infrared (IR) range. This effect takes place because the high scattering phenomena cannot be fully and efficiently tackled by means of conventional spectrophotometer and more advanced testing is necessary to fully characterise the behaviour of the PCM layer in the solid state.

Taking into account the previously published works [6] & [7] and analysis of results presented above it seems to be justified to analyse optical properties of PCM windows more realistically. The main goal of the presented study was to determine experimentally the optical properties of PCM-triple-glazing windows in solid, liquid and intermittent state. It was assumed that material can be placed in the inner and outer cavity of glazing unit. The effect of phase change was induced by constant heat radiation flux (at the level of 1000 W/m^2) emitted by halogen light source. The total and visual radiation in different configurations was measured by professional pyranometers (Hukseflux – LP02) and illuminance meters (Konica Minolta – T-10MA). The specific experimental set-up was developed for the purpose of experiments to provide non disturbance values for the whole glazing unit. The measurement results were compared to the standard – empty glazing unit and finally the transmissivity was determined.

2. Materials and methods

2.1. Experimental set-up

The experimental set up consists of the three separate elements combined together: artificial sun, sample holder equipped with triple glazing unit filled PCM and data acquisition system. As a light and radiation source an artificial sun Gunt HL 313.01 was used. This light source provides radiation flux similar to natural sunlight and enables experiments under stable, repeatable irradiation conditions. The artificial sun comprised of eight halogen lamps each with a total power of 8000 W, arranged in two vertical rows. Dimensions of this device were L×W×H: 1340×810×2100mm. The intensity of radiation was adjusted by the distance between lamps and measurement devices – the pyranometers or illuminance meters.

The main element of the installation was the sample holder in which the triple glazing unit and sensors were located. Dimensions of glazing were L×W×H: 150×45×150 mm. One cavity was constantly filled with argon, while the second one for the first series of measurements was treated as a reference one and filled with air. For the purpose of the further, primary measurements the second cavity was fully filled up with Rubitherm RT 21HC.

The distance between the glass and the sensors has been calculated and adjusted to eliminate reflections to sensors from the aluminum distance piece. The diameter of the outer hole was 120 mm while the internal was 51.5mm. Schematically it was shown in Fig. 1. Experiments were carried out with the simultaneous use of two calibrated sensors where one was the reference (without covering from outside).

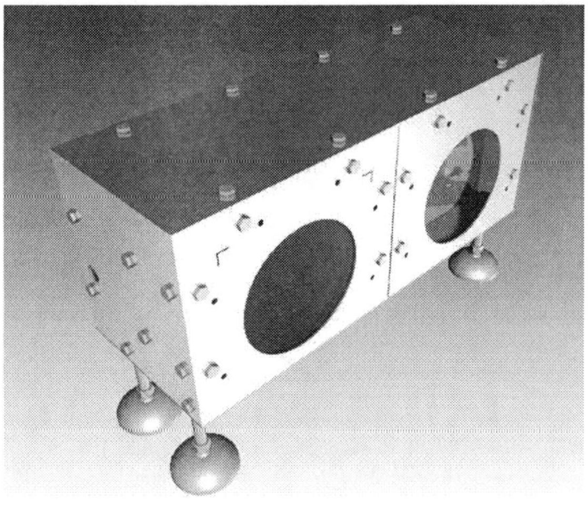

Figure 1. Sample holder with sensors.

2.2. Control and data acquisition system
To measure luminous flux per unit area Konica Minolta T-10MA Illuminance Meter with two mini receptor heads was used. The data was transmitted to a computer via the RS232 interface. The measurement was performed with a 30-second interval. The second class (specifications of the ISO 9060) solar radiation sensors Hukseflux LP02 were used. This sensors measure the solar radiation received by a plane surface from a 180° field of view angle therefore the interior of the sample holder was covered with a black light absorber. The measurements were carried out continuously using National Instruments cDAQ-9181 with the 9219 module. The data was collected using program written in the LabView.

2.3. Experimental procedure
The experiment was divided on calibration part and real measurements. Calibration was used to unify output data from sensors as well as to stabilize the radiation source. It was assumed that experiments will be conducted under the intensity of irradiation on the glazing surface of 1000 W/m^2. The spectral radiation of the artificial sun slightly differs from the real sun and is the same as for halogen lamps. It means that the maximum relative irradiance occurs for 750 nm. Moreover:

- the radiation in ultraviolet range is lower,
- the radiation in near infrared range is higher.

in comparison with spectral characteristic of solar radiation.

The samples were prepared by conditioning the whole glazing with PCM in 8°C during 24 hours before the experiment started. It means that material fully solidified and was cooled down 5 degree below the point when phase change begins. The air temperature in the experimental zone was 20°C. Any other light or heat radiation sources were turned off. The halogen lamps were preheated for 5 minutes before each set of measurement.

The experiment was done for standard value of irradiance 1000 W/m^2 in two window configurations. The first one, fig. 2a, when cavity with PCM was on the outer side of the window (was exposed directly to radiation). The second one, fig. 2b, when PCM layer was located on the inner side. All dimensions are given in millimetres. Each experimental procedure lasted for more than 2 hours from the beginning to fully melting of PCM. During experiments the irradiance and illuminance were registered every 30 seconds. Additionally, the image of the material was registered in 60 seconds interval using digital camera.

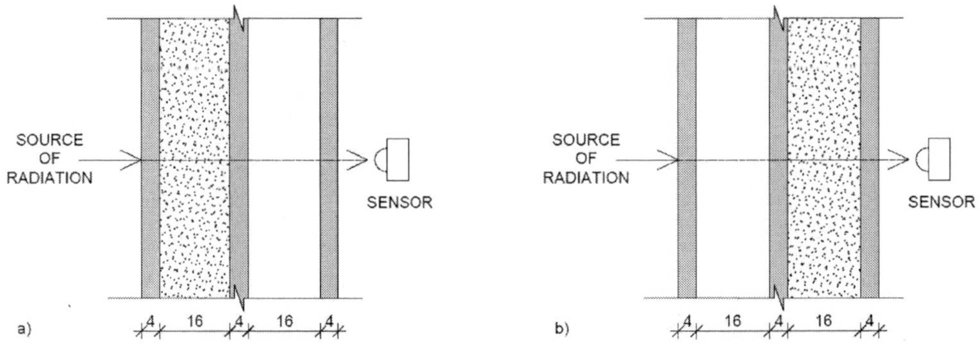

Figure 2. Two configurations of PCM layer towards solar radiation, a - case A (from the front side of radiation), b - case B (from the back side of radiation).

3. Results and analysis

3.1. Total solar energy transmittance
As it was stated in the introduction, the PCM layer in transparent construction affects the overall solar energy performance dependently on the physical state. Using traditional techniques and laboratory equipment it is very difficult or almost impossible to analyse the optical properties as a function of

temperature of material. Therefore, the main goal of this research was to use a standard pyranometer to monitor transmitted solar radiation during change of phase under standardized and stable conditions.

The results of measured solar radiation were presented in figure 3 (case A) and 4 (case B), respectively. The analysis of presented results shows that position of PCM layer determine the amount of radiation transmitted through triple glazing unit. When material is located on the outer side (the first cavity from radiation source), the radiation is higher in both, solid and liquid states. It can be explained by the transmittance processes in complex glazing structures where PCM plays a role of additional, reflective surface. When PCM is located behind the two sheets of glass the inter-reflection in the unit enhanced the effect of reemission. The irradiation values measured for solid state were 5 W/m^2 in case A and 0 W/m^2 in case B. For liquid state the irradiation were 142 and 118 W/m^2 respectively. Zero value registered for a solid PCM can be also a result of sensor sensitivity. Nevertheless, the melting process was also shifted in time due to the lower radiation incident on the PCM external surface and transferred to heat. Registered delay was around 25 minutes until the material achieved fully mushy state.

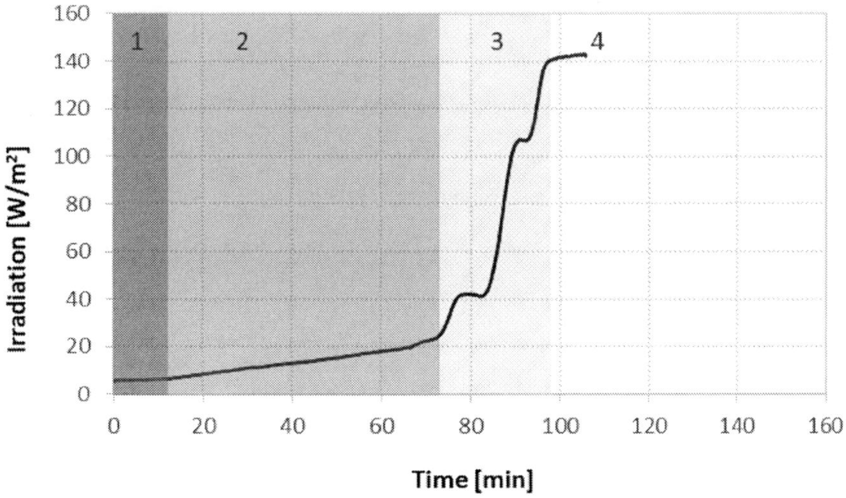

Figure 3. Radiation transmitted through PCM-triple glazing vs. time during change of phase – case A (PCM located in the outer cavity of the window).

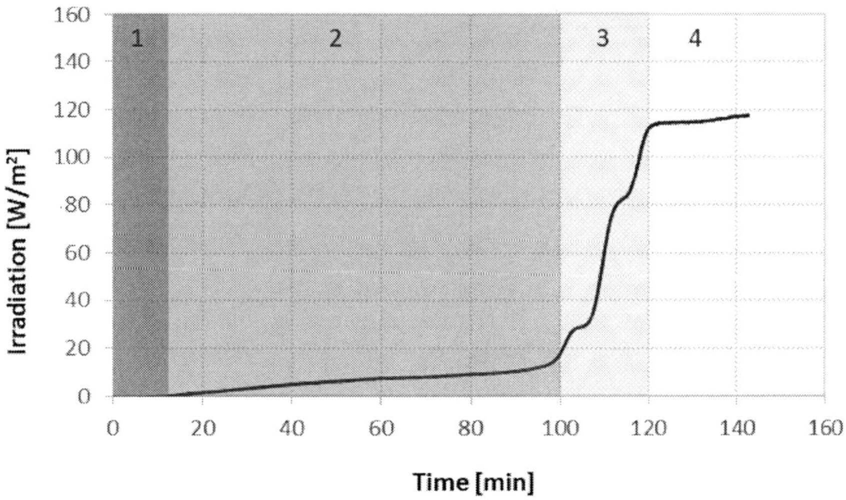

Figure 4. Radiation transmitted through PCM-triple glazing vs. time during change of phase – case B (PCM located in the inner cavity of the window).

Analysis of the presented results allowed to specify the four characteristic periods when material was in a specific state. These periods were highlighted in figures 3 and 4 and indicated from 1 to 4. In period 1 the material was fully solid and its transmissivity was constant. In period 2 material melted and became more translucent but still homogenous. Its transparency was slowly rising. In period 3 the material was partly liquid (from upper side) and partly mushy (from the bottom). The mushy part was moving down letting more radiation to be transmitted through the glazing unit. This period ends up relatively quickly. The characteristic plateau, which is visible two times in that period can be explained as an artefact caused by uncovering of the sensor by the mushy slice of PCM moving down. In period 4 the material was fully melted and transmissivity became constant again. This effect was confirmed by the sets of images registered by digital camera and presented in 3.3.

3.2. Light transmittance

The total irradiation presented in figures 3 and 4 corresponds to light transmittance measured by illuminance meter. The experimental procedure presented in 2.3 was repeated for the second sets of the sensors. The illuminance was measured for the same sample and under the same conditions as irradiance was. The results for case A and B were presented in figure 5. The characteristic of the curves are similar. The effect of PCM location reduced the light transmittance in similar ratio as total radiation transmittance. The differences between solid and liquid phase were around 10 times, for both cases. The maximum illuminance for case A was above 16 klx when for case B was slightly below 14 klx. The luminance efficacy of the light transmitted through liquid PCM was around 115 lm/W.

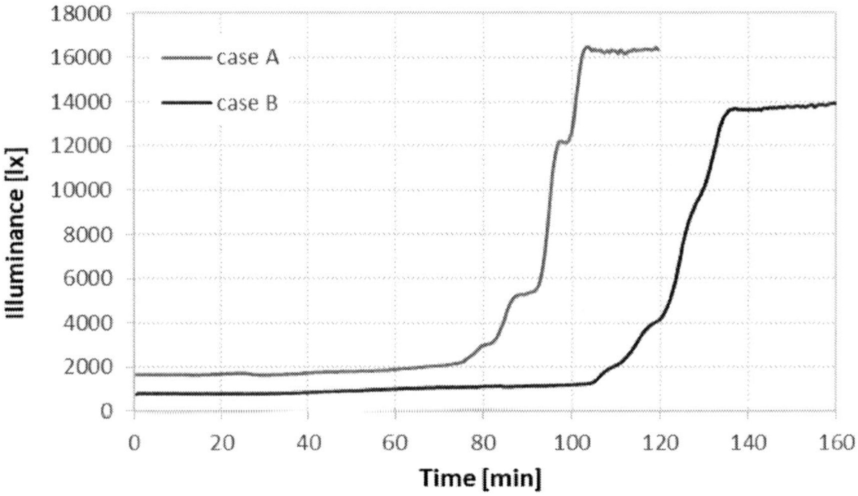

Figure 5. Light transmitted through PCM-triple glazing vs. time during change of phase – case A (PCM located in the outer cavity of the window) and B (PCM located in the inner cavity of the window).

3.3. Overall aspects of PCM glazing in different physical states

The results presented in table 1 were registered during experiment and divided into four characteristic periods. The photos show the colour changes as well. When material was in a liquid phase the material was fully transparent and colouristically neutral. In intermittent phase material became rather translucent than transparent, while when it was fully solid it became less transparent and light grey-white. The darker colour means more melted state and higher transmissivity of the PCM layer. The interesting remark is that material was melted non homogeneously during whole period. Completely liquid phase appeared on the top and finally the boundary front between melted and mushy parts moved down of the glazing.

IOP Conf. Series: Materials Science and Engineering **415** (2018) 012041 doi:10.1088/1757-899X/415/1/012041

Table 1. The view of glazing surface with PCM layer from outside in selected periods of transition.

Periods			
1	2	3	4

4. Conclusions

The experimental results of radiation transmission through triple glazing PCM structure were presented in the paper. The authors developed the original experimental set up and used artificial sun made of the halogen lamps to conduct the experiment under constant irradiation.

The results showed that total and visual part of solar energy transmitted through glazing is very low when PCM is in a solid or even mushy state. The thickness of the PCM layer (16 mm) caused that the glazing unit become almost fully blind, especially when PCM is located on the back side from the radiation source. It can cause the serious problem in daylighting the building interior during that time. When material was fully melted the transmitted radiation was much higher. In case of illuminance these values can be even around ten times higher.

Acknowledgement
This work was funded in a framework of ERANet-LAC 2nd Joint Call on Research and Innovation by NCBiR as part of the project entitled: Solar hybrid translucent component for thermal energy storage in buildings (acronym: SOLTREN).

References

[1] Heim D 2011 *Modification of thermo-optical properties of transparent elements of building envelope* (Warsaw: Polish Academy of Sciences)
[2] Goia F, Zinzi M, Carnielo E and Serra V 2012 Characterization of the optical properties of a PCM glazing system *Energy Procedia* vol 30(Elsevier)pp 428–37
[3] Gowreesunker B L, Stankovic S B, Tassou S A and Kyriacou P A 2013 Experimental and numerical investigations of the optical and thermal aspects of a PCM-glazed unit *Energy Build.* **61** 239–49
[4] Bianco L, Komerska A, Cascone Y, Serra V, Zinzi M, Carnielo E and Ksionek D 2018 Thermal and optical characterisation of dynamic shading systems with PCMs through laboratory experimental measurements *Energy Build.* **163** 92–110
[5] Goia F, Zinzi M, Carnielo E and Serra V 2015 Spectral and angular solar properties of a PCM-filled double glazing unit *Energy Build.* **87** 302–12
[6] Fokaides P A, Kylili A and Kalogirou S A 2015 Phase change materials (PCMs) integrated into transparent building elements: a review *Mater. Renew. Sustain. Energy* **4** 6
[7] Silva T, Vicente R and Rodrigues F 2016 Literature review on the use of phase change materials in glazing and shading solutions *Renew. Sustain. Energy Rev.* **53** 515–35

IOP Conf. Series: Materials Science and Engineering **415** (2018) 012042 doi:10.1088/1757-899X/415/1/012042

National and European requirements concerning acoustic insulation from air sounds for internal walls

H. Nowak[1] and T. Kania[1]

[1] Faculty of Civil Engineering, Wroclaw University of Science and Technology, Wybrzeze Wyspianskiego 27, 50-370 Wroclaw, Poland

Abstract. Neighbourhood noises in newly erected residential buildings are problems which are more and more frequently raised by their inhabitants. Pursuant to the binding European norms, the way of testing acoustic insulation properties has been standardized, however both required values and indices used to express them remain at the discretion of individual countries of the European Community. In the article, single-number acoustic insulation indices for air sounds used in Europe are presented and the methods of converting them are described. Current national requirements concerning acoustic insulation properties for interior walls within the apartment and between apartments in multi-family buildings as well as their comparison with requirements applicable in selected countries of the Western, Central and Southern Europe are presented. Based on the provided data, the evaluation of national requirements regarding acoustic insulation properties for interior walls was carried out. It was concluded that Polish acoustic requirements for interior walls are at a low level not only compared to the countries of the Western Europe, but also Central and Southern Europe.

1. Introduction

Noise is one of the most essential factors negatively affecting health and quality of life of the contemporary man. Both European [1] and national [2, 3] regulations mention protection from noise as one of seven fundamental mandatory requirements which should be fulfilled by building structures. The European regulations do not specify the level of the requirements but their scope, parameters used to formulate them, principles of checking them, whether the basic requirement is fulfilled as well as technical specifications and guidelines for European technical approvals. Buildings should be protected from the following noise [4]:

- external air noise;
- air noise penetrating between interior spaces;
- impact noise;
- noise generated by technical equipment;
- noise caused by excessive reverberation.

Requirements regarding acoustic insulation properties in the residential building industry exist in more than 30 European countries. In some of them, they have been binding since the nineteen fifties [5], in Poland the first standard concerning soundproof protection of buildings was published in 1954 [6].

Current methods regarding testing of acoustic insulation properties are included in the European standards. However, even though the methods of measuring acoustic insulation properties have been standardized within the European Community, requirements concerning single number indices and their minimal values are the responsibility of individual member states. The international standard valid in Poland and other European countries ISO 717-1 allows application of various single number indices for acoustic insulation properties and spectral adaptive indices for various frequency ranges [7].

2. Single number acoustic insulation indices for air sounds

Requirements concerning acoustic insulation from air sounds refer to standard single number indices. The regulations binding within this scope in Poland are expressed as minimum single number acoustic insulation indices [8]. Requirements for interior partitions within the apartment are specified using the R_{A1R} index. In order to evaluate the acoustic insulation properties of partitions between apartments, the R'_{A1} and $D_{nT,A1}$ indices are used.

Apparent evaluation rate of sound reduction R'_{A1} (taking into account flanking transmission of sound) is described using the formula (1):

$$R'_{A1} = R'_{w} + C\ [dB]. \tag{1}$$

In the presented formula, R'_w is a weighted apparent sound reduction index determined by comparing the reference calibration curve to the characteristics of approximate specific acoustic insulation properties R' within the considered frequency range. C is A-weighted pink noise spectrum adaptation term which takes into account the spectrum of sounds having an effect on the partition, characteristic for noise occurring inside residential buildings. This index can have negative values or values equal to zero.

The index indicating the evaluation of reference difference of levels $D_{nT,A1}$ is used when neighbouring spaces are shifted versus each other in such a way that the common part of the partition constitutes only a fragment of the considered wall or if the surface area of the common part of the partition is lower than 10 m^2. It is determined using the formula (2):

$$D_{nT,A1} = D_{nT,w} + C\ [dB], \tag{2}$$

where $D_{nT,w}$ is a weighted index of the reference difference of levels obtained by comparing the reference curve to the characteristics of the reference difference of acoustic pressure levels D_{nT} within the considered frequency range. The reference difference of acoustic pressure levels is acoustic insulation between spaces converted to the reference conditions determined by reverberation time T_0 = 0.5 s (specific for furnished spaces).

In European Union countries, various parameters regarding acoustic insulation from air sounds are used. Figure 1 presents types of indices according to EN ISO 717-1 and the number of European countries in which they are used to formulate requirements concerning acoustic insulation from air sounds [9].

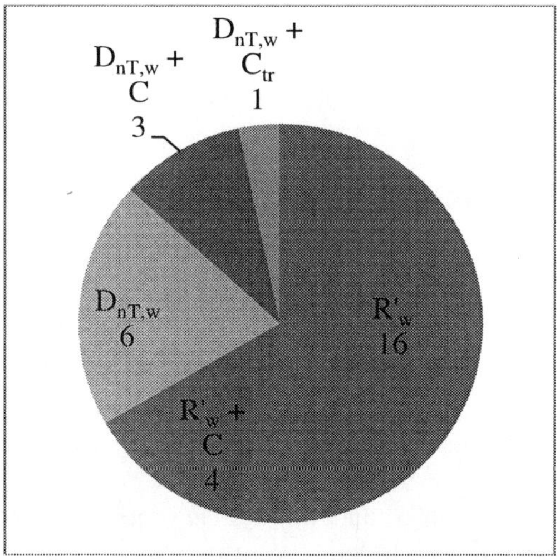

Figure 1. Breakdown regarding the use of single number indices of acoustic insulation from air sounds for 30 selected European countries

IOP Conf. Series: Materials Science and Engineering **415** (2018) 012042 doi:10.1088/1757-899X/415/1/012042

Not only single number indices of acoustic insulation but also requirements concerning their minimum values show considerable differentiation between individual European countries.

3. Proposals for standardizing European indices of acoustic insulation from air sounds

As a part of work aimed at integration and harmonization of regulations regarding acoustic insulation in Europe, it was proposed to standardize the indices and requirements concerning acoustic insulation from air sounds. According to literature available in this field [10], the index for evaluation of the reference difference of levels described by the following formula is regarded as the preferred one (3):

$$D_{nT,50} = D_{nT,w} + C_{50\text{-}3150} \text{ [dB]}. \tag{3}$$

The advantage of accepting the reference difference of acoustic pressure levels $D_{n,T}$ as the basic measurement parameter is, among other things, the considerable availability of data and experience in using it in many European countries, and the lack of any necessity to measure the volume of spaces and surface area of tested elements.

The lower threshold of measurement frequencies extended versus the existing one in 1/3 octave bands up to 50 Hz (currently it is equal to 100 Hz) was accepted as essential mainly for the acoustic characteristics of light building structures. The upper threshold of measurement frequencies in 1/3 octave bands equal to 3150 Hz was regarded as sufficient for the characteristics of acoustic properties of building partitions [10].

4. Dependence between indices used in Europe and requirements concerning acoustic insulation from air sounds

According to [11], in order to convert the specific acoustic insulating properties R' to the reference difference of acoustic pressure $D_{nT,}$ the dependence described by the formula (4) can be used:

$$D_{nT} = R' + 10lgV \cdot \frac{0.16}{T_0 \cdot S} \tag{4}$$

where: V is the volume of the receiving space [m^3], T_0 is the reverberation time [s], and S is the surface area of the partition between the transmitting and receiving space [m^2].

In the case of conversion between weighted indices of acoustic insulation from air sounds, their differentiated frequency ranges and calculation procedures should be taken into account. The measurement frequency range commonly used so far for 1/3 octave bands is 100 – 3100 Hz.

Direct comparison of requirements and single number indices used in various countries of the European Union is very difficult. This results from the fact that there are no simple mathematical dependencies between them. Such estimation can be made only based on existing experience and obtained statistical dependencies. Results of these types of dependencies vary for different types of partitions, for example light framing structures and massive brick walls. The attempt to convert values of various single number sound reduction indices to $D_{nT,50}$ values was undertaken by Dunbavin and Gerretsen [11] using the formulas (5-8):

$$R'_w: \; D_{nT,50} = R'_w - 1.0 + (5.2 - 0.12 \cdot R'_w), \tag{5}$$

$$D_{nT,w}: \; D_{nT,50} = D_{nT,w} + 5.2 - 0{,}12(D_{nT,w} + 1.0), \tag{6}$$

$$D_{nT,A1}: \; D_{nT,50} = D_{nT,A1} + 6.2 - 0.12(D_{nT,A1} + 2.0), \tag{7}$$

$$D_{nT,A2}: \; D_{nT,50} = D_{nT,A2} + 10.2 - 0.12(D_{nT,A2} + 6.0). \tag{8}$$

The presented dependencies were used to compare the requirements of acoustic insulation for walls inside residential buildings in Poland and in selected European countries. The calculation results are shown in table 1.

It can be concluded from the presented data that Polish requirements concerning acoustic insulation from air sounds for internal walls are relatively low. This concerns especially vertical partitions between apartments. Among countries for which requirements were quoted in table 1, lower values were found only in the case of the Romanian standard (calculated value of index $D_{nT,50} \geq 49$ dB) and lowest category of dwellings in Belarus standard (calculated value of index $D_{nT,50} \geq 48$ dB). Higher requirements were recorded in all other analysed countries.

Table 1. Requirements concerning acoustic insulation of walls inside residential buildings in Poland and in selected European countries.

Country		Required acoustic insulation [dB]			
		Walls between rooms in dwelling		Walls between dwellings	
		Sound Reduction Index	$D_{nT,50}$	Sound Reduction Index	$D_{nT,50}$
POLAND [8]		$R_{AIR} \geq 35\text{-}38$	32-37	$R'_{A1} \geq 50$	50
LITHUANIA – (in dependence on the category of building) [12]	category A	$R`_w \geq 40$	39	$R`_w \geq 63$	60
	category B	$R`_w \geq 35$	35	$R`_w \geq 58$	55
	category C*	$R`_w \geq 30$	31	$R`_w \geq 55$	53
	category D	$R`_w \geq 25$	26	$R`_w \geq 52$	50
	category E	$R`_w \geq 20$	22	$R`_w \geq 48$	46
LATVIA [13]		$R`_w \geq 45$	44	$R`_w \geq 54$	52
ESTONIA [14]		$R`_w \geq 43$	42	$R`_w \geq 55$	53
BELARUS [15]	category A	$R`_w \geq 45$	44	$R`_w \geq 54$	52
	category B	$R`_w \geq 43$	42	$R`_w \geq 52$	50
	category W	$R`_w \geq 43$	42	$R`_w \geq 50$	48
CZECH REPUBLIC, SLOVAKIA [16, 17]		$R`_w \geq 42$	41	$R`_w \geq 53$	51
HUNGARY [18]		$R'_{A1} \geq 37$	39	$R'_{A1} \geq 51$	51
ROMANIA [19]		$\Gamma'_a \geq 32$	32	$\Gamma'_a \geq 51$	49
GERMANY [20]		-	-	$R`_w \geq 53$	51

* Lowest acoustic class for the new buildings

Tests regarding evaluation of insulation acoustic properties for walls in residential apartments conducted in Poland in the seventies and the eighties of the previous century showed dissatisfaction in approximately 20 % of their inhabitants which also indicates the need to increase requirements regarding acoustic insulation of internal walls in residential buildings [4, 21-23]. Furthermore, it should be taken into consideration that walls erected in the seventies of the previous century were characterized by higher density and surface mass compared to many currently common light brick structures. Tests carried out in other countries of the European Community, often with higher acoustic requirements, also indicate that inhabitants of multi-apartment buildings experience the negative effects of noise caused by co-inhabitants. Based on the conducted surveys and statistical research, it is estimated that more than 50 million Europeans are exposed to neighbourhood noise which negatively affects their quality of life [5].

In spite of relatively low national requirements concerning the acoustic insulation properties of partitions between apartments, it should be stated that their fulfilment is not an obvious thing and requires use of materials having high acoustic parameters both with respect to horizontal and vertical partitions and appropriate selection of the method of bonding. In design calculations regarding acoustic insulation of walls between apartments, it should be borne in mind that it is impossible to derive a direct dependence between values of acoustic insulation obtained in laboratory tests and their values in actual structures.

5. Calculation of acoustic insulation using the simplified method

In order to obtain an appropriate forecast regarding acoustic insulation of walls between apartments from air sounds, the effect of transmitting sound via the direct route (through the calculated partition) and all side routes should be determined. The effect of flanking transmission of sound on acoustic insulation properties of partitions in the building is determined using the resultant of acoustic insulation for flanking transmission of sound via material routes. The scheme of air sound transmission routes is illustrated in figure 2.

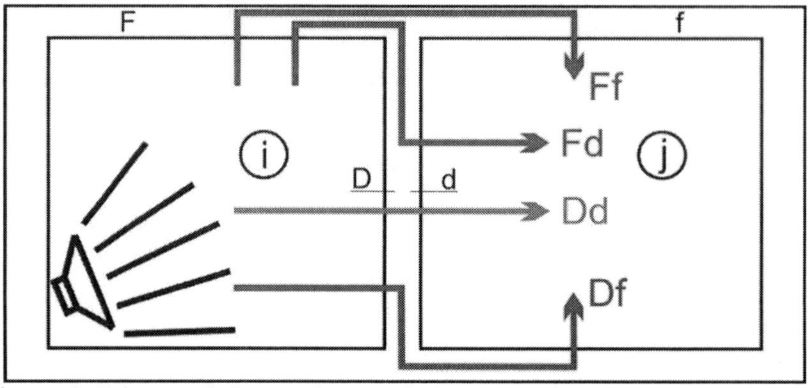

Figure 2. Scheme of air sound transmission routes between spaces [22, 24].

According to the Sound Reduction Indices that are valid in Poland, the calculations regarding acoustic insulation from air sounds using the simplified method given in EN 12354-1 [24] are based on apparent sound reduction indices R_{A1} [dB]. For the calculations, an equation given in formula (9) is used,

$$R'_{A1} = -10\lg\left[10^{-0,1R_{Dd,A1}} + \sum_{F=f=1}^{n} 10^{-0,1R_{Ff,A1}} + \sum_{f=1}^{n} 10^{-0,1R_{Df,A1}} + \sum_{F=1}^{n} 10^{-0,1R_{Fd,A1}}\right] \quad [dB], \tag{9}$$

where:

R'_{A1}- apparent evaluation rate of sound reduction [dB] described in equation (1);
$R_{Dd,A1}$ – specific acoustic insulation properties of the direct route (through the partition separating spaces) [dB];
$R_{Ff,A1}$, $R_{Df,A1}$, $R_{Fd,A1}$ – acoustic insulation properties of side routes [dB] according to markings in fig. 2;
n – number of side partitions in the room; usually $n = 4$.
Flanking apparent sound reduction index ($R_{ij,A1}$) for the paths of sound given in figure 2 are described by the formulas (10-12). They depend on the sound transmission routes and on material and structural data of building partitions:

$$R_{Ff,A1} = \frac{R_{F,A1} + R_{f,A1}}{2} + \Delta R_{Ff,A1} + K_{Ff} + 10log\frac{S_s}{l_0 l_f} \; [dB], \tag{10}$$

$$R_{Fd,A1} = \frac{R_{F,A1} + R_{s,A1}}{2} + \Delta R_{Fd,A1} + K_{Fd} + 10log\frac{S_s}{l_0 l_f} \; [dB], \tag{11}$$

$$R_{Df,A1} = \frac{R_{s,A1} + R_{f,A1}}{2} + \Delta R_{Df,A1} + K_{Df} + 10log\frac{S_s}{l_0 l_f} \; [dB]. \tag{12}$$

$R_{F,A1}$ is an apparent sound reduction index for element F in the transmitting space [dB],
$R_{f,A1}$ is an apparent sound reduction index for element f in the transmitting space [dB],
$\Delta R_{Ff,A1}$ is a total apparent sound reduction index of the growth of specific acoustic insulation of the system on the side element on the transmitting and/or receiving side [dB],
$\Delta R_{Fd,A1}$ is a total apparent sound reduction index of the growth of specific acoustic insulation of the system on the side element on the transmitting and/or of separating element on the receiving side [dB],

$\Delta R_{Df,A1}$ is a total apparent sound reduction index of the growth of specific acoustic insulation of the system on the separating element on the transmitting side and/or on the side element on the receiving side [dB],
K_{Ff} is a vibration reduction index of the wall junction on sound path *Ff* [dB],
K_{Fd} is a vibration reduction index of the wall junction on sound path *Fd* [dB],
K_{Df} is a vibration reduction index of the wall junction on sound path *Df* [dB],
S_s is an area of separating element [m^2],
l_f is a length of the junction of separating element with side elements *F* and *f* [m],
l_0 is a reference length, $l_0 = 1m$.

It should be taken into consideration that according to the Polish demands, apparent sound reduction indices of materials (R_{A1}) should be decreased by 2 dB to so called design values of apparent sound reduction indices (R_{A1R}).

As an example of determining airborne acoustic insulation, the calculations for the aerated concrete wall between apartments, density 700 kg/m^3 and thickness 400 mm, plastered on both sides using lime-cement mortar 2 x 10 mm thick, dimensions 4.61 x 2.70 m are presented. Autoclaved aerated concrete in density class 700 kg/m^3 and thickness 400 mm is not currently produced in Poland. The material was chosen in this way to examine the possibility of fulfilling acoustic requirements for this type of wall, using its relatively high surface mass. The design apparent sound reduction index R_{A1R} of such a partition is equal to 53.0 dB. It has been calculated as partition between full reinforced concrete ceilings, thickness 160 mm, with a floating floor made of mineral wool, thickness 30 mm, and with cement screed 40 mm, for which the design index for evaluation of acoustic insulation properties R_{A1R} + ΔR_{A1R} = 54.0 + 2.0 dB. Material data for all partitions, calculated ratios of surface masses of the calculated partition (m_s') for circumferential partitions (m_f') and vibration reduction indices for rigid cross junctions on all flanking sound transmission routes are presented in table 2.

Table 2. Material data for the wall between apartments and circumferential partitions together with the summary of acoustic insulation properties, surface masses and vibration reduction indices K_{ij}.

Wall/Ceiling	Materiał	Surface mass m' [kg/m^2]	R_{A1R} [dB]	m_s'/m_f' [kg/kg]	K_{Ff} [dB]	K_{Fd} [dB]	K_{Df} [dB]
Wall between dwellings	Aerated concrete, class 700 kg/m^3	280	53.0	-	-	-	-
Ceiling between dwellings No 1 (F=f=1)	Reinforced concrete slab	384	54.0 +2.0	0.73	6.5	8.8	8.8
Ceiling between dwellings No 2 (F=f=2)	Reinforced concrete slab	384	54.0	0.73	6.5	8.8	8.8
Wall between rooms in dwellings (F=f=3)	Aerated concrete, class 600 kg/m^3	72	38.0	3.89	20.8	10.7	10.7
External wall (F=f=4)	Aerated concrete, class 500 kg/m^3 + ETICS	125	40.0	2.24	11.3	6.4	6.4

On the grounds of the provided input data, calculations regarding acoustic insulation properties were carried out for all sound transmission routes according to the formulas (10-12) and resultant acoustic insulation properties were calculated for a wall between apartments made of autoclaved aerated concrete, class 700, according to the formula (9). The calculation results are shown in table 3.

On the basis of the calculations it was concluded that the autoclaved aerated wall with above-average density 700 kg/m^3 and thickness 400 mm, characterized by a relatively high design index for evaluation of acoustic insulation properties R_{A1R} = 53.0 dB (R_{A1} = 55.0 dB), after taking into consideration flanking transmission in analysed set of compartments achieved the value of apparent evaluation rate of sound reduction index R'_{A1} = 49.4 dB. Correction for flanking transmission of this wall reached the value of 3.6 dB. According to the presented calculations and bearing in mind the assortment of aerated concrete that is currently being produced in Poland it can be stated that this kind of walls does not meet Polish requirements for acoustic insulation of partitions between dwellings ($R'_{A1} \geq 50$ dB).

Table 3. Component values of indices for evaluation of acoustic insulation properties for all sound transmission routes and calculated approximate index for evaluation of specific acoustic insulation properties for an aerated concrete wall between apartments in the layout of aerated concrete walls and reinforced concrete ceilings

COMPONENT VALUES OF INDICES FOR EVALUATION OF ACOUSTIC INSULATION							
Wall between dwellings:	[dB]	**Ceiling No 1:**			**Wall between rooms in dwellings:**		
$R_{1,d}$ =	68.6	R_{D1} =	68.6	dB	R_{D3} =	62.8	dB
$R_{2,d}$ =	66.6	R_{11} =	67.8	dB	R_{33} =	65.4	dB
$R_{3,d}$ =	62.8	**Ceiling No 2:**			**External wall:**		
$R_{4,d}$ =	59.5	R_{D2} =	66.6	dB	R_{D4} =	59.5	dB
-		R_{22}=	64.8	dB	R_{44} =	58.0	dB
APPARENT EVALUATION RATE OF SOUND REDUCTION FOR THE CALCULATED WALL BETWEEN DWELLINGS:					**R'_{A1} =**	**49.4**	**dB**

The use of walls of great thickness for the construction of partitions between dwellings is also not economically justified due to the occupancy of a larger residential area than thinner walls of high density and higher acoustic parameters.

It has to be stated that aerated concrete blocks with volume density of 700 kg/m^3, and thickness of 250 mm, that are from time to time being used as the material for walls between dwellings cannot meet Polish requirements of acoustic insulation. Apparent evaluation rate of sound reduction R'_{A1} for this type of walls in most of the cases vary from 44 to 47 dB.

6. Conclusions

The paper presents the comparison of requirements for acoustic insulation of walls inside residential buildings in Poland and selected European countries. Based on the calculations and comparisons made, it has been concluded that currently, in the European countries considerable differentiation appears with respect to requirements and indices used for evaluation of acoustic insulation of walls inside the buildings. In order to compare the requirements, diversified single-number indices were brought to the value of the weighted standardized level difference with spectrum adaption term $D_{nT,50}$. Standardization of the evaluation of acoustic properties is essential not only due to international exchange of working experience but also due to problems connected with the trade in building products between counties belonging to the European Union.

As a result of conducted comparison it has been concluded that the Polish acoustic requirements for internal walls introduced in 2015 are at a low level not only compared to the countries of Western, but also to Eastern and Southern Europe. Lower required values were recorded only in Romania and for the lowest category of apartments in Belarus. Pursuant to the UE [1] and national associated regulations [2, 3] '*the building and equipment associated with it should be designed and erected in a way that ensures that the level of noise to which users or people staying in its neighbourhood will be exposed does not cause any danger to their health, and also allows their work, rest and sleep in satisfactory conditions'.*

The acoustic comfort survey conducted in the previous years in Poland pointed on high dissatisfaction of residents with the acoustic insulation of walls between dwellings. It should be noted that the traditional, massive walls made in the buildings under evaluation are characterized by significantly higher acoustic insulation than many of the currently used material solutions. As a negative example of such technologies, the calculations of sound insulation of 400 mm thick cellular concrete wall were presented, for which the sound reduction index of 49 dB has been obtained (Polish requirements has not been met). Although the requirements and principles of building acoustics calculations are included in relevant documents, they are often neglected and omitted in the practice of designing residential buildings. One should remember that the commonly used solutions of light partitions have a negative impact on their acoustic properties which often does not meet the required values.

7. References

[1] Regulation (EU) No 305/2011 of the European Parliament and of the Council of 9 March 2011 laying down harmonised conditions for the marketing of construction products and repealing Council Directive 89/106/EEC Text with EEA relevance pp 33-34
[2] Journal of Laws from 1994 No. 89, item 414 with all changes, Poland
[3] Journal of Laws from 2002 No. 75, item 690 with all changes, Poland
[4] Szudrowicz B 2004 Polish acoustics demands against the UE regulations [in Polish] *Mat. Bud.* **4** 2-6
[5] Rasmussen B 2010 Sound insulation between dwellings– Requirements in building regulations in Europe *Applied Acoustics* **71**(4) 373-385
[6] PN-54/B-02151 Building Acoustics – *Protection of buildings against sound* (in Polish)
[7] ISO 717-1 2013 *Acoustics -- Rating of sound insulation in buildings and of building elements -- Part 1: Airborne sound insulation*
[8] PN-B-02151-3 2015-10 Building Acoustics – Protection against noise in the buildings – Part 3: Requirements for sound insulation of walls, ceilings and building components in buildings
[9] Rasmussen B and Rindel J H 2010 Sound insulation between dwellings – Descriptors in building regulations in Europe *Applied Acoustics* **71**(3) 171-180
[10] Gerretsen E. and Dunbavin P. 2014 *Proposal of Harmonized Sound Insulation Descriptors. Building acoustics throughout Europe. Volume 1: Towards a common framework in building acoustics throughout Europe.* (Madrid: DiScript Preimpresion, S. L.) pp 55-64
[11] Dunbavin P and Gerretsen E 2014 *How to Translate Sound Insulation Descriptors and Requirements. Building acoustics throughout Europe. Volume 1: Towards a common framework in building acoustics throughout Europe* (Madrid: DiScript Preimpresion, S. L.) pp 65-79
[12] STR 2.01.07: 2003 Building environmental and nuclear environmental protection. Ministry of the environment of Lithuania [in Lithuanian]
[13] LBN 016-03 2004 Construction acoustics. Cabinet of Ministers Regulations No 468 [in Latvian]
[14] EVS 842: 2003 Sound insulation requirements for buildings. Noise protection. Estonian Centre for Standarization [in Estonian]
[15] SNIP II-12-77 Protection against noise. Russian federation building regulations [in Russian]
[16] ČSN 73 0532 Acoustics - Protection against noise in buildings and related acoustic properties of construction products – Requirements [in Czech]
[17] STN 73 0532 Acoustics. Assessment of soundproofing properties of buildings and building structures. Requirements [in Slovakian]
[18] MSZ 15601-1:2006 An acoustic building. Part 1: Soundproofing requirements for buildings. Hungarian Standards Board [In Hungarian]

IOP Conf. Series: Materials Science and Engineering **415** (2018) 012042 doi:10.1088/1757-899X/415/1/012042

[19] Revision and completion P 122-1989 2010 Normative acoustics in construction and urban areas Part III - Measures to protect against noise at residential, social, cultural and administrative buildings [in Romanian]
[20] DIN 4109 Sound insulation in building construction. German Institute for Normalization [in German]
[21] Szudrowicz B 2003 Acoustic insulation of products for inner partitions execution according to the researches of Acoustic Department of Building Research Institute *Quarterly Journal Works of Building Research Institute* **3** (127) 17-36
[22] Szudrowicz B 2008 Airbourne sounds acoustic insulation of building partitions [in Polish] *Mat. Bud.* **3** 72-75.
[23] Szudrowicz B 1998 Acoustic problems of masonry walls [in Polish] Conference Masonry Constructions (Puławy) pp 181-198
[24] ISO 12354-1 2017 Estimation of acoustic performance of buildings from the performance of elements -- Part 1: Airborne sound insulation between rooms

IOP Conf. Series: Materials Science and Engineering **415** (2018) 012043 doi:10.1088/1757-899X/415/1/012043

Acoustical renovation of beam-and-block Ackermann floor in Cracow residential building from 1940

A K Kłosak[1,2], B Ziarko[1] and K Nering[3]

[1] Institute of Building Materials and Structures, Cracow University of Technology, Cracow, Poland
[2] archAKUSTIK, Cracow, Poland
[3] Cracow University of Technology, Cracow, Poland

Abstract. During the renovation works one of the apartments, in a tenement house in the center of Cracow, we had the opportunity to conduct measurements of airborne and impact sound insulation of existing and newly designed floors between apartments (before and after replacing the floor layers). Article presents results of these measurements along with the analysis of floor solutions used in the interwar period and the comparison of their effectiveness with solutions used nowadays.

1. Introduction

In a tenement building built in 1940 in Cracow, the owner of one of apartments intended to perform a general renovation. One of the authors was asked to provide guidelines for the execution of new floor layers in this apartment. During the renovation works, the authors had the opportunity to conduct measurements of airborne (weighted sound reduction index in the field R'w(C;Ctr)) and impact (weighted normalized impact sound pressure level in the field L'nw) sound insulation. The results of these measurements, along with the analysis of floor solutions used in the interwar period and the comparison of their effectiveness to solutions used today are the subject of this article.

2. Building

The discussed apartment is located on the fourth floor of a multi-family, five-story residential building located at Sereno Fenna str. in the center of Cracow. The building was designed in 1938 on behalf of the company "Henryk Franck Sons - Fabryka Środków Kawowych Spółka Akcyjna" from Skawina. The company began construction of a tenement building in 1939, which was interrupted by the outbreak of war. The construction was completed by Germans in 1940 and German officers settled in the building. The building has been designed as a mixed structural system, with load-bearing (or stiffening) brick walls of thickness equal to 51cm and reinforced concrete columns and girders. A bomb shelter was located in the basement. The building was equipped with an elevator, central heating and central hot water with a recirculation system from its own boiler house, delivered to each of the flats.

3. Before renovation state

Individual residential floors were separated from each other by structural ceilings made beam and block floor system using Ackermann bricks [1, 2, 3]. Based on the measurements of the thickness of structural ceilings, it was found that structural floors consisted of Ackermann blocks of 20cm height, with Griffel 25mm steel reinforced ribs and approx. 4cm layer of concrete topping. Floor thickness was increased by 3cm on the 120cm wide area centered along non load-bearing dividing walls. Figure 1 shows the floor layers and the longitudinal and transverse cross-section through the system in the discussed apartment before renovation. Layout of the floor layers in the discussed apartment shown in Figure 1 remained unchanged from 1940 until the renovation in 2017.

The floor layers had total thickness of approx. 15 cm, which consisted of floor joists, with cross-section of 7x7 cm, spaced every 80 cm, to which subfloor planks, with thickness approximated 20 mm, spaced approximately 20mm, were nailed. The oak parquet, of thickness about 20mm, was laid on the boards. Space between the joists was filled with a sand mixed with brick debris.

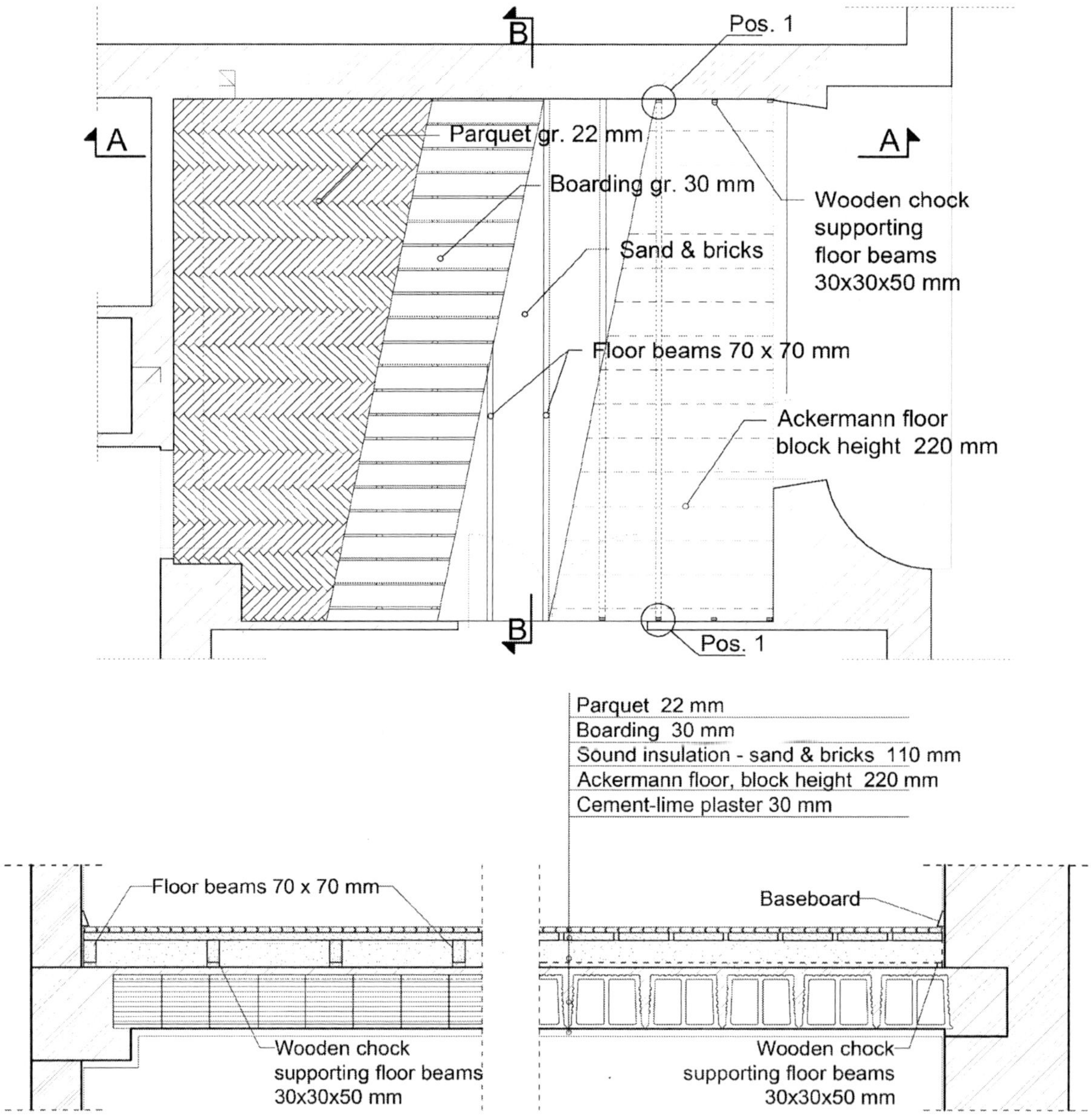

Figure 1. Plan and cross-section of floor slab before renovation (1940).

Noteworthy is the method of supporting the floor joists. Floor joists were not supported on the top of the Ackermann floor along its entire length. They were supported only on both ends by small wooden chocks with dimensions of section about 30x30mm, in such a way that they worked as simply supported beams (deflecting), not as joists. As a result, the contact area of these beams with the Ackermann floor was limited only to the surface of wooden chocks. Floor beam periodically deflected under the weight, compacting the 30mm layer of the sand, which was under the joist, additionally minimizing the contact area of the beams with the ceiling. Joist scheme is shown in Figure 2.

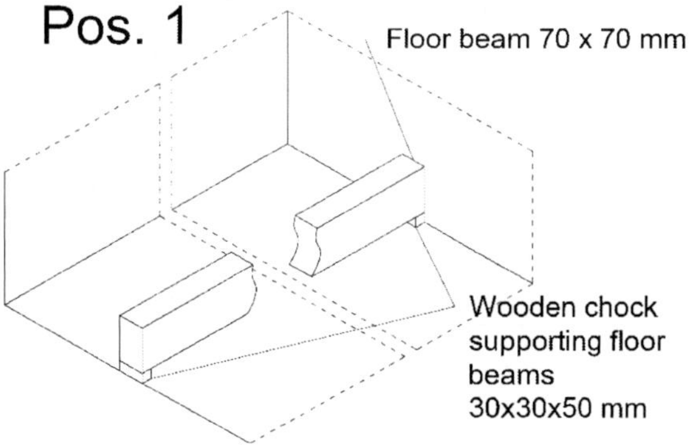

Figure 2. Schematic drawing of joist supported by wooden chocks of height 30mm (1940).

In Figure 3, an axonometric diagram of the cross-section of the structural ceiling in the discussed building is shown in the before renovation state.

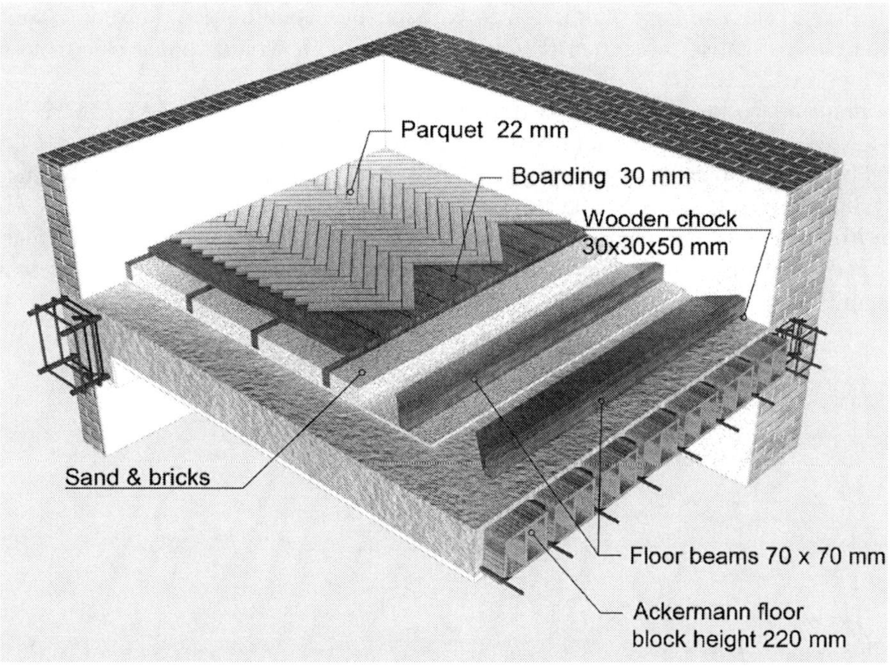

Figure 3. Axonometric view with cross-section of floor slab with floor layers before renovation (1940).

4. After renovation state

The owner of the apartment described earlier repaired the second apartment, which was on the same floor, but on the opposite side of the staircase. In second apartment, the renovation company replaced existing floor layers (similar as described in chapter 2) with OSB floor, based on joists placed directly on the Ackermann floor, with mineral wool filling space between joists. The use of such a floor system was dictated (according to renovation company opinion) by the ease of assembly and economic factors. Unfortunately, such floor design resulted in a distinct feeling of impact sounds in the flat below. The Authors were able to perform additional measurements of impact sound insulation just recently, therefore the analysis of this solution was not included in this article. However, the very fact of problems reported by the renter, the owner of the apartment above, prompted the latter to consult one of the authors on the acoustic issue.

Analyzing the available possibilities of floor renovation, two solutions were considered. A "lightweight" solution, similar to that described above, using OSB boards, but with the support of joists on the ceiling through the elastic elements. And a "heavyweight" solution using a cement screed as a pressure layer and underlying resilient layer made of mineral wool or elastic expanded polystyrene. The owner and contractor's arguments for the "lightweight" solution were low cost, quick assembly and no need to wait for screed to dry out, which is necessary to start the installation of the parquet. One of the authors arguing against the "lightweight" solution was significant reduction in the weight of the existing system, resulting from the removal of approx. 10 cm layers of sand and brick, which surface mass was estimated at approx. 140 kg / m2 with a sand and debris density of approx. 1,400 kg / m3. With a relatively small surface mass of Ackermann floors (approx. at 300-330 kg / m2), acoustic performance considered as airborne sound insulation of Ackermann floor together with layers of debris and sand was estimated at approx. RW = 59dB. The lack of these additional 140kg / m2 could significantly reduce the roof sound insulation, especially in the low frequency range. This reduction was estimated at approx. ΔRW = -6 dB, which would translate into the RW value = 53 dB for the Ackermann floor alone. According to [4], beam and block floors of similar surface mass amounting to 305-314kg / m2 are characterized by acoustic insulation from airborne sounds at RW = 51dB, and therefore slightly lower than RW = 52 ~ 53dB, as it would appear it is from the mass law [in 4, Table 4, point 1].

The expected impact sound level for a floor with a mass of approx. 300kg / m2 is approx. 82dB. According to [4], beam and block floors of similar surface mass from 305-314kg / m2 are characterized by impact sound level LnW, eq = 77 ~ 78dB, which is measured according to standards from 1980-1990, not exactly in the same wasy as values currently used.

A comparison of the estimated airborne sound insulation for a floor with a surface mass of 300 and 440 kg / m2 and a typical reinforced concrete slab with a thickness of 20 cm and a weight of 470 kg / m2 is shown in Figure 4.

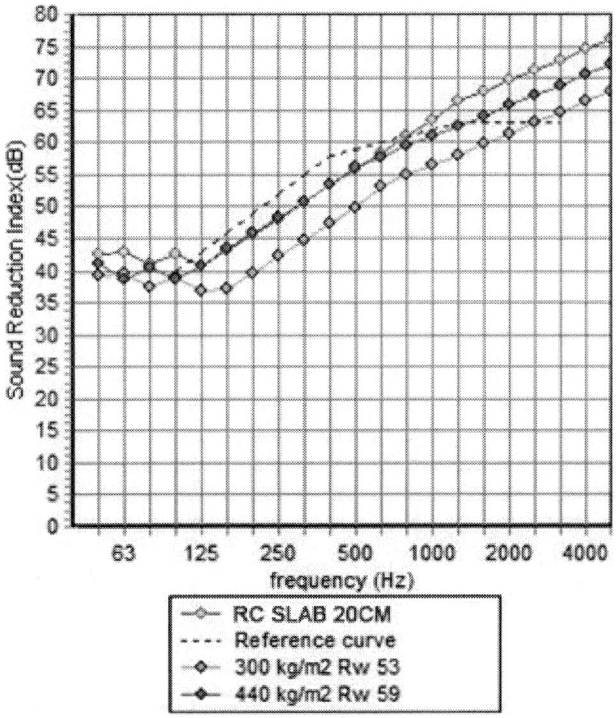

Figure 4. Comparison of estimated airborne sound insulation for Ackermann floors with surface mass of 300kg/m2 (red curve) with floor of surface mass 440kg/m2 (blue curve) and typical reinforced concrete floor 20cm thick and with surface mass 440kg/m2 (green curve).

Second argument against the use of the "lightweight" floor layer solution was fear of "rumbling" of a single layer of OSB boards when users walk and jump on them.

Finally, the "heavyweight" solution was chosen, with mineral wool acoustic insulation and a pressure coating made of cement screed. Due to the need to distribute installation in the floor layers, acoustic insulation in two layers, 5cm + 2cm was applied, with the assumption that the installations will be carried out in the bottom 5cm layer of mineral wool. Choice of mineral wool instead of elasticized expanded polystyrene resulted mainly from desire to maximize airborne sound insulation and the use of wool as a resilient material in a layered system (ceiling-screed). In the areas where Ackerman floor topping was thicker, 2cm layer of acoustic insulation was ommited.

The area of the renovated apartment was approx. 90m2, so it was estimated that approximately 100 m2 of 5cm thick mineral wool panels will be needed and the same number of 2cm thick panels. For the total sound insulation thickness of 70mm used in the discussed solution, dynamic stiffness should amount to approximately 8.5 MN / m3 [according to 5, see pg. 84].

As a pressure layer, a cement screed was used, with density of approx. 2000 kg / m3 and thickness of approx. 70 mm, which translated into a surface mass of approx. 140 kg / m2, the same as the sand and brick pugging weighed. Maintaining the mass of ceilings during the renovation of old tenement buildings is an important point of view of the safety of the structure.

For a combination of insulating material with dynamic stiffness of approx. 8.5 MN / m3 and pressure layer with a surface mass of approx. 140 kg / m2, based on EN 12354-2 [6], Annex C predicted reduction of impact sound level (Weighted impact sound reduction improvement index) should be approx. $\Delta Lw = 36$dB, at the resonance frequency $f0$ of the pressure layer, depending on the calculation method, approx. 39Hz [in 6, Annex C1] or 47Hz [in 7, Annex D2]. At the same time, calculation of increase in insulation from airborne sound (Weighted sound reduction improvement index) due to the use of additional massive layer on the ceiling should amount to approx. $\Delta Rw = 9dB$. In order to maximize the impact sound insulation of designed floor layers, an additional layer of so-called "rubber-cork", 4.5 mm thick was glued between the screed and the parquet.

Figure 5 shows plan of floor layers and the longitudinal and transverse cross-section through the structural ceiling in the discussed apartment after renovation. In Figure 6, an axonometric diagram of the cross-section of the floor in the discussed building is shown in the current state and after renovation.

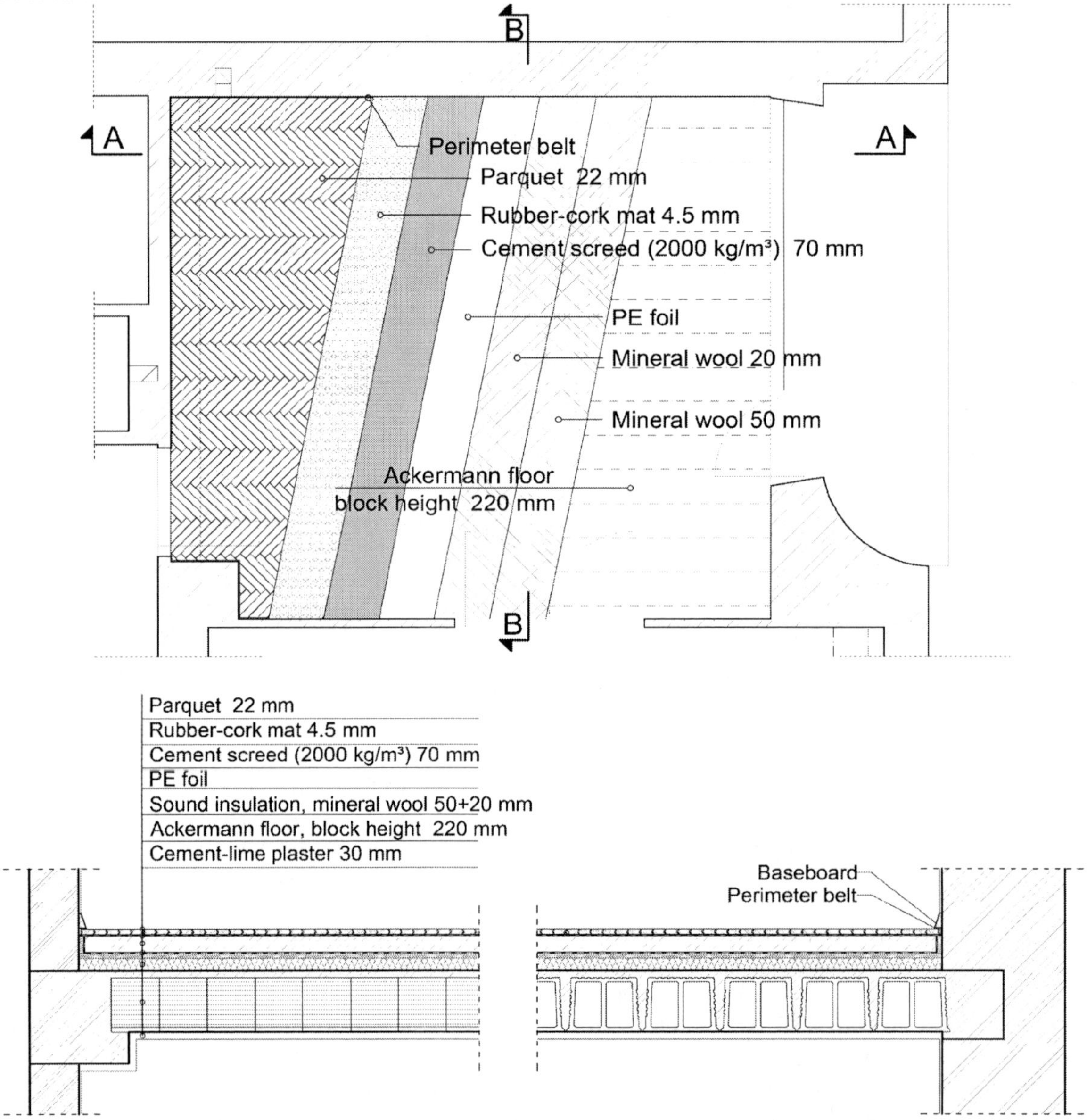

Section A-A Section B-B

Figure 5. Floor layers plan and cross-section of floor after renovation (2017).

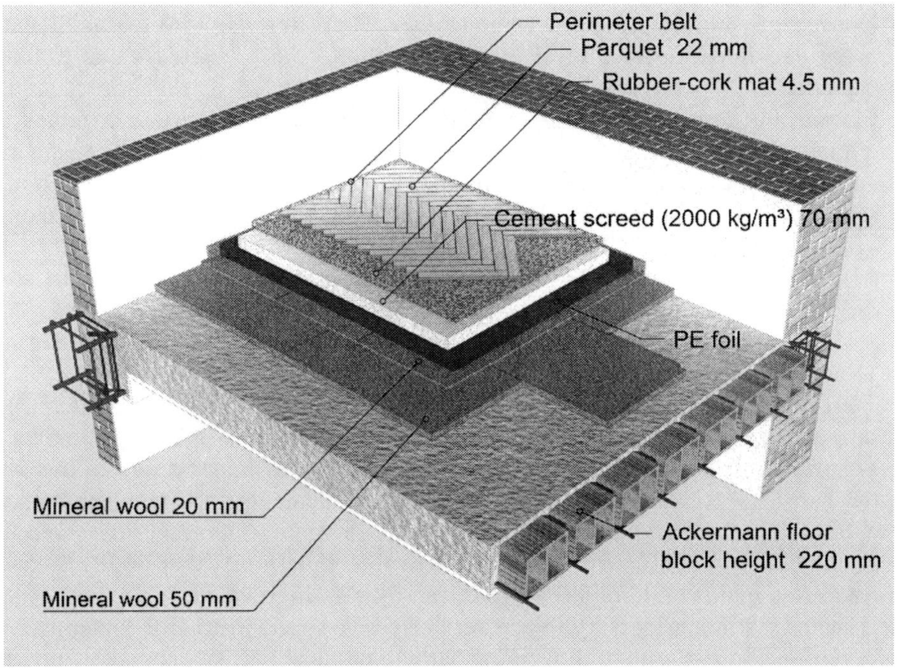

Figure 6. Axonometric view of structural ceiling and floor layers in existing state after renovation (2017).

5. Results of acoustic insulation measurements

Before renovation and during the renovation works, authors made impact and airborne sound insulation measurements of the ceiling between the renovated apartment on the fourth floor and the apartment located directly under it. The results of these measurements together with the comparison to the requirements of PN-B 02151-3 from 2015 [9] are presented in Table 1.

Table 1. Results of acoustic insulation measurements carried out before, during and after renovation compared to requirements of PN-B 02151-3:2015 [9].

Lp.	Floor system	R'_W [dB]	R'_{A1} [dB]	R'_{A2} [dB]	$L'_{n,W}$ [dB]
	PN-B 02151-3:2015-10 [9] requirements	-	*≥ 51*	-	*≤ 55*
1.	Ackermann floor (without floor layers)	50	49	47	86
2.	With floor layers 1940	57	55	51	53
3.	With floor layers 2017r (no parquet)	59	58	55	57
4.	With floor layers 2017r (with parquet and rubber-cork mat)	62	61	57	47

6. Discussion

As can be seen from the analysis of the results given in Table 1, tested Ackerman floor without any additional layers (Table 1, item 1) does not meet the requirements in the field of airborne sound insulation (1dB deficit) or impact sound insulation (deficit 31dB). After considering 2dB corrections for the differences between the laboratory and the construction, the measured airborne sound insulation is consistent with the predicted value (R'W = 50dB in relation to the predicted RW = 51 ~ 53dB). The measured impact sound level is very similar to the predicted value (L'nW = 86dB in relation to the predicted LnW = 83dB).

The most interesting result of measurements is fact that Ackermann floor together with wooden floating floor made in 1940 (Table 1, item 2) met the present (2015) [9] acoustic requirements. Airborne sound insulation requirement is fulfilled with 4dB surplus, and impact sound insulation - with a reserve of 2dB, both in relation to the requirements of the 2015 standard [9]. Compared to the requirements of the 1999 standard [8], impact sound level is 5dB better than required by the standard. Requirements in field of air sounds for floors between flats are the same in both standards. The measured airborne sound insulation is in line with predicted value including a 10cm layer of sand and brick pugging (R'W = 57dB compared to predicted RW = 59dB). The measured impact sound level is very similar to predicted value (L'nW = 86dB in relation to predicted LnW = 83dB).

Another interesting result is fact that floating floor without top layers (Table 1, item 3), i.e. without rubber-cork and parquet floor, did not meet the modern requirements in field of impact sound when laid on the Ackermann floor. The deficit is not large (-2dB) and may be due to the fact that when measuring, the screed was not loaded with furniture or elements that simulated it, which could contribute to lowering the measurement result. What is more important measured impact level reduction was only $\Delta L'w$ = 29dB, with the predicted value equal to ΔLw = 36dB. It can be seen that Ackermann floor used in this building is characterized by very high values of the impact level (L'nW = 86dB), which means that it is necessary to use very effective sound insulation materials. Nevertheless, this result is a certain surprise, although weaker than the predicted impact sound level reduction could also result from a partial thinning of screed in one of ends of analyzed room, resulting from the need to compensate for inequalities of the existing ceiling. The measured airborne sound insulation is consistent with the calculated value, taking into account the additional weight of the screed, (R'W = 59dB in relation to the predicted RW = 59dB). In addition, the predicted increase of 9 dB airborne sound insulation of ceiling after using floating floor proved to be accurate (R'W = 50dB for Ackermann floor itself in relation to R'W = 59dB for ceiling with floating floor).

The last of the results are measurements made on a fully finished floor, together with an oak parquet stuck to the screed through the rubber-cork layer and a typical load of furniture. Designed and made floating floor (Table 1, item 4) fulfilled with great reserve all modern requirements in terms of both airborne sound (11dB reserve) and impact sound (8dB reserve). However, what is more important for users of flat below, new floor is better soundproof than the one from 1940, by 6dB in both airborne and impact sound.

In order to better analyze changes in the acoustic insulation of subsequent examined systems, Fig. 7 shows diagrams of from airborne sound insulation (a) and impact sound level (b) of individual examined layer systems.

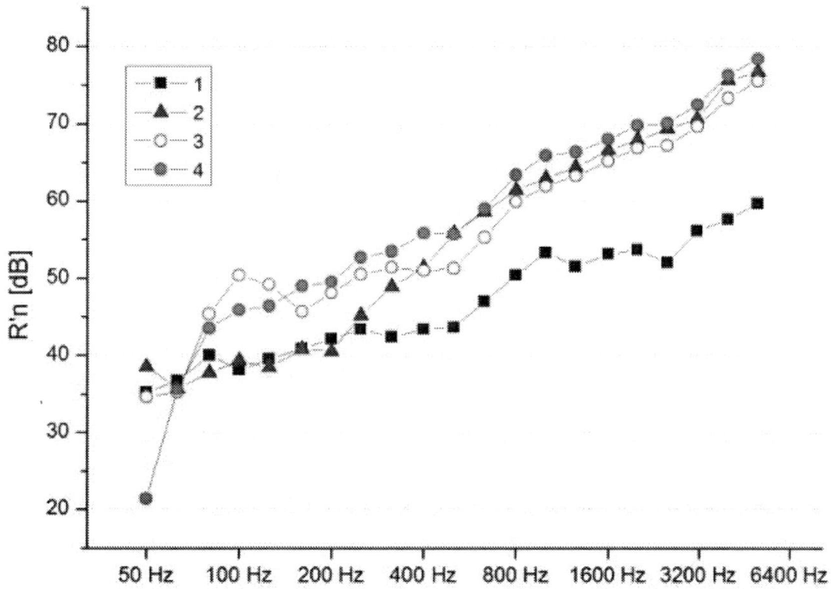

a) airborne sound insulation

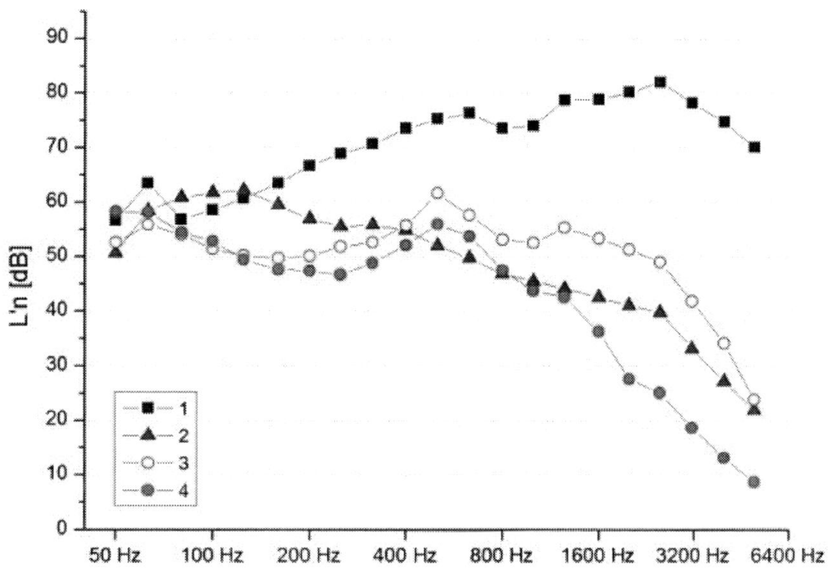

b) impact sound level

Figure 7. Comparison of measured airborne sound insulation (a) and impact sound level (b) for 4 examined layer systems. Numbering of systems according to Table 1.

As it can be seen from comparison of measurement results shown in Figure 7a, the airborne sound insulation of Ackermann floor with original layers from 1940 (No. 2 in the graph) is almost equal to

insulation of floor itself without floor layers (No. 1 in the graph) in terms of 50 Hz to 200 Hz one-third octave bands. Above these bands "lightweight" floor layers begin to affect the insulation of the entire system. In Figure 7a it can also be seen that "heavyweight" floor layers begin to affect acoustic insulation from much lower frequencies, above the resonant frequency of the entire system (about 50 ~ 80Hz). It can also be seen that furniture load on the screed (No. 4 in the graph) compensated slightly the course of the insulation curve in relation to the value for screed itself (No. 3 on the graph).
Comparison of results of impact sound insulation measurements shown in Figure 7b shows that "lightweight" floor from 1940 (No. 2 in the graph) is effective as impact sound insulation for frequencies above 125 Hz one-third octave band. In contrast, the "heavyweight" solution currently designed (No. 3 and No. 4 in the graph) reduces impact sound for frequencies above 50Hz band, above the resonant frequency of the floating floor system. The increase in impact level of the "heavyweight" floor in frequency band 500Hz is clearly visible. This is probably related to the critical frequency of the screed, which for the 50-70mm screed ranges around 400-500Hz, which is probably the cause of the deterioration of results shown in Figure 7b (No. 3 and No. 4 in the graph).

Finally, it is worth paying attention to the subjective aspect of receiving changes in impact sound insulation of original solution from 1940, and its contemporary version. Despite clear improvement of acoustic parameters of new solution in relation to original floor from 1940, subjective reception of changes is not unambiguous. The subjective tests of receiving footsteps of a person walking on both floors in shoes and barefoot showed that new solution causes a strengthened feeling of low frequency penetration, audible in form of low-frequency rumbling of the screed layer around 40-50Hz. This frequency is marked on the Figure 8 with an arrow. Paradoxically, the increase of impact sound insulation in above the resonant frequency - visible in Figure 8 from about 100Hz, where impact sound insulation after a renovation is lower, marked in brown, than analogous level, marked in blue, for floor solution before renovation - caused that after refurbishment to the dwelling below, the audible noises of footsteps in higher frequencies penetrate to a much lesser extent, which masked the lower frequencies to some extent. Currently, for solution after renovation, due to the lack of footsteps from frequencies above 100Hz, people conducting subjective tests heard only low-frequency rumbling, which was perceived by some as more onerous. This is shown in Figure 8.

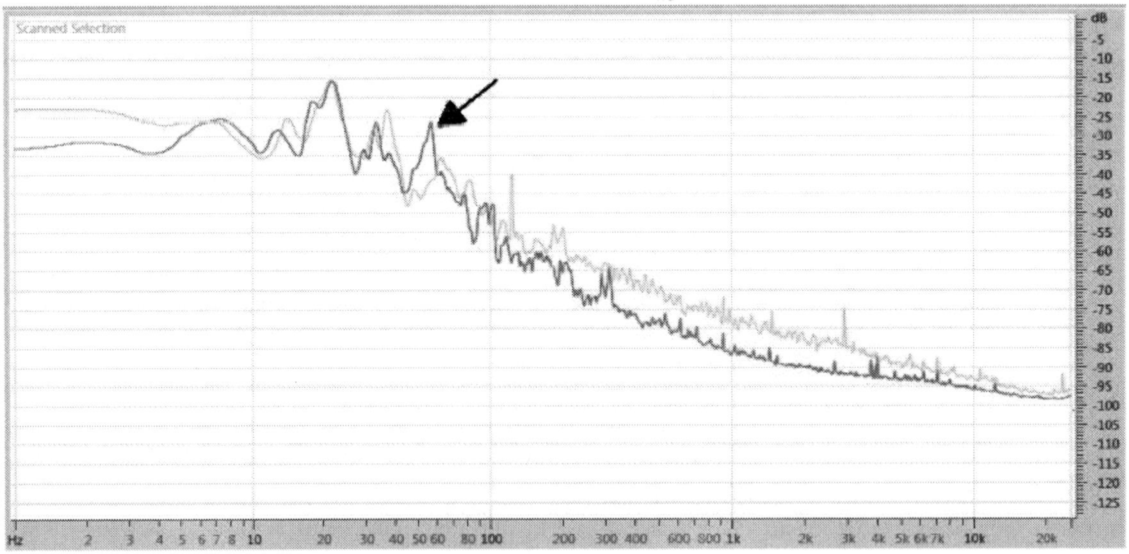

a) walking in shoes

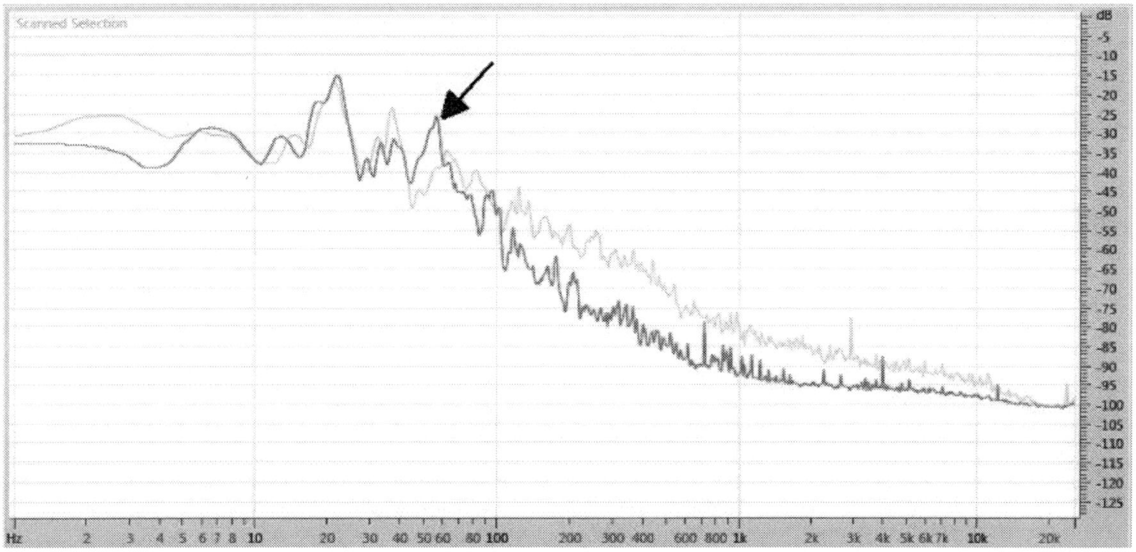

b) walking bare feet

Figure 8. Comparison of perceptible impact sound level in the flat below tested ceiling, for person walking in shoes (a) and bare feet (b) for floor layers before renovation (item 2 in Table 1, blue in the graph) and after renovation (item 4 in Table 1, brown in the graph).

Another possible reason for appearance of such rumble is difference between a locally reactive layer of a light floor in the solution before renovation and a resonant reactive layer of a heavy screed in the solution after renovation. This means that in first case, force from impact of tapping machine or the steps of walking person is transmitted from top layer to ceiling only around point of impact, while in second case a reverberating bending wave field is created [10]. Due to lack of dilatation in screed layer on a large area of the renovated flat (due to later possibility of changing the partition walls) the resonance of the screed layer due to impacts can be more clearly heard.

7. Summary

This paper discusses the process of renovation of floor layers in an apartment from the interwar period. It has been shown that solutions used in residential buildings of last century meet modern requirements in the field of sound insulation. What is more modern solution presented in this article has not fulfilled impact sound level requirement. Additional layers had to be done to provide satisfying results. Awareness of connections between solutions in the field of general construction and their acoustic function is a key element in the process of acoustic renovation of existing buildings. This applies especially to older constructions, such as the 1940 tenement building described in this article.

Acknowledgement

The authors would like to thank owners of flats in the described building for making their flats available for acoustic measurements. The research shown in this article was financed, inter alia, as part of the research work L-1/158/2016/DS "Revitalization of existing buildings".

IOP Conf. Series: Materials Science and Engineering **415** (2018) 012043 doi:10.1088/1757-899X/415/1/012043

Literature:

[1] Nicer, T. *Flat ceilings in historic buildings*. Budownictwo i Architektura 5 (2009) 85-100
[2] Drobiec, Ł., Pająk Z., *Ceilings with small-sized elements*. Gliwice, 2013, ISBN: 978-83-7880-023-1
[3] Mielnicki S., *Construction systems*, Katowice, 1938
[4] ITB Instruction no. 369, Sound insulation properties of building partitions, 2002
[5] Rockwool SeaRox Marine & Offshore Insulation, Technical Guidlines 08/2012
[6] EN ISO 12354-2:2017 Building acoustics — Estimation of acoustic performance of buildings from the performance of elements — Part 2: Impact sound insulation between rooms.
[7] EN ISO 12354-1:2017 Building acoustics -- Estimation of acoustic performance of buildings from the performance of elements -- Part 1: Airborne sound insulation between rooms.
[8] PN-B 02151-3:1999: Building acoustics. Noise protection in buildings. Sound insulation in buildings and of building elements. Requirements.;
[9] PN-B 02151-3:2015-10: Building acoustics - Noise protection in buildings - Part 3: Requirements for sound insulation in buildings and building elements.;
[10] Vigran, T.E. „Building Acoustics", Taylor & Francis, New York, 2008, ISBN 978-0-415-42853-8

IOP Conf. Series: Materials Science and Engineering **415** (2018) 012044 doi:10.1088/1757-899X/415/1/012044

Occupant behavior modeling based on migration registration technique

J W Dziedzic[1], V Novakovic[1]

[1]Norwegian University of Science and Technology,
Department of Energy and Process Engineering,
Kolbjørn Hejes v 1B

jakub.w.dziedzic@ntnu.no

Abstract. The building occupant energy related behavior modelling is a key factor for building performance simulation that supports realistic forecasting of energy consumption. The current state of development of this topic delivers partial knowledge about occupant's behavior, but information about transient thermal comfort of each occupant is still, beyond the reach. Access to such a data would visualize outcome of the proposed HVAC design. With a use of proper measurement technique, it is possible to develop virtual bot that will recreate typical energy related behavior scenarios inside designed building and gather consistent information regarding thermal properties of the indoor environment. Such a tool will display potential faults of the design regarding occupants thermal comfort and point out potential paths of adjustment during design phase

1. Introduction

Re-creation of a building's occupant behavior has always been a challenge. The main reasons are the diversity of personal preferences among building users and various acceptance conditions. The above factors make modeling of occupant's behavior a troublesome subject. Since there are no significant input data, they have to be replaced by individual behavior approximations. Behavior models can be applied to all human-related scientific topics. They may even be used as a tool for monitoring and forecasting of energy consumption inside buildings. Several methods aiming at description of energy-related human behavior were developed in the past [1] [2]. They opened a new chapter in the research on indoor environment quality since they allowed evaluating sensation of comfort on a fundamental level. It was an essential step towards improvement of indoor air quality (IAQ) control and optimization of energy consumption for heating, ventilation and air conditioning (HVAC) systems.

The next step should lead towards personalized comfort detection. Reaching this goal will open a new space for automation in buildings. Without synergy of accuracy and reliability, it is impossible to reach the level of personalized simulations that could support preferential HVAC control. The current state-of-the-art instruments enable detecting occupants inside offices/zones with a low error margin [3]. It was possible to distinguish between two types of office/zone users: average daily users and guests. This simulation approach ensures accurate results, but it can only define behavior of occupant groups and does not offer a precise picture for a personalized indoor air model. This obstacle concerns both office and residential buildings. Occupants are described as a mass. It is difficult, within this solution, to distinguish individual actions of an occupant from occupants group. Insufficient accuracy has been emphasized in Yan's general review of occupant behavior modeling in buildings [4]. The results of his research indicate that if higher simulation accuracy is to be achieved, it must be connected with detection of occupants' personal needs.

To improve energy simulation accuracy, input information about occupants behavior has to be more accurate/transparent [5]. With registration of each occupant's actions, it is possible to develop a pattern of their behavior. Organized chronologically, it will tell a "story" of their thermal sensation. Such data will generate a range of possible interpretations concerning an occupant's thermal needs. The collected feedback can later be interpreted as a habit or standard practice of each occupant. In order to reach the required level of detail in description, there is a need to improve the measurement accuracy. The collected data has to allow distinguishing each occupant's position and action in time. At the same time, it may not go as far as to cross the important boundary of recording data that can be considered sensitive from occupants' point of view. It has to provide sufficiently meaningful information so that all the future recordings could be attributed to a given person (for example Occupant A). The collected data has to provide specific metadata that enables interpretation of all the occupant's activities.

If all of the above requirements are fulfilled, the obtained data will allow taking the next step in development for personalized control for IAQ. With direct identification, it will be possible to build models that will support identification of occupants' needs based on actual thermal conditions. With this holistic approach, the accuracy of building performance simulation (BPS) results may reach the acceptable level. It may minimize the difference between the simulated and measured energy consumption of the building. This goal may be achieved with the use of the transient occupant behavior model. The first step towards its goal should aim at discovering patterns of occupants' movements that could later on be used to create simple yet robust movement model.

Improvement of the gathered data quality requires selection of a new measurement device or/and technique. To make them more reliable, measurements have to be conducted with the use of an in-situ method. This means that all solutions that require using a device, such as a wristband or smartphone, have to be excluded from this study as potentially leading to measurement errors with reference to some occupants. The errors may occur if some occupants lose their device or leave it in the monitored zone. Similar effects of under/overestimation of some occupants may appear if the system counts door opening as input data. It is hard to link the number of times when the door was opened to the number of occupants using it. Moreover, this solution give us no information on the direction of migration.

Other solutions – like infra-red (PIR) detectors – provide information on crossing its monitoring point. This measurement method can sometimes produce results that come close to the description of occupants' activity. However, this method does not allow capturing anticipated information. Using these detectors produces binary output. It can be enough for the lighting control or presence detection but the accuracy and resolution are not significant enough to support the development of personal thermal comfort sensing. The method has a great potential when it comes to the detection of CO2 concentration. Unfortunately, due to the response delay in this measurement technique, it can only be used for areas, like an auditorium, where occupants stay for a longer period [6]. For residential solutions, such measurements can deliver only brief information regarding the occupancy state. Occupants activity and their degree of freedom is higher in residential buildings.

In order to obtain higher quality data, it seems that it is necessary is use video recording data. Supported by Kalman filtering or pre-trained model, it can map the paths of movement. The registered movement will be only a projection on an observed surface. The absence of perspective from the recording can be re-calculated. Unfortunately, applicability of video recording is limited due to ethical issues. It is most unlikely to be used in monitoring of residential buildings, but it can be used to track movement in wide, open areas, such as halls or lobbies. The only other solution for the measurement of occupant behavior that bypasses all of the obstacles is depth registration camera. Collected data do not allow direct identification of the monitored occupant, but it can provide information about the movement of each limb. Additionally, it can detect the outline of the observed body and thus identify individual persons. The main issue related to this type of observed depth recording is its limited range. The range boundaries are defined by the limitation of the power of signal amplifier and the price of the detector. There are many different producers currently on the market who are focusing on manufacturing this kind of devices. Most of them find applicability in the automotive, personal aviation (drones) and entertainment sectors. Previous investigation into these measurement technique has shown promising results, and that is why using it will be explored further. [7] [8]

2. Aim

The aim of the presented work is to investigate the use of data collected in previous studies. The collected data offers high-resolution information, and its potential has not been fully used. With high-quality data, it is possible to produce an occupant behavior transition model that accurately recreates their movement. The developed model will become a foundation for building an occupant behavior simulator for model predictive control or annual simulations during the design phase. However, it is necessary to discover patterns of transition to reach this goal.

It will have to be seen if the already available data is sufficient to build a behavioral model. The first step towards development of a building occupant transient agent-based model (BOT-ABM) has to focus on observation regarding occupants' movement itself. The goal is to develop a self-sufficient model that will recreate pathways of transitions inside the investigated area. In the current state of development, drivers of occupants' actions regarding their transitions will be not investigated.

3. Methods

3.1. Measurements

Measurement data will be obtained with the use of a depth registration camera. For the purposes of this research, the Microsoft Xbox One Kinect device will be use. The choice of the measurement device was determined by the desired quality of the data to be collected, the sampling time, price and availability in the market. The selected device is capable of registering up to six people at one time and has the detection distance limitation of five meters. Regardless of these obstacles, this type of measurement can be successfully used in close spaces like corridors, entrances, small offices or residential buildings. Access to the full measurement device potential was obtained from the software development kit published by its manufacturer. Collection of this data is done with the use of a grid projected by a beam in an infra-red spectrum (to avoid discomfort of the observed occupants). Once someone crosses the observation spectrum, his or her transition/activity is registered. The obtained data is a skeleton model (SM) that has representation of twenty-five common points in three dimensions. Each SM is connected to an identification number, from one to six, by a semi-random procedure. During measurement, the device registers movements of each joint with the frequency up to thirty frames per second. A more detailed description of the measurement technique may be found in manufacturer webpage [9].

The whole recording was conducted with the use of self-developed software in the Matlab environment. The choice of the programming language was based on the availability of the existing library, the available third-party support and personal preference. The registered data will recreate paths of movement taken by each monitored occupant. Two Kinect devices were used during the measurement procedure. To make data more coherent, it was necessary to establish one common global coordination system .One device has to be selected as a 'master' device, and rest as a 'salves'. All recordings from the 'slave' devices has to be recalculated via referential 'master; device. It was done by measuring distances between both devices and setting up a translation vector to recalculate readings. Streamed information from SM provides three-dimensional data. The devices were placed so as to cover the whole office area, from one end to the other.

It was detected that there were certain specific points within the monitored area that were more actively used. All of these points on this surface can be treated as nodes of communication or points of interest (POI). POIs have to be understood as being limited by boundary spaces/zones which can be defined as points. POI location was detected with the use of the Pareto filter. While "typical" enterers were exposed, it could be used to generate a hidden Markov chain model. This part of study regarding occupancy modeling was put aside for future investigation. At the present current stage, movement inside the measurement zone and discovery of its patterns was treated as a priority.

3.2. Raw data

The recording process produces a vast amount of data. Each minute of recording generates an 1800x457 matrix. Each row of the matrix represents one frame, 30 frames per second multiplied by 60 seconds give 1800 rows. In one exposition, the Kinect device can capture up to six occupants standing within the device's range. Each registered occupant provides information about the movement of 25 points in three dimensions, 75 columns in the registration matrix. Based on this information, the recording device has to reserve recording space for six people – six multiplied by 75 gives 450.The last seven columns are reserved for the clock data that gives information about the year, month, day, hour, minute, second and frame number.

The software built into the device is unable to distinguish occupants directly, and it does not link any observed occupant with the same SM number. After it loses track of a specific person, it may link it to the same SM, but the odds of this happening are one to six. The process of connecting the SM with an observed human is semi-random. It links an occupant with the next available SM number. Once the software reaches the SM number six, it resets numbering and starts from SM one. That is why it is necessary to record all the possible SMs simultaneously. During the registration procedure, each of device's clocks was synchronized. Recorded data packages provide an overwhelming amount of data, far too much to process it manually, so it was necessary to develop software to do it automatically. It is essential to develop filtering and linking software that will transform the acquired data into a suitable form enabling its correct interpretation

3.3. Processing

The first step of data processing is initialized during recording. To save the hard drive space and improve performance of the recording software, minimization of the file size was required. To discard rows with zero activity, all empty rows have to be deleted before saving the data on the hard drive. The developed software was checking the sum of the first 450 columns from a row and deleted it if it was equal to zero (no movement detected), so that it would not be included in the saved matrix. Because the reporting (saving) protocol is resolved every minute, most of the saved files are entirely blank, due to the absence of movement inside the observed field. If there is any detected action, it immediately provides a significant data stream. It makes the files with recorded activity stand out, regarding its file size, from the rest of the files, which are blank. Simple filtering of files by their size allow quick determination which file has data with registered activities.

Afterwards, the data is successfully connected and calculated by the transition vector to one global coordinate system. The collected records require extra processing steps to identify whether there exists any POIs. The merged data still requires connecting pathways because of the existing discontinuities caused by the following two factors:

- The devices lose track of an occupant;
- The second source of discontinuities in the recorded data is dependent on some characteristics of the device used in the experiment.

Each device binds SM data by itself, with no communication between them. It is highly unlikely that both devices will link a monitored occupant to the same identification number of SM. To connect detached paths, it is necessary to sort records by time and calculate the velocity of each step. The next step requires investigation of the SM body ratio.

Each SM provides a complete set of information about the movement of each monitored limb. It can be used to distinguish each occupant regardless of the connected SM identification number. Registered occupants may differ from each other regarding their body dimensions, height, width etc. They can be distinguished by their body ratio factor. For example, the value of height divided by the horizontal distance between shoulders. The choice of the representative ratio or ratios will be based on its reading stability and transparency. If this layer of filtering does not provide a reliable sorting criterion, limb data can be applied to use its swing (pendulum) movement analysis. Each limb has its own specific harmonic move during walking. Observation of its repeatability can also provide some "marker" information. Frequency and deviation angle of limb movement can provide significant support in the personal \\\identification process.

3.4. Moment knowledge

Post-process data allows displaying all chosen pathways separately. All of the pathways were projected

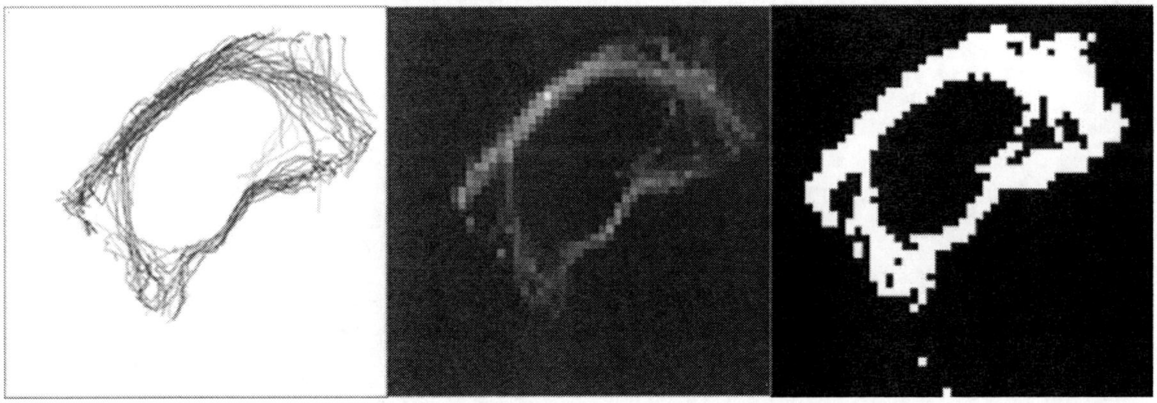

Figure 1. Pathway registration, the Heat map of space usage, transition outline plots

on a heat map to discover the frequently used tracks. Displaying these results allows swift identification of the monitored area layout (Figure 1). Based on ground truth, the uncrossed area represents the position of the table and chairs. It shows that the transition model has to take into account all possible connotations of the layout of the building. It is impossible to achieve it with just an overview of the transition and that is why the next part of the analysis has to focus on each separate step of the transition. As has already been mentioned, all monitored pathways have to be plotted on a normalized shape, where it is possible to discover patterns of transition from point A to B. Conceptual analyses of the movement were presented in Figure 2. It shows the possibilities of transition from point 1 to point 2. Transition steps and the angularity of consecutive steps are unequal. The probability of the step selection was illustrated by a gradient cloud, which represents the density of the selected steps from the starting to the end position. To understand the whole transition phenomenon, all of the movement transition has to be observed separately..

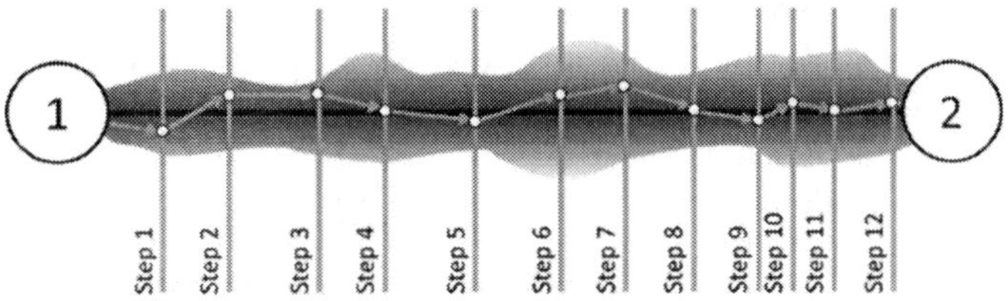

Figure 2. Movement pathway distribution illustration

3.5. Movement model

Investigation into the transition patterns requires to include all of the recoding inputs in the scope. The most interesting observation, regarding the model development potential, was the distribution of step lengths and its angularity, which is understood as an angle between the previous step and the next step turn. The distribution of the step lengths and angularity was displayed in histograms (Figure 3). Both of the graphs show that each value is characterized by natural distribution in a limited range. It demonstrates the existence of actual patterns that can be explored further. The displayed data shows a

Gaussian semi-bell-shaped curve. Both distributions were investigated, and distribution data, such as range and peek, were used to develop a movement.

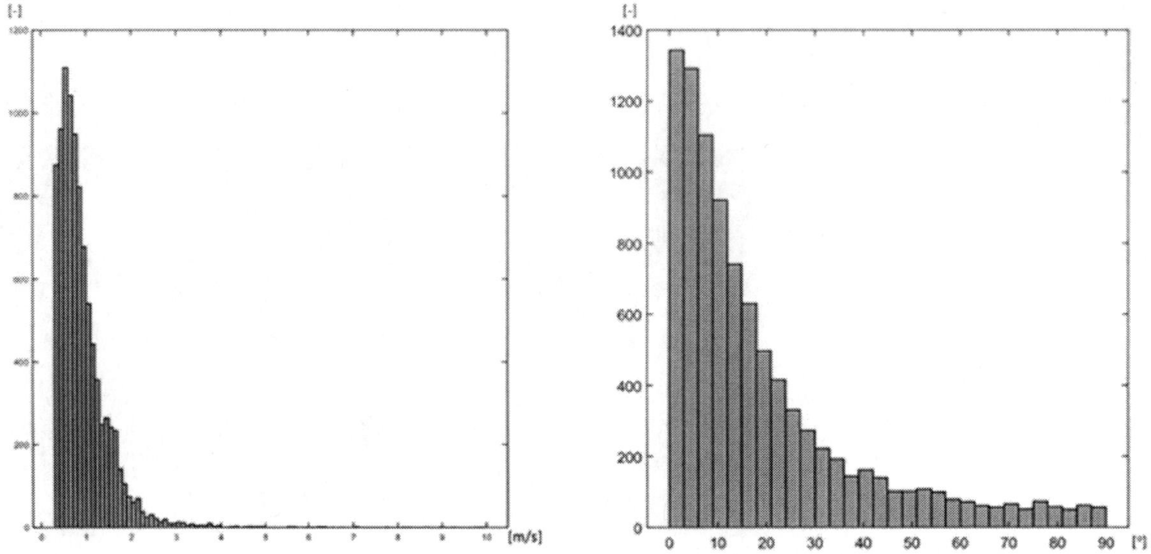

Figure 3. Histogram of movement speed and angle difference between steps

For the development of the Agent-Based Model that will recreate a selected pathway, all of the necessary data has been obtained. The essential model parts are the current location, the step length distribution, the step angle distribution and the turn vector (Figure 4). The step length was directly translated into the natural distribution model, where its metadata, such as the position of the semi-bell peek, range and standard deviation, were used in its description. Each step has been resolved with a random number that indicated the step deviation from the natural model. The step turn, was resolved in the nearly the same fashion as the step length. Data from angular distribution was re-calculated to absolute values to take into account all possible changes of the angle. It produced a semi-bell shaped natural distribution curve which includes data on step angle differences and the maximum difference between all the necessary statistical values. The maximum angle difference allowed locating the left and right arc ends. The turn vector always followed the general turn of the simulated agent from the previous step, and that is why the whole distribution arc was continually updating its step turn per time. It seems that the angular distribution was split into two semi arcs (the left with a red outline, the right one with a blue outline). For each step, it is necessary to decide which part of it will be taken into account. Selection of the semi arc part was made by measuring the distance from the intended goal to the arc ends. The smaller distance has always been selected. If both distances were even, selection of the semi-arc was random.

The designed movement model has been tested, and it has produced similar pathways in a normalized transition scenario. That is why it has been decided to test its capabilities in a virtual environment. For this purpose, four different layout cases were investigated. The developed movement model requires information on the transition towards the intended goal. If the randomly generated start and end point

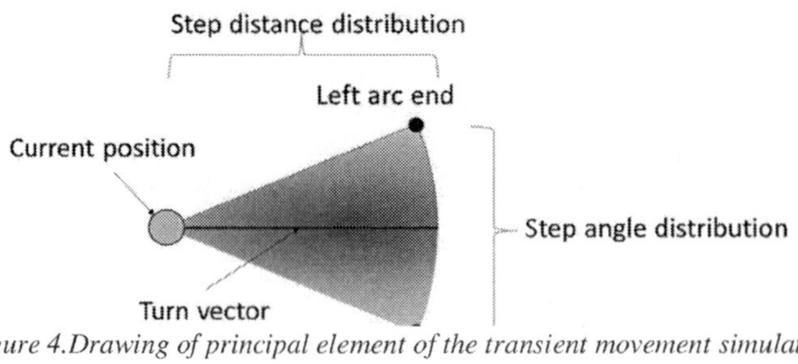

Figure 4.Drawing of principal element of the transient movement simulator

have some obstacle between them, the developed model will not be able to compile it. The model requires more information regarding the specific position of the investigated case. That is why test layouts have to be supplemented with additional information on local transition points. The points play the role of navigation points inside the tested cases. Location of navigation points was detected with a use of self-designed software. Each corner of the layout has been taken into account. During that phase, it was tested if other currents are also visible. If so, each pair provides one middle point between the visible corners. All middle points on the layout outline were discarded. Next, all the middle points identify connections between them. From now on these points are called orientation points. The identification process uses once again visibility analysis. Once the matrix of connection is established, the layout is ready to be tested for movement simulations. Each layout had one designed purpose of the testing. Scenarios naming is adequate to the figure included in results (Figure 5).

- Scenario A was investigating behavior in close corners;
- Scenario B was a testing transition between rooms in cross-like connections;
- Scenario C was focused on testing transition in an advanced room connection;
- Scenario D has tested the reaction of the model to the occurrence of possible circular patterns in the simulated space.

Once the location of beginning (A) and ending (B) of the pathway is established, the shortest pathway algorithm is used to describe the step of the agent. The procedure prescribes a schedule of the points that the agent should reach to achieve the end position. For example, it describes that to reach goal (B), the agent has a scheduled vector [A,1,3,6, B]. It means that from point (A) it should be focused on aiming towards and reaching orientation point no.1. First of all, in each step, the algorithm checks if end point (B) is reached. If not, the software movement procedure is initialized. While progressing towards this orientation point, it will detect that orientation point no. 3 is visible. From that event, the aim of the agent is updated to reach orientation point no. 3. This procedure continues until the end of the pathway has been reached. To avoid extension of the simulation time of the agent movement, if endpoint (B) is within reach of the agent, it finishes the movement simulation with a transition towards it. If during the simulation procedure the next step taken by the agent will cause over-crossing the space outline, it resolves the step selection once again. In critical condition, if it is not possible to resolve a movement without crossing the layout border, the simulation software cancels the last step, and it is resolved once again. To check the frequency of appearance of such situation, a specific counter has been applied in the code. Once all of the monitoring systems have been set up to track all the potential misconducts of the movement model, it is possible to initialize simulation procedure.

4. Results

For the model test trials, four different layout scenarios were designed in a black and white bitmap. Once all the necessary assumptions had been formulated, each layout was tested for generation of fifty pathways. The start and end points were generated at random. Due to that reason, it was impossible to predict if all of the selected areas layouts will be visited by the designed agent. The results of the simulated pathways were displayed in Figure 5.

Figure 5.Movement simulation results on selected floor layouts

5. Discussion & Conclusion

The performed simulation test shows that the suggested modeling approach has its potential applicability in building performance simulation. This study has to be treated as the first step towards transient occupant behavior modeling. The results of simulated pathways show that a few parts of occupant transition have to be improved. The next improvement steps should aim at a transition from the raster to vector data. Re-calculation of the whole matrix into a raster matrix can be overwhelming, for CPU or GPU. That is why it is important to deliver this calculation switch. Additionally, all of the designing software uses vector drawing as its basis. If the potential designers are to support their work with such a tool, it is critical that their workflow will not be disturbed by additional unnecessary tasks. Simulation of the transient occupant behavior has to reach the level of a complementary feature. During the simulation trials, none of the monitored errors were observed. The movement simulator respected the limitations imposed on all the four test scenarios.

With continuous work aimed at improvement of this method, it will eventually be able to obtain more precise data regarding occupant thermal sensation. With the use of this method, it may be coupled with a converging CFD simulation. If it holds information of the indoor air thermal condition, it is possible to track the history of an occupant's exposition to local conditions. Such asset can be used for evaluation of the already existing local thermal comfort models. Once the accuracy of the previously developed comfort models is taken into account, it will be discovered what their transient accuracy is. It can indicate the potential direction for future improvement towards development of the personal transient thermal comfort model. To reach that level of confidence, the proposed approach has to be put to more extensive trial. To do so, it is necessary to conduct a new series of measurements with wider spectrum of occupants included in the study. It has to be checked what are the pathway decision-making factors which influence the selection of a specific pathway. This knowledge can be obtained from investigating simple movement scenarios. Each future scenario will have to include the choice of at least two different pathways. Additionally, it will have to be checked how the beginning position influences the selection of a pathway. Once the above-mentioned parameters have been investigated, data collected in those experiments may be used for improvements in the movement model. The current state of development shows promise of a potential breakthrough in building performance simulations. If all the missing parts of the knowledge are verified and receive positive feedback, it will mean that this model could be treated as the first milestone in development of the Building Occupant Transient Agent-Based Model (BOT-ABM).

References

[1] A K. Parsons, 2014, *Human Thermal Enviroments*, (CRC Press)
[2] P. Fanger, 1970, *Thermal comfort*, (Danish Technical Press)
[3] J. Hviid and M. Baun Kjærgaard, 2018, *Activity-Tracking Service for Building Operating Systems*, (IEEE International Conference on Pervasive Computing and Communications Workshops)
[4] D. Yan, W. O'Brien, T. Hong, X. Feng, H. B. Gunay, F. Tahmasebi and A. Mahdavi, 2015, *Occupant behavior modeling for building performance simulation: Current state and future challenges*, (Energy and Buildings)
[5] T. Hong, J. Langevin and K. Sun, 2018, *Building simulation: Ten challenges*, (Building Simulation)
[6] I. B. Arief-Ang , M. Hamilton and F. D. Salim, 2018, *RUP: Large Room Utilisation Prediction with carbon dioxide sensor*, (Pervasive and Mobile Computing)
[7] J. Dziedzic, Y. Da and V. Novakovic, 2017, Occupant migration monitoring in residential buildings with the use of a depth registration camera, (Procedia Engineering)
[8] J. Dziedzic, Y. Da and V. Novakovic, 2018 , *Real Time Measurement of Dynamic Metabolic Factor (D-MET)*, (Springer Proceedings in Energy)
[9] Micorsof Xbox One, "http://www.xbox.com/," [Online], Available: http://www.xbox.com/en-US/xbox-one/accessories/kinect. [Accessed 10 1 2018].

IOP Conf. Series: Materials Science and Engineering **415** (2018) 012045 doi:10.1088/1757-899X/415/1/012045

The influence of building zones with different dynamic properties on the building energy performance

H Jędrzejuk

Institute of Heat Engineering, Faculty of Power and Aeronautical Engineering, Warsaw University of Technology, Nowowiejska 21/25, 00-665 Warsaw, Poland

Hanna.Jedrzejuk@itc.pw.edu.pl

Abstract. The paper presents an analysis of the energy performance of single-family buildings with two zones differing in dynamic properties. The procedure currently applied for determining the energy performance distinguishes several cases in which a division into separate calculation zones is required, whereas none of them sanctions dissimilarities of dynamic properties between parts of buildings. In order to determine the differences resulting from the separation into zones, e.g. the ground floor with a heavy structure and a lightweight usable attic, the monthly balance method was used as legally binding for establishing energy performance of buildings. Heat demand for heating and cooling a building was determined. The results obtained for a building treated as one zone and divided into two zones were compared. The analysis was carried out in a variant manner, i.e. assuming varying internal thermal capacity, usable space, surface of glazed partitions, and air exchange rate, for 59 locations in Poland. It was shown that, under assumption of the same useful area of upper and lower storey, in cases taken into consideration the maximal difference in energy demand for heating only is 5.41% (Nowy Sącz), but in cooling can reach as much as 46.75% (Zakopane).

1. Introduction

The principles of determining the energy performance of buildings currently in force in Poland are based on the PN-EN ISO 13790: 2009 standard [6] which specifies conditions for dividing a building into calculation zones. Generally, the partitioning of a building into thermal zones is required in several independent cases:

- set-point temperatures for heating or cooling of spaces differ by more than 4 K,
- spaces are serviced by different heating systems and different cooling systems (if any),
- less than 80% of the floor area is serviced by the same ventilation system, and
- the ventilation air flow relative to the floor area differs by more than four times in 80% of the floor area or doors are not likely to be opened frequently between spaces.

Among the above listed conditions, no special case is distinguished where a building consists of parts with different dynamic properties.

The dynamic properties of a building are influenced by, among others, internal thermal capacity. Internal heat capacity (or: thermal mass) is a heat capacity of building materials, which has an effects on fluctuation of the internal temperature [3].

This property enables storage a heat in building construction to reduce peak heating or cooling loads [2, 4] as well as energy needs [1, 9].

2. Mathematical model

The demand for useful heat for heating and cooling in each of the two assumed zones as well as in the entire building was determined using the monthly balance method according to the formula provided in PN-EN ISO 13790: 2009 [6]. In Poland, this algorithm is required to determine energy performance of a building [8]. Climate statistics published on the website of the Ministry of Investment and Development [5] were used in the calculations.

If the energy assessment takes account only of the building construction, i.e. without heating and cooling systems, then the building energy performance can be expressed by an indicator of the annual useful energy demand (EU_H) for heating and cooling (EU_C). This indicator is determined on the basis of the annual useful demand for heating (Q_H) and conditioned floor area of the space (A_f):

$$EU_{H,x} = \frac{Q_{H,x}}{A_{f,x}}, \text{ kWh m}^{-2}\text{ K}^{-1} \tag{1}$$

The additional subscript x is *1* for the first floor zone, *2* for the second floor zone, and *B* for the whole building treated as one conditioned zone.

Similarly, on the basis of the annual useful heat demand for cooling (Q_C), the annual useful heat demand indicator for cooling was determined as:

$$EU_{C,x} = \frac{Q_{C,x}}{A_{f,x}}, \text{ kWh m}^{-2}\text{ K}^{-1} \tag{2}$$

To evaluate the influence of different internal heat capacities on the energy efficiency of the building, the indicators were compared:

- for heating $EU_{H,Tot} = \frac{Q_{H,1}+Q_{H,2}}{A_{f,B}}$ and $EU_{H,B}$,
- for cooling $EU_{C,Tot} = \frac{Q_{C,1}+Q_{C,2}}{A_{f,B}}$ and $EU_{C,B}$.

The obtained values of the indicators were used to determine:

- the relative difference between annual useful energy demand indicators for heating:

$$\delta EU_H = \frac{EU_{H,Tot}-EU_{H,B}}{EU_{H,B}}, \%, \tag{3}$$

- the relative difference between annual useful energy demand indicators for cooling:

$$\delta EU_C = \frac{EU_{C,Tot}-EU_{C,B}}{EU_{C,B}}, \%, \tag{4}$$

- the relative root mean square of the annual useful energy demand for heating:

$$dRMS_H = \sqrt{\frac{\left(Q_{H,B}-Q_{H,1}\right)^2+\left(Q_{H,B}-Q_{H,2}\right)^2}{2}} \Big/ Q_{H,B}, \%, \tag{5}$$

- the relative root mean square of the annual useful energy demand for cooling:

$$dRMS_C = \sqrt{\frac{\left(Q_{C,B}-Q_{C,1}\right)^2+\left(Q_{C,B}-Q_{C,2}\right)^2}{2}} \Big/ Q_{C,B}, \%. \tag{6}$$

3. Case study

The impact of different dynamic properties of two zones on the value of the annual energy demand indicator for heating and cooling of a building was analysed by performing a number of variant calculations. The assumed data are provided in Table 1.

Maximum values that indicate differences in the energy performance of the analysed building due to different dynamic properties of the two zones are provided in tables 2 ÷ 3.

Due to specific metrological data, the maximum values the relative difference in heating mode occur in Nowy Sącz (Table 2), and in cooling mode occur in Zakopane (Table 3).

The analysis of the mathematical model applied leads to a general conclusion that the largest differences in the energy performance of a building, as described by the useful heat demand indicator for heating, will occur in the case of:

- the largest internal thermal heat capacity of the first storey and the lowest internal heat capacity of the second storey,
- the height standard of thermal insulation of external partitions,
- the lowest air change rate,
- the smallest conditioned space, and
- large heat gains, especially from solar radiation.

Similarly, while the useful heat demand indicator for cooling is evaluated the largest differences of the indicators can be expected when:

- the largest internal thermal heat capacity of the first storey and the lowest internal heat capacity of the second storey,
- the highest standard of thermal insulation of external partitions,
- the highest air change rate,
- the smallest conditioned space, and
- large heat gains, especially from solar radiation.

However, the calculations performed indicate how large these differences can be. For example, in the case shown in Table 2, the actual useful heat demand for heating, determined as the sum of the demand in each zone, is more than twice as high as that calculated for the entire building.

Table 1. Dataset for calculations.

Description	Specification and variants
Location	59 weather stations [5]
Shape of the building	Cuboid
Net height of a storey	2.80 m
Conditioned (e.g. usable) space of one storey (in brackets: the whole building)	50 (100) m^2; 75 (150) m^2; 100 (200) m^2
Internal heat capacity of the upper storey per conditioned area ($C_{mu,2}$)	260,000 J K^{-1} m^{-2}; 370,000 J K^{-1} m^{-2}; 400,000 J K^{-1} m^{-2}
Internal heat capacity of the lower storey per conditioned area ($C_{mu,1}$)	500,000 J K^{-1} m^{-2}; 600,000 J K^{-1} m^{-2}
Area of external partitions facing S and N	25% larger than the surface of external partitions facing E and W
Area of the glazed external partitions	The minimum variant: 1/8 of the conditioned (used) area The intermediate variant: the arithmetic mean of the minimum and maximum variants The maximum variant: the total surface of external partitions
Set-point temperature for space heating	20.0 °C
Set-point temperature for space cooling	24.0 °C
Air exchange rate	0.5 h^{-1}; 1.5 h^{-1}; 3.0 h^{-1}
Thermal insulation standard	The 2002, 2008, 2017, and 2021 requirements [7]

Across Poland, the relative differences in annual useful energy demand indicators for heating (δEU_H) vary between 5.41% (Nowy Sącz) and 1.82% (Suwałki) (Figure 1). The main data for Nowy Sącz location are included in Table 2. It is worth noting that these relative differences are always positive, which means that the useful heat demand for heating calculated to assess the energy performance of a building with two zones differing in dynamic properties is lower than the actual one.

The smallest yearly useful heat demand indicator for heating is observed in Nowy Sącz ($EU_{H,B}$ = 42.99 kWh m^{-2} a^{-1}). In that case the largest relative change of δEU_H (5.41%) is present (Figure 2). The highest yearly useful heat demand indicator for heating is noticed in Suwałki ($EU_{H,B}$ = 138.35 kWh m^{-2} a^{-1}), but a relative change δEU_C is equal 1.82%, when the absolute difference is 2.52 kWh m^{-2} a^{-1}. The relative root mean square of the annual useful energy demand for heating varies from 48.15% (Nowy Sącz) to 49.18% (Suwałki) (Figure 2).

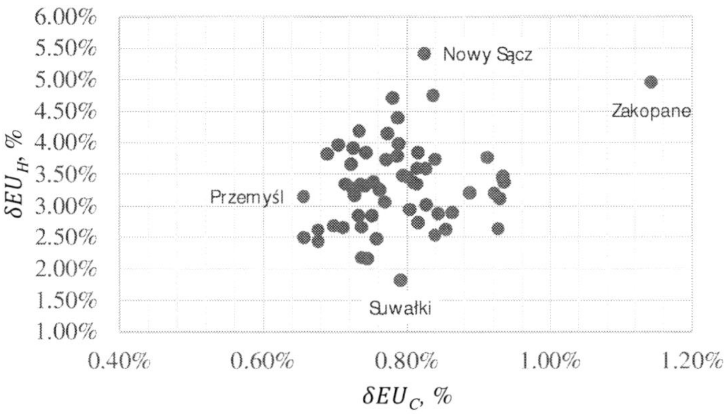

Figure 1. Maximal values of the relative difference between annual useful energy demand indicators for heating and the corresponding values in cooling mode.

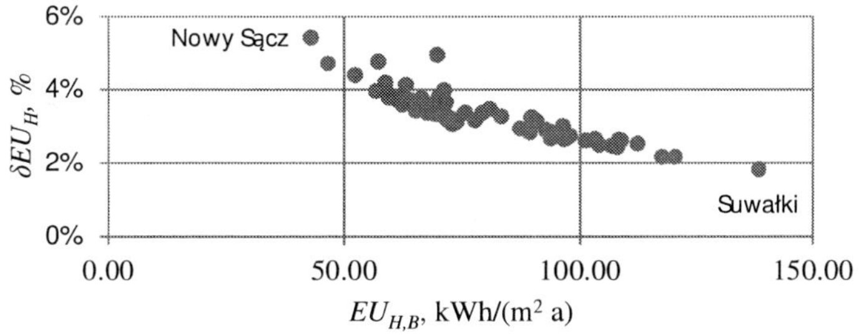

Figure 2. Relative difference between annual useful energy demand indicators for heating in relation to annual useful energy demand indicator for heating.

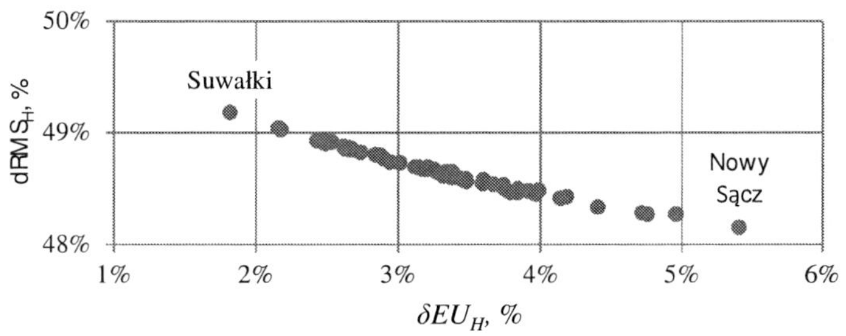

Figure 3. Relative root mean square of the annual useful energy demand for heating in relation to relative difference between annual useful energy demand indicator.

Monthly changes of the useful heat demand for heating for the Nowy Sącz location are presented in Figure 4. The monthly useful heat demand for heating of the two zones building is higher than one zone building. Due to the heat solar gains in January the useful heat demand is lower than in December.

A comparison of monthly heat demand for heating per conditioned area (Figure 5) shows that heating is noticed in the lower storey in 7 months, in the upper story in 10 months, while in the entire building it is equal 8 months.

Table 2. Main data for the chosen case for the heating mode of a building located in Nowy Sącz.

Results	Variant description
$EU_{H,1}$ = 27.32 kWh m^{-2} a^{-1}	$C_{mu,1}$ = 600,000 J K^{-1} m^{-2}
$EU_{H,2}$ = 43.95 kWh m^{-2} a^{-1}	$C_{mu,2}$ = 260,000 J K^{-1} m^{-2}
$EU_{H,Tot}$ = 71.27 kWh m^{-2} a^{-1}	The 2017 thermal insulation standard
$EU_{H,B}$ = 33.13 kWh m^{-2} a^{-1}	$A_{f,B}$: 150.00 m^2
$dRMS_H$ = 49,89%	Air exchange rate: 0.5 h^{-1}
δEU_H = 5.41%	Share of window surfaces in the total area of external partitions: N – 0%, S – maximum, W – maximum, E – maximum

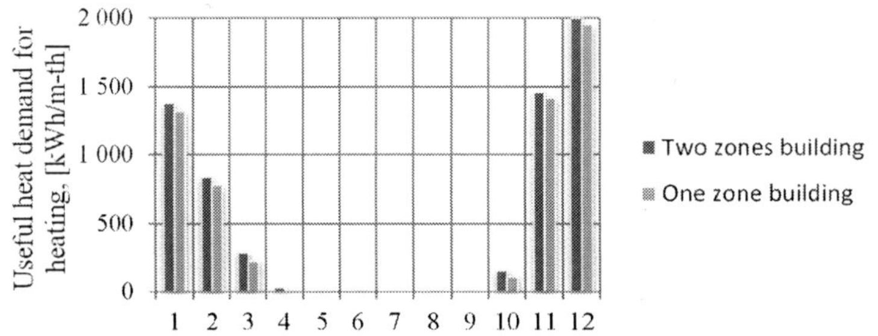

Figure 4. Monthly useful heat demand for heating for the Nowy Sącz location.

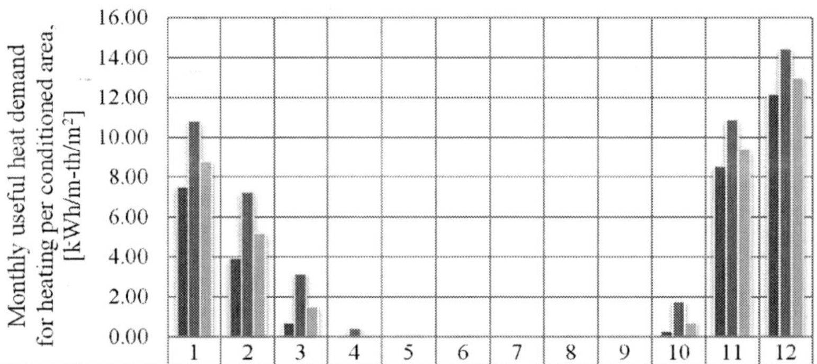

	1	2	3	4	5	6	7	8	9	10	11	12
Lower storey	7.51	3.93	0.72	0.02	0.00	0.00	0.00	0.00	0.00	0.30	8.54	12.17
Upper storey	10.85	7.26	3.16	0.44	0.04	0.01	0.00	0.00	0.07	1.76	10.91	14.48
One zone building	8.79	5.21	1.50	0.09	0.00	0.00	0.00	0.00	0.01	0.72	9.42	13.00

Figure 5. Monthly heat demand for heating per conditioned area for the Nowy Sącz location.

In the cooling mode, the largest differences occur in the case of:

- the largest internal thermal capacity of the first storey and the lowest internal heat capacity of the second storey,
- the worst standard of thermal insulation of external partitions,
- the largest air change rate,
- the smallest conditioned space, and
- the largest heat gains, especially from solar radiation.

Across Poland, the relative differences in annual useful energy demand indicators for cooling (δEU_C) vary between 46.75% (Zakopane) and 12.93% (Krosno) (Figure 6).

The smallest yearly useful heat demand indicator for cooling is observed in Zakopane ($EU_{C,B}$ = 1.39 kWh m^{-2} a^{-1}), then a small change of the absolute value causes a largest relative change of δEU_C (48.75%) (Figure 7). The highest yearly useful heat demand indicator for cooling is noticed in Krosno ($EU_{C,B}$ = 5.89 kWh m^{-2} a^{-1}), but a relative change δEU_C is equal 12.93%, when the absolute difference is 0.76 kWh m^{-2} a^{-1}.

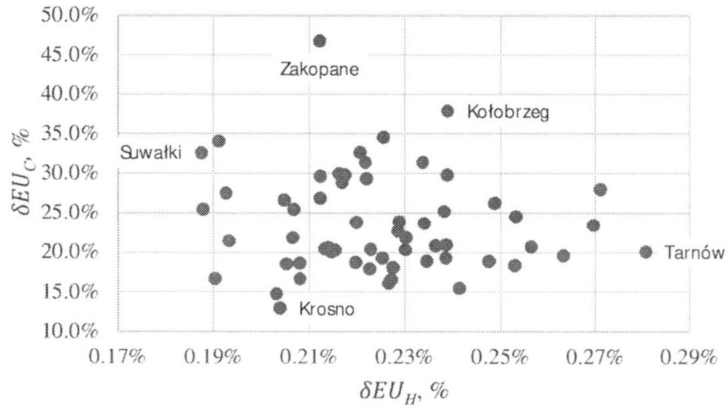

Figure 6. Maximal values of the relative difference between annual useful energy demand indicators for cooling and the corresponding values in heating mode.

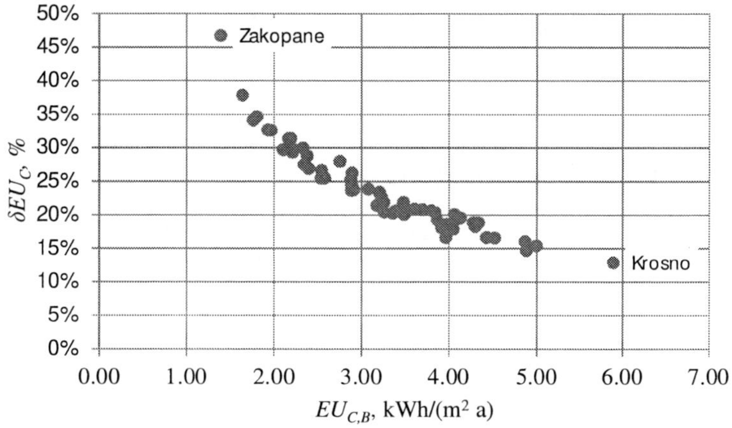

Figure 7. Relative difference between annual useful energy demand indicators for cooling in relation to annual useful energy demand indicator for cooling.

. In the extreme case at the Zakopane location, the useful heat demand indicator for cooling calculated as the sum of relevant indicators determined in the zones is 46.75% of the indicator characterizing the entire building (Table 3).

Monthly changes of the useful heat demand for cooling for the Zakopane location is presented in Figure 9. The sum of the useful energy demand for cooling of two zones exceeds analogical value calculated for one zone.

In Zakopane (Figure 10) lower storey requires cooling for 6 months, while for upper storey cooling is needed for the whole year. The entire building needs cooling in eight months.

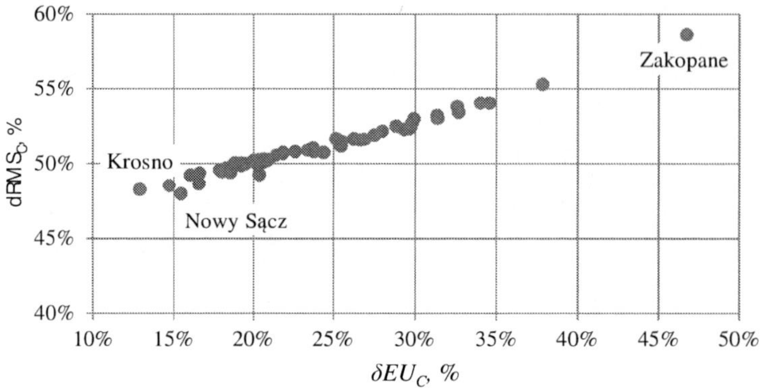

Figure 8. Relative root mean square of the annual useful energy demand for cooling in relation to relative difference between annual useful energy demand indicator.

Table 3. Main data for the chosen case for the cooling mode of a building located in Zakopane.

Results	Variant description
$EU_{C,1}$ = 0.59 kWh m^{-2} a^{-1} $EU_{C,2}$ = 3.50 kWh m^{-2} a^{-1} $EU_{C,Tot}$ = 2.04 kWh m^{-2} a^{-1} $EU_{C,B}$ = 1.39 kWh m^{-2} a^{-1} δEU_C = 46.75%	$C_{mu,1}$ = 600,000 J K^{-1} m^{-2} $C_{mu,2}$ = 260,000 J K^{-1} m^{-2} The 2021 thermal insulation standard $A_{f,B}$: 100 m^2 Air exchange rate: 3.0 h^{-1} Share of window surfaces in the total area of external partitions: N – 0%, S – minimum, W – minimum, E – minimum

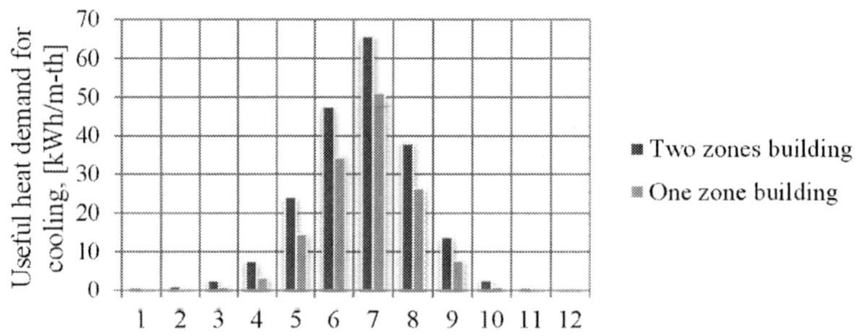

Figure 9. Monthly useful heat demand for cooling for the Zakopane location.

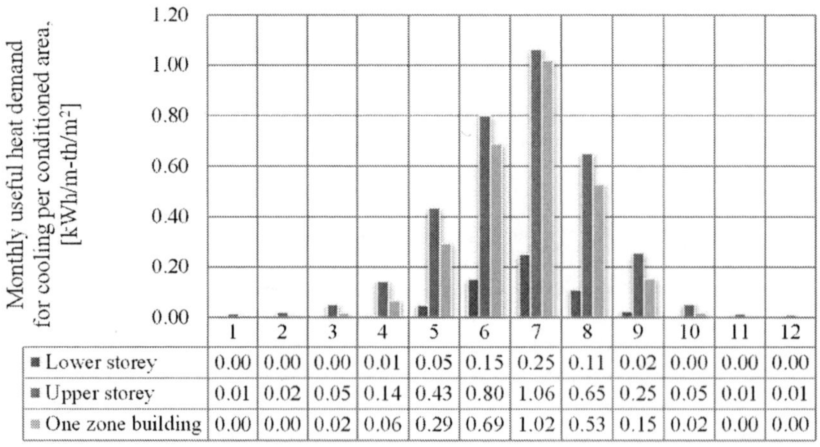

	1	2	3	4	5	6	7	8	9	10	11	12
Lower storey	0.00	0.00	0.00	0.01	0.05	0.15	0.25	0.11	0.02	0.00	0.00	0.00
Upper storey	0.01	0.02	0.05	0.14	0.43	0.80	1.06	0.65	0.25	0.05	0.01	0.01
One zone building	0.00	0.00	0.02	0.06	0.29	0.69	1.02	0.53	0.15	0.02	0.00	0.00

Figure 10. Monthly heat demand for cooling per conditioned area for the Zakopane location.

4. Conclusions

Different energy characteristics of the distinguished building zones result not only from different dynamic properties but also from a different thermal insulation of external partitions, for example between the roof and the ceiling above the unheated basement or ground floor.

In buildings with a lower thermal insulation standard, a smaller influence of the internal thermal capacity on the differences in the energy performance indicators is observed than in the case of very well insulated buildings.

Generally in the Polish climatic conditions, in the case of the same useful area of upper and lower storey, the influence of internal heat capacity on useful heat indicator for heating can be omitted, because the relative difference of that do not excide 5.41% (Nowy Sącz), and in absolute value is a maximum of 3.46 kWh m^{-2} K^{-1} (Zakopane).

The relative root square mean in the heating mode varies between 48.15% (Nowy Sącz) and 49.91% (Suwałki) (Figure 2), and in cooling mode changes from 47.99% (Nowy Sącz) to 58.64% (Zakopane) (Figure 8), and the analysed effect of internal heat capacity is more important in the cooling mode.

Disregarding the division into two zones due to the different dynamic properties slightly improves the value of the building assessment indicator.

In heating mode the division into two zone due a to different internal heat capacity can be omitted because of a neglected effect on the useful heat demand indicator (maximum: 5.41%). In cooling mode that effect can be neglected because of the small absolute value (maximum: 3.46 kWh m^{-2} K^{-1}), even if the relative difference between useful heat demand indicator is equal to 46.75%. However, noticeable differences appear in the number of months influencing the value of heat demand for heating and cooling.

It can be expected that in the case of a building with two zones differing significantly in terms of conditioned area the division into zones should be introduced due to their different dynamic properties.

References

[1] Aste N., Angelotti A., Buzzetti M. 2009 The influence of the external walls thermal inertia on the energy performance of well insulated buildings *Energ Buildings* **41** pp 1181–1187

[2] Balaras C.A. 1996 The role of thermal mass on the cooling load of buildings. An overview of computational methods *Energ Buildings* **24** pp 1-10

[3] Haase M., Andresen I. 2007 Thermal Mass Concepts. State of the art. SINTEF Report No. SBF BK A07030

[4] Kosny J., Petrie T., Gawin D. Childs P., Desjarlais A., Christian J. 2006 Thermal Mass -Energy Savings Potential in Residential Buildings. (https://www.researchgate.net/publication/268411404_Thermal_Mass_-Energy_Savings_Potential_in_Residential_Buildings)

[5] Ministry of Investment and Development. Data for energy calculations of buildings, in Polish (http://www.miir.gov.pl/strony/zadania/budownictwo/charakterystyka-energetyczna-budynkow/dane-do-obliczen-energetycznych-budynkow-1/)

[6] PN-EN ISO 13790:2009 Energy performance of buildings -- Calculation of energy use for space heating and cooling, in Polish

[7] Regulation of the Minister of Infrastructure of 12 April 2002 on technical specifications for buildings and their location. Journal of Laws 2015, Item 1422, consolidated text, amended in Journal of Laws 2017, Item 2285, in Polish

[8] Regulation of the Minister of Infrastructure and Development of 27 February 2015 on the methodology for calculating the energy performance of the building or part of the building and the preparation and design of energy performance certificates. Journal of Laws 2015, Item 376, amended in Journal of Laws 2017 Item 1100, in Polish

[9] Reilly A. Kinnane O. 2017 The impact of thermal mass on building energy consumption, *Appl Energ* **198** pp 108-121

IOP Conf. Series: Materials Science and Engineering **415** (2018) 012046 doi:10.1088/1757-899X/415/1/012046

Using CFD software for the evaluation of hygrothermal conditions at wall-window perimeters

A Stolarska[1], J Strzałkowski[1], Halina Garbalińska[1]

[1] Department of Building Physics and Building Materials, Faculty of Civil Engineering and Architecture, West Pomeranian University of Technology Szczecin, al. Piastów 50, 70-311 Szczecin, Poland

jstrzalkowski@zut.edu.pl

Abstract. This article evaluates the results of heat and moisture analyses for areas adjacent to a window, mounted in various ways to an external wall. The parameters for the window were assumed to be constant. The calculations were focussed on evaluating the effect of installing a window in a wall and the influence of the wall's construction on the quality of the microclimatic conditions in the contact zone of these two components. The following variables were analysed: a partition without insulation, a partition with thermal insulation, with additional expanding tape and with additional aerogel mat. Each the variable was additionally diversified with respect to the location of the window within the wall. The influence of the material used for the load-bearing wall layer was also analysed in the simulations. The variables were modelled using CFD (computational fluid dynamics) software. For all variables, the cumulative value of heat flux and linear thermal transmittances were calculated. Additionally, distributions of the 0 isotherm were determined and the f_{Rsi} temperature factors were calculated. These results have been thoroughly analysed to determine the most beneficial solution that provides optimal thermal insulation and reduced risk of mould. Our findings can help ensure the maintenance of an ideal microclimatic conditions and reduced energy wastage in the contact zone of windows with external walls.

1. Introduction

One of the requirements of energy-efficient buildings is the airtightness of the building envelope and any possible contact zones of the components. Ensuring proper airtightness and thermal insulation requires, among other things, the appropriate joining of the windows with the wall. Insufficient amount of attention given to these details may reduce the benefits of installing well-insulated windows in walls meeting the requirements for thermal transmittance *U*. Wrong product installation can also result in the occurrence of air leakages and thermal bridges. In the energy balance of buildings, in addition to heat losses due to air leakages, thermal bridges play an important role. While the modern construction and material solutions for building partitions and window frames contribute to the reduction of heat loss, their joining in sensitive areas still remains vulnerable. Therefore, more attention should be paid to the methods of joining windows with partition walls in order to minimize its effect on heat loss.

A similar problem was already discussed in article [1]. The article assessed the influence of the window frame location in the window opening on the reduction of the effects of thermal bridges. Five different wall constructions with two windows of different thermal transmittance *U* values were studied. The results of the thermal simulations demonstrated that the window location has a decisive influence on the amount of cold bridge energy loss. For each type of wall, optimal window location was found. In some cases, placing the windows in the most energy efficient location reduced the linear thermal transmittance by more than 50%. Within the studied window locations, the temperature on the inner surface of the windowpane groups was weakly dependent of the window location.

Variant installation of windows in ceramic wall also evaluated in [2], using simulations generated by the THERM software. The results showed a consistent reduction of up to 75% of linear thermal transmittance as a result of moving the window from the internal to the external position.

In turn, the paper [3] presents the results of studies on the airtightness of the window-wall interface in 13 different installation variants. The obtained results proved that it is possible to obtain very high performance using the polyurethane foam, hermetic membranes, plywood frame and plaster and seal. The authors conclude that in locations where good airtightness is required, mineral fibre, partial polyurethane filling and plaster without seal should be avoided.

The work [4], on the other hand, analyses the influence of windows located in the lintel area. Six installation variants were investigated. The resulting ψ values and the f_{Rsi} were assessed. The study confirmed the importance of the window position in the improvement of thermal properties and the avoidance of condensation in this area.

The authors of [5] focused on the influence of the material of the lintel beam and its design on the heat flow through exterior wall with a window. Such beams are used as supporting elements in the installation of windows in the thermal insulation layer. The study analysed temperature distribution in the wall, the 0°C isotherm gradient and the possible occurrence of condensation inside the wall. Based on the results of numerical analysis and experimental tests, it was found that moving the window towards the insulation layer using a mounting element made of high-performance materials and increasing the thickness of the insulating layer in the outer wall eliminates the risk of condensation.

An important element of the study is also the solution applied in the window frame construction. Numerical CFD simulations, this time only for the window frame, were presented in paper [6]. The study demonstrates the two-dimensional numerical method for the evaluation of the thermal behaviour solely of the framework.

Other research concerning minimization of windows thermal brides were described in [7] and [8]. In the research [7] a new innovative insulating coating which can be used to limit thermal bridge effects was evaluated. In addition, the authors computed the cooling/heating load coming from the windows offset thermal bridges of a typical French house before and after adding the insulating coating.

The authors of [9] demonstrated an application of the quantitative infrared thermography technique to evaluate the heat losses through windows thermal bridges. This kind of research can be a proper way to confirm CFD simulations. Other thermal imaging measurements analysis were presented in [10].

2. Materials and methods

This article focuses on the solution for the installation of windows in the sill area. The analysis of the solution of this critical zone was performed using the numerical CFD modelling which, as a rule, is used for the simulation of stationary and non-stationary thermal processes [11]. The analysed interface was modelled two-dimensionally. 10 cases were considered presented in figure 1 and figure 2.

Options were diversified in terms of window location in the wall, i.e. in the external face of the wall (variants a) and in the centre of the wall (b). All variants assumed that the window had thermal transmittance equal to 0.830 (W/m^2·K). The coefficient U of the concrete 24 cm thick wall was 3.310 (W/m^2·K), while total coefficient U of the partition with an extra layer of 15 cm thick foam amounted to 0.218 (W/m^2·K).

The simulation included the following variants:

- variant 1 – wall without insulation, windows installed using assembly foam,
- variant 2 – wall without insulation, windows installed using expanding tape, application of aerogel mat on the outside,
- variant 3 – wall with insulation, windows installed using assembly foam,
- variant 4 – wall with insulation, windows installed using expanding tape, application of aerogel mat on the outside,
- variant 5 – wall with insulation, windows installed using styrofoam mould.

IOP Conf. Series: Materials Science and Engineering **415** (2018) 012046 doi:10.1088/1757-899X/415/1/012046

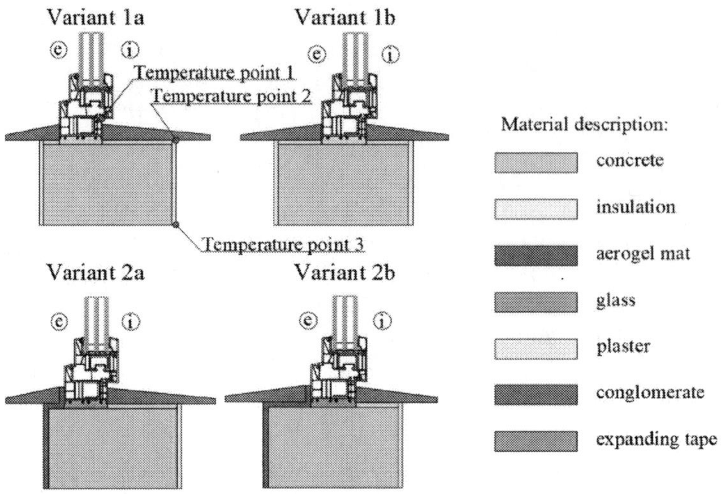

Figure 1. The diagrams of walls variants without thermal insulation.

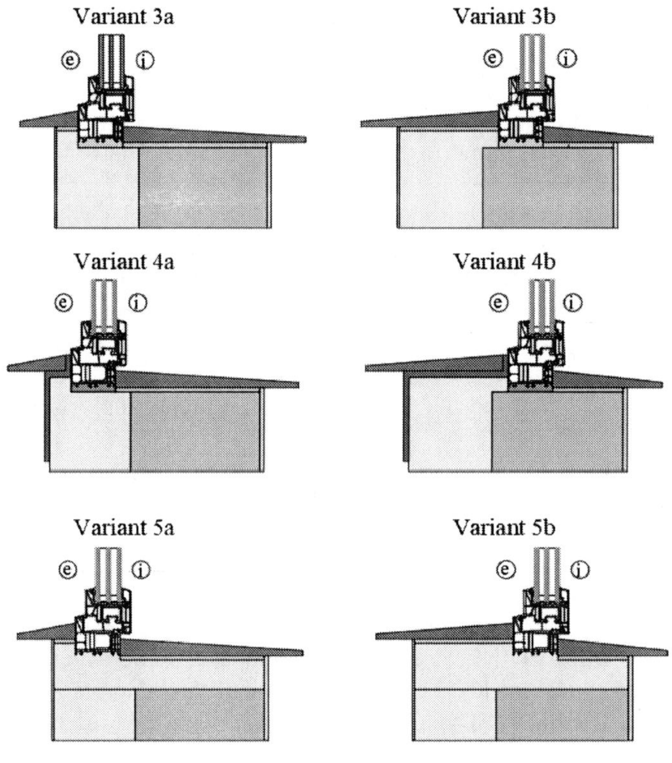

Figure 2. The diagrams of walls variants with thermal insulation.

The computational model was dimensioned in accordance with the [12] standard. In all variants, the same internal θ temperature of +20°C and external θ_e temperature equal to -16°C were assumed. The adopted surface resistance of the boundary layer for the outer surface was 0.04 $m^2{\cdot}K/W$ and 0.13 $m^2{\cdot}K/W$ for interior surfaces, while for determining temperatures necessary for the calculation of the temperature factor f_{Rsi} the R_{si} resistance was increased to 0.25 $m^2{\cdot}W/K$. The horizontal boundaries of the model were identified as adiabatic.

3. Results and discussion

The results obtained were: total heat flux Q [W], thermal coupling coefficient L^{2D} [W/m·K] and temperatures in all three set points of the partition wall shown in figure 1. The third point was placed 15 cm from the bottom of the window sill. The obtained results are collected in table 1 and in figure 5.

Table 1. The total energy of each variant and the thermal coupling coefficients.

Variant	Q [W]	L_{2D} [W/m·K]
Variant 1a	0.1384	3.843
Variant 1b	0.1365	3.791
Variant 2a	0.1348	3.745
Variant 2b	0.1345	3.736
Variant 3a	0.0274	0.760
Variant 3b	0.0294	0.818
Variant 4a	0.0272	0.756
Variant 4b	0.0286	0.795
Variant 5a	0.0276	0.768
Variant 5b	0.0281	0.781

The Q values point to clear differences in the variants without wall insulation (variants 1 and 2) compared to insulated walls (variants 3, 4, 5). A more than fivefold reduction in heat flux in favour of the thermally insulated walls was observed. This had its impact on the values of the linear thermal transmittance ψ [W/m·K]. Figure 3 shows the extent of the influence of the window location and the methods applied for the insulation of this zone.

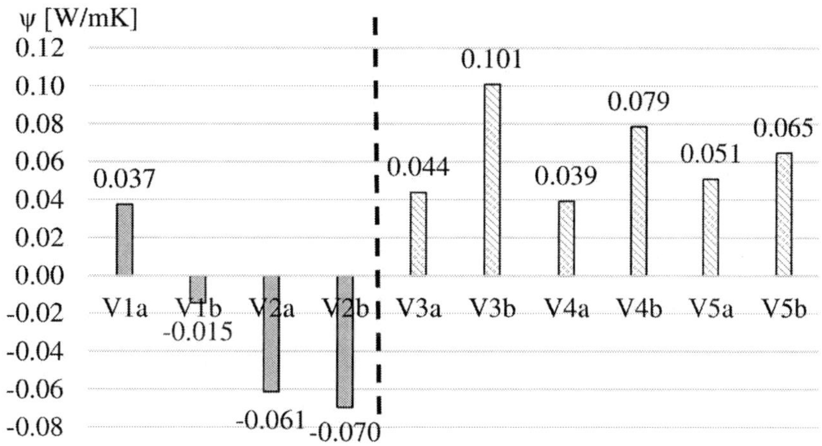

Figure 3. The linear thermal transmittances of the tested variants.

In the case of the first two variants with non-insulated walls and very well thermally insulated windows, it was shown that the use of a short strip of aerogel mat contributes to obtaining better linear thermal transmittance values (negative values of the coefficient ψ). In these cases we observe larger losses of heat due to transmittance through walls than in the case of thermal bridges.

IOP Conf. Series: Materials Science and Engineering **415** (2018) 012046 doi:10.1088/1757-899X/415/1/012046

In variants 3, 4 and 5, the installation of the window in the insulation layer (variants a), resulted in a reduction of the coefficient ψ compared to the values obtained in case of centrally positioned windows (b). The most evident influence of the window location was demonstrated in variant 3, where the use of foam more than doubled the values measured. The application of the styrofoam mould under the window sill also proved to be a beneficial option (variant 5). However, the best solution was the use of expanding tape and aerogel mat which, in case of the window being installed in the insulation layer, provided the lowest value of the parameter ψ compared to solutions proposed in variants 3 and 5.

As a result of the simulations performed, charts showing temperature distribution in the analysed variants were obtained. The example temperature distribution for variants 3a and 3b is shown in figure 4.

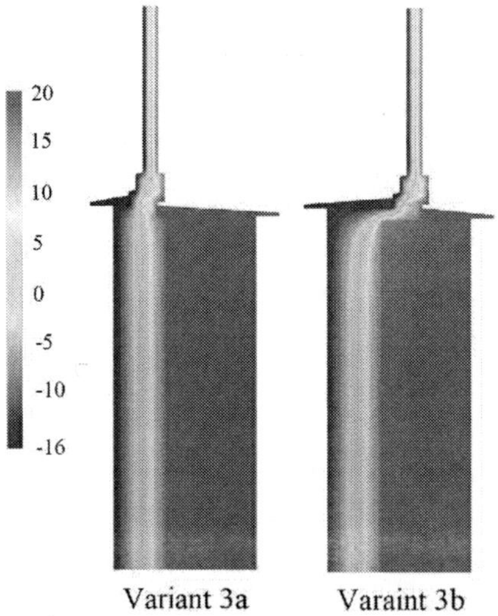

Figure 4. The example temperature distribution for variants 3a and 3b.

In turn, figure 5 shows the temperature values for the three critical points in the window-wall interface in the area of the window sill and the wall. In the case of the first two variants, notably the lowest temperatures in all three analysed areas were obtained. In the worst case, the wall surface temperature was even close to 0°C (V1a and V1b). It was demonstrated once again that the application of aerogel mat and expanding tape contributes to the increase of temperatures in all three evaluated points.

In other variants, the obtained temperature values for the second (under the sill) and the third point (on the wall surface) were similar. The effect of window location on the obtained temperature was observed. While the installation of the window in the insulation layer is conducive to the increase of temperatures in these two points, in the first point (on the frame surface) it leads to lower temperature values in relation to the values of the variants where the window was installed in the wall layer.

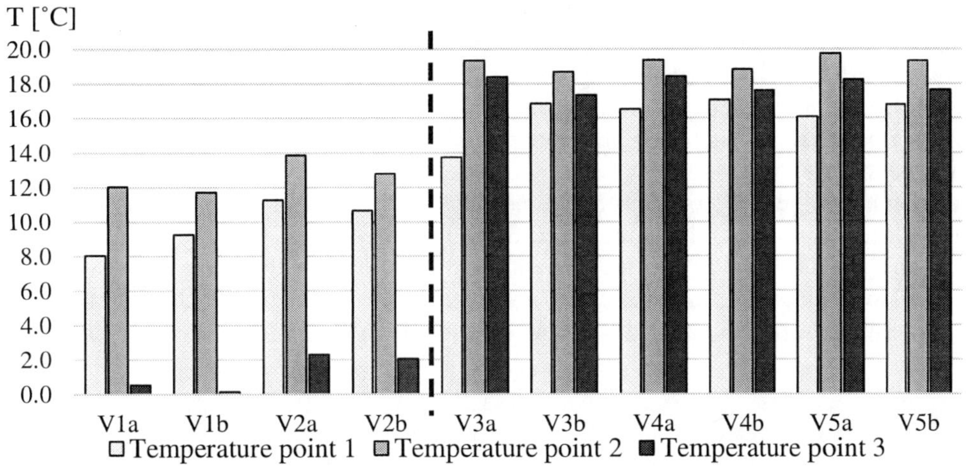

Figure 5. The temperature values in the three specified points.

Another important aspect in the diagnosis of the interface elements of buildings is also the assessment of the risk of surface condensation and the development of mould. Table 2 shows the obtained temperature values in point 1 (according to figure 1a). On this basis, a boundary value for humidity was determined that meets the dew point condition and the value of the f_{Rsi} temperature factor. The results show that the use of adequate insulation in the window installation area by the windowsill increases the temperature in point 1: in the best scenario, even by 9°C. At the same time, it ensures maintaining the temperature factor at a sufficiently high level and above the critical value of 0.72 [13]. This condition is not met solely by the variant 1, in which no insulation in the area of the thermal bridge was used.

What is more, table 2 shows the specified permissible humidity values below which there is no risk of occurrence of condensation based on the dew point temperature. The best results were obtained for the variant 4b, for which the relative humidity would have to exceed 83 % in order for the condensation to occur in point 1 within the specified temperature conditions.

Table 2. The comparison of hygrothermal parameters.

Variant	The temperature point 1 [°C]	Boundary humidity that meets the requirements of dew point [%]	f_{Rsi} [-]
Variant 1a	8.04	45.92	0.67
Variant 1b	9.25	49.84	0.70
Variant 2a	11.27	57.05	0.76
Variant 2b	10.67	54.82	0.74
Variant 3a	13.74	67.09	0.83
Variant 3b	16.84	81.85	0.91
Variant 4a	16.52	80.21	0.90
Variant 4b	17.06	83.00	0.92
Variant 5a	16.08	77.99	0.89
Variant 5b	16.77	81.49	0.91

Another important issue is the depth of frost penetration of walls. Data concerning this aspect can be seen in the charts in figures 6 and 7. They present the distribution of the zero isotherms for all the five variants – subdivided to non-insulated (figure 6) and insulated walls (figure 7). Out of all the variants without wall insulation, the most profitable was the variant 2 insulated by means of aerogel mat and expanding tape. The use of such type of insulation significantly shifts the 0°C isotherm towards the negative temperatures.

On the other hand, the 0°C isotherms in the insulated wall variants in all cases show that the freezing area is in the insulation layer. The comparison demonstrates the difference between the variants with windows in the insulation layer and the variants with windows in the central area of the wall. The window installation place has, in these cases, a significant influence on the characteristics of the zero isotherm in the window sill area. For the variant 5b using the styrofoam mould, the values are the least profitable.

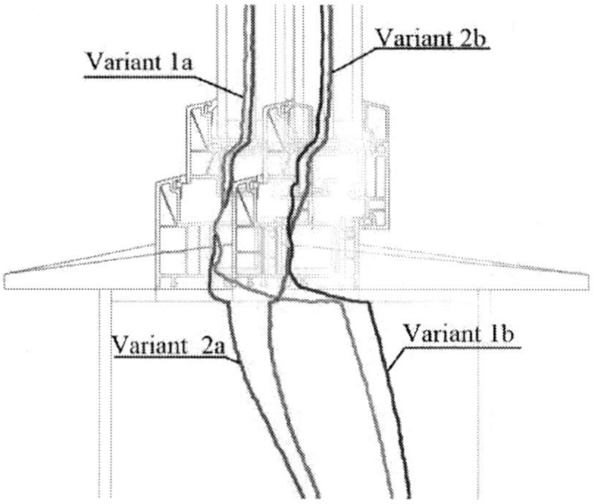

Figure 6 The isotherm 0°C distributions of variants without thermal insulation.

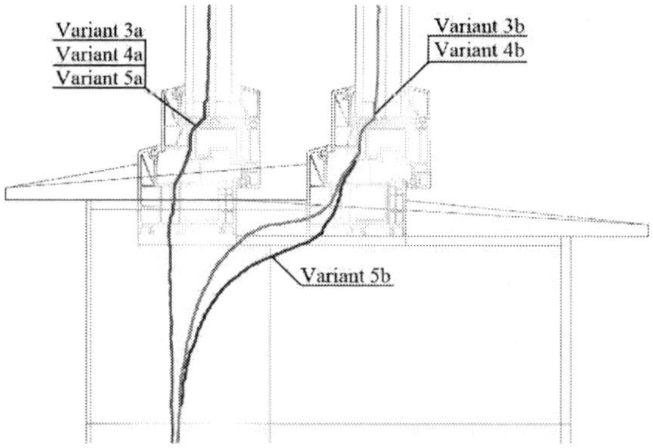

Figure 7. The isotherm 0°C distributions of variants with thermal insulation.

4. Conclusions

The article discusses the problems concerning the occurrence of thermal bridges located in the window-wall interface at the height of the window sill. The results of the simulation confirmed the importance of finding an appropriate solution to this issue with regard to the location of window frames in the wall and the application of proper insulation materials. The most profitable method of window installation among the analysed cases is the location of the window within the insulation layer. The paper also shows the positive effect of the application of aerogel mats and expanding tapes which leads to a decrease in the value of the linear heat transfer coefficient. It was demonstrated that the correct location and insulation of windows in the sill area is important for the prevention of condensation and therefore should be subject to assessment. Moreover, attention was drawn to the problems that arise in the installation of windows in walls with insufficient insulation parameters. In this case, improvements can be obtained by using the aforementioned aerogel mats and expanding tapes.

Another important issue for persons creating energy saving plans for buildings is also the source of the adopted values of the linear thermal transmittance ψ. The linear thermal transmittance of the thermal bridge should be determined according to the standard [12] (numerical calculations) or, by approximation, using the tabulated values specified in the standard [14]. However, the tables included in the standard [14] do not feature many cases relevant to the analysed variants as they do not take into account the newest insulation solutions available nowadays for this critical zone. As a result, this can lead to significant inaccuracies in the calculations of heat loss. It is therefore reasonable to perform simulations for such kinds of thermal bridges and to include in the calculations the exact values reflecting the actual solutions applied.

References

[1] Misiopecki C, Bouquin M, Gustavsen A and Jelle B P 2018 Thermal modeling and investigation of the most energy-efficient window position *Energy and Buildings* **158** 1079–86

[2] Cappelletti F, Gasparella A, Romagnoni P and Baggio P 2011 Analysis of the influence of installation thermal bridges on windows performance: The case of clay block walls *Energy and Buildings* **43** 1435–42

[3] Van Den Bossche N, Huyghe W, Moens J, Janssens A and Depaepe M 2012 Airtightness of the window–wall interface in cavity brick walls *Energy and Buildings* **45** 32–42

[4] Sierra F, Gething B, Bai J and Maksoud T 2017 Impact of the position of the window in the reveal of a cavity wall on the heat loss and the internal surface temperature of the head of an opening with a steel lintel *Energy and Buildings* **142** 23–30

[5] Adamus J and Pomada M 2017 Analysis of heat flow in composite structures used in window installation *Composite Structures*

[6] Malvoni M, Baglivo C, Congedo P M and Laforgia D 2016 CFD modeling to evaluate the thermal performances of window frames in accordance with the ISO 10077 *Energy* **111** 430–8

[7] Ibrahim M, Biwole P H, Wurtz E and Achard P 2014 Limiting windows offset thermal bridge losses using a new insulating coating *Applied Energy* **123** 220–31

[8] Ben-Nakhi A E 2002 Minimizing thermal bridging through window systems in buildings of hot regions *Applied Thermal Engineering* **22** 989–98

[9] O'Grady M, Lechowska A A and Harte A M 2018 Application of infrared thermography technique to the thermal assessment of multiple thermal bridges and windows *Energy and Buildings* **168** 347–62

[10] Garrido I, Lagüela S, Arias P and Balado J 2018 Thermal-based analysis for the automatic detection and characterization of thermal bridges in buildings *Energy and Buildings* **158** 1358–67

[11] Stolarska A and Strzałkowski J 2017 Modelling of Edge Insulation Depending on Boundary Conditions for the Ground Level *IOP Conf. Ser.: Mater. Sci. Eng.* **245** 042003

[12] PN-EN ISO 10211:2017 Thermal bridges in building construction. Heat flows and surface temperatures. Detailed calculations

[13] Rozporządzenie Ministra Infrastruktury i Budownictwa z dnia 14 listopada 2017 r. zmieniające rozporządzenie w sprawie warunków technicznych, jakim powinny odpowiadać budynki i ich usytuowanie, Dz. U. poz. 2285 (in Polish)

[14] PN-EN ISO 14683:2017 Thermal bridges in building construction. Linear thermal transmittance. Simplified methods and default values

Mathematical modeling of the building thermal state taking into account the heat and energy impact of the environment

A Malinowski, A Muzychak

Department of electrisity and control systems, Lviv Polytechnic National University, 29a Bandery str., Lviv, 79013, Ukraine

E-mail: malinowski@polynet.lviv.ua

Abstract. There are some difficulties in analysing the actual thermal state of a building under changing conditions of the environment, particularly solar insolation. The methodological approach to overcoming these difficulties and the corresponding mathematical models of the thermal state of a building are proposed. The base of the models is a power circuit in which there are described all components of heat losses and heat revenues. Controlled and uncontrolled heating systems are considered by representing heating devices as a source of effort or as a source of flow. The proposed models as a result of the analysis of the thermal state allow obtaining energy indicators of the building and an objective picture of the thermal comfort in the rooms simultaneously.

1. Introduction

Nowadays there are two problems in the focus of dwelling building engineers, the first problem is a microclimate quality improvement and the second one is an energy efficient increasing. Since the improving of microclimate causes the increasing of energy consumption, these two problems are combined in essence. The searching for these problems' solutions has led to a new conception of ecological building [1] which purpose is to provide high air quality at minimum energy consumption. There are different approaches and software for resolving the particular problem, for example, one of which is specialized program EnergyPlus [2]. The approach considered in the article is common for different kinds of energy circuits and is a prerequisite for the establishment of methodological bases for analyzing the state of the building and its energy supply systems.

According to the system approach the mathematical model of a building thermal state as a single heat energy system consists of three interrelated submodels [3]:

- the mathematical submodel of heat transfer through an enclosure of a building,
- the mathematical submodel of heat accumulation characteristics of a building enclosure and radiation and convective heat exchanging in building rooms,
- the mathematical submodel of heat and energy impact of the environment on a building.

Such approach allows us to divide the complex task of thermal steady-state analyzing into simpler parts and to introduce them into a general model as needed. There are several approaches to build the mathematical submodel of heat transfer. For analyzing of steady-state it is worthwhile to use the submodel of heat transfer in the form of an array of two-terminal resistive components that allows us to represent thermal characteristics of enclosure constructions with varying degrees of detail [4]. The next step is formation an equivalent circuit (a power circuit) of the thermal state in a building which allows using the network approach of the theory of power circuits for further analysis.

According to the theory of power circuits the power circuit can be described by following system of equations [4, 5]

$$\begin{cases} \mathbf{\Pi} \boldsymbol{X} = -\boldsymbol{F}, \\ \mathbf{\Gamma} \boldsymbol{Y} = 0, \\ \boldsymbol{Y} = \boldsymbol{E} - \mathbf{R}\boldsymbol{X}, \end{cases} \tag{1}$$

where $\mathbf{\Pi}$, $\mathbf{\Gamma}$ are incidence matrix and circle matrix, $\boldsymbol{X}, \boldsymbol{Y}$ are column vectors of flows (serial variables) and efforts (parallel variables) of brunches, $\boldsymbol{E}, \boldsymbol{F}$ are column vectors of effort sources and flow sources and $\mathbf{R}$ is resistance matrix of resistive components.

Such mathematical model of the thermal state in a building allows us to analyze a thermal steady-state for the buildings in which there are heated rooms, unheated premises and spaces with forced cooling. This corresponds to the requirements of the current normative documents in particular [6] and [7], which require the division of a building into separate zones and the calculation energy demands for heating and cooling of each zone.

The mathematical submodel of heat transfer should be complemented with the mathematical submodel of heat and energy impact of the environment and it is the purpose of the article. The mathematical submodel of heat accumulation characteristics of a building has an important role in transient processes analyzing however it is possible to omit it in steady-state analyses.

2. The mathematical model of the thermal state in a building

Applying of the proposed model will be shown on an example of a representative building. The building consists of three floors (height of each floor is 3 m) on each floor there are four flats. There is a basement in the building.

Plan of a typical floor (except a ground floor) and main dimensions are shown in figure 1.

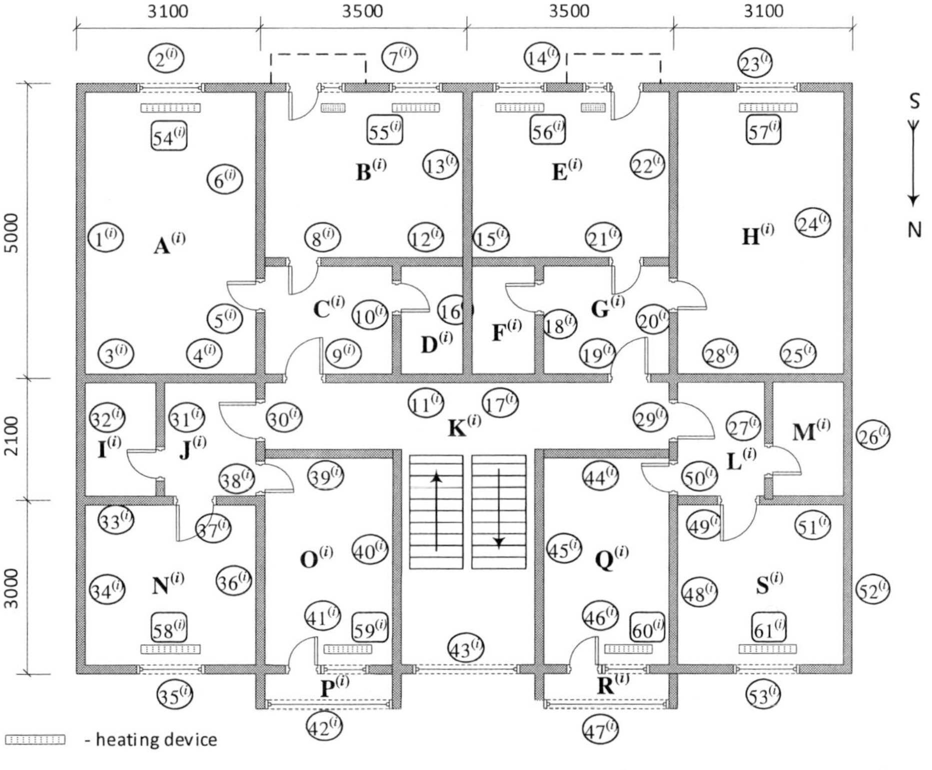

Figure 1. The plan of a typical floor of the building.

The numbers in circles indicate the main enclose constructions and the numbers in rounded rectangles indicate the heating devices. Bold capital letters indicate rooms' names of the building. Index in all cases indicates the number of a floor. The orientation of the building relative to the world is also shown in figure 1.

The analysis of a thermal state in the building is performed from simpler case to more complicated one. At first, an only mathematical model of heat transfer including the heating system is considered. It is necessary for the reason that such kind of model is the basis for determining rated (peak) mode of a heating system.

2.1. The mathematical model of a thermal state in the building that does not take into account heat and energy impact of the environment

According to the [4] each element of an enclose construction should be represented by at least one separate two-terminal component in the submodel of heat transfer. In order to simplify the presentation of the material the whole possible array of resistive components for each room is simplified to a seven-terminal network (figure 2(b)). First six equivalent resistive components $R_{e.s1}$, $R_{e.s2}$, $R_{e.s3}$, $R_{e.s4}$, $R_{e.c}$, $R_{e.f}$ reproduce equivalent thermal characteristics of main enclose constructions (walls S1, S2, S3, S4, celling C and floor F in figure 2(a)) and seventh equivalent resistive component $R_{e.i}$ reproduces heat losses through air infiltration (ventilation).

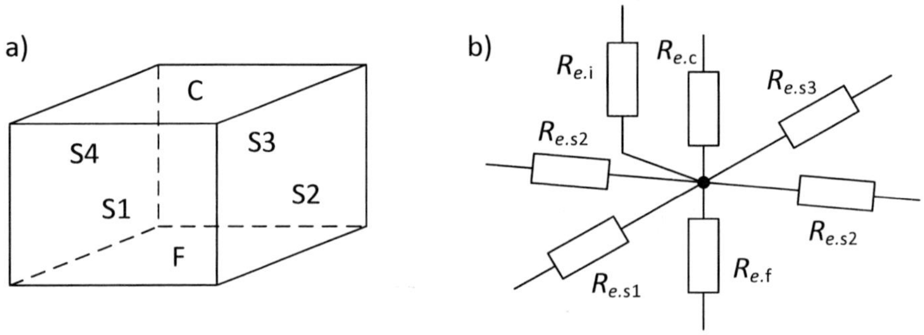

Figure 2. A separate room of the building (a) and its equivalent circuit (power circuit) as a part of an energy circuit of a building (b).

There are two possible systems of flow and effort dimensions [4]. The approach in which flow X is in W and effort Y is in K is accepted. Therefore resistive components dimension is K/W and their values should be defined as

$$R = R_{ht} / S, \tag{2}$$

where R_{ht} is heat transfer resistance in m^2·K/W and S is square in m^2. There are such values of R_{ht}: outside wall – 2.4 m^2·K/W, side wall – 3.2 m^2·K/W, interior flat room – 0.6 m^2·K/W, interior wall – 0.2 m^2·K/W, celling – 0.76 m^2·K/W, roof – 3.2 m^2·K/W, floor – 2.8 m^2·K/W, window – 0.82 m^2·K/W, flat door – 0.6 m^2·K/W, interior door – 0.24 m^2·K/W, balcony door – 0.94 m^2·K/W.

In order to reproduce the thermal state of a building in the model, should take into account the heat introduction by heating devices.

A building in relation to the thermal state is considered as an uninsulated system that is separated from the universe by an imaginary boundary. The part of the universe that we are interested in is usually called "the system" and everything else is called "the surroundings". The first law of thermodynamics for the system can be expressed as the sum of the energy flows across its boundary is equal to the change in energy content of the system ΔE, i.e.

$$\Delta E = E_{in} - E_{out}, \tag{3}$$

where E_{in} and E_{out} are the energy flows in and out respectively.

Energy can flow across the boundary as work done either to the system or by the system, as heat transfer either in or out of the system or it can be carried in a flow of matter across the boundary. Hence,

$$\Delta E = (W_{\text{in}} - W_{\text{out}}) + (Q_{\text{in}} - Q_{\text{out}}) + (E_{\text{mass.in}} - E_{\text{mass.out}}), \tag{4}$$

where W, Q and E_{mass} are the energy carried by work, heat transfer and matter moving in and out of the system, as indicated by the subscripts.

Regardless of how heat energy is introduced into rooms, all sources of the power circuit are divided on two kinds. If a source provides a constant temperature in the room it should be represented as a source of effort E and if a source provides a constant amount of thermal energy regardless of the room temperature it should be represented as a source of flow F. In general, an introduction of heat energy into building rooms is shown in figure 3.

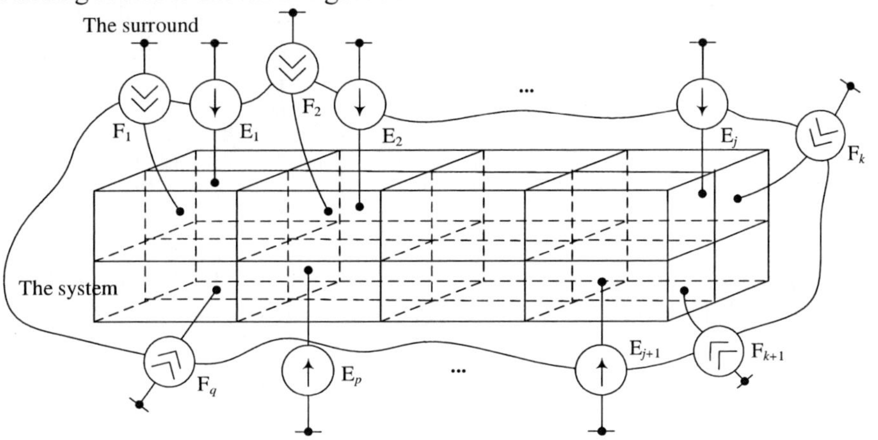

Figure 3. Introduction of heat energy into building rooms.

In a power circuit each source of effort is represented by a separate branch. The part of the power circuit of the building for the southern facade of a typical floor is shown in figure 4.

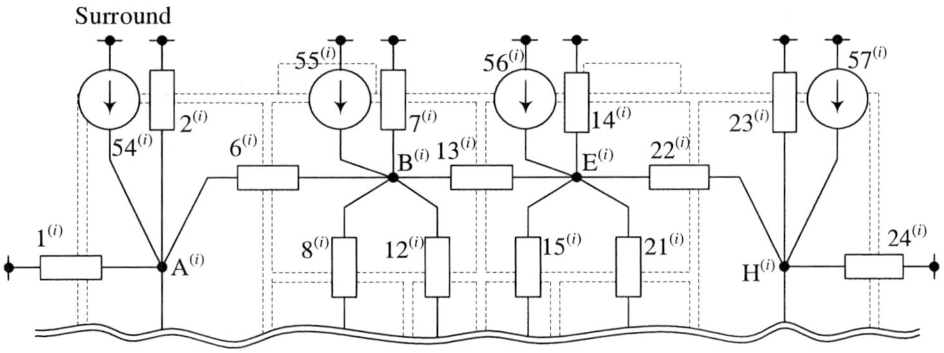

Figure 4. The part of the power circuit of the building for the southern facade of a typical floor.

Each room in the power circuit corresponds to the separate node (the nodes' names in figure 4 correspond to the rooms' names in figure 1). The numbering of resistive components (passive branches) in figure 4 corresponds to the numbering of enclosure constructions in figure 1 and the numbering of sources of flow corresponds to the numbering of heating devices.

The next stage is to construct a directed graph (digraph) for the power circuit and to choose a graph tree. In the case of a linear mathematical model of heat transfer (linear power circuit), there are no additional recommendations how to form digraph tree. In the case of a nonlinear mathematical model of heat transfer (nonlinear power circuit), the graph tree should be formed from edges whose resistance is the largest. This provides the highest convergence rate of numerical methods. Thus it is proposed to introduce edges into a graph tree in such sequence:

- the edges that correspond to outside enclosure constructions,
- the edges that correspond to interior flat constructions,
- the edges that correspond to interior constructions.

In figure 5 it is shown a directed graph for a part of the power circuit of the building that corresponds to a typical floor. The edges of the digraph tree are highlighted by thick lines.

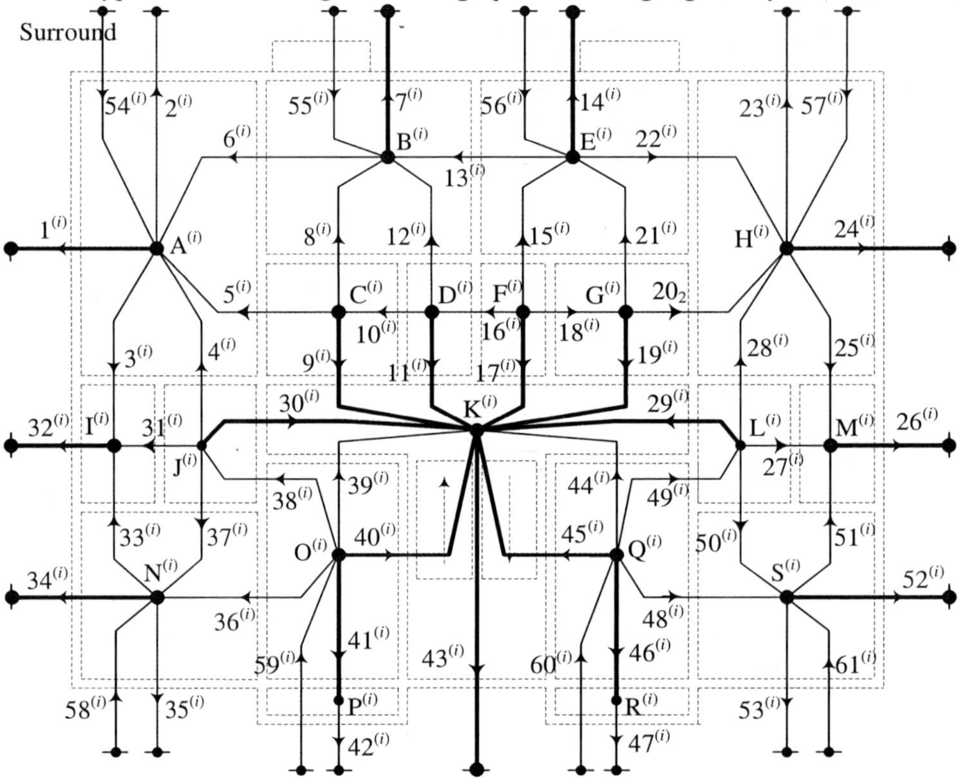

Figure 5. The directed graph of a typical floor of the building.

The digraph of the whole building is non-planar and consists of the digraphs of all floors that are connected with each other by edges which correspond to the inter-floor ceilings. These edges should be introduced into the set of graph chords.

The calculation of the mathematical model obtained in a form of equations set (1) could be performed by any of known formalized methods. The mesh-flows method (analog of the mesh-current method in the theory of electrical circuits) was applied in the work. As a result, there are obtained the main parameters of thermal steady-state such as temperatures and heat flows for separate rooms.

The calculated parameters of thermal state for rooms of the second and third (upper) floors are shown in table 1. The values of calculated temperatures (T) are given for all rooms and the calculated heat removal from heating devices (Q) are given for heated rooms. The calculations are performed at an ambient temperature –1.5°C that corresponds to the average temperature for the day part of the representative day in January for a latitude of Lviv [7]. The obtained results give a detailed picture of the thermal state and they are the basis for constructing an energy balance of a building according to current normative documents of Ukraine [6, 7] and international documents [8].

In addition, it is possible to obtain temperatures of internal and external surfaces for enclosure constructions. For example, for the room A (figure 1) there are such temperatures of inner surfaces for the walls: wall 1 – 20.1°C, wall 2 – 19.8°C, wall 3 – 20.8°C, wall 4 – 20.9°C, wall 5 – 20.9°C and wall 6 – 21°C. It allows us to better estimate the thermal comfort in rooms of the building.

Table 1. The calculated parameters of thermal state for rooms of the second and third (upper) floors of the building (not taking into account the heat and energy impact of the environment).

Room	$A^{(2)}/A^{(3)}$ Heated	$B^{(2)}/B^{(3)}$ heated	$C^{(2)}/C^{(3)}$ unheated	$D^{(2)}/D^{(3)}$ unheated	$E^{(2)}/E^{(3)}$ heated	$F^{(2)}/F^{(3)}$ unheated	$G^{(2)}/G^{(3)}$ unheated	$H^{(2)}/H^{(3)}$ heated	$I^{(2)}/I^{(3)}$ unheated	$J^{(2)}/J^{(3)}$ unheated
T, °C	21/21	21/21	20.6/18.9	20.4/18.1	21/21	20.4/18.1	20.6/18.9	21/21	17.8/16.2	19.8/18.6
Q, W	281/399	337/427	–	–	337/427	–	–	281/399	–	–
Room	$K^{(2)}$ unheated	$L^{(2)}/L^{(3)}$ unheated	$M^{(2)}/M^{(3)}$ unheated	$N^{(2)}/N^{(3)}$ heated	$O^{(2)}/O^{(3)}$ heated	$P^{(2)}/P^{(3)}$ unheated	$Q^{(2)}/Q^{(3)}$ heated	$R^{(2)}/R^{(3)}$ unheated	$S^{(2)}/S^{(3)}$ heated	Surround
T, °C	18.4	19.8/18.6	17.8/16.2	21/21	21/21	9.5/8.9	21/21	9.5/8.9	21/21	–1.5
Q, W	–	–	–	245/318	139/230	–	139/230	–	245/318	9146[a]

[a] the heat extraction from the heating system of the whole building.

Further development of the mathematical model of a thermal steady-state is to complement it by a mathematical submodel of heat and energy impact of the environment. There some features for cases of controlled and uncontrolled heating systems.

2.2. The mathematical model of a thermal state in the building that take into account heat and energy impact of the environment (for the case of a controlled heating system)

The most important component of heat end energy impact of the environment is a solar insolation which brings a certain amount of heat into rooms through the translucent elements of enclosure and heats the external surfaces of opaque enclosure (figure 6(a)).

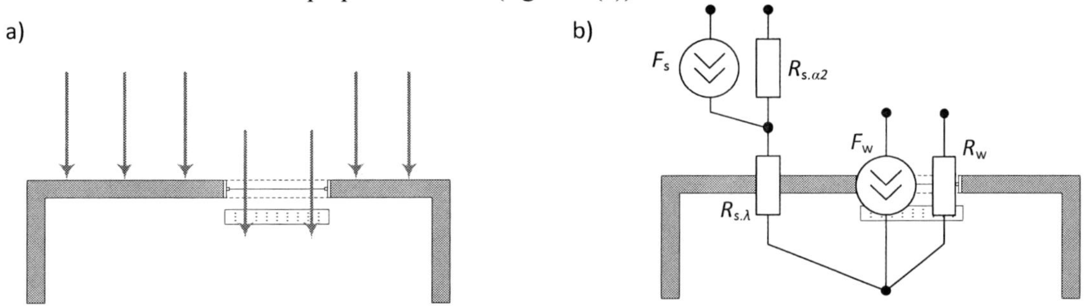

Figure 6. Heat and energy impact of the environment (a) and its power circuit (b).

The equivalent resistance of main enclosure construction $R_{e.s}$ (figure 2(b)) has to be divided on a resistance of translucent elements (R_s) and a resistance of opaque elements (R_w). Moreover, the heat transfer resistance for opaque constructions of an enclosure has to be divided into two parts $R_{s.\lambda}$ (includes the resistance of heat transfer between the wall surface and the air inside and the resistance of thermal conductivity) and $R_{s.\alpha 2}$ (the resistance of heat transfer between the wall surface and the air outside) (figure 6(b)). The source of effort E_h corresponds to the heating device and the sources of flow F_s and F_w correspond to a solar insolation.

The dimension of flow X corresponds to the power hence flow of a source of flow F_w which brings a certain amount of heat into rooms should be defined as

$$F_w = k_{sh} I_{sol} g_{gl} S_w, \tag{5}$$

where k_{sh} is shadow coefficient, I_{sol} is solar insolation per unit area (43 W in January for a southern facade for a latitude of L'viv and 30 W for a horizontal surface [7]), g_{gl} is transmission coefficient of solar insolation through translucent elements of the enclosure (0.75 for triple-glazed window [7]) and S_w is square of translucent elements.

Flow of a source flow F_s which reproduces the heating of the external surface of opaque enclosure should be defined as

$$F_s = k_{sh} I_{sol} S_s, \tag{6}$$

where S_s is square of opaque elements.

The part of the power circuit of the building including heat and energy impact of the environment for the southern facade of a typical floor is shown in figure 7.

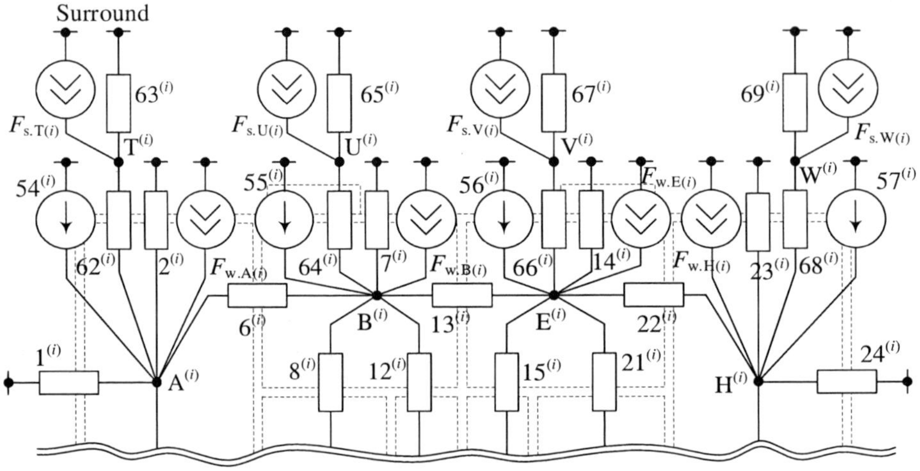

Figure 7. The part of the power circuit of the building including heat and energy impact of the environment for the southern facade of a typical floor.

Following given rules, the power circuit and the mathematical model of a thermal steady-state for the whole building is formed. The calculated parameters of thermal state for rooms of the second and third (upper) floors are shown in table 2.

Table 2. The calculated parameters of thermal state for rooms of the second and third (upper) floors of the building (for the case of a controlled heating system).

Room	$A^{(2)}/A^{(3)}$ heated	$B^{(2)}/B^{(3)}$ heated	$C^{(2)}/C^{(3)}$ unheated	$D^{(2)}/D^{(3)}$ unheated	$E^{(2)}/E^{(3)}$ heated	$F^{(2)}/F^{(3)}$ unheated	$G^{(2)}/G^{(3)}$ unheated	$H^{(2)}/H^{(3)}$ Heated	$I^{(2)}/I^{(3)}$ unheated	$J^{(2)}/J^{(3)}$ unheated
T, °C	21/21	21/21	20.6/19.1	20.4/18.3	21/21	20.4/18.3	20.6/19.1	21/21	17.8/16.4	19.8/18.8
Q, W	238/343	290/370	–	–	290/370	–	–	238/343	–	–
Room	$K^{(2)}$ unheated	$L^{(2)}/L^{(3)}$ unheated	$M^{(2)}/M^{(3)}$ unheated	$N^{(2)}/N^{(3)}$ heated	$O^{(2)}/O^{(3)}$ heated	$P^{(2)}/P^{(3)}$ unheated	$Q^{(2)}/Q^{(3)}$ heated	$R^{(2)}/R^{(3)}$ unheated	$S^{(2)}/S^{(3)}$ heated	Surround
T, °C	18.4	19.8/18.8	17.8/16.4	21/21	21/21	9.4/9.1	21/21	9.4/9.1	21/21	–1.5
Q, W	–	–	–	245/310	139/219	–	139/219	–	245/310	8172

For a controlled heating system the temperatures in the heated rooms are the same as in table 1 but heat removal from heating devices has decreased. Decreasing of heat consumption for the whole building equals 10.6%.

2.3. The mathematical model of a thermal state in the building that take into account heat and energy impact of the environment (for the case of an uncontrolled heating system)

In the case of an uncontrolled heating system the heat removal from the heating devices does not change (or these changes could be neglected). To reproduce this in the mathematical model of a thermal state in the building, the heating devices should be represented by the sources of flow instead of the sources of effort. It means that the sources of effort $54^{(i)}$, $55^{(i)}$, $56^{(i)}$ i $57^{(i)}$ in figure 7 should be replaced by the sources of flow whose flows are equals to powers of heating devices in table 1.

Following given rules, the power circuit and the mathematical model of a thermal state for the whole building are formed. The calculated parameters of thermal state for rooms of the second and third (upper) floors are shown in table 3.

Table 3. The calculated parameters of thermal state for rooms of the second and third (upper) floors of the building (for the case of an uncontrolled heating system).

Room	$A^{(2)}/A^{(3)}$ heated	$B^{(2)}/B^{(3)}$ heated	$C^{(2)}/C^{(3)}$ unheated	$D^{(2)}/D^{(3)}$ unheated	$E^{(2)}/E^{(3)}$ heated	$F^{(2)}/F^{(3)}$ unheated	$G^{(2)}/G^{(3)}$ unheated	$H^{(2)}/H^{(3)}$ Heated	$I^{(2)}/I^{(3)}$ unheated	$J^{(2)}/J^{(3)}$ unheated
T, °C	25.2/24.8	25.8/25.0	24.6/24.0	24.4/23.8	25.8/25.0	24.4/23.8	24.6/24.0	25.2/24.8	19.4/19.8	20.4/20.8
Q, W	281/399	337/427	–	–	337/427	–	–	281/399	–	–
Room	$K^{(2)}$ unheated	$L^{(2)}/L^{(3)}$ unheated	$M^{(2)}/M^{(3)}$ unheated	$N^{(2)}/N^{(3)}$ heated	$O^{(2)}/O^{(3)}$ heated	$P^{(2)}/P^{(3)}$ unheated	$Q^{(2)}/Q^{(3)}$ heated	$R^{(2)}/R^{(3)}$ unheated	$S^{(2)}/S^{(3)}$ heated	Surround
T, °C	21.1/21.6	20.4/20.9	19.4/19.9	21.1/21.5	21.2/21.7	9.6/10.4	21.2/21.7	9.6/10.4	21.1/21.5	–1.5
Q, W	–	–	–	245/318	139/230	–	139/230	–	245/318	9146

In the case of an uncontrolled heating system, the heat removal from the heating system does not change, but the environmental impact is manifested in changing the temperatures in the rooms.

3. Conclusions

The mathematical models of a thermal state in a building both without and with taking into account heat and energy impact of the environment are proposed in the paper. The first model is relevant for the analysis of the rated (peak) mode of a heating system and the second model is relevant for the analysis of building thermal states in different conditions.

The proposed models as a result of the thermal steady-state analysis allow obtaining energy indicators of the building and an objective picture of the thermal comfort in the rooms simultaneously.

The addition of a mathematical model of thermal state with a mathematical submodel of heat and energy impact of the environment requires the changes in the mathematical submodel of the heat transfer of the building. The solar insolation could be adequately reproduced by the sources of flow in the power circuit.

References

[1] Bandurski K, Mielczynski T and Koczyk H 2014 Thermal comfort and energy consumption of the ecological house – simulation analysis of domtrzon. *Technical transactions Civil engineering* **3-B(8)** 11-22

[2] The official website EnergyPlus Energy Simulation Software. Available: https://energyplus.net /sites/all/modules/custom/nrel_custom/pdfs/pdfs_v8.6.0/InputOutputReference.pdf

[3] Tabunshchykov Y and Brodach M 2002 *Mathematical modeling and optimization of thermal efficiency of buildings* (Moscow: Avok-Press) 194

[4] Malinowski A, Turkowski W and Muzychak A 2014 Thermal conditions of buildings: mathematical modelling by power circuit theory *Technical transactions Civil engineering* **3-B(8)** 299-309

[5] Sauh S 2011 Mathematical modeling of the power circuits *Electronic modelling* **33-3** 3-12

[6] DBN V.2.6-31 2016. *Construction of buildings and facilities. Thermal insulation of buildings* (Kyiv: Ministry of construction, architecture and housing and communal services of Ukraine)

[7] DSTU B A.2.2-12 2015 *The energy efficiency of buildings. Method of calculating energy consumption for heating, cooling, ventilation, lighting and hot water supply* (Kyiv: Ministry of construction, architecture and housing and communal services of Ukraine)

[8] ISO 52016-1:2017 *Energy performance of buildings. Energy needs for heating and cooling, internal temperatures and sensible and latent heat loads* (Technical Committee: ISO/TC 163/SC 2)

IOP Conf. Series: Materials Science and Engineering **415** (2018) 012049 doi:10.1088/1757-899X/415/1/012049

CFD modelling of transient thermal performance of solar chimney used for passive ventilation in a building

A Lechowska[1], N Szczepanik-Ścisło[1], J Schnotale[1], M Stelmach[2], T Pyszczek[2]

[1] Institute of Thermal Engineering and Air Protection, Cracow University of Technology, 24 Warszawska St., 31-155 Krakow, Poland
[2] Architektura Pasywna Pyszczek i Stelmach Sp. J., 65 Szlak str, 31-153 Krakow

agnieszka.lechowska@pk.edu.pl

Abstract. The paper analysed the 24 hour performance of a solar chimney placed on a building roof in Krakow, Poland. The solar chimney was an element of as a passive ventilation system allowing energy efficient night ventilation during the summer season. The chimney was built as an air duct. Three walls of the chimney were made of glass and faced toward the sun, the opposite wall was made of a concrete absorber. The system worked in two stages: day and night. The absorber located in the chimney was heated during the day, and cooled during the night. During the night phase, the released heat provided a draft contributing to passive ventilation.
Absorbers with different thermal capacities were studied. Simulations were conducted for three different absorber thickness 10 cm, 20 cm or 30 cm. The simulations were carried out using the Ansys Fluent simulation software. The hourly results of temperature distributions in the chimney allowed the authors to determine the optimal absorber thickness for the passive ventilation system. Calculations showed that the chimney with the absorber that was 10 cm thick was the optimal among analysed solutions. The absorber is warmed up to the highest temperature, up to 82°C, and induces air flow which lets removing even 3820 m^3 of air during the night.

1. Introduction

Directives of the European Parliament introduce the idea of lowering energy consumption in Europe for heating purposes in winter, as well as for cooling in summer. One of the ways of lowering energy demand is the application of cooling systems that are based on natural ventilation systems. They are called passive cooling systems and help to improve thermal comfort without the use of an energy source. In summer, the walls of buildings accumulate heat, causing the building to maintain a higher inside temperature during the night than the outdoor air. Because of this, the indoor air does not cool down to the required temperature during the night which means that, during the day, more energy will have to be used to maintain indoor air quality.

To provide proper ventilation and night cooling, a system based on the application of a solar chimney can be used thanks to which the airflow used to cool the building is forced. It is constructed in the form of a duct with one or more sides made of glazing that allows the sun to heat an absorber on the opposite side of the duct. During the day phase the air inlet and outlet in the lower and higher chimney walls are shut. The absorber accumulates heat during the day and after opening the air inlet and outlet, it releases the air during the night phase. The air flow is created by the pressure difference in the night – air is induced out of the building through the chimney and fresh, cool air flows in from the opened windows through the building and to the chimney.

The intensity of the ventilation caused by the chimney is strictly connected to its design and the solar radiation that heat it during the day. The best effects are gained when the radiation is the highest, which is why it is recommended that they are located in hot climates were the sun shines for long periods of time [1].

Solar chimneys in hot climates have been studied around the world and have proved to provide passive ventilation. The studies focused on the location of the absorber and of the chimneys' geometry, as well as on its maximum effectiveness. Analysis of solar chimneys in Singapore [2] and Spain [3,4] showed that the location of the absorber being heated during the day is an important factor. Moreover, the studies proved that shorter chimneys with the absorber located on the roof of a building and inclined in the right angle were more effective, even though the chimney length was shorter and the pressure difference that drives the process should have been lower. Many other authors have shown that such chimneys have a positive effect on lowering energy usage. Jianliu and Weihua [5] designed a mathematical model of such a chimney used in South-East China. Their study showed that their model reflected the existing chimney. The same conclusions were made by Bassiouny and Koura [6], in whose study computer simulations were used to provide a detailed view of the airflow within the chimney. Neves et al. [3] analysed the angle of the absorber in the chimney and its influence on its effectiveness in Brasilia, Mahdavinejad et al. [7] studied the optimal angle of the chimney in Iran and Sakonidou et al. [8] did a similar analysis for the conditions in Greece.

The presented studies show that the matter of passive ventilation using solar chimneys is important and ongoing. What is more, computer simulations can reflect the conditions within a solar chimney. This is why in this paper it was decided to conduct a simulation for an absorber in such a chimney located on the roof of a building in Krakow, Poland. The purpose of the study was to see if it was possible to use a passive ventilation system in Poland's cooler climate. The aim of the analysis was to gain the largest airflow through the system, so that the cooling would be as intense as possible. Because the thermal capacity of the absorber is the crucial parameter in the solar chimney design, the thickness of the absorber was adjusted to determine, which was the optimal for the ongoing conditions. Each case was simulated for the period of 24h to reflect the heating and cooling process throughout the day and night.

The simulations were conducted using the Ansys Fluent modeling software, widely used to conduct simulations on the topic of airflow, turbulence, heat transfer in complicated geometries. After the assumptions of boundary and initial conditions and solver settings, transient numerical calculations were performed.

2. Geometry of the solar chimney

The solar chimney was designed is such a way that maximum radiation could reach the absorber. It was designed as a duct with three glass walls and one wall with a concrete absorber. The glass walls were placed so that solar radiation could reach the absorber, the concrete wall was opposite the radiation. Figure 1 shows the detailed 3D geometry of the chimney designed in the Ansys geometry tool.

An inlet and outlet at the bottom and top chimney walls from were designed as throttling dampers that close and open automatically according to the day and night cycle. Each position was maintained for 12 hours. During the day, when the absorber was heated the inlet and outlet dampers were closed to prevent heat loss, while during the night, when passive ventilation occurred, they were both open. Thanks to this arrangement, the heated absorber released the accumulated energy and created a draft caused by the difference of temperature/density between the air in the chimney and inside the building.

Fresh air was provided by window vents located inside the building. Thanks to this, fresh, cool air entered and lowered the building's temperature and improved indoor air quality. This was done without the use of a mechanical ventilation system meaning that such a system is a passive one. The only energy used was to provide the impulse for the dampers to change position between cycles.

Three different cases were studied; in each the thickness of the absorber differed. In figure 1, the thickness is shown as 20 cm, the other thicknesses include 10 cm and 30 cm. In each case the inlet and outlet sizes were the same. The thickness of the absorber determines its thermal capacity and the study will show which absorber is optimal.

Figure 2 shows the cross-section of the chimney along with the materials used and their dimensions.

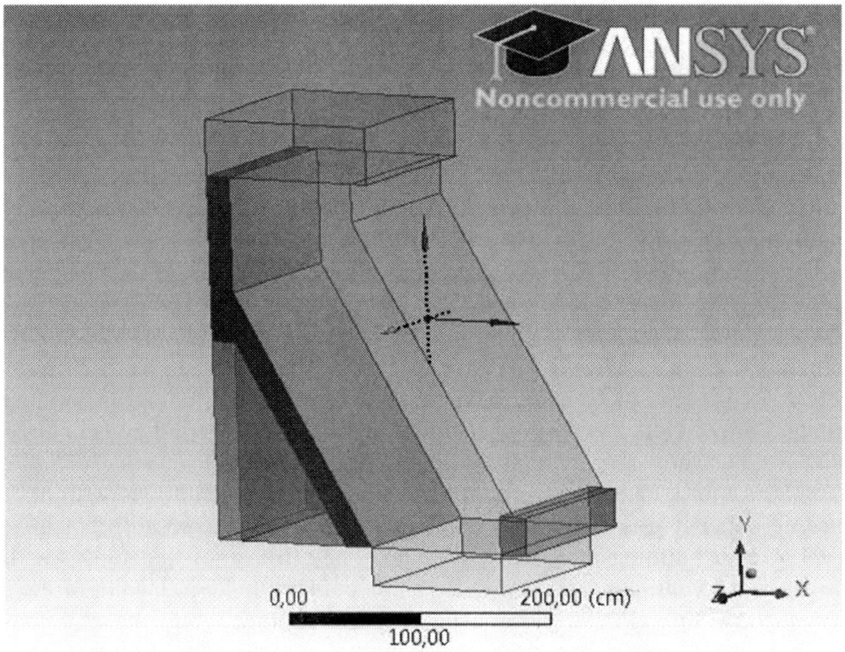

Figure. 1. 3D geometry of the solar chimney

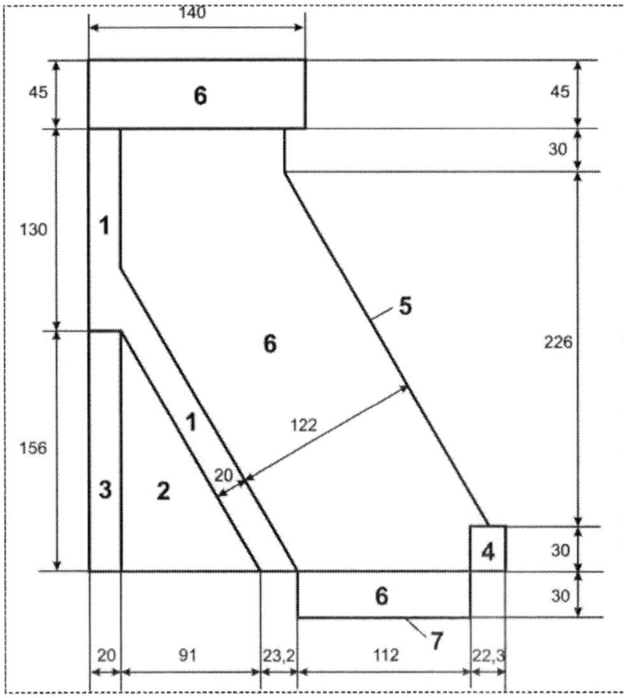

Figure. 2. Vertical cross section through the solar chimney – the absorber thickness 20 cm. Dimensions in centimetres. 1 – absorber, 2 – insulation of the absorber made of polystyrene, 3 – concrete support of the absorber, 4 – concrete support of the glazing, 5 – glazing, 6 – air in the chimney, 7 – inlet damper.

3. Mesh

The mesh created to conduct the calculation is an important step in the simulation process. It must be of high quality to prevent calculation errors. The mesh was created in Ansys meshing tool. Each analysed case had a mesh consisting in tetrahedral elements. The amount of elements in each mesh was around 223 000, 191 000 and 196 000 for the cases in which the absorber was 10 cm, 20 cm and 30 cm thick respectively.

The quality of the mesh was determined by two parameters: the aspect ratio and skewness. Their value was not larger than 1:5,5 and 0,85 respectively for each case. The more regular the mesh is the lower these parameters are, where the maximum values are 1:35 for the aspect ratio and 0.95 for the skewness. Both of these parameters are sufficient and the quality of the mesh is adequate for the case study.

4. Boundary conditions and simulation setup

To conduct the simulation, a series of input parameters had to be defined. One of them was the materials used for the study that created the solar chimney. The materials and their thermal properties shown in table 1 were used in the simulation.

Table 1. Material properties used in the study.

Material	Thermal conductivity λ (W/(mK))	Density ρ (kg/m^3)	Specific heat c_p (J/(kgK))
concrete	1.3	2200	840
polystyrene	0.04	30	1460
glazing	1.0	2500	840

Another important aspect was the air turbulence model, as it determines how the air behaves throughout the heating and cooling process. The k-ε turbulence model was used to simulate air movement. It was also assumed that the physical properties of air changed in a linear way along with temperature change. Table 2 shows the air properties used in the simulation. Because the chimney's 24-hour cycle was divided into two stages (the day stage, when the absorber was heated, and the night stage, when the absorber cooled and night ventilation took place) it was decided to divide the cycle into two sub-cycles that lasted 12 hours each. The day cycle lasted from 8:00 a.m. until 8:00 p.m. while the night cycle lasted from 8:00 p.m. to 8:00 a.m. of the next day. During the day, the inlet and outlet dampers were shut to maximize heat gain in the chimney, while during the night, they were open. To simulate solar radiation heating up the absorber, an hourly distribution of external air temperature as well as the daily sun route on the horizon were assumed for both for day and night stages. The outdoor temperature was chosen for a sunny day in Krakow according to hourly meteorological data. The external air temperatures for each hour are shown in table 3.

During the simulations it was also assumed that the absorber was facing south. To simulate the moving sun, the solar ray tracing model that reflects the movement of the sun during the day was used.

The value for each phase was calculated with a time step equal to 300 s (5 min).

Table 2. Air thermal properties used in the study.

Temperature (°C)	Thermal conductivity λ (W/(mK))	Density ρ (kg/m^3)	Dynamic viscosity μ (Ns/m^2)	Specific heat c_p (J/(kgK))
10	$2.51 \cdot 10^{-2}$	1.247	$1.76 \cdot 10^{-5}$	1005
20	$2.59 \cdot 10^{-2}$	1.205	$1.81 \cdot 10^{-5}$	
30	$2.67 \cdot 10^{-2}$	1.165	$1.86 \cdot 10^{-5}$	
40	$2.76 \cdot 10^{-2}$	1.128	$1.91 \cdot 10^{-5}$	
50	$2.83 \cdot 10^{-2}$	1.093	$1.96 \cdot 10^{-5}$	
60	$2.90 \cdot 10^{-2}$	1.060	$2.01 \cdot 10^{-5}$	

Table 3. External air temperatures for each hour.

Temperature during the day phase		Temperature during the night phase	
Hour	Temperature [°C]	Hour	Temperature [°C]
8:00	17.4	20:00	26.5
9:00	17.4	21:00	23.0
10:00	24.2	22:00	20.6
11:00	26.9	23:00	19.2
12:00	29.2	00:00	15.2
1:00	30.8	1:00	14.5
2:00	32.6	2:00	13.8
3:00	34.2	3:00	13.2
4:00	35.3	4:00	12.6
5:00	36.1	5:00	12.5
3:00	35.8	6:00	12.2
7:00	33.5	7:00	12.4
8:00	30.2	8:00	12.4

5. Results for 24h work of the solar chimney

This chapter shows the results for the 24-hour phase in which the absorber is heated from 8:00 a.m. to 8:00 p.m. (12-hour day phase) with inlet and outlet dampers shut and then with passive ventilation working when the absorber is cooled (12-hour night phase). In the night the dampers were open from 8:00 p.m. to 8:00 a.m.

The average temperature of the absorber volume for 24-hour period is shown in figure 3. Figure 3 shows that the absorbers gain their maximum temperature around 4 p.m., the 8th hour after shutting the dampers. The absorber with the thickness of 10 cm heats up the quickest and also releases the most heat during the night stage due to the lowest thermal capacity. It reaches the temperature of 82°C, which is twice as hot as the temperature of 30 cm thick absorber.

The temperature contours on the cross section in the middle of the chimney for each absorber thickness at 4 p.m. are shown in figures 4 to 6.

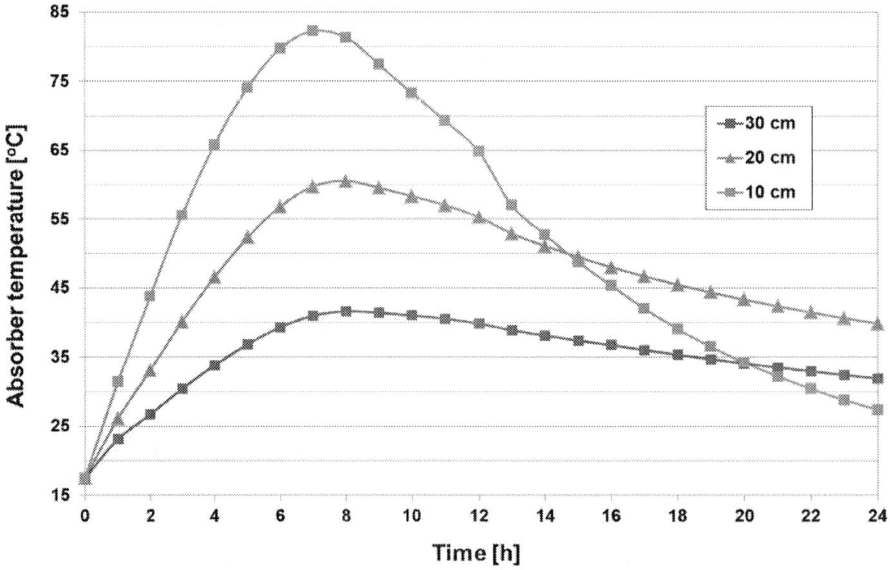

Figure 3. Average absorber temperature (10 cm, 20 cm and 30 cm thick) for the 24-hour period.

IOP Conf. Series: Materials Science and Engineering **415** (2018) 012049 doi:10.1088/1757-899X/415/1/012049

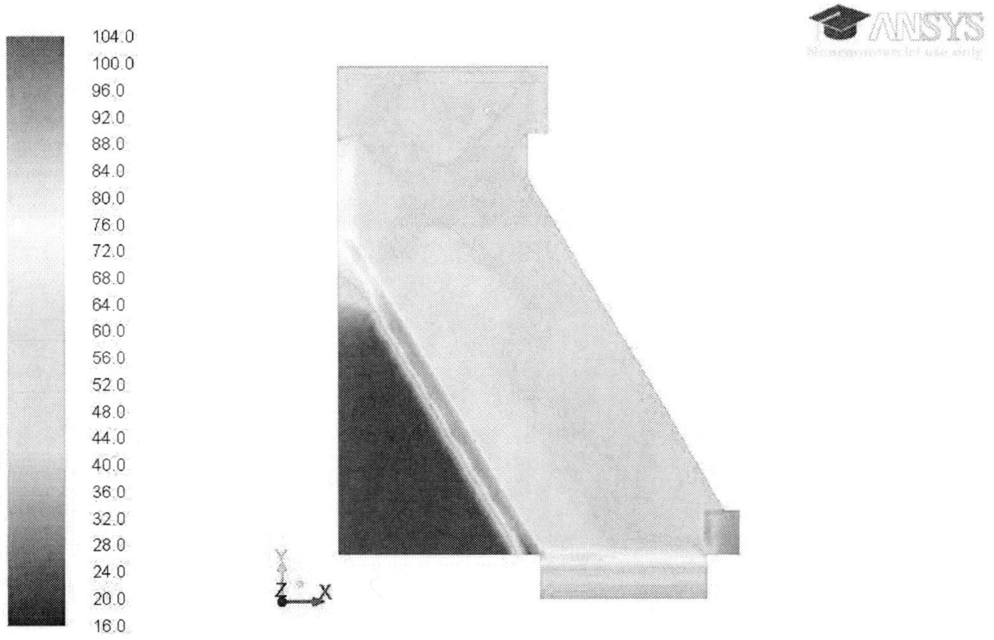

Figure 4. Temperature distribution in the chimney (°C) with 10 cm thick absorber at 4:00 p.m.

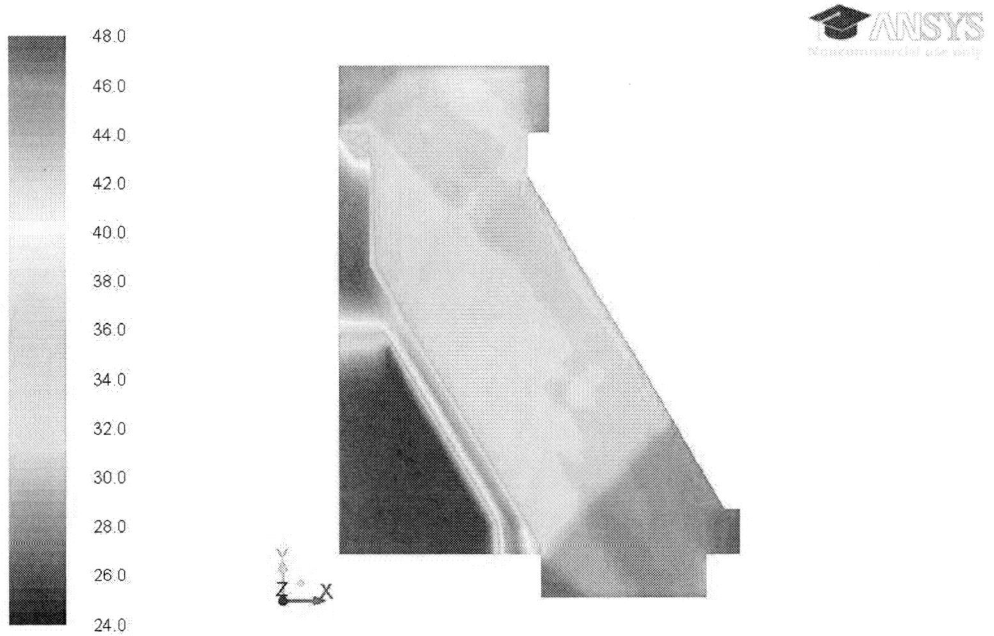

Figure 5. Temperature distribution in the chimney (°C) with 20 cm thick absorber at 4:00 p.m.

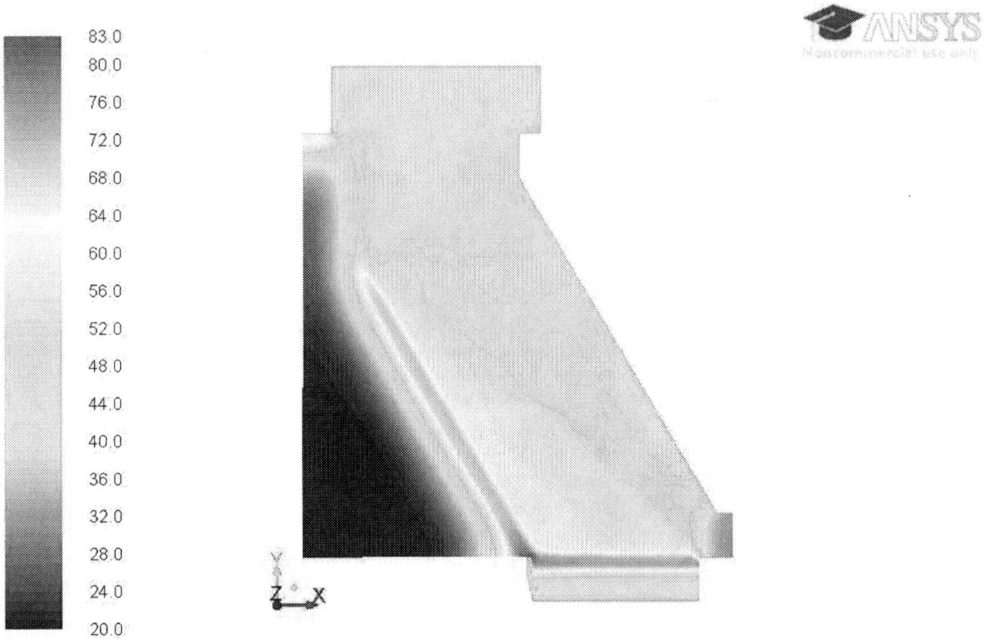

Figure 6. Temperature distribution in the chimney (°C) with 30 cm thick absorber at 4:00 p.m.

In addition to the temperature contours, it was possible to achieve airflow for each case during the 24-hour period. The airflows are shown in figure 7.

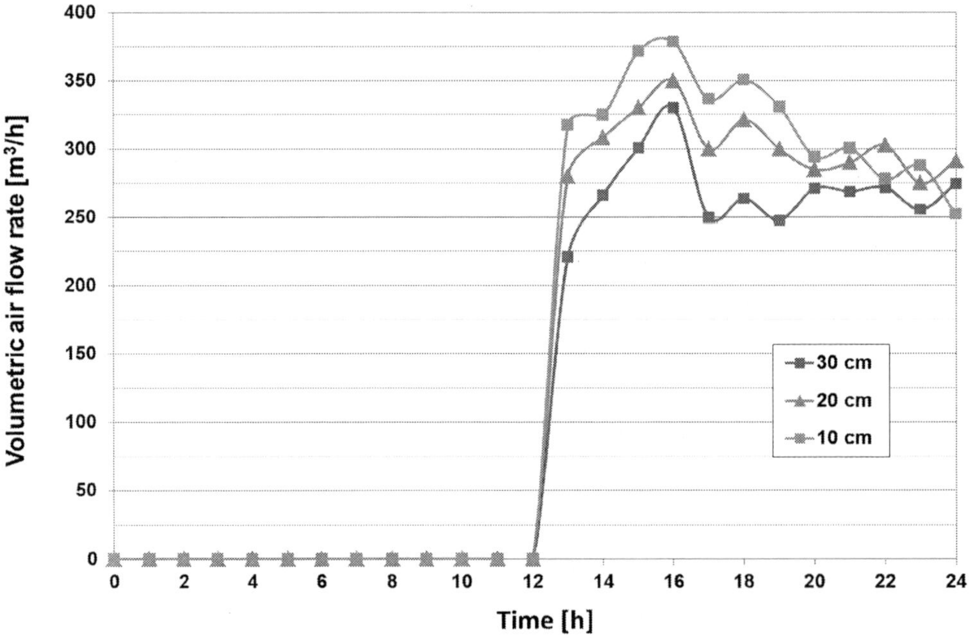

Figure 7. Volumetric air flow rates through the solar chimney for each case for a 24 h cycle.

On the basis of the results shown in figure 7, the total amount of air that was removed from the building during the 12-hour period was the following:

- 3821 m^3 of air in 12 h for the chimney with an absorber that was 10 cm thick,

- 3634 m^3 of air in 12 h for the chimney with an absorber that was 20 cm thick,
- 3222 m^3 of air in 12 h for the chimney with an absorber that was 10 cm thick,

The results show that the most air was removed through the system when the absorber was the thinnest (10 cm). It can be then stated that it is the best for a passive cooling system in Poland and can reduce the cooling energy consumption of the building.

6. Summary

The aim of this paper was to see if night passive ventilation applying a solar chimney would be possible for a building in Krakow, Poland, because the most studies that analysed such structures are mainly on cases for hot climates closer to the Equator.

The analysed chimney was located on the roof of a building and designed in such a way that the absorber, which purpose was to accumulate solar heat gains, was facing south. The rest of the structure was made of glazing. This was done to accumulate the maximum amount of solar radiation.

The chimney work was simulated using Ansys Fluent software in order to determine which thickness of the absorber is the optimal for the Central European climate. Three different cases that differed in the thickness of the absorber were studied: 10 cm, 20 cm and 30 cm. The calculations were conducted for two phases: the day phase, in which the absorber was heated up by the solar energy and the night phase, in which the absorber released the accumulated heat which contributed to the passive ventilation effect. The outside air temperature and solar radiation differed on an hourly basis.

The results of the simulations showed that the solar chimney with the thinnest absorber (10 cm) was the most efficient. It heated up the quickest and achieved the highest temperature – up to 82°C. It also released the most heat during the night phase, which caused a draft to occur and generated the highest air flow through the chimney, 3820 m^3 in 12 h, meaning that the building would be ventilated more efficiently. The thicker absorbers (20 cm and 30 cm) heated up slower and accumulated less heat.

It must be highlighted that the results are dependent on weather conditions. This study was conducted for a sunny and warm day with no clouds that could lower the solar radiation received by the absorber. Thus the results are maximal to be reached for Krakow.

References

[1] Neves L Roriz M Da Silva F M *Modelling a solar chimney for maximum solar irradiation and maximum airflow, for low latitude locations* Proceedings of Building Simulation – 12th Conference of International Building Performance Simulation Association, Sydney, 14-16 November, 2011

[2] Tan A Y K Wong N H *Natural ventilation performance of classroom with solar chimney system* Energy and Buildings, 53, pp. 19–27, 2012

[3] Arce J Jimenez M J Guzman J D *Experimental study for natural ventilation on a solar chimney* Renewable Energy, 34, pp. 2928–2934, 2009

[4] Macias M Mateo A Schuler M Mitre E M *Application of night cooling concept to social housing design in dry hot climate* Energy and Buildings, 38, str. 1104–1110, 2006

[5] Jianliu X Weihua L *Study on solar chimney used for room natural ventilation in Nanjing* Energy and Buildings, 66, pp. 467–469, 2013

[6] Bassiouny R Koura N S A *An analytical and numerical study of solar chimney use for room natural ventilation* Energy and *Buildings, 40, pp. 865-873, 2008*

[7] Mahdavinejad M Fakhari M Alipoor F *The study on optimum tilt angle in solar chimney as a mechanical eco concept* Frontiers of Engineering Mechanics Research, 2, pp. 71-80, 2013

[8] Sakonidou E P Karapantsios T D Balouktsis A I Chassapis D *Modelling of the optimum tilt of a solar chimney for maximum air flow* Solar Energy, 82, pp. 80-94, 2008

IOP Conf. Series: Materials Science and Engineering **415** (2018) 012050 doi:10.1088/1757-899X/415/1/012050

Use of vegetation and natural management of rainwater and grey water resources in the external space of the building

A Bucka[1], B Ziarko[1]

[1] Institute of Building Materials and Structures, Cracow University of Technology, Cracow, Poland

Abstract. The purpose of the publication is to focus attention on the potential of vegetation application and revitalisation of non-industrial sewage and rainwater in contemporary architectural design. Use of plants, along with the natural water resources management system in the external zone of the building and in the public space, has an ecological justification, characterised by numerous advantages, making it easier for us to live in an urbanised area. These are: simplicity of implementation, thermal and acoustic benefits, energy saving, the possibility of using local plants, health and aesthetic values, as well as low maintenance costs.

"Trouble in nature appears only there, where there has been human interference. "
Shaun Ellis

1. Introduction

Drinking water is definitely scarce these days, which is why it is so important not to use it only for activities such as watering plants, flushing toilets, washing or doing laundry. In the 21st century, in the era of strong urbanisation, when water became an indispensable element of the civilised world, we completely forgot how valuable this life-giving substance is and how much we contribute to the deterioration of its quality, treating its resources as unlimited. The amount of drinking water is constantly decreasing due to the rapidly growing industry, increasing number of areas with hardened surface, as well as the uneconomical management of water in our daily life. The system designed to manage the collected rainwater could be one of the solutions to the aforementioned problems, making it an alternative source of water.

2. Recycling of rainwater and grey water

Advanced rainwater collection systems have been known and have already been present around 3,000 years BC in Jordan [1]. In Poland, rainwater was used mainly for washing in the 1960s [2], mostly in rural areas. Rainwater collected from roofs, paved areas such as driveways, car parks, squares, etc. is collected by a system of gutters and downpipes, then goes to a tank (most of the time located underground). Subsequently it is filtered and pumped to the place of water demand. Rainwater cannot be used interchangeably with tap water in a completely arbitrary way, because their parameters differ from each other. It is obvious that rainwater cannot be used for food purposes, but due to the lower sanitary quality it is a good solution for watering lawns and vegetation in a garden, a patio, living walls or a green roof. An additional advantage is that rainwater has less calcium compounds, so it is more easily absorbed by plants than tap water. For the same reason, rainwater as soft water has no destructive effect on the elements of washing machines and it facilitates the dissolution of detergents (which reduces their amount). Rainwater can be successfully used for flushing the toilet bowls, which enlarge savings of drinking water. Moreover, rainwater can be also used for various activities such as washing a gardening equipment, a terrace or driveways. Undoubtedly there is a possibility of making significant savings of tap water in a daily consumption. The water used to rinse a toilet and urinals is called black water (black sewage), while the water that does not contain faeces and urine named grey water is a significant part of the general sewage as reported in the summary (Table 1). To sum up:

grey water makes up about 51% of the total sewage and this is the amount that we can recover and re-use-water for this type of use does not have to be drinking water. It is also very important that the use of a rainwater management system makes it possible to avoid including given estate in "rain tax" (fee for maintenance of rainwater and snowmelt infrastructure in proper technical efficiency to remove or minimise any hazards associated with its deficit and excess, such as floods and droughts).

Table 1. The structure of the consumption of water intended for living and household needs [3].

Water consumption structure	Usage rate [dm3/Mxd]		%
	min.-max.	average	
Drinking and cooking	05-mar	4	3
Washing dishes	15-paź	12	10
Washing the body	15-paź	12	10
Bathing in the shower or in the tub	25-40	33	26
Rinsing the toilet bowl	30-45	38	30
Laundry	16-20	18	15
Cleaning and other needs	10-cze	8	6
Total	100-150	125	100

Education about how to properly dispose water in the world is an important aspect of human life, especially due to the fact that the space around us is more and more filled with concrete and asphalt (especially in cities), limiting the natural resources of the Earth (both flora and fauna are connected with it). In the era of technological progress, ways of water purification can be more advanced without a doubt, unfortunately they involve higher costs and require well-trained and qualified staff. These reasons alone, often make it impossible to apply these methods and technologies in practice, which is why development of simple, cheaper and more cost-effective natural methods of cleaning water arouses interest. Among the natural water treatment systems, we can distinguish: phyto-purification (Figure 1), lagoons and storage in reservoirs.

3. Features of natural water treatment systems

Undoubtedly, the priority aspect of natural grey and rainwater treatment systems is the pro-ecological aspect, which is environmentally friendly and does not harm the ecological balance of natural resources. Another important features are as follows: the simplicity of the system construction, the ability to conduct care treatments even by unqualified staff, the ease of designing vegetation and its application in practice. Vegetation is a cost-effective undertaking because it does not require high construction, maintenance or energy costs. The lack of use of mechanical devices reduces maintenance costs. The only limiting factor is the land cost on which the establishment of a natural wastewater treatment system is designed. Storage of rainwater increases the independence of the owner from the water supply. The rainwater collection system reduces the amount of rainwater discharged to the network and thus prevents the collector to overflow, it is also a sustainable investment that increases the value of the land. Most of the contaminants can be removed, so natural water treatment systems are very efficient (however, climatic conditions determine its effectiveness - at lower temperatures the efficiency is lower). Another feature of the discussed systems is high reliability even under extreme operating conditions (the ability to absorb a wide range of organic compounds).

4. Impact of the use of vegetation on human health, urban climate and the environment

A natural environment conducive to human health is a rich ecosystem, that we are fully dependent on and it is crucial to keep us alive. Therefore, in the era of strong urbanisation affecting the impoverishment of the natural living space, it is very important to pay special attention to the use of vegetation in a city.

The most important benefits [4] of plants are the reduction of carbon dioxide and, of course, oxygen production. Bushes and trees absorb CO_2 and "build it" into the wood structure, which is not possible for herbaceous plants. These plants rots in the winter hence they emit carbon dioxide in the same year or the next one, while their organic material decomposes and creates organic fertiliser. Plants also contribute to the filtration of the air from industrial dust, keeping it on their surface, acting as a self-renewing biological filter rinsed by rainfall.

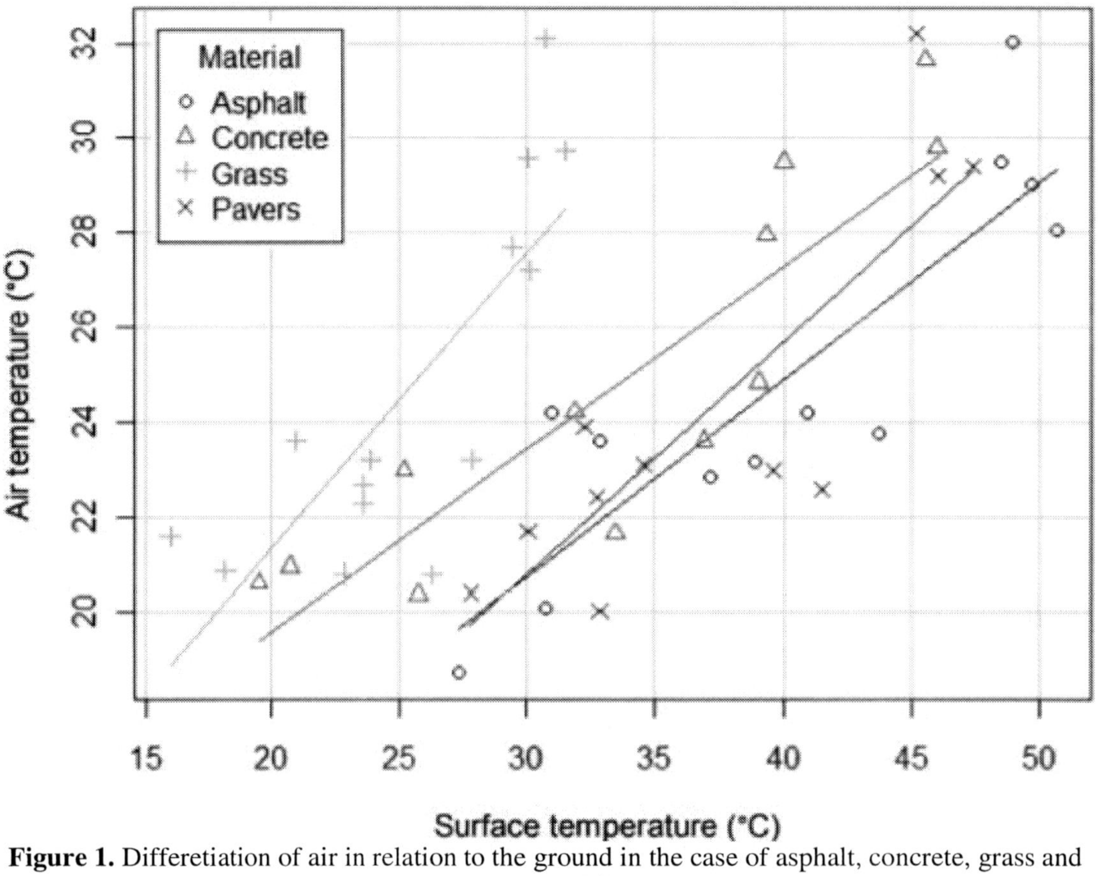

Figure 1. Differetiation of air in relation to the ground in the case of asphalt, concrete, grass and pavement [5].

Greenery has a positive effect on the climate in urban space. It was found that over the area of an ordinary lawn, the temperature at noon is even 20 degrees lower than above the asphalt surface [Figure 1]. Results of study carried out by Katharine Gaun [5] showed no significant difference in surface temperatures between asphalt, concrete, asphalt and pavers, but there was a significant difference between all artificial materials and grass. During the summer months when the sun's radiation is highest, the materials absorb more heat and the surface temperature increases. The ambient temperature will also increase due to convection from materials. However, the temperature remains relatively lower compared to the natural cover due to evapotranspiration. Similarly, the humidity of the air, where even a single tree produces its irreplaceable microclimate, raising the humidity of the air in the immediate area, also lowering the temperature preventing the formation of "urban heat islands". As it is written [6] water in the environment helps to reduce heat adsorption by: encouraging evaporation and transpiration; keeping pervious surfaces cool; and enhancing the effectiveness of vegetation for shading.

Heat Island is a term for the phenomenon that involves a significant absorption of heat during the day by extensive artificial surfaces such as concrete, asphalt, pavement, located mainly in a city. The most reliable indicator of the intensity of the urban heat islands is the temperature difference between a city and its suburban areas. In large,mostly in European and American agglomerations, in time of the highest intensity such as at night, the difference can reach 10-15^{0}C.

Green roofs and their vertical equivalents so-called living walls, set an example and inspire local councils to some groundbreaking activities in favour of saving energy and maintaining a healthy climate. For example, every new building in France, an office building, a hotel or a warehouse will require installation of greenery or solar cells on the roof. This is due to the law adopted in France [7] in March 2015, which requires the development of roof surfaces in harmony with climate and natural environment.

In Berlin, neighbouring buildings covered with green roofs with a total area of 40,000 m^2 were built on Potsdamer Platz [8]. Such a large area of the green roof together with a reservoir storage of a total of 3500 m^2 and an artificial lake of a total of 13,000 m^2 constitute an effective natural element of the rainwater collection and cleaning system by landscape management. This is a perfect example of a conscious policy in this area.

Figure 2. Green roofs at the Potsdamer Platz in Berlin [8] photo. K. Wolańska.

A research [9] conducted in New York shows that the surface temperatures on the regular roofs in hot summer days can be up to 40^{0}C higher than the roof surface covered with properly selected vegetation. Relatively the surface temperature of the standard roof was 19^{0}C higher during the day and it was 8^{0}C lower at night comparing to the green roof (measurements were carried out in July 2003). However, the temperature inside the building with the green roof was on average 2^{0}C lower during the day and 0.3^{0}C higher at night.

To avoid the occurrence of heat islands, we should seriously approach the subject and rationally plan the urban space taking into account also the creation of green roofs with large areas. Measurable results of reducing the effects of the urban heat island are possible to achieve with significant green areas located close to each other. Installations that are scattered over a large area may have no effect on lowering the air temperature.

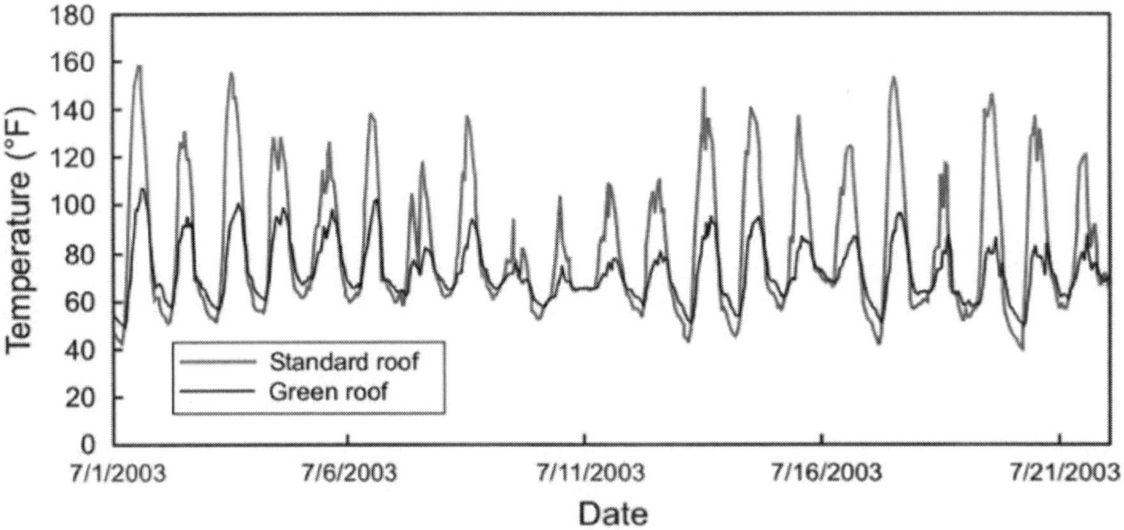

Figure 3. Example of the average surface temperatures on green roofs and standard roofs [9].

Green roofs can also be ideal for bee breeding as well as for vegetable growing. Located in the city, they could ensure production at the level of 3/4 of the vegetables consumed in it, as suggested by the Bologna case study in Italy [11]. In 2010, Bologna became the first city in Italy testing roof gardens on public buildings as a part of a project run by local authorities, universities and non-profit organisations. The specialists observed the process for three years in the period from 2012 to 2014. At that time, gardeners, working on the roof, cultivated lettuce, cabbage, chicory, tomatoes, aubergines, chilli peppers, melons and watermelons in the structures of plastic pipes and in recycled pallets filled with compost.

The researchers estimated that such a hypothetical garden can produce more than three tons of vegetables a year. Then the researchers estimated potential efficiency of a network of vegetable gardens, occupying all potential flat roof surfaces throughout the city of Bologna. This means that, based on current consumption data of the city, roof gardens could satisfy 77% (12,500 tons of vegetables) of the demand for fresh vegetables grown locally.

In addition, a map of the spaces located in the area of 500 meters from each other was made, showing that thanks to this type of solution, it is possible to create a network with a length of 94 kilometres of corridors for wildlife, including pollinating insects. A distance of 500 meters has been considered by scientists as adequate because most of the popular pollinating insects have a flight over a distance of 750-1500 m.

5. Examples of the use of vegetation with the possibility of using purified grey water and rainwater in the city space

The above features make these systems competitive to the conventional methods of wastewater treatment, and in addition can be part of a landscape, urban space or green architecture in themselves, as well as fulfil the function of irrigating vegetation. The cities and areas with dense buildings are a problematic environment for rainwater. This is the reason why more and more emphasis is placed on the design of sustainable rainwater management systems. Their advantage is the management of rainwater at their place of installation [12].

The first example of using greenery along with the possibility of using managed rainwater and grey water in the architecture of the building and its immediate surroundings is the Pure SPA facility in Vietnam[15] (figure 4, figure 5, figure 6, figure 7). The openwork building is flooded with greenery, its inner courtyard is filled with pools, plants are present in almost every corner of it, and translucent walls with hanging vegetation contribute to the wonderful play of light and shade. Greenery and water not only have aesthetic value but allow you to fully use the character of the building as a place of peace, stimulating all the senses and relaxation of both the body and the mind. A healthy, natural environment was created using local vegetation, which has become a part of architecture.

Figure 4. Pure SPA, Vietnam [12]. Photo: Oki Hiroyuki

Figure 5. Pure SPA, Vietnam [12]. Photo: Oki Hiroyuki

Figure 6. Pure SPA, Vietnam [12]. Photo: Oki Hiroyuki

Figure 7. Pure SPA, Vietnam [12]. Photo: Oki Hiroyuki

Another example of rational management of greenery and water in the city is a small building with a vertical herb garden on the façade in the centre of Lisbon [13]. It delights not only with its appearance, but also with the scent of plants, which are a compilation of 25 different species of herbs found on the Iberian Peninsula. The most original aspect of the design and implementation is the fact that each floor has a different smell - the aroma of lavender in the bedroom, rosemary in the living room, etc.

Figure 8. A vertical herb garden in the center of Lisbon [13].Photo: Fernando Guerra

Figure 9. A vertical herb garden in the center of Lisbon [13].Photo: Fernando Guerra

Something from "our yard" – objects presenting the green roof as examples of green infrastructure in cities, contributing to the reduction of pollution contained in the air, as well as fulfilling the above mentioned pro-ecological tasks. As we can see in the illustrations below - the green roof can be found on objects of different functions.

Figure 10. A flower meadow on the terrace of an office building in Cracow [14]. Photo: Piotr Wolański APK Dachy Zielone

Figure 11. Green roof as terrace by the pool – a private house in Cracow [14], Photo: Piotr Wolański APK Dachy Zielone

Figure 12. Roof available for visitors at the Copernicus Science Center in Warsaw [14] Photo: Katarzyna Wolańska .

Figure 13. Extensive roof near the office building in Cracow [14]. Photo: Piotr Wolański APK Dachy Zielone

Figure 14. Shopping gallery Tarasy Zamkowe in Lublin [14], Photo: Katarzyna Wolańska

Figure 15. Polish sky-high gardens at the Centre for the Meeting of Cultures in Lublin [14], Photo: CSK Dorota Bielak

6. Conclusion

The management of rainwater and its reuse for specific purposes leads to savings in tap water and is rational, safe and most importantly - compatible with nature. The use of complete rainwater management systems is undoubtedly a pro-ecological activity that brings many benefits not only to current generations, but also to future ones and is consistent with the message of Native Indians, for whom nature is very close: "*We do not inherit our earth from our parents, we borrow her from our children* " (Chief Washaki).

Literature and other sources:

[1] Jones P M, Hunt F W : *Performance of rainwater harvesting systems in the southeastern United States.* „Resources, Conservation and Recycling" 54/2010

[2] Błażejewski R : *Kanalizacja wsi.* Wydawnictwo Polskie Zrzeszenie Inżynierów i Techników Sanitarnych. Poznań 2003

[3] Sakson G : *Wykorzystywanie wód opadowych w budynkach.* „Rynek Instalacyjny" 5/2010

[4] J. Johnston, J. Newton, "Building green. A guide for using plants on roofs, walls and pavements", The London Ecology Unit, London 1993.

[5] Guan K.K., "Surface and ambient air temperatures associated with different ground material: a case study at the University of California, Berkeley", Surface and Air Temperatures of Ground Material, Spring 2011;

[6] Coutts A. M., Beringer J., Jimi S., Tapper N.J., "The urban heat island in Melbourne: drivers, spatial and temporal variability, and the vital role of stormwater.", STORMWATER 2009 Conference paper.

[7] www.spcleantech.pl/smart-cityzrownowazony-rozwoj-miast/zielone-dachy-i-zyjace-sciany/ [access date: 27.03.2018]

[8] Wolański P., Wolańska K., „Dachy zielone i ich wpływ na klimat miasta", IZOLACJE 7/8/2015;

[9] Rosenzweig, C., S. Gaffi n, and L. Parshall (Eds.), "Green Roofs in the New York Metropolitan Region: Research Report.", Columbia University Center for Climate Systems Research and NASA Goddard Institute for Space Studies. New York. 2006;

[10] Orsini, F., Gasperi, D., Marchetti, L., et al. (2014). Exploring the production capacity of rooftop gardens (RTGs) in urban agriculture: the potential impact on food and nutrition security, biodiversity and other ecosystem services in the city of Bologna. Food Security 6(6): 781-792;

[11] Burszta-Adamiak E , *Projektowanie systemów zrównoważonego gospodarowania wodami opadowymi*, „RynekInstalacyjny" no 10/2014, p. 71–74

[12] www.miadesignstudio.com [access date: 27.03.2018]

[13] www.plataformaarquitectura.cl [access date: 27.03.2018]

[14] Wolański P., „Zielone dachy jako przykład innowacyjnego sposobu tworzenia zielonej infrastruktury w mieście", Budujemy hale, Poradnik inwestora, www.budujemyhale.com, 2018;

IOP Conf. Series: Materials Science and Engineering **415** (2018) 012051 doi:10.1088/1757-899X/415/1/012051

Potential Influence of Carbon Fixation by Wood Carbonization on Atmospheric CO_2 Concentration and Temperature

Y Baba[1] and S HOKOI[2,3]

1) Ministry of Land, Infrastructure, Transport & Tourism of JAPAN, Tokyo, Japan
2) Southeast University, No.2, Sipailou Road, Nanjing, China
3) Kyoto University, Yoshida-honmachi, Sakyo, Kyoto 606-8501, Japan

hokoi@maia.eonet.ne.jp

Abstract. Out of greenhouse gases causing global warming, CO_2 has the largest direct contribution after industrial revolution. To reduce CO_2 concentration, it is essential to collect CO2 in the atmosphere and store it in separate places from the atmosphere. As a method of CO_2 capture and fixation, we studied a method using carbonization of wood. Based on the statistics of domestic wood production and import/export, we estimated possible amount of fixed CO_2 as 8 to 10% of annually emitted CO_2 in Japan. Influence of wood carbonization on CO_2 concentration in the atmosphere was also examined based on a prediction model for the atmospheric CO_2 concentration.

1. Introduction

CO_2 comprises approximately 93% of the current greenhouse gas emissions in Japan [1]. Therefore, significant reductions in atmospheric CO_2 concentrations are important to mitigate global warming.

Although previous research has focused on reducing the rate of CO_2 emissions, additional mechanisms for atmospheric CO_2 reduction should be considered due to the uncertainty associated with the consequences of the current CO_2 concentration levels. One mechanism to reduce atmospheric CO_2 is to collect it from the atmosphere and store it in non-atmospheric carbon reservoirs, leading to what is known as negative emissions [2]. Previous studies have investigated the transformation of CO_2 into useful materials by microbial activities, hydrogenation, or by direct storage of CO_2 in terrestrial and oceanic reservoirs [3]. However, the long-term behavior of the stored CO_2 and its influence on ecological systems has not been fully understood. Our study focuses specifically on using wood carbonization as a carbon capture method.

Trees grow by absorbing CO_2 from the air and converting it into organic materials. After a tree is fully grown, it can be utilized as a building material, fuel, or left unused. Within several decades, wood will naturally decompose or burn, returning the organic material to the atmosphere as CO_2 gas. Consequently, the cycle of tree growth and harvest does not generally influence the long-term concentration of atmospheric CO_2. However, if a harvested tree is carbonized, organic carbon is transformed into inorganic carbon and can be stored as charcoal. In this manner, atmospheric CO_2 can be fixed as stable inorganic carbon indefinitely, or until the charcoal is burned [4].

Charcoal is a result of the thermal decomposition process of inorganic matter. The carbon contained in organic matter is liberated as CO_2 by burning or decaying, but the carbon contained in charcoal produces inorganic polymer compounds, which are chemically stable and can be maintained long-term, unless they are burnt. Since this carbon originates from the CO_2 in the atmosphere and is absorbed by trees through photosynthesis, the carbonization of organic matter can be regarded as a CO_2 capture method. The carbonization of wood interrupts the cycle where the CO_2 absorbed by trees returns to the atmosphere, and reduces the amount of CO_2 in the atmosphere, as shown in Fig. 1. Trees are reproducible, and CO_2 can be continuously captured. Furthermore, most of the energy required for carbonization can be supplied from the process of carbonizing the organic matter themselves. The essential point in this system is to store the carbonized charcoal, and not use it as a fuel.

Our aim in this research is to estimate the potential for CO_2 fixation by carbonization of the available wood base products (WBP) in Japan and demonstrate how this process could potentially contribute to the reduction of atmospheric CO_2 and prevent global warming. Here, WBP refer to the wood produced in the forest, and the wood and paper products produced using this wood as a raw material.

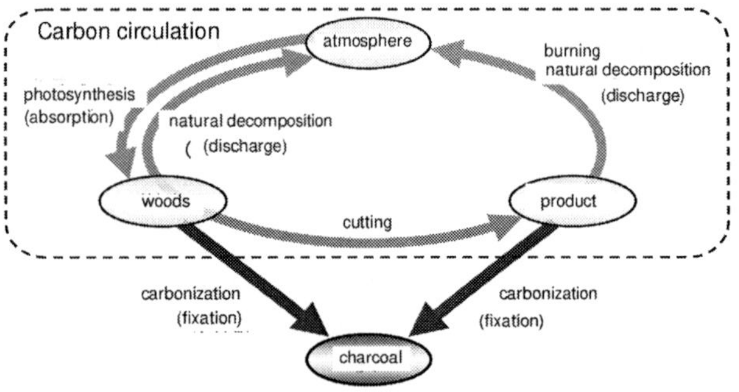

Fig. 1 Carbon circulation and wood carbonization.

2. Amount of Raw Materials Available for Carbonization

To evaluate CO_2 reduction by carbonization of WBP, the available amount of raw material for carbonization must first be estimated. In this study, the available WBP in Japan was considered, assuming that carbonization is performed only in Japan.

Both the WBP produced in Japan and the imported resources are discussed, and the available amount of wood was estimated. Figure 2 shows a simplified flow chart of the WBP, where 'WBP produced in Japan' is the growth rate of trees, exemplified by their weight increase, caused by the fixation of atmospheric CO_2. The total forested area of Japan will not decrease by this process. The term 'left' is the WBP that are neither thinned nor cut, but decay naturally.

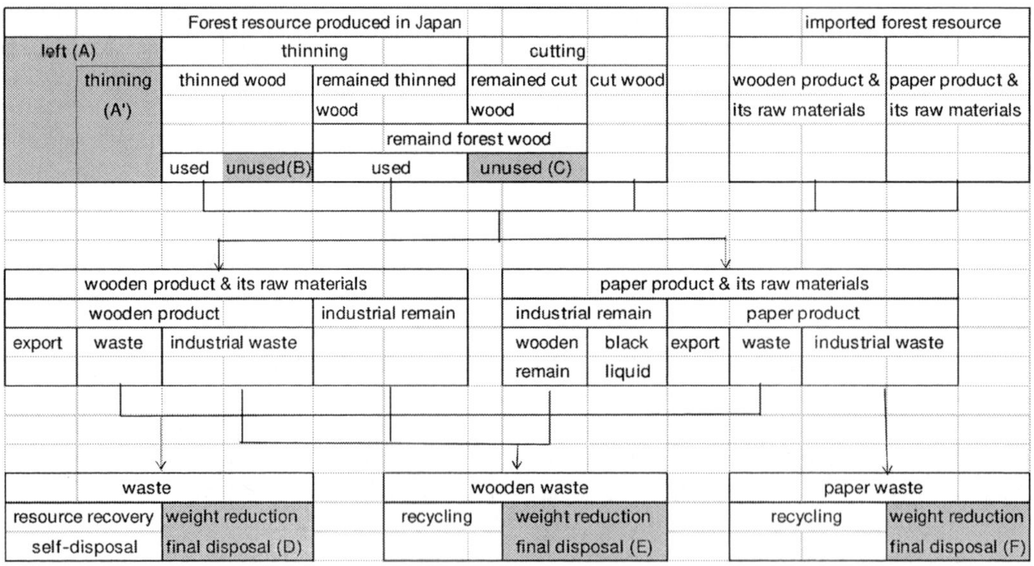

Fig. 2 Flow chart of wood base products.

'Thinned wood' is the wood processed into logs, and the remaining small diameter wood and slender twigs are called 'remaining thinned wood.' 'Cut wood' and 'remaining cut wood' in the cutting process are also defined in the same manner. The remaining thinned wood and remaining cut wood are altogether called 'remaining forest wood', and most of them are left in the forest. It is important to use the thinned wood and remaining wood in forests effectively.

The final use of the cut and thinned WBP can be classified into wooden and paper products. All wooden and paper products other than those that are exported are consumed or disposed of in Japan. The waste material can be classified into 'waste' [5] and 'industrial waste.' In the category of industrial waste, 'chips' and 'waste-paper' were considered in this study.

As available resources, (A) left forest, (B) unutilized thinned wood, (C) unutilized resources remaining in woodland, (D) wooden and paper waste products, (E) chips from industrial waste, and (F) waste-paper from industrial waste were examined based on available statistics. This model estimated the maximum amount of the WBP currently available in Japan and gave an estimate of approximately 122 million t/y (air-dry weight).

3. Relationship between Carbon Fixation Rate and Carbonization Temperature.

The production efficiency of charcoal[A1)] as a function of carbonization temperature was determined, and the carbon fixation rate[A2)] and carbon remaining ratio[A3)] in the carbonization process of wood and paper were calculated based on the results of the references [6] and [7]. Charcoal and paper are primarily composed of carbon, hydrogen, oxygen, and ash. Figure 3 shows the charcoal weight [%] with respect to the air-dried wood weight as a function of carbonization temperature. The fraction of C drastically decreases between 200 and 400 °C because of pyrolysis and remains almost constant at higher temperatures. A carbon fixation rate of 20.9% was used in this study, and was taken from the average between 400 and 1100 °C. The remaining ratio of carbon was also nearly constant between 400 and 1100 °C with an average value of 49.1%.

Figure 4 shows the charcoal weight [%] of the waste paper. The carbon fixation rate and carbon remaining ratio of the waste paper were determined to be 19.6 and 45%, respectively, and were taken from the averages of the values between 700 and 1000 °C.

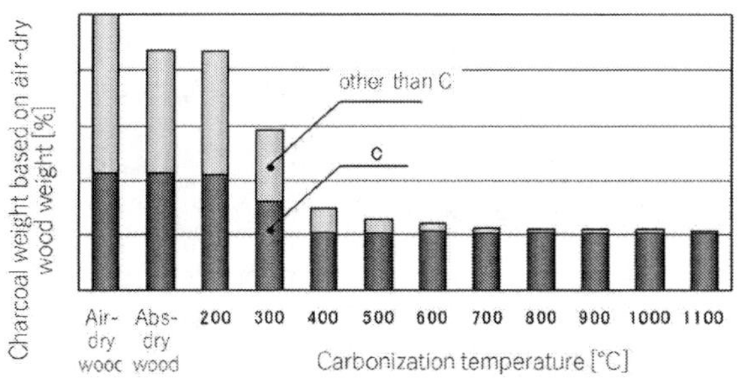

Fig. 3 Charcoal weight from the wood as a function of carbonization temperature.

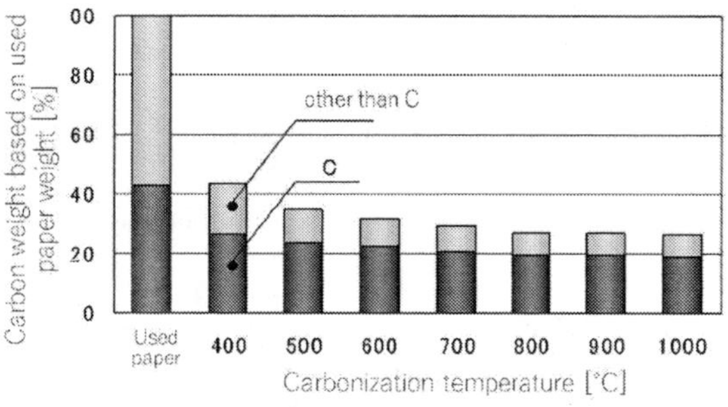

Fig. 4 Charcoal weight from the waste paper as a function of carbonization temperature.

4. Carbon Dioxide emissions from the carbonization system.

The amount of CO_2 emitted from the carbonization of WBP, which originates from the use of fossil fuel, was calculated relative to the unit weight of the air-dried raw charcoal material. The following five processes were examined; (1) cultivation and cutting of the left forest, (2) transportation of the raw charcoal materials to the carbonization factory, (3) production and operation of the carbonization facility, (4) construction and operation of the carbonization factory, (5) transportation of the charcoal to storage. The raw charcoal materials were classified into 4 categories from I to IV, as listed in Table 1, and the unit CO_2 emission rates were calculated for each material.

IOP Conf. Series: Materials Science and Engineering **415** (2018) 012051 doi:10.1088/1757-899X/415/1/012051

4.1 Forest cultivation and cutting of the left forest

Forest cultivation, which is necessary to produce the raw carbonization material I, includes afforestation, brushing, thinning, cutting, and take-out. To evaluate the amount of CO_2 emissions, the sum of overseas and domestic emissions during production, and the division 'Raw materials (domestic)' was used. This is given in the table of CO_2 emission rates per unit production, which was estimated based on the input-output table [8].

4.2 Transportation of the raw charcoal material

CO_2 emissions were calculated as a product of units of CO_2 emitted per unit loading capacity of the truck per unit distance to be 0.08 kg-C/t/km [8][9], and the transported distance was assumed to be 30 km. Here, the weight was corrected for moisture content.

4.3 CO_2 emissions from the production and operation of the carbonization facility

A continuous self-combusting carbonization system consisting of a crusher, dryer, and carbonization furnace was assumed [10]. The units CO_2 emission during the construction process was evaluated based on the weight of the raw materials used such as steel, stainless-steel, and cement; their respective unit CO_2 emission rates; depreciation of machines; and annual production rate. The unit CO_2 emissions for plant operation were calculated based on the consumed energy, unit energy use, and operation processing capacity. Since the raw charcoal materials III and IV do not require drying, only the unit CO_2 for crusher and carbonization furnace operation were summed.

4.4 CO_2 emissions from the construction and operation of the carbonization factory

The CO_2 emissions from the construction and operation of carbonization factory was evaluated by multiplication of unit Life Cycle CO_2 ($LCCO_2$) emission rate per floor area per year, 390 kg-CO_2/m^2/y, with the total floor area of the carbonization factory assumed to be 800 m^2/ward, taking the annual production rate into consideration. The unit $LCCO_2$ emission was obtained from the 'steel division' in the LCA guideline for buildings [11].

4.5 Transportation of the charcoal

In the same manner as described in Section 4.2, the unit CO_2 emission per unit loaded weight was evaluated assuming a 30 km distance by truck and corrected considering the load weight ratio against the air-dried weight of the raw charcoal material.

Table 1 Classification of the raw charcoal material

	forestation・cutting	Original material	carbide	Wood base Products
I	necessary	raw wood	charcoal	A
II	unnecessary	raw wood	charcoal	B, C
III	unnecessary	waste	charcoal	wood product of D, E
IV	unnecessary	waste	paper charcoal	paper product of D, F

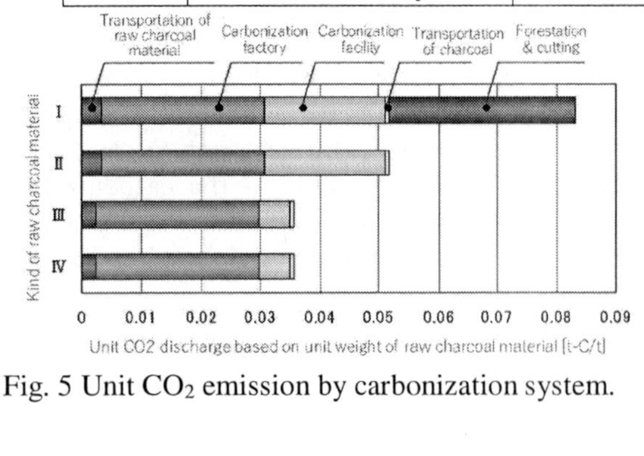

Fig. 5 Unit CO_2 emission by carbonization system.

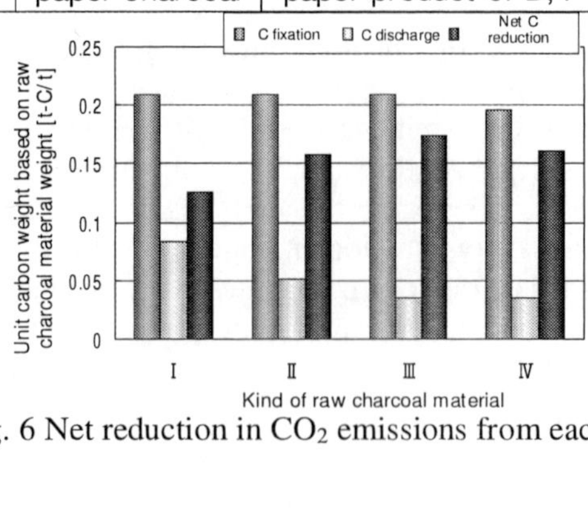

Fig. 6 Net reduction in CO_2 emissions from each raw

charcoal material using the carbonization system.

4.6 Calculated results

The unit CO_2 emission rate was the largest for the raw charcoal material I, which requires forestation and cutting, and the smallest for type III, which does not require the construction and operation of the dryer. A significant amount of CO_2 is emitted during the process of construction and operation of the carbonization factory and facilities, while the emission during transportation is relatively small.

5. Net Reduction in CO_2 Emissions Using the Carbonization System

Figure 6 shows the net reduction in CO_2 emissions from each raw charcoal material[A4)], which was calculated based on carbon fixation rate estimated in Chapter 3 and the unit CO_2 emission calculated in Chapter 4. It shows that the carbonization system has a CO_2 reduction effect, although the amount of reduction differs depending on the type of raw material. The CO_2 reduction rate from the raw charcoal material III is the highest, while the raw charcoal material I exhibited the lowest rate.

Using the available WBP estimated in Chapter 2, the net reduction in CO_2 emissions when carbonization is performed amounts to 16.34 million t-C/year. This corresponds to 5% of the anthropogenic CO_2 emission in Japan, 328 million t-C/y (as of 2002) [12], and 31% of the Japanese reduction goal as per the Kyoto protocol, 53 million t-C/y (as of 2002).

6. Prediction of the Reduction in Atmospheric CO_2 Concentration

To understand the influence that the CO_2 reduction by carbonization would have on the atmospheric CO_2 concentration, an extended model that integrates the carbonization process into the well-known CO_2 prediction model [13], [14] was proposed. The calculated atmospheric CO_2 concentration when carbonization is performed was compared to a scenario without carbonization analyzed using the proposed model.

6.1 Surface Biomass Carbon Circulation Model Combined with Carbonization

The previously proposed model [13], [14] predicts the CO_2 concentration from time 0 to t, using a convolution integral considering the net CO_2 emission at time t and CO_2 absorption by the ocean. The net emission is calculated using CO_2 emission data from industrial activities and changes in CO_2 accumulation of surface biomass. The latter is given by a carbon circulation model of the surface biomass[A5)].

Since the proposed carbonization process involves biomass resources as a raw material, it will influence the carbon circulation of the surface biomass. Therefore, the carbon circulation model was extended to consider CO_2 fixation by carbonization by adding the model (Fig. 7).

Because the raw charcoal materials I and II, which were to be resolved into humic substances or soil without utilization, are removed from the forestland, the corresponding amount of carbon, R_w, is deducted from decomposition flux of plants. The amount of CO_2 from the raw charcoal materials III and IV was represented by W_w and W_p, respectively, and deducted from the carbon flux from the plants to the atmosphere caused by the changes in land use (mainly deforestation), because these materials were utilized before incineration and reclamation. The carbon content of the raw charcoal materials, R_w, W_w, and W_p, was evaluated based on the carbon contents of the air-dried materials (types I to III: 42.7%, type IV: 43.1%). The carbon flux from charcoal box to the atmosphere, E_c, in the form of emitted carbon gas which was transformed from the carbon contained in the raw charcoal material without being changed into charcoal is given by the following equation.

$$E_C = (R_W + W_W)(1 - CR_W) + W_P(1 - CR_P)$$

where CR_w and CR_p represent the remaining carbon fractions [t-C/t-C] of the charcoal and paper charcoal, respectively, and $CR_W = 0.491$ and $CR_P = 0.456$.

Based on these considerations, a new prediction model of atmospheric CO_2 concentration incorporating carbonization was proposed by changing the surface biomass carbon circulation model and adding the CO_2 emissions from the carbonization process to those emitted from industrial activity (Fig. 7).

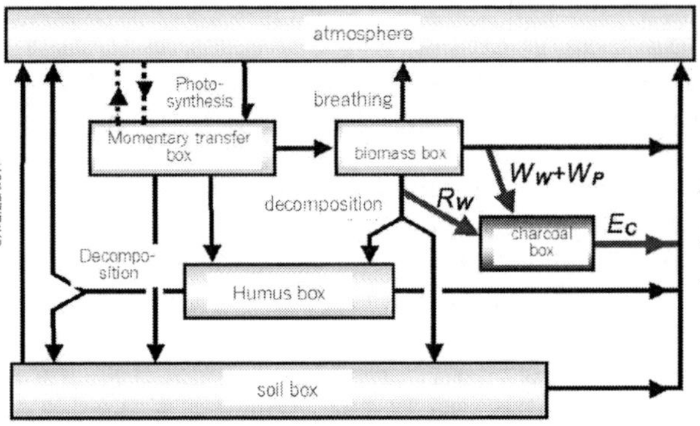

R_W : carbon weight of raw charcoal material (I and II) directly taken out from forestland [Gt-C/yr]

W_W : carbon weight of carbonized wooden waste (III) [Gt-C/yr]

W_P : carbon weight of carbonized paper waste (IV) [Gt-C/yr]

E_C : carbon weight liberated from raw charcoal material without being carbonized [Gt-C/yr]

Fig. 7 Surface biomass carbon circulation model incorporating carbonization.

6.2 Predicted reduction in CO_2 concentration

The predicted reduction in CO_2 concentration caused by carbonization, i.e., the difference in CO_2 concentration between the basic (without carbonization) and carbonization scenarios, is shown in Fig. 8. It was assumed that carbonization started in the year 2000, and that the annual amount was constant. The annual decrease rate of CO_2 concentration was determined to be approximately 0.0032 ppmv/y, and the total reduction amounts to 0.338 ppmv by 2100. These values represent a global reduction in CO_2 concentration when carbonization is performed only in Japan.

Figure 9 shows the unit CO_2 concentration reduction for each raw charcoal material[A6]. Similar to the unit net CO_2 reduction of each raw charcoal material in Fig. 6, it is largest for the type III materials, and smallest for the type I materials. The reduction efficiency is high 5 to 10 years after carbonization started in types I and II (raw wood), while it immediately jumps for carbonization of types III and IV (wastes).

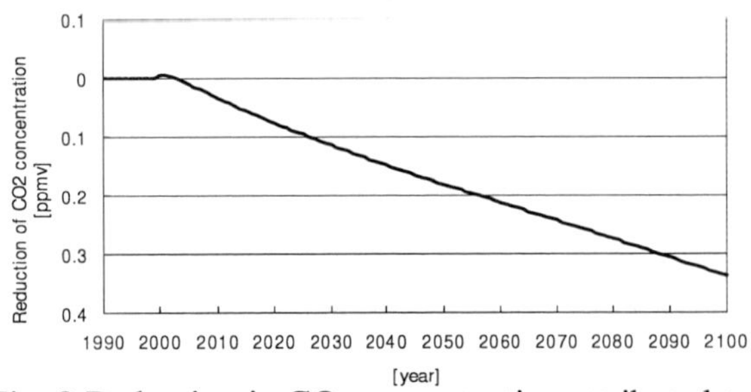

Fig. 8 Reduction in CO_2 concentration attributed to carbonization.

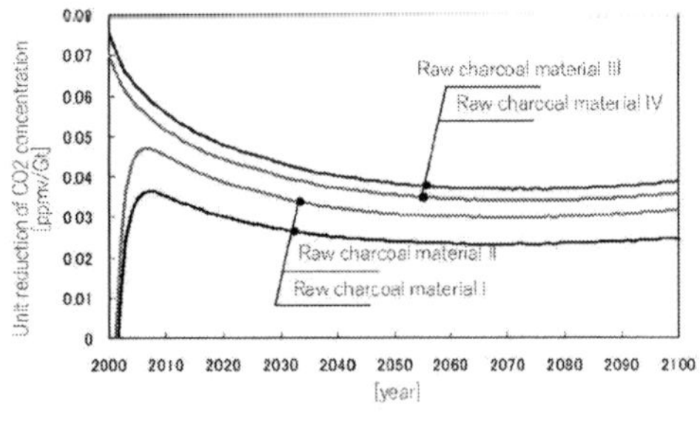

Fig. 9 Unit reduction of CO_2 concentration for each type of raw charcoal.

7. Prediction of the Decrease in Air Temperature

7.1 Calculation Method

The influence of CO_2 concentration on global climate has been discussed by comparing the temperatures at equilibrium for different atmospheric CO_2 concentrations, which are computed by climate models [15] assuming constant CO_2 concentrations. The temperature increase when the CO_2 concentration is doubled (temperature sensitivity) is predicted to be 2.0±0.5 °C by the one-dimensional radiation-convection equilibrium model, and 3.0±1.0 °C (global average) by three-dimensional atmosphere-ocean combined global circulation model. In this study, the temperature reduction effect caused by carbonization was roughly estimated using these models.

Since the temperature in the troposphere is nearly proportional to the logarithm of the CO_2 concentration [16], the temperature increase ΔT [°C] when the CO_2 concentration changes from C_0 [ppmv] to C [ppmv] is described by the following equation using a temperature sensitivity ΔT_{2X} [°C] factor.

$$\Delta T = \Delta T_{2X} \log_2 (C/C_0)$$

This is the predicted asymptotic value of the temperature increase at a constant CO_2 concentration, C. Because it takes several decades for this transient change to equilibrate, the temperature response is complicated when the CO_2 concentration continuously changes. To examine this response delay, the temperature change calculated using CO_2 concentration history was compared with the observed values. It was concluded that the above mentioned prediction model assumes a response delay of 30 years when $\Delta T_{2X} = 2$ [°C], 60 when $\Delta T_{2X} = 3$ [°C], and 80 years when $\Delta T_{2X} = 4$ [°C].

7.2 Predicted temperature decrease

Figure 10 shows the temperature decrease caused by carbonization and the difference between the temperatures calculated using the CO_2 concentrations estimated for the basic and carbonization scenarios in Chapter 6. When $\Delta T_{2X} = 3$ [°C], the temperature reduction appears in 2060, 60 years after the start of carbonization. The reduction of the global average temperature was estimated as 0.0022 °C by 2160. Although with larger temperature sensitivity, larger temperature reductions were observed along with larger response delays.

Figure 11 shows the estimated unit temperature reduction per unit of raw charcoal materials when $\Delta T_{2X} = 3$ [°C] [A7]. The unit temperature reduction increases between 2065 and 2070 for raw material types I and II (raw wood) due to the CO_2 concentration reduction effect 5 to 10 years after the start of carbonization. For types III and IV (waste) materials, a large effect is observed at 2060, because the carbon reduction appears immediately after the start of carbonization, and subsequently, the effect decreases and approaches 0.

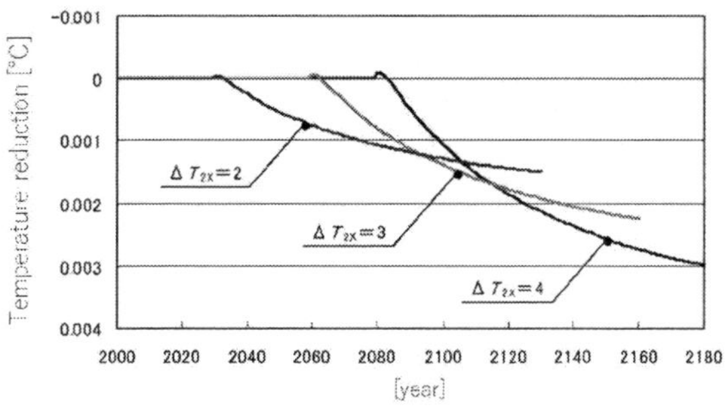

Fig. 10 Temperature decrease attributed to carbonization.

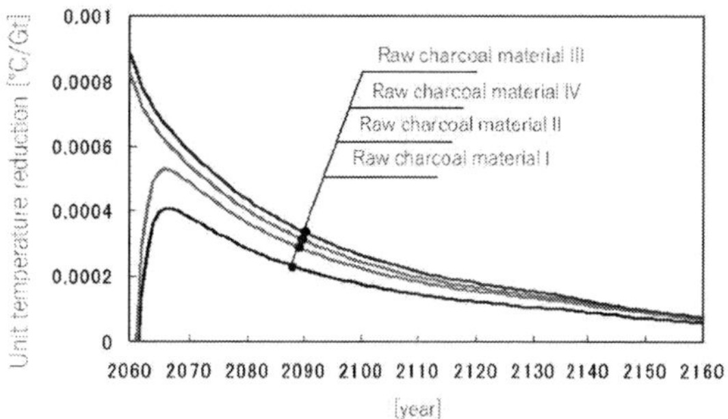

Fig. 11 Unit temperature reduction for each raw charcoal material when $\Delta T_{2X} = 3$.

8 Conclusions

(1) Carbonization fixes CO_2 corresponding to approximately 20% of the weight of the raw charcoal material (carbon-based weight). Since CO_2 corresponding to 4 to 8% of the raw materials is emitted during the carbonization process, the net reduction of CO_2 emissions amounts to 13 to 17% of the raw materials. The reduction efficiency strongly depends on the moisture content of the raw charcoal materials, and wooden wastes with low moisture content show the highest efficiency.

(2) To examine the influence of the carbonization of wood base products on the carbon circulation of surface biomass, a prediction model for the atmospheric CO_2 concentration integrated with

carbonization was developed. The reduction in the atmospheric CO_2 concentrations caused by continuous carbonization amounts to 0.024–0.037 ppmv/Gt per air-dried weight of raw charcoal material. The reduction efficiency becomes the highest 5 to 10 years after the start of carbonization for raw wood, while it occurs immediately after the start for the waste materials.
(3) If the available forest resources in Japan are fully utilized, 16 million t-C/y of net CO_2 can be removed, which corresponds to 5% of the anthropogenic CO_2 emissions of Japan. This corresponds to 31% of the Japanese goal of CO_2 reduction declared in the Kyoto protocol. This would decrease the global CO_2 concentration at a rate of approximately 0.0032 ppmv/y. By continuing this carbonization process for 100 years, the global CO_2 concentration would decrease by 0.338 ppmv, and the temperature by 0.0015–0.0030 °C. The effect on temperature reduction would appear 30–80 years later.
It should be noted that these conclusions are based on a limited consideration using models with uncertain assumptions, and the carbonization has its merits and demerits from a wider point of view.

Appendix
A1) Production efficiency of charcoal: air-dried weight of charcoal/air-dried weight of raw charcoal material
A2) Carbon fixation rate: carbon content of the charcoal/air-dried weight of the raw charcoal material
A3) Remaining ratio of carbon: carbon content of the charcoal/carbon content of the raw charcoal material
A4) Unit net CO_2 reduction = (fixed CO_2 − emitted CO_2) /air-dry weight of raw charcoal material
A5) 4-box model, whose surface biomass is composed of momentary transfer box, biomass box, humus box, and soil box
A6) Unit CO_2 concentration reduction = reduction rate of CO_2 concentration/air-dry weight of raw charcoal material
A7) Unit temperature reduction = reduction rate of temperature/air-dry weight of raw charcoal material

References
[1] Ministry of the Environment Government of Japan: 2003 White Paper, CLd. Gyosei, PP.90, 2003.
[2] P. Williamson: Scrutinize CO2 removal methods, Nature, Vol. 530, PP. 153-155, Feb. 2016.
[3] Research Institute of Innovative Technology for the Earth: Scenario saving the earth – Strategy for reducing CO2, Nikkan Kogyo Shinbun, PP.151-161, 2000.
[4] S. Kishimoto: Misteries of Charcoal, DHC CLd., 1994.
[5] Ministry of Internal Affairs and Communications, Administrative Management Bureau: "Waste Management and Public Cleaning Law (Dec. 25, 1970 Act No. 137)", http://law.e-gov.go.jp/htmldata/S45/S45HO137.html, (referred 2005/12/13). d (referred 2005/12/13).
[6] Seiichi Satonaka: Fundamental Study on Wood Carbonization, Research Bulletins of the College Experiment Forests, Hokkido University, PP. 257-277, Vol. 22, No. 2, 1963.
[7] Shimada, M., T. Iida, K. Kawarada, T. Okayama, and M. Fushitani. Pore structure and adsorption properties of carbonized material prepared from waste paper. Journal of Material Cycles and Waste Management. 3 (2), pp. 135-143, 2001.
[8] LCA Guideline Subcommittee, Architectural Institute Japan, "LCA Database Ver. 3.0", http://news-sv.aij.or.jp/tkankyo/s0/news.htm, (referred 2006/1/26).
[9] Special Policy Office, General Planning Bureau of Mie Prefecture, Vision of Biomass Energy Utilization, Mie Prefecture, March 2004, http://www.pref.mie.jp/TOKUTEI/plan/se/nev/bev/bev-2.pdf, (referred 2006/1/27).
[10] Research Institute of Innovative Technology for the Earth, Development of Integrated Technologies for Environment Policy Preventing Global warming by Carbonization of Trees (II), Result report, March 2002, Database of NEDO result report.
[11] Architectural Institute Japan, LCA Guideline of Buildings, Maruzen, 2003.
[12] Marland, G., T. A. Boden, and R. J. Andres. Global, Regional, and National CO2 Emissions. In Trends: A Compendium of Data on Global Change. Carbon Dioxide Information Analysis Center, Oak Ridge National Laboratory, U.S. Department of Energy, Oak Ridge, Tenn., U.S.A. 2005. http://cdiac.esd.ornl.gov/trends/emis/meth_reg.htm (referred 2006/2/2).
[13] Wigley, T. M. L.: A simple inverse carbon cycle model, Global Biogeochemical Cycle, 5 (4), PP. 373-382, 1991.
[14] Wigley, T. M. L.: Balancing the carbon budget. Implications for projections of future carbon

dioxide concentration changes, Tellus, 45B (5), PP. 409-425, 1993.
[15] Taro Matsuno: Influence of CO2 on climate, Research Notes, No. 160, PP. 161-185, The Meteorological Society of Japan, 1987 (in Japanese).
[16] K. Itoh: Global Warming - Last pieces of Jigsaw Puzzle, Nihon Hyoronsha, PP. 21-90, 2003.

IOP Conf. Series: Materials Science and Engineering **415** (2018) 012052 doi:10.1088/1757-899X/415/1/012052

Innovations in shaping the residential and retail buildings. Functional and pro-environmental potential of shipping containers in architecture

A M Berbesz[1], I M Szefer[1]

[1] Wroclaw University of Science and Technology, Faculty of Architecture, Department of Housing and Industrial Architecture, Interior Design, Rural and Landscape Planning and Visual Arts, W1/K6

E-mail: anna.berbesz@pwr.edu.pl, ilona.szefer@pwr.edu.pl

Abstract. Shipping containers are currently one of the innovative trends of shaping architectural objects. They are used to build portable and permanently set in the ground residential and public buildings. Realization with the use of shipping containers appear all over the world, i.a. single- and multifamily houses in the Netherlands, Spain, Denmark and France, the bus station in the Netherlands, the art gallery in Tokyo. Particularly, many realization with the use of shipping containers relate to housing environment due to ease of transport, speed of installation and assembly, shorten the construction time and pro-environmental potential. In contrary, conventional housing is responsible for ¼ forest logging area, absorbs 1/6 of fresh water resources and consumes 2/5 of liquid fuels and materials. The respect for the environment in the aspect of the global direction of shaping residential and retail buildings is particularly important. This paper focuses on pro-environmental, usably and economical aspects of architectural objects with the use of shipping containers on the example of Polish and world-wide realization.

1. The concept of container architecture – ideology

'The essential element in architecture is the manipulation of space. It is essence which separates it from all other arts'

Paul Rudolph

Container architecture constitutes a contemporary type of architecture making use of steel maritime shipping containers as the primary construction element. The appearance of containers in international trade around the year 1956 facilitated transport, reduced costs [1] and enabled the global economic growth [2]. Given the modular construction allowing easy structure shaping in a horizontal and vertical direction, this solution was adopted in architecture. An ever increasing interest in containers in architecture led to a new term - cargotecture. Objects built of shipping containers offer an innovative approach to shaping the human environment, including ecological and economic aspects. The first ideas and realizations based on shipping modules date back to 1989 when Phillip C. Clark issued a patent application described as: 'Method for converting one or more steel shipping containers into a habitable building at a building site and the product thereof'[3].

The popularity of universal container modules as building blocks comes largely from their durability and accessibility. Frames and transverse elements of containers are made of 4 mm-thick steel profiles and outer walls from trapezoidal metal sheet. Flooring is most often made of wood plates or plywood having a thickness of 28 mm. A container gate is equipped with bolt locks and the whole construction is perfectly sealed. Thanks to high-quality materials, containers can withstand any atmospheric conditions, heavy load and transportation. Their average life expectancy in extreme conditions is about 30 years, but a vast majority of typical sea containers is used for maximum 10 years. A special advantage of shipping containers is also an unlimited choice of available sizes (and easy modification): from universal/standard, to HC (High Cube), to PW (Pallet Wide), to Open Top containers, and thus they are ideal for designing building development dedicated to housing, service and commercial, office, recreational, cultural and even medical functions.

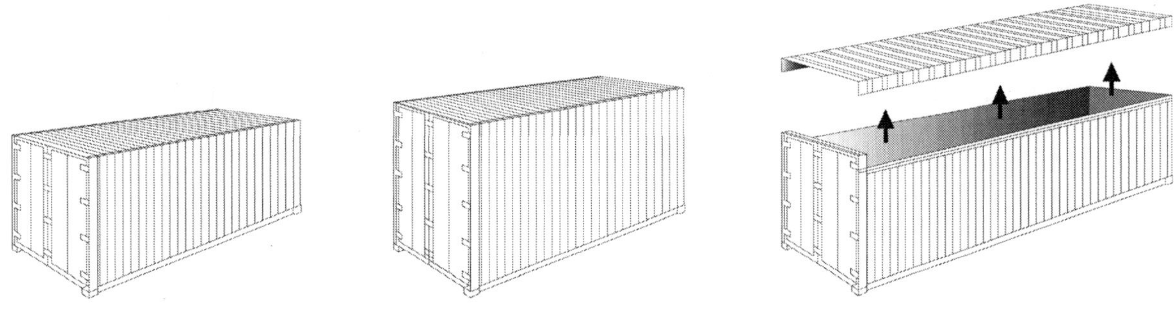

Figure 1. Scheme of available sizes of shipping containers that can be used in construction

2. Legal and economic aspects of container solutions

From the legal perspective, according to the Polish regulations, a container is classified as a temporary building object 'dedicated to temporary use over a period shorter than the container's technical durability, designed to be moved to another location or deconstructed, not connected to the ground permanently'[3]. Such objects (meant to be used temporarily during construction works) do not require building permit, yet their deconstruction or relocation has to take place no later than 180 days after the beginning of the construction defined in the building notification. The new location of the relocated object has to be reported to a competent authority again. The current regulations do not ban a relocation within the same plot.

In the case of a permanent connection between shipping containers and the ground (provided that the anchorage method is reliable enough to guarantee the building's durability and ability to withstand external conditions, which could change its position or even lead to its destruction), in accordance with the Polish building code, a notification or - in the case of an impact of the object on neighbouring plots - building permit is required. Based on the information obtained from competent municipal authorities, the building procedures are the same as in the case of traditional development. Similarly, as far as planning offices are concerned, there are no obstacles related to the object construction proposed.

From the economic point of view, an undeniable advantage of container-based constructions is lower building costs, easy transportation as well as a short assembly time. According to SG Blocks, a company dealing with deep-sea container adaptation on a wide scale, an application of this technology helps to shorten the construction time by up to 40% and reduce the costs even by 70% [4].

Harbours around the world store over 30 million containers since, for financial reasons, their transport without cargo to the home-port is not cost-effective. Finding a right container is not so difficult. An average cost of a single standard 20' module (6.1 m long x 2.4 m wide x 2.6 m high) and a 40' module (12.2 m long x 2.4 m wide x 2.6 m high) is only about 2,000 $, with the price depending on the technical condition and delivery costs. In Poland, prices are around ~1,640 $ (6,000 zł) per piece. Maritime container adaptation in the world is popular now and yet in Poland this is a narrow building niche. One of the companies that specialize in this kind of development is Flexicube. The smallest single-storey apartment in a container costs around ~ 15,030 $ (55,000 zł), a large multi-storey building ~ 48,920 $ (179,000 zł) (net prices not including transportation and anchorage costs - established individually depending on the location). On the Polish territory, transport and anchorage of containers is in the cost range of ~ 273 – 957 $ (1,000-3,500 zł). In contrast, the cost of building a house in the traditional system with an area of 135 m^2 with an attic is, according to the web page *Kalkulatory Budowlane*, about ~114,786 – 125,718 $ (420,000-460,000 zł).

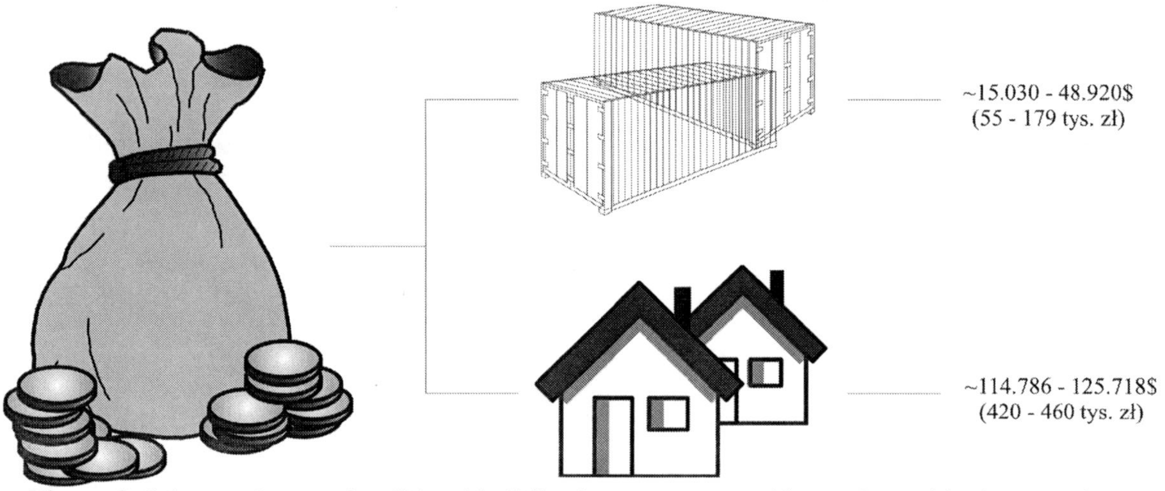

Figure 2. Scheme of costs of traditional building house system and house from shipping container

3. An outline of technological and material solutions in container-based objects

The contemporary container architecture, thanks to a use of modern technologies, materials, installation solutions and financial resources, allows designing objects respecting the natural environment based on the idea of sustainable growth. According to SG Blocks, each year hectares of forests are cut down for building industry purposes, 1/6th of fresh water and 2/5ths of liquid fuel resources and materials produced are used. Shipping containers substantially balance CO_2 emissions and decrease the demand for natural resources. Disposing of one container costs about 8,000 kWh of energy, whereas recycling and upcycling consume only 500 kWh. A substantial energy surplus may support the object's operation regardless of its function.

Adaptation of shipping containers as living and service modules requires adjustments to pre-existing climate and geological conditions in a given place. Detailed information about the adaptation, insulation and installation solutions applied in the adaptation of container units is available within Residential Shipping Containers Primer (RSCP ™), which constitutes a set of references to building objects based on ISO containers. It is managed by architects and designers with experience in designing and producing residential and service buildings and pop-up pavilions on the basis of containers as basic units [5].

One of the first stages of adaptation of containers as architecture objects is, apart from cutting out openings for windows and doors and adjusting the functional arrangement of the structure, proper insulation. A choice of good insulation will improve the operation of the object and ensure a proper user's comfort. One of the options is internal batt insulation using e.g. glass wool. In this technology, it is important to choose materials with the highest density. According to NAIMA (North American Insulation Manufacturers Association), glass fibers make up over 50% of the insulation material market in the USA. Producers use 25-40% of recycled glass in the production process. Another solution is sprayed insulation using low-toxicity materials, e.g. polyurethane foams (PUR). The blown-in technique is also used in applications of cellulose. This method minimizes the occurrence of thermal bridges in an object. Moreover, recycled cellulose allows retrieving 75% of the material and cellulose fibers exhibit resistance to biodegradation. An alternative to traditional insulation materials these days is SIP (Structural Insulated Panel) whose construction comprises facings made of OSB plates with a polymer foam insulation core [6].

Anchoring a shipping container construction to the ground may be based on various kinds of foundation. The most frequently used solution is traditional cast-in-place foundation walls and point-like foundations in the form of foundation footings. In locations where the frost line runs not deep in the ground, it is recommended to use shallow foundations in the form of a slab on the ground (slab-on-grade foundation). A use of traditional foundation no longer allows considering a container object as a temporary building object.

When adapting deep-sea containers so that they can serve as residential buildings, one may also consider alignment to the cardinal points analogous to shaping passive buildings. Then, longer walls should be situated along the east-west axis. The southern wall should face the Sun between 9 A.M. and 3 P.M. Spaces requiring illumination should be situated in the southern part of the object. Container structures can be located in places equipped with installation terminals, but they can also function as off-grid units or fully autonomous ones which requires a use of installation solutions independent from the pre-existing terminals.

4. Examples of technological objects and their functional division

4.1. Single-family housing development

An interesting example of an application of shipping containers is a house designed and built in 2012 for an art collector and his family in the Andes near Santiago in Chile. The house created by the architect Sebastián Irarrázaval consists of five 40', six 20' and one 40' open-top container dedicated to serve as a swimming pool. The total area of the object is 350 m^2. While designing, the author's priority was to integrate the new object with its nearest vicinity strongly influenced by the presence of the Andes, both visually and tectonically. Another key issue was to secure free air flow through the building (avoiding mechanical cooling installation) [7]. The use of cheap, durable materials which do not require maintenance has given the owner a possibility to live in a unique object and ensured low operation costs as well as a capacity of the building to age and increase the value of its materials and itself [8].

Figure 3. Scheme of single-family house in Chile, proj. Sebastián Irarrázaval, author: Szefer I, based on http://www.disenoarquitectura.cl/casa-oruga-sebastian-irarrazaval/, access: 15.03 2018.

4.2. Multi-family housing development

Drivelines Studios is an object opened in September 2017 aimed at multi-family housing function, whose main construction elements are adapted deep-sea containers. The realization took place in Johannesburg (The Republic of South Africa) in the Maboneng district. It constitutes one of the largest enterprises in that region of the world utilizing container architecture. At the same time, this development direction is justified by the fact that the city plays host to one of the largest inland ports with an industrial shipping park. The realization was handled by an office specializing in widely-understood cargotecture - LOT-EK (founded by Giuseppe Lignano and Ada Tolla). The building is V-shaped with a free-access gallery and a view of the inner yard. All housing units make up 100 studio-type apartments, each having an area of 40-60 m^2. Additionally, the object has been equipped with three service units. Each housing unit has been designed to allow for alternative energy sources, e.g. solar panels, rainwater-based systems. The idea of the city officials is to convert the heart of the Maboneng district into a culture and commercial hub of the city. The container-based housing development is meant to act like a magnet attracting new residents through its innovative character [9]. The current apartment rent cost is 4,900 R (about 1,400 zł per month) [10].

Figure 4. Scheme of multi-family house in Johannesburg, proj. LOT-EK, author: Szefer I, based on http://www.skyscrapercity.com/showthread.php?t=542573&page=88, access: 15.03 2018..

4.3. Collective housing buildings

- Dormitory

The dormitory building Urban Rigger designed by the architecture office BIG (Bjarke Ingels Group) in Kopenhagen in 2016 is an equally inspirational example. The project aims at securing affordable halls of residence for students located in the central part of the city and docked in the harbour. The object consists of 9 maritime shipping containers situated on a floating foundation, the so-called buoy. In order to protect the structure against rising sea levels, the containers were angled so that they create an inner patio as a common space for residents [11]. Flat roofs serve three functions: one of them is a terrace, the second roof is a green area and the third has been equipped with solar panels [12]. In total, the object's rooms, each having an area of 23-27 m^2, may accommodate 12 people. There is a plan according to which by 2025 next units will have been built [13]. Urban Rigger is a long-term urban project aspiring to create revolutionary and innovative system of floating student accommodation units. From the perspective of its founder and originator Kim Loudup, Kopenhagen seemed to be an ideal city given its harbour spaces and at the same time being one of the most expensive cities to live in for young people at the beginning of their higher education period [14].

Figure 5. Scheme of Urban Rigger dormitory, proj. Bjarke Ingels Group, author: Szefer I, based on http://www.urbanrigger.com/?da_image=25, access: 15.03 2018.

- Hotel

Noteworthy is also an atypical and eco-friendly hotel Quadrum Ski, using modularity of containers in Gudauri in Georgia, opened in 2017. Built at an altitude of 2,200 m above sea level, it has been named a new ski resort. Thanks to its unique construction, the hotel offers amazing views of the Caucasus mountains both in single and double rooms as well as in deluxe apartments. The architects who undertook the designing work of this demanding project were Sandro Ramishvili and Irakli Eristavi [15]. The proposed interiors constitute a combination of minimalistic and luxurious solutions [16].

Figure 6. Scheme of Quadrum Ski Hotel in Georgia, proj. Sandro Ramishvili and Irakli Eristavi, author: Szefer I, based on http://quadrum-gudauri.com, access: 15.03 2018.

4.4. Public buildings

- Library

An example of a public building in which deep-sea containers have been used is a library in Batu (Indonesia) being a project of Dpavilion Architects. Eight deep-sea containers 820 USD each were utilized in the object. Batu is a small city with a growing wealth disparity. The idea of the project was to fill the education gap by providing a public library space equipped with 6,000 books and creating a clinic with a basic public healthcare service. Colours, corresponding to various functions, play an important role in the object. The blue volume is dedicated to entertainment and popular books, the red containers offer a terrace and scientific books, the yellow unit is reserved for women and the green one serves as a lobby space [17].

IOP Conf. Series: Materials Science and Engineering **415** (2018) 012052 doi:10.1088/1757-899X/415/1/012052

Figure 7. Scheme of library in Batu, proj. Dpavilion Architects, author: Szefer I, based on http://www.morethangreen.es/el-pabellon-social-por-dpavilion-architects/, access: 15.03 2018.

- Mobile healthcare objects

One of the contemporary applications of container structures is their use as temporary objects relocatable to places stricken by natural disasters and serving as mobile medical units. Here, apart from aesthetic considerations and eccentric likes of the Investor, humanitarian aspects are of great importance. The first clinic based on container solutions was invented by Michał Wawrzewski, the author of the concept Clinics in a Can in 2005. M. Wawrzewski was also nominated to AAPA 2005 Physician Assistant of the year for his work on boosting the quality, optimization and accessibility of healthcare [18]. Clinic in a Can is now a firm providing medical objects adapting shipping containers to, among others, Ecuador, Haiti, Liberia, Nigeria, Sierra Leone, Kenya, the USA, the United Arab Emirates and the Philippines. The units can be delivered in the form of three functional groups. The first is connected to intensive medical care where the units have been equipped with an X-ray system, an ultrasound device, a suction machine, a portable ventilator, a multi-parameter screen and laboratory diagnostic devices, among other things. The second group is related to primary healthcare units enabling medical treatment for all people from the newborn to the elder. The third functional group is a solution for a time of an epidemic or infectious disease and provides proper quarantine spaces [19].

Figure 8. Scheme of mobile healthcare Clinic (Clinic in a Can), author: Szefer I, based on http://www.clinicinacan.org, access: 15.03 2018.

5. Summary

Shipping containers, as a type of modular architecture, are now a commonly used element prefabricated in the building industry. Many realizations with various functionalities have been created these days. In Poland, there is a number of firms adapting shipping containers, i.e. Carodesign, Flexicube, Weldon, etc. Currently, shipping containers are frequently used to shape the residential environment, especially single-family units. Apart from economic aspects, more extravagant objects are also produced. A pioneer of such solutions is Adam Kalkin, the author of numerous projects and books concerning adaptation of shipping containers, among other works: *Quick Build: Adam Kalkin's*

ABC of Containers Architecture and *Architecture and Hygiene* [21]. Given the economic and technological advantages, such structures may be also used in places touched by natural disasters and military conflicts providing temporary habitable spaces as well as primary medical care. Such an implementation of container structures gives them a humanitarian aspect, opens new perspectives for applications of this type of objects in architecture and can be a solution to housing problems in the future [22,23].

References

[1] Levinson M 2006 *The Box. How the Shipping Container Made the World Smaller and the World Economy Bigger* (Princeton University)
[2] Palma Oliveras A A 2010 *Sustainability in prefabricated architecture, A comparative life cycle analysis of container architecture for residential structures* Victoria University of Wellington Faculty of Architecture and Design School of Architecture
[3] Clark P C 1989-08-08 *Method for converting one or more steel shipping containers into a habitable building at a building site and the product thereof,* (patent no.: US4854094A) https://patentimages.storage.googleapis.com/48/c5/f8/9183a02b6bbad5/US4854094.pdf
[4] The Act of July 7, 1994 Building Law (Ustawa z dn. 7 lipca 1994 *Prawo budowlane* (art.3 pkt 5))
[5] https://www.sgblocks.com/, access: 06.03 2018
[6] http://residentialshippingcontainerprimer.com/what%20is%20RSCP, access: 03.03 2018
[7] http://residentialshippingcontainerprimer.com/insulation, access: 04.03 2018
[8] https://www.archdaily.com/394846/caterpillar-house-sebastian-irarrazaval-delpiano, access: 01.03 2018
[9] http://sirgear.com/hilltop-shipping-container-house-in-chilean-the-inside-is-so-cool/, access: 27.02 2018
[10] http://www.lot-ek.com/DRIVELINES-STUDIOS, access: 04.03 2018
[11] https://www.hippo.co.za/news/johannesburgs-new-container-homes-open-in-maboneng/, access: 06.03 2018
[12] https://www.big.dk/#projects-con, access: 05.03 2018
[13] https://www.dezeen.com/2016/09/22/big-bjarke-ingels-shipping-containers-floating-student-housing-urban-rigger-copenhagen/, access: 04.03 2018
[14] http://www.urbanrigger.com/urban-rigger/, access: 26.02 2018
[15] http://www.urbanrigger.com/about-us/our-history/, access: 02.03 2018
[16] https://www.georgianjournal.ge/discover-georgia/33407-extraordinary-container-hotel-built-in-gudauri-georgia.html, access: 04.03 2018
[17] http://quadrum-gudauri.com/, access: 04.03 2018
[18] https://www.designboom.com/architecture/dpavilion-architects-amin-shipping-container-library/, access: 05.03 2018
[19] http://www.hospitalsofhope.org/images/stories/Documents/news/aapa7-30-05.pdf, access: 01.03 2018
[20] http://www.clinicinacan.org/#about, access: 04.03 2018
[21] Kalkin, A., *ABC of Containers Architecture* and *Architecture and Hygien,*
[22] Cherenet Z Sewnet H 2012 *Building Ethiopia: Sustainability and innovation in architecture and design* Ethiopian Institute of Architecture Building Construction and City Development
[23] Martinez-Garcia M 2014 *Alternative Housing: The Shipping Container Home* Center for Realtor Technology,NATIONAL ASSOCIATION OF REALTORS

GIS modeling of green roofs allocation potential – case study (Kosice)

Z Poorova[1*], Z Vranayova[2]

[1,2] Faculty of Civil Engineering, Technical University of Kosice, Vysokoskolska 4, Kosice 040 01

*zuzana.poorova@tuke.sk

Abstract. Green infrastructure needs to be where people live and work – everywhere in the city. Current city challenges such as hot, dry, pollution, traffic, housing, urban sprawl etc. can be solved simply with nature. Improving stormwater management, changing microclimate, windbreaks, reducing energy use, biodiversity, applying green roof are few examples of solving problems in the cities these days. Ecosystem services must come to towns and people need to learn to use them. Sponge city is not only new term. It provides ecological, economic and social benefit through natural solutions. It is a key to sustainable spatial planning, urban design and development. The aim of this paper is to explain a case study of Košice city how green infrastructure can solve climate petrubation.

1. Introduction

The ideal way to measure any city´s heat island effect would be to examine regional weather patterns with and without the city in the place. Measuring heat island´s effects on regional climate is useful, but it cannot tell how effective mitigation measures would be at reducing a heat island´s effect. This is where modeling becomes necessary. Models are used to predict how well mitigation measures can reduce urban temperatures, energy use, and air pollution and retain water [1].

The simplest and the most common way to analyze a heat island is to compare existing weather data from two or more fixed locations. Fixed station data are used in three different ways:

- comparing data from a single pair of urban and rural weather stations;
- studying data from multiple stations to find regional, two-dimensional impacts;
- investigating a large set of historical data to evaluate heat island trends over time as region was developing.

More rigorous analysis of a heat island includes data from numerous fixed stations in and around the city. If enough stations are available, two-dimensional contour map of the city´s temperatures can be generated [1]. Aim of this article is to show that using available data creating of map of heat island of any city is possible using two main parameters – minimum humidity and maximum temperature.

2. Methodology

According to five year period analysis, World Meteorological Organization (WMO) stated that between 2011 and 2015 was the warmest five year period in the history of observations [2].During this period there have been many cases of extreme weather, particularly heat waves. Situation of the global climate was historic for many reasons. The level of greenhouse gases in the atmosphere has reached new highs in the northern hemisphere during spring 2015 [3]. For the first time it exceeded 3 month average of global concentrations of CO_2. 2015 was the warmest year in the history records in terms of the surface temperature of the oceans since the beginning of observations [4].

In 2016 there were 2-7 more days of summer than in extremely warm year 2015. On the other hand, significantly fewer tropical days were experienced than in the previous year 2015, approximately 15 to 30 days less [5,6].

Summer of 2016 was in all the main features quite different than the summer of 2015. While the year 2015 was extremely warm at the same time it was also very dry. Summer 2016 was much wetter than summer 2015. This resulted in a high number of days and nights, when people feel stuffy. It also resulted in intense thunderstorms. An essential feature of summer 2016 was also part of the transition significant cold front, which also initiated significant convective systems that brought abundant rainfall. That is why in 2016 we did not observe similar long periods of high daily maximum air temperatures exceeding 35 °C as in summer 2015. Characteristic features of the summer 2016 were extremely high temperatures and high frequency of intense storms and torrential rainfalls that are associated with them [5,6].

Following above mentioned information, data from July 01-31, 2015 [7] were used to create map of heat island of Košice city and map of green roofs of Košice city - map of a sponge city using green infrastructure [8]. The concept of a water-sensitive built environment is already established in many urban areas around the world on various scales through approaches such as sustainable drainage systems and low impact development. The traditional idea of managing surface water runoff in an urban environment is to funnel that water away from cities as fast as possible. Recently, however, city planners and governments are realizing that this design concept is flawed as it means that they are, to all intents and purposes, throwing away a highly valuable resource: water. The new mode of thinking is to design a city in such a way that it retains all the surface water runoff that occurs within the confines of the city and that it can be reused at a later date, thereby creating an urban environment that absorbs the water then releases that water when required, in a similar manner to a sponge [9].

3. Results

The mitigation effect of the green roof on urban heat island therefore retaining water in the original area of rainfall therefore lowering temperature and raising humidity is accessible using meteorological measurements. The measurements were collected using automatic weather stations that are installed on roofs of the buildings in Košice.

This research, completing Košice heat island map and map of green roofs of Košice uses only weather stations listed in Tab.1. The study takes into account only parameters given by these stations. For further research of this topic, more parameters will be needed to complete it.

3.1. Data

Table 1. List of max. temperatures and min. humidity

address	max. temperature [°C]	min. humidity [%]
Berlínska 3	**41.1**	**22**
Branisková 25	**36.6**	**25**
Brnenská 41	39.6	21
Bukovecká 43	42.7	20
Bukovecká 68	38.4	22
Cesta pod Hradovou 6	**31.4**	**29**
Čingovská 15	**32.4**	**41**
Čingovská 72	36	25
Magnezitárska 2	**37.8**	**22**
Fatranská 39	37.9	24

Hrabová 8	**43.4**	**19**
Kmeťova 51	35.4	26
Komenského 56	33.8	28
Obrody 63	38.4	22
Park angelium 58	35.1	26
Park mládeže 4	34.2	27
Severné Nábrežie 30	45.1	18
Sokolovská 23	**47.6**	**17**
Starozagorská 5	46.5	18
Šuhajova 49	34.3	27
Šuhajova 70	40.5	21
Turgenevova 66	37	24
Zinková 9	**37.8**	**22**
Zupkova 7	**43**	**19**

3.2. Maximum temperature

The studies on urban heat islands in other cities over the world are focusing on air temperatures rather than focusing on surface temperatures as a first factor. There are several types of heat islands depending on the climate and topography [10].

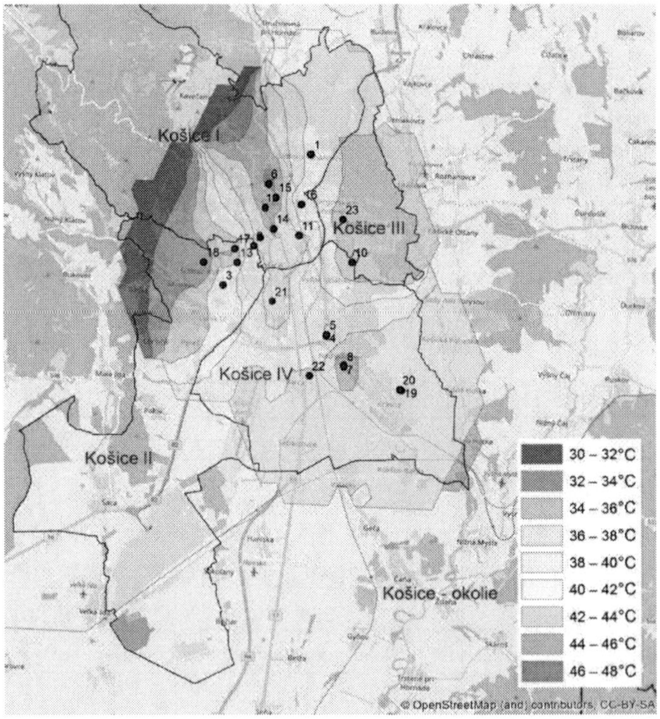

Figure 1. Spatial distribution of temperature trends in July 2015 in Košice (black dots are representing weather stations)

This study of heat island in Košice is focusing also on air temperatures, following study of efficiency of green roofs to mitigate urban heat island effect in Rio de Janeiro [10]. Online data from period July

01-31, 2015 (bold text in Tab.1) were supplemented with calculated data (regular text) caused by gaps in measurement, using data from period July 01-31, 2016, mathematic calculation and taking into account the location of weather station.

The study uses maximum temperatures, because it takes the worst case scenario- extreme into account. Spatial distribution of temperature trends in July 2015 in Košice is shown in Fig.1, detail in Fig. 2. The distribution of temperature trends used only data (max. temperature [°C]) listed in Tab.1. The urban microclimate is influenced by urban form and surfaces. City is characterized by impervious surface with high concentration of anthropogenic activities leading to significant increases in the air temperature and the surface temperature. The urban heat island effect occurs in cities all around the world and it is a result of different thermal properties of surfaces in urban areas. In Košice during summer 2015, up to 15.2°C air temperature differences (minimum 32.4°C, maximum 47.6°C) occurred comparing one housing estate to another housing estate. The biggest extremes are observed in Nad Jazerom and Krásna housing estates. The differences are due to presence of heat station, lake and park near weather stations used for this research.

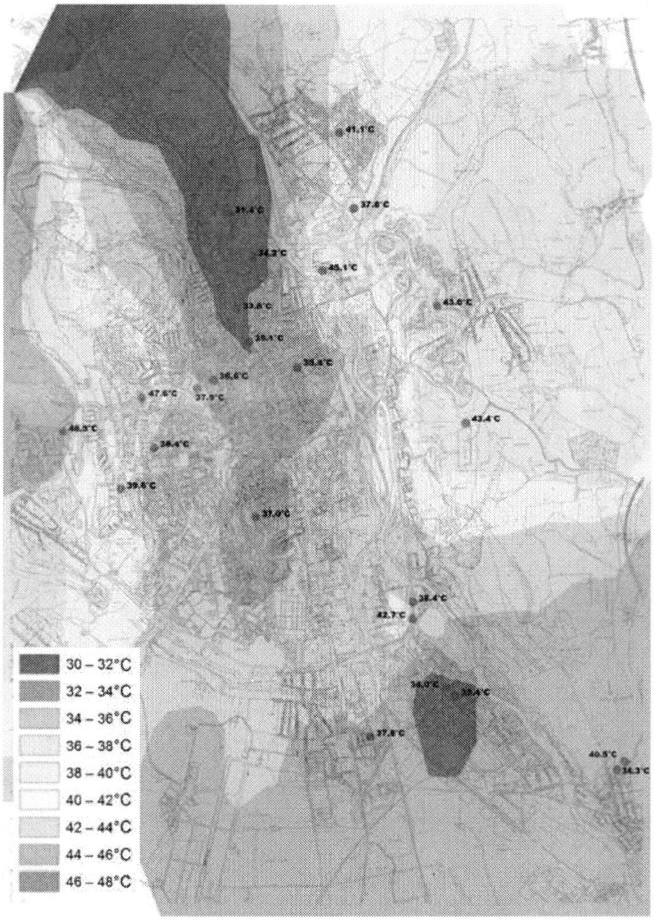

Figure 2. Spatial distribution of temperature trends in July 2015 in Košice

3.3. Minimum humidity

Spatial distribution of humidity trends in July 2015 in Košice is shown in Fig.3. The distribution of humidity trends used only data (min. humidity [%]) listed in Tab.1. The study uses minimum humidity because it takes the worst case scenario - extreme into account. The first radius of minimum humidity of each weather station represents 250 meters, minimal distance between weather stations used for this research. The second radius represents 500 meters.

The detail of spatial distribution of humidity trends in July 2015 in Košice focusing on city centre and surrounding housing estates is shown in Fig.3. The distribution used only data listed in Tab.1. 22 self-governing neighborhoods have more than just one humidity zone. The biggest extremes are observed in Nad Jazerom and Krásna housing estates. These differences are caused because of presence of heat station, lake and park near weather stations used for this research. Important is that these measurement results are related to the temperature measurements. The weather stations with higher temperatures have lower humidity and the weather stations with lower temperatures have higher humidity. In Košice during summer 2015 up to 24% humidity differences (minimum 17%, maximum 41%) occurred in comparison one housing estate to another housing estate.

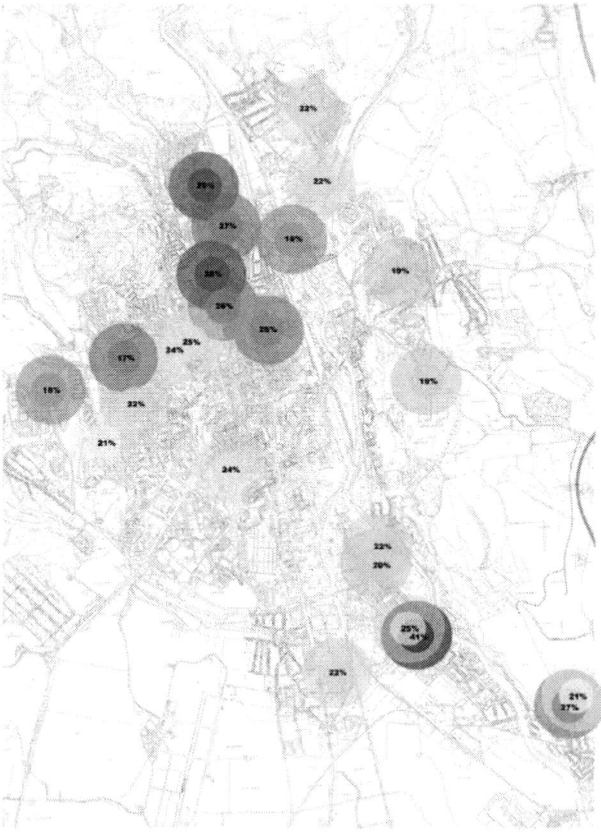

Figure 3. Spatial distribution of temperature trends in July 2015 in Košice - detail

4. Analysis

4.1. Maximum temperature

Modelling and analysing tools of ArcGIS – Geostatistical Analyst – in modelling of spatial distribution of temperature trends were used. Geostatistics is based on the regionalization of random variable in a given area. A set of random variables generated random function. Random function model is based on a study of the spatial variability of the studied phenomenon in different directions – experimental variogram. The result of this study is a mathematical model of variogram defined by changing the spatial variability in different directions of space by anisotropy and autocorrelation. Calculation of empirical semivariogram is written in the form [11]:

$$\gamma(h) = \frac{1}{2n(h)} \sum_{i=1}^{n(h)} [z(s_i) - z(s_i + h)]^2 \tag{6.1}$$

where: $\gamma(h)$ is estimated semivariation for the distance h; $n(h)$ is the number of pairs of measured points separated by a distance h; $z(si)$ is a measured value in point (si).

The first step is the calculation of empirical semivariogram. Followed by the transfer of empirical semivariogram by its theoretical model and determining its parameters. Model found for a given set of data depends on the experimental and theoretical assumptions.

The determining of semivariogram parameters is followed by the actual process of estimating the phenomenon of unknown values based on known data – Kriging [11].

The detail of spatial distribution of temperature trends in July 2015 in Košice focusing on city centre and surrounding housing estates is pictured in Fig.2.

4.2. Minimum humidity

The detail of spatial distribution of humidity trends in July 2015 in Košice focusing on city centre and surrounding housing estates is shown in Fig.3.

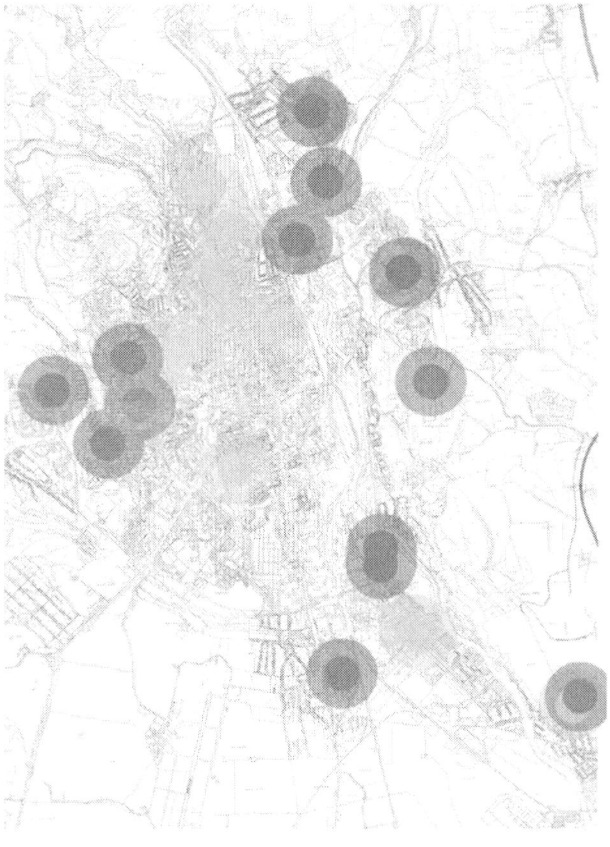

Figure 4. Allocation of green roofs in Košice following spatial distribution of humidity trends

5. Discussion

Košice was chosen as a study area to demonstrate the possibility to mitigate low humidity and therefore heat island in the city center and housing estates implementing green roofs and designing sponge city. The paper deals with the possibility to use green roof structures to manage stormwater runoff delay, its reduction into the public drainage system and its storage for later reuse.

Green roofs in urban areas are significant due to their capacity of water storage in the layers of the roof. This capacity, retention quality, allows minimizing the effect of extreme rainfall in short period. Retaining water in the layers of the green roof means reducing the flood peak in the urban drainage system and reducing the risk of water distress.

6. Conclusion

The area with low humidity (17-22%) needs green roof construction with higher retention qualities (Fig. 4, "dots in blue color"). The city area with higher humidity (24-41%) needs green roof construction with lower retention qualities ("dots in green color").
The first radius of suggested green roof construction represents 250 meters. The radius represents minimal distance between weather stations used for this research. The second radius of suggested green roof construction represents 500 meters.
The map of green roofs in Košice (Fig. 4) is modeled using data in Tab. 1. Allocation of two different major types "green color roofs" and "blue color roofs" of green roofs follows these data. More input data is needed to complete more precise and more accurate map of green roofs of Košice city.

7. References

[1] Gartland, L. (2011) Heat Islands. Undrestanding and Mitigating Heat in Urban Areas. London, Earthscan
[2] Coffel, E. D., Horton, R. M., Sherbinin, A. (2017) Temperature and humidity based projections of a rapid rise in global heat stress exposure during the 21st century.
[3] Gutro, R. (2015) NASA - What's the Difference Between Weather and Climate? online: https://www.nasa.gov/mission_pages/noaa-n/climate/climate_weather.html Retrieved: June 11, 2018
[4] Bochníček, O., Faško, P., Švec, M. (2016) Rok 2015 - druhý najteplejší v histórii meteorologických meraní v Európe (2015: second warmest in history of meteorological measurements in Europe) http://www.shmu.sk/sk/?page=2049&id=681
[5] Šťastný, P., Faško, P., Pecho, J., Labudová, L., Ivvaňáková, G., Kajaba, P., Turň, M,
[6] Bochníček, O. (2017) Rok 2016: mimoriadne teplý a veľmi bohatý na atmosférické zrážky (Year 2016: extremely warm year with very rich in precipitations)
[7] Climate: average monthly weather in Kosice, Slovakia (2018) online: www.weather-and-climate.com/average-monthly-Rainfall-Temperature-Sunshine,Kosice,Slovakia Retrieved: June 11, 2018
[8] Yeang, K., Spector, A. (2011) Green design. From theory to practice. London: Black dog publishing. ISBN 978 1 907317 12 5
[9] Atmospheric Humidity And Precipitation (2018) online: https://www.britannica.com/science/climate-meteorology/Atmospheric-humidity-and-precipitation Retrieved: June 11, 2018
[10] Siman, C., Pecho, J. (2017) Extrémne úhrny zrážok v závere júla 2016 (Extreme rainfall at the end of July 2016) http://www.shmu.sk/sk/?page=2049&id=736
[11] Konasova, S. (2016) The efficiency of green roofs to mitigate urban heat island effect in Rio de Janeiro Advances and Trends in Engineering Sciences and Technologies II Edited by Mohamad Al Ali and Peter Platko, Taylor & Francis Group, 6000 Broken Sound Parkway NW, Suite 300, Boca Raton, FL 33487-2742 CRC Press 2016 SBN: 978-1-138-03224-8 http://www.crcnetbase.com/doi/abs/10.1201/9781315393827-80

Acknowledgments

This work was supported by project VEGA n. 1/0202/15: Sustainable and Safe Water Management in Buildings of the 3rd Millennium

IOP Conf. Series: Materials Science and Engineering **415** (2018) 012054 doi:10.1088/1757-899X/415/1/012054

Vertical greenery systems for the energy retrofitting of buildings in Mediterranean climate: a case study in Catania, Italy

S Cascone[1], G Evola[2], C Leone[1] and G Sciuto[1]

[1] Department of Civil Engineering and Architecture, University of Catania, Via Santa Sofia 64, 95123, Catania, Italy
[2] Department of Electrical, Electronics and Computer Engineering, University of Catania, Viale Andrea Doria 6, 95125, Catania, Italy

E-mail: stefano.cascone@unict.it

Abstract. The increasing awareness of issues like climate change and reduction of available fossil resources moves scientific research into the development of new technological solutions in order to reduce the energy and environmental impact of buildings. Hence, in recent years, key concepts such as energy saving and environmental sustainability affected both the design of new buildings and the retrofitting of existing buildings.
The present study aims to evaluate the energy behavior of an existing building in Mediterranean area, and to quantify the potential contribution provided by retrofitting solutions implying the adoption of green roofs and green façades. Dynamic simulations, validated by comparison with experimental results, show that a green façade with dense foliage allows a reduction by 1.7 °C in the indoor air temperature and by 1.5 °C in the inside surface temperature, if compared to a traditional retrofit solution for the envelope. Furthermore, the outside surface temperature is reduced by 2.9 °C, and the cooling load is reduced by 32%.
These results suggest that green façades are a viable solution to mitigate current environmental issues in buildings. Lower energy consumption corresponds to lower pollutants emissions and to economic savings in the use of the building.

1. Introduction

Scientific research moves forward the development of new technological solutions in order to reduce the energy and environmental impacts of buildings [1]. In recent years, green systems are becoming widespread thanks to numerous studies confirming their multifunctional benefits at both building and urban scale, such as energy savings, mitigation of the urban heat island effect, biodiversity improvement, carbon dioxide sequestration and building acoustic insulation.

Traditionally, vertical greenery systems (VGS) are made up of climber plants growing directly on the exterior surface of buildings [2]. In recent years the construction and technological characteristics of VGS have been improved by separating the vegetation from the façade in order to avoid functional problems related to linking the building with vegetation.

Whilst green roofs are clearly distinguished into extensive and intensive, a classification for vertical greenery systems results more complex due to the different constructive systems, the variety of vegetal species, the different thermal behavior, etc. [3]. Previous research [4] proposed a classification distinguishing VGS into two categories: the green façade and the living (or green) walls.

Green façade consists of climber plants growing directly on the building vertical surface or indirectly by using support systems, such as cables or trellis made of different materials (steel, wood, plastic, aluminum), thus creating a double skin façade. Living wall systems are based on hydroponic crops in which vegetation can grow in absence of soil thanks to nutrient solutions. These systems are usually made up of geotextile felts or prefabricated panels fixed to a vertical support or the wall structure [5].

VGS contribute to different benefits both at buildings and urban scale. Different studies [6] confirmed that greening the building envelope is a valid passive systems for energy savings thanks to the following main effects: the shadow produced by the vegetation, the thermal insulation provided by vegetation and substrate, the evaporative cooling due to the evapotranspiration process and the protection from wind action [5]. Pérez et al. [7] analyzed the shading effect of a green façade on the wall surface temperature, measuring a maximum difference of 15.2 °C by using the green façade. The air in the intermediate space between the green façade and the building wall had a lower temperature and higher humidity during summer, proving the evapotranspiration effects of plants.

The vertical greening systems, in addition, contribute to the reduction of noise pollution in urban areas thanks to the sound-proofing and sound-absorbing power provided by vegetation. Azkorra et al. [8] evaluated the acoustic impacts of eight different VGS. The results showed a stronger attenuation at low and middle frequencies due to the absorbing effect of substrate, while a smaller reduction was observed at high frequencies due to vegetation scattering. Moreover they found higher values of the sound absorption coefficient of the VGS compared with other building materials.

The urban heat island (UHI) phenomenon determines an urban microclimate characterized by higher temperatures than the surroundings rural areas. Thanks to the higher values of albedo and thermal inertia compared to the usual artificial materials, urban green plays a fundamental role in the mitigation of the UHI [9]. The research conducted by Song and Wang [10] verified an increase in air temperature in several large U.S. cities, and estimated that green areas could mitigate the UHI effect, reducing the national energy demand in air conditioning by 20% and improving the urban air quality.

Furthermore, vertical greenery systems improve the economic value of the building. Des Rosiers et al. [11] estimated an increase in the property value of 3.9%. Gao and Asami [12] found that greening walls, streets and pedestrian spaces would provide an increase in land price by 1.4% in Tokyo and by 2.7% in Kitakyushu (Japan).

However, only a few studies have experimentally quantified the benefits of green façade used for energy redevelopment of existing buildings. The present study aims to evaluate the energy behavior of an existing building in Mediterranean area and to quantify the contribution provided by the shading effect of green façade in terms of both surface temperature and energy savings. Using dynamic simulation software, various technological solutions are analyzed in order to assess the effectiveness of the green façade for the retrofitting of existing buildings, comparing the performance of different envelope solutions.

2. Material and methods

2.1. The building case of study

The experimental site is located in the city of Catania (37°31′ N, 15°05′ E), south of Italy. Catania is characterized by a Mediterranean climate with hot and dry summer and mild winter (Csa according to Köppen and Geiger climate classification). The hottest months are July and August, when the air temperature usually exceeds 30°C. The yearly average precipitation is around 500 mm, mainly from October to March.

The existing building has a rectangular perimeter measuring 7 x 15 m, 4 m high, and is made up of wooden structure. The main elevation of the building has an optimal exposure as it is south facing and free from obstructions (Figure 1). Internally the building consists of a single large room. A large covering shades the entire building roof. The building hosts a showroom. The aim of retrofitting is to convert the structure into offices. Table 1 shows the stratigraphy of the existing building envelope.

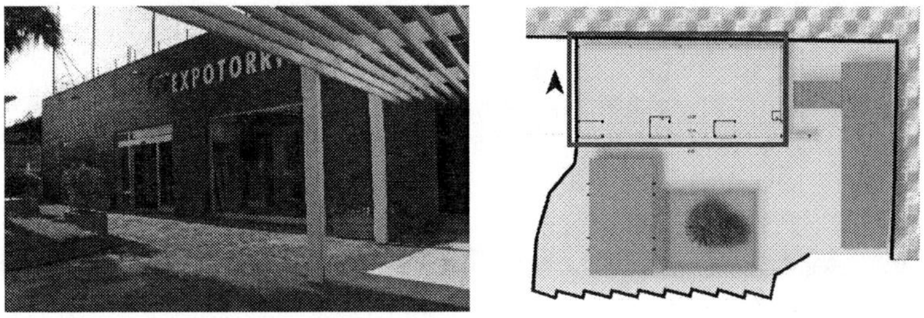

Figure 1. South elevation and plant view of the building

Table 1. Stratigraphy and thermal properties of the current envelope.

Component	Description (from ext. to int. side)	Conductivity l [W/(mK)]	Specific heat *c* [J/(kgK)]	Dry density r [kg/m³]	Thickness *s* [cm]	*U*-value [W/(m²K)]
Walls	Plywood	0.15	2700	550	1.8	2.86
	Plasterboard	0.21	1000	900	1.25	
Windows	16-mm Low-e double glazing + Aluminum frame with thermal break					3.1
Roof	Laminated fir wood	0.12	1380	510	1.4	3.16
	Plasterboard	0.21	1000	900	1.25	
Foundation	Concrete slab	1.13	1000	2000	29	2.22
	Parquet	0.14	1200	650	1.5	

2.2. *Experimental set-up*

The internal environmental parameters were recorded using TESTO 480 data logger, placed in the center of the room. Table 2 reports the sensors used and the parameters monitored. A weather station located less than one kilometer away from the building was used to measure the outdoor air temperature. The indoor air temperature was used to validate the Energy Plus building model. Data collection began on the afternoon of 18th July 2017 until the afternoon of 21st July 2017, at 5-minute intervals. During the measurements no activities were carried out inside the building, and the entrance door was opened only in the afternoons of 19 and 20 July, from 2.30 to 6.00 pm.

Table 2. Sensors used with TESTO 480

Probe	Measure	Range
Globe Temperature (D = 150 mm)	Mean radiant temperature	0 to +120 °C
IAQ	Air temperature, humidity, CO_2	0 to +50 °C
Thermocouple PT100	Air temperature	-50 to + 250 °C

2.3. *The envelope retrofitting*

In order to evaluate the energy behavior of the building and identify the solution that allows the greatest energy savings, six scenarios have been supposed for the retrofitting of the existing building envelope (Figure 2).

Scenario S1 considers a traditional retrofitting based on the insulation of the envelope components, aimed to comply with the limits of thermal transmittance defined by the Italian regulations. Scenario S2 introduces a green roof. In scenario S3, a green façade consisting of a wooden support system at 20 cm from the wall was placed to shade the south-facing wall. Furthermore, a horizontal and vertical shading of 70 cm depth protects the entrance window from direct solar radiation. The creeper supposed for the green façade is *Virginia creeper* (Parthenocissus quinquefolia), a deciduous plant with a great density of foliage. In scenario S3.A a low density of the foliage was supposed, which implies a light transmission coefficient of 0.4 over the walls and 0.8 over the windows in order to ensure a better lighting of the indoor space. Conversely, in the scenario S3.B, a higher density of the foliage was supposed, defined by a light transmission coefficient of 0.2 over the walls and 0.6 over the windows. This scenario is more realistic and is closer to Pèrez et al. results [7], where the authors measured a 0.15 light transmission coefficient for *Virginia Creeper*.

Finally, Scenario S4 combines green roof and green façade. Table 3 provides the detailed stratigraphy of all components; in all the scenarios proposed, the existing covering above the roof is removed.

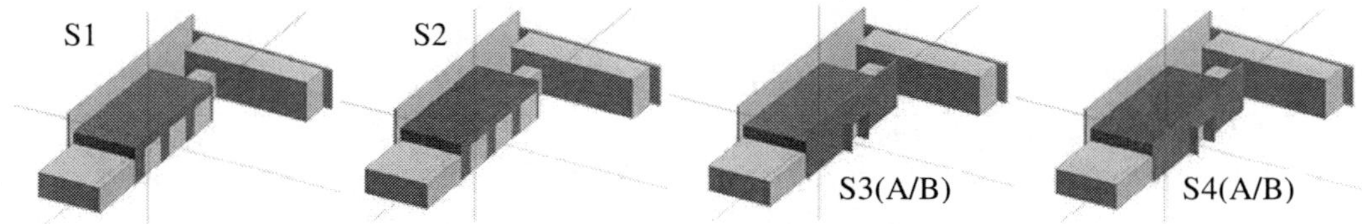

Figure 2. The different retrofitting hypothesis

Table 3. Stratigraphy and thermal properties of the different scenarios

Scenario	Component	Descriptions (from ext. to int. side)	Conductivity λ [W/(mK)]	Specific heat *c* [J/(kgK)]	Dry density δ [kg/m³]	Thickness *s* [cm]	Total *U*-value [W/(m²K)]
S1	**Wall**	Gypsum fiberboard	0.32	1100	1150	1.25	0.26
		Wood wool insulation	0.065	1810	360	3.5	
		OSB panel	0.1	1700	530	1.5	
		Wood fiber insulation	0.038	2100	160	10	
		OSB panel	0.1	1700	530	1.5	
		Vapor retarder					
		Non-ventilated air gap	0.55	1000	1000	2	
		Gypsum fiberboard	0.32	1100	1150	1.25	
	Roof	Sheet metal covering	45	420	7680	0.05	0.31
		Ventilated air gap	0.55	1000	1000	5	
		OSB panel	0.1	1700	530	2	
		Wood fiber insulation	0.038	2100	160	10	
		OSB panel	0.1	1700	530	2	
		Gypsum fiberboard	0.32	1100	1150	1.25	
S2	**Wall**	S1					0.26
	Green roof	Sedum					0.31
		Substrate	0.44	880	950	10	
		Filter layer + Drainage					
		Waterproof membrane					
		Wood fiber insulation	0.038	2100	160	10	
		Vapor barrier					
		OSB panel	0.1	1700	530	2	
		Gypsum fiberboard	0.32	1100	1150	1.25	
S3.A-S3.B	**Wall**	Green façade + S1					0.26
	Roof	S1					0.31
S4.A-S.4.B	**Wall**	S3.A-S3.B					0.26
	Green roof	S2					0.31

2.4. Simulation set up

The environmental parameters analyzed are the indoor air temperature, the outside and inside surface temperature of the south façade and the energy needs for space cooling. These are estimated by using Energy Plus, a dynamic simulation software tool. Temperatures are evaluated in free running conditions and the simulations are performed from 18th to 21st July. On the other hand, the energy consumption is obtained from 1st June to 30th September by means of an ideal thermostat control that keeps the indoor temperature at 26 °C and a controlled mechanical ventilation system according to the work schedule (Monday to Friday, 9.00 h to 13.00 h and 16.00 h to 19.00).

3. Results and discussion

3.1 EnergyPlus model validation

Figure 3 shows that the indoor air temperature rises rapidly from 7:00 a.m., when solar radiation hits the building and the outdoor air temperature increases, reaching a peak value at 3:00 p.m. The slight irregularities of the thermal gradient observed in the afternoon are due to the opening and closing of the entrance door. During the night the temperature keeps decreasing until dawn. Moreover, the peak of the indoor air temperature is almost in phase with the peak of the outdoor air temperature, due to the reduced thermal inertia of the existing envelope, which is typical in northern Europe but not suitable for hot climates of southern Europe. The mean radiant temperature is very close to the indoor air temperature, indicating that the inside surfaces are not significantly overheated.

The first step of the research concerned the validation of the EnergyPlus model of the building. To this aim, Figure 4 shows good accordance between the measured and simulated indoor air temperature for the current state (CS1 in Figure 4), with only small differences when the air temperature decreases.

Then, in order to compare the different proposed scenarios starting from the same boundary conditions, the behavior of the building was assessed without the shading contribution due to the fiber cement roof (CS2 in Figure 4), that must be in any case replaced. As expected, this cover plays a fundamental role: without it, the internal temperature increases by about 4 °C, now exceeding 38 °C.

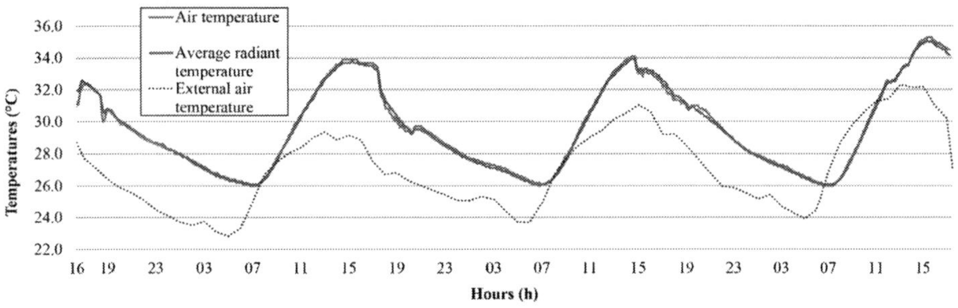

Figure 3. Temperature measured during the monitored period

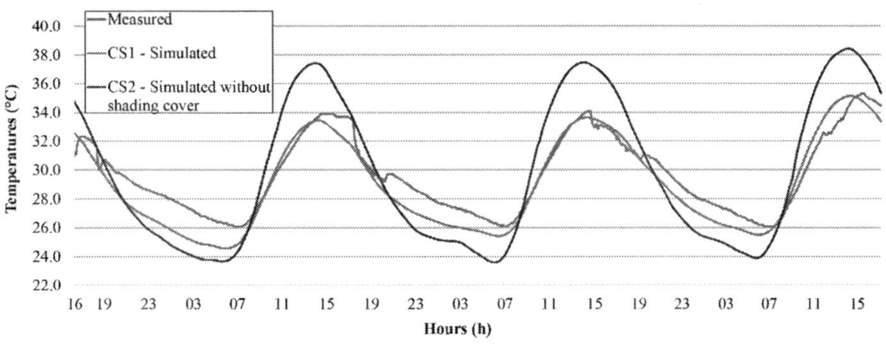

Figure 4. Energy Plus model validation – measured vs. simulated air temperature

3.2 Air temperature

Table 4 reports the maximum indoor air temperatures for the different retrofitting scenarios. In comparison with case CS2 discussed in the previous section, better results are achieved in scenario S1, i.e. by adopting a retrofitting solution based on a traditional building envelope without greenery systems. Indeed, the maximum indoor air temperature decreases from $T_{max,CS2}$ = 38.4 °C to $T_{max,S1}$ =

34.9 °C. Moreover, this scenario shows higher thermal inertia, and the indoor air temperature is less sensitive to fluctuations in the external temperature (Fig. 5).

Moreover, by using a green roof (scenario S2), the indoor temperature does not vary significantly if compared to scenario S1, and $T_{max,S2}$ = 34.6 ° C. This small difference between the traditional roof retrofit (S1) and the green roof (S2) highlights the effectiveness of the micro-ventilation of the sheet roof used in scenario S1, which prevents the overheating of the lower layers. Therefore, the installation of the green roof cannot be justified only from an energy point of view.

On the other hand, the best results pertain to solution S4.B, i.e. when the green roof is coupled with the VGS having a high density of the foliage; in this case the maximum indoor air temperature decreases to around 33 °C. However, eliminating the green roof (solution S3.B) or adopting a smaller density for the foliage (solution S4.A) do not change the outcomes significantly.

Table 4. Air temperatures of the different scenarios analized

Case study	CS1	CS2	S1	S2	S3.A	S3.B	S4.A	S4.B
T_{max} (°C)	35.1	38.4	34.9	34.6	33.7	33.2	33.4	32.9

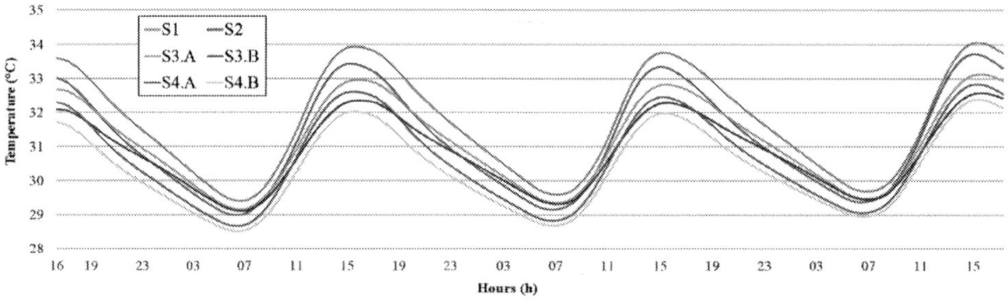

Figure 5. Indoor air temperature of the proposed scenarios

3.3 Outside and inside surface temperature

The results in terms of outside surface temperature are reported only for scenario S3.A and S3.B and compared with scenario S1 (reference building retrofitting). As Figure 6 shows, the green façade reduces outside surface temperatures by about 3 °C, from $T_{max,S1}$ = 43.3 °C without the shielding to $T_{max,S3.A}$ = 40.5 °C with the climber. The higher foliage density of scenario S3.B does not provide further reductions on the outside surface temperature (Table 5). This reduction, compared to traditional finishing materials, can contribute to mitigate the Urban Heat Island effect.

Table 5. Maximum outside surface temperature

Scenarios	S1	S3.A	S3.B
Temperature (°C)	43.4	40.5	40.5

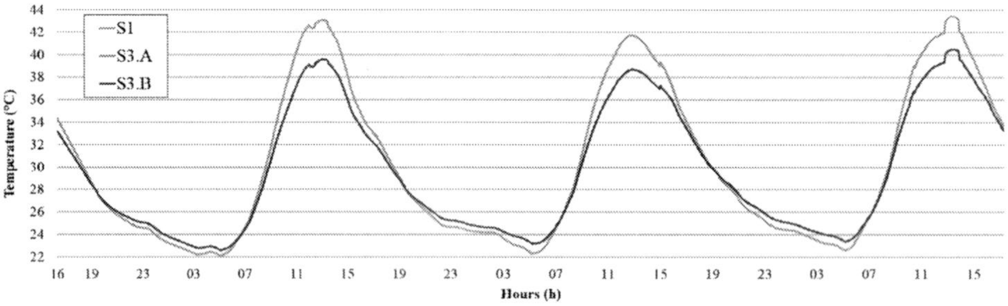

Figure 6. Outside surface temperature of the proposed scenarios

As found for the indoor air temperature, the inside surface temperature is higher in scenario S1 ($T_{max,S1}$ = 34.1 °C) than in all other scenarios. Temperatures decrease for scenarios S2, S3.A and S3.B, with a T_{max} of 33.7 °C, 33.1 ° C and 32.6 ° C respectively (Table 6).

Figure 7 shows that the lower surface temperatures are achieved by the combination of the green roof and facade ($T_{max,S4.A}$= 32.8 °C, $T_{max,S.4.B}$ = 32.4 °C). The maximum difference between the internal surface temperature of scenario S4.B (green facade with high density of foliage) and the reference scenario S1 (traditional building envelope) is 1.7 ° C.

Table 6. Maximum inside surface temperature

Scenarios	S1	S2	S3.A	S3.B	S4.A	S4.B
Temp (°C)	34.1	33.7	33.1	32.6	32.8	32.4

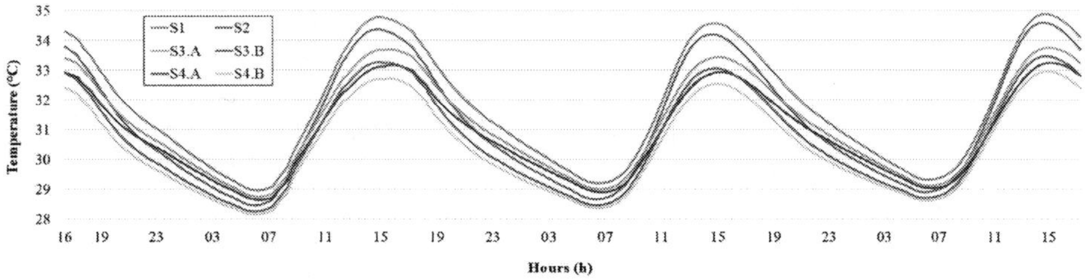

Figure 7. Inside surface temperature of the proposed scenarios

3.4 Energy cooling load

Figure 8 shows the energy needs for space cooling in kWh/m^2, estimated by EnergyPlus under thermostatic control. Currently, the building has a cooling energy consumption of 27.9 kWh/m^2 (CS1) during the summer season. In case CS2 (without the shading of the canopy), the energy needs would be almost twice as high, which shows once again the energy benefits coming from shading the building roof from direct solar radiation. As shown in Table 7, for all the proposed retrofitting scenarios the cooling needs decrease significantly: the reduction ranges between 56% and 72%, if compared to case study CS1 (current state of the building envelope).

However, if we take scenario S1 (traditional building retrofitting solution) as a basis for comparison, the energy savings obtained for all the other retrofitting scenarios are shown in Table 8. The scenario that combines the green roof with the green façade characterized by a higher index of foliage (S4.B) has the greatest energy saving potential.

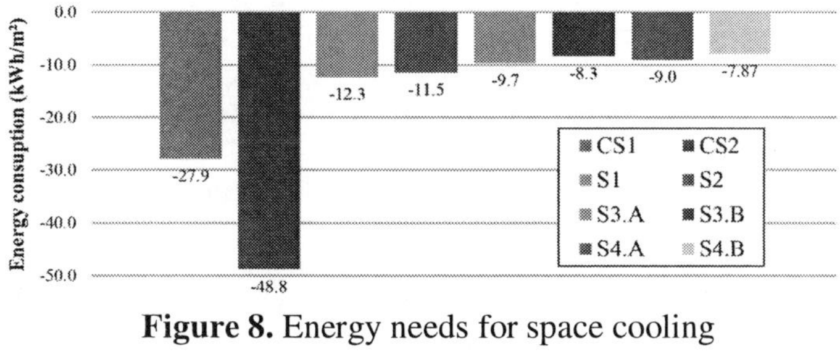

Figure 8. Energy needs for space cooling

Table 7. Percentage variation of the energy needs with respect to CS1

Scenarios	CS2	S1	S2	S3.A	S3.B	S4.A	S4.B
Variation (%)	+75%	-56%	-59%	-65%	-70%	-68%	-72%

Table 8. Percentage of energy savings with respect to S1

Scenarios	S2	S3.A	S3.B	S4.A	S4.B
Energy reduction (%)	-7%	-21%	-32%	-27%	-36%

4. Conclusions

The present study assessed a vertical greening systems as a possible passive strategy for the reduction of summer cooling load. Through dynamic simulations, the energy behavior of different building envelope solutions was evaluated for the retrofitting of an existing building. The parameters evaluated are the internal air temperature, the inside and outside surface temperature of the wall shaded by the green façade and the summer thermal load. Among the proposed scenarios the green facades have higher energy performance. The results obtained with a dense leaf cover of the creeper showed a reduction in the air temperature by 1.7 °C as well as a reduction by 1.5 °C for the inside surface temperature with respect to the traditional envelope. Furthermore, the outside surface temperature was reduced by 2.9 °C. Finally, the thermal load is reduced by 32%. These results show that shading the wall has allowed high energy savings for summer air conditioning.

Future studies should aim at a cost-benefit analysis of green facades, taking into account both construction costs and energy savings. Finally, in addition to the shading effect, the evapotranspiration process needs to be evaluated as further benefit from the green envelope.

Reference

[1] Cascone S, Catania F, Gagliano A and Sciuto G 2018 Energy performance and environmental and economic assessment of the platform frame system with compressed straw *Energy Build.* **166** 83–92

[2] Wong N H, Kwang Tan A Y, Tan P Y, Chiang K and Wong N C 2010 Acoustics evaluation of vertical greenery systems for building walls *Build. Environ* **45** 411–20

[3] Pérez G, Coma J, Martorell I and Cabeza L F 2014 Vertical Greenery Systems VGS for energy saving in buildings: A review *Renew. Sustain. Energy Rev.* **39** 139–65

[4] Köhler M 2008 Green facades-a view back and some visions *Urban Ecosyst.* **11** 423–36

[5] Pérez G, Rincón L, Vila A, González J M and Cabeza L F 2011 Green vertical systems for buildings as passive systems for energy savings *Appl. Energy* **88** 4854–9

[6] Dahanayake K W D K C and Chow C L 2017 Studying the potential of energy saving through vertical greenery systems: Using EnergyPlus simulation program *Energy Build.* **138** 47–59

[7] Pérez G, Rincón L, Vila A, González J M and Cabeza L F 2011 Behaviour of green facades in Mediterranean Continental climate *Energy Convers. Manag.* **52** 1861–7

[8] Azkorra Z, Pérez G, Coma J, Cabeza L F, Bures S, Álvaro J E et al. 2015 Evaluation of green walls as a passive acoustic insulation system for buildings *Appl. Acoust.* **89** 46–56

[9] Tan C L, Wong N H, Jusuf S K and Chiam Z Q 2015 Impact of plant evapotranspiration rate and shrub albedo on temperature reduction in the tropical outdoor environment *Build. Environ* **94** 206–17

[10] Song J and Wang Z H 2016 Diurnal changes in urban boundary layer environment induced by urban greening *Environ. Res. Lett.* **11**

[11] des Rosiers F, Thériault M, Kestens Y and Villeneuve P 2007 Landscaping attributes and property buyers' profiles: Their joint effect on house prices *Hous. Stud.* **22** 945–64

[12] Gao X and Asami Y 2007 Effect of urban landscapes on land prices in two Japanese cities *Landsc. Urban Plan.* **81** 155–66.

IOP Conf. Series: Materials Science and Engineering **415** (2018) 012055 doi:10.1088/1757-899X/415/1/012055

Polycarbonate multi-wall panels integrated in multi-layer solar façade concepts

M Čekon[1] and K Struhala[1]

[1] Centre AdMaS, Faculty of Civil Engineering, Brno University of Technology, Czechia

cekon.m@fce.vutbr.cz

Abstract. Recent technological advances in transparent insulation materials´ (TIMs) production may have opened ways for new integration of these materials in buildings. This paper presents a study analysing several variants of façades incorporating TIMs on solar façade principles. The analysis is based on ongoing long-term full-scale experiments in Brno, Czech Republic. The paper introduces the test platform which is used for evaluation of specific aspects of integrated polycarbonate TIMs. The paper describes the experiment results, especially long-term thermal response and passive solar gain data. These data are applied for analysis of: influence of various polycarbonate-based TIMs on the real performance of the façades; influence of implementation of different solar absorbers on the performance of the proposed facades; effectivity of application of latent thermal energy storage (based on PCMs) as a part of heat accumulation layer and coupling of TIM with prismatic glass to enhance optical selectivity aspect. Presented results demonstrate for example significant influence of the type of solar absorber on the thermal performance of tested solar facades: the difference is up to 35% to 54%. Also the integration of prismatic glass coupled with simple two-wall polycarbonate panel can reduce solar penetration through components at the level comparable with the most complex six-wall panel.

1. Introduction

Commonly, the efficient way to improve energy consumption of buildings is based on reduction of the heating and ventilation energy losses. This can be achieved through addition of thermal insulation on the building envelope and installation of efficient HVAC system with heat recovery [1, 2]. However the potential for further savings in future building designs is limited by increasing investment costs, embodied energy and other environmental impact [3, 4, 5]. Therefore the building sector is looking for solutions that would not only minimize the energy losses [6], but also utilize the energy gains available on-site. These solutions integrate renewable energy sources (RES) such as photovoltaics (PV) or solar thermal collectors for on-site energy generation [7, 8, 9]. Although the field of RES is already well covered in literature, there are still some directions that need to be carefully investigated. Therefore this paper focuses on one of more straightforward ways for exploitation of available renewable energy gains: solar (heat) energy gains through TIMs based on Polycarbonate multi-wall panels integrated in building envelopes, particularly in multi-layer solar façade.

Integration of TIMs is already proven in case of solar thermal collectors [10, 11]. Existing TIMs [12] could be also integrated into solar thermal facades providing both transparency and thermal insulation. Several studies in the last two decades for example applied TIMs as replacement of the original glazing elements in solar wall concepts [13]. However, their involvement in building envelope is so specific that application in contemporary building practice is rather rare, even though commercial products already exist [14]. One of the main drawbacks is that specialized TIMs are rather expensive. Therefore application of cheaper materials with similar properties (such as polycarbonate sheets and panels) is studied. Various types of multiwall panels with improved thermal properties (with optical efficiency remaining the same) are being developed. Their application in building envelopes (facades or roofs) can provide cost-effective way for achieving passive energy gains in building design [15]. This paper introduces eight experimental polycarbonate solar façade (PSF) concepts utilizing different types of TIMs (based on multiwall co-extruded polycarbonate). Components representing these PSFs are investigated as a part of long-term experiment in Brno, Czechia. Especially the influence of varying structure of TIMs (and related thermal and optical parameters) on the performance of the components is studied in detail. However, other issues like selection of proper solar absorber or incorporation of phase-change materials (PCMs) for better heat accumulation are also investigated. The paper presents comparative analysis of thermal, optical and solar properties of individual materials and illustrates performance of the whole PSF concepts on selected summer and winter days (with clear sky) measurements.

2. Experimental polycarbonate solar façade concepts

At first the properties of materials applied in the proposed concepts were tested. This was followed by full-scale testing of eight experimental PSF components in real climate conditions (see Figure 1). The full-scale tests are taking place at AdMaS research centre belonging to Brno University of Technology, Czechia [16]. The PSF components are based on four polycarbonate TIMs. The TIMs are installed in pairs of components opposite or next to each other. This limits the number of tested materials. On the other hand, it allows comparative measurements and increased accuracy of the experimental results. The test cell's longitudinal axis is oriented towards South-East / North-West. One of the walls with the test components is oriented to the South-East to receive as much solar radiation as possible. The other (North-Western) wall is adjacent to one of the AdMaS center's buildings. Therefore, the test components there are not exposed to the solar radiation directly. The space inside the test cell is divided into eight compensating boxes (TCBs) made of particleboard. They are constructed around individual PSF components of tested materials – four components on each side. The TCBs separate individual testing components from each other and the internal climate of the test cell to dampen the temperature amplitudes of the internal air. Internal climate in the test cell (not in the individual test boxes) is controlled by an air-conditioning unit.

Figure 1. Outside view of the test cell (left), inside view of the test cell with individual TCBs (middle), test cell cross section (right).

Eight full-scale PSF components (A to H) are long-term tested in real climate conditions. Components A to D are oriented to South-East, while components E to H are oriented to North-West. Thermal and optical performance of all components are described in sections 3 and 4, while real outdoor performance in section 5 is described only for components A to D oriented to the South-East (due to direct solar radiation). Schemes of the components are in Figures 2 and 3. Each tested component fulfils several different functions thanks to four functional layers. First layer of tested components is made of clear polycarbonate multi-wall panel in the role of TIM. All the TIMs are installed with air cavities parallel to the longer side of the components. Components A and H integrate simple 10mm thick two-wall panel (PC10). In case of component A the PC10 panel is covered by prismatic glass pane rayWall90 to improve its overall optical performance.

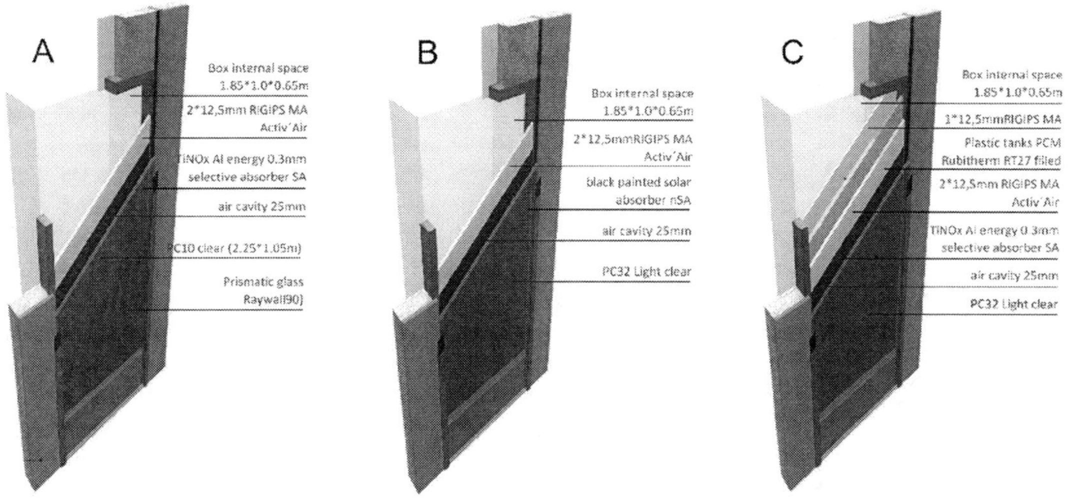

Figure 2. Test components A, B and C, key material layers, TIM and solar absorber types.

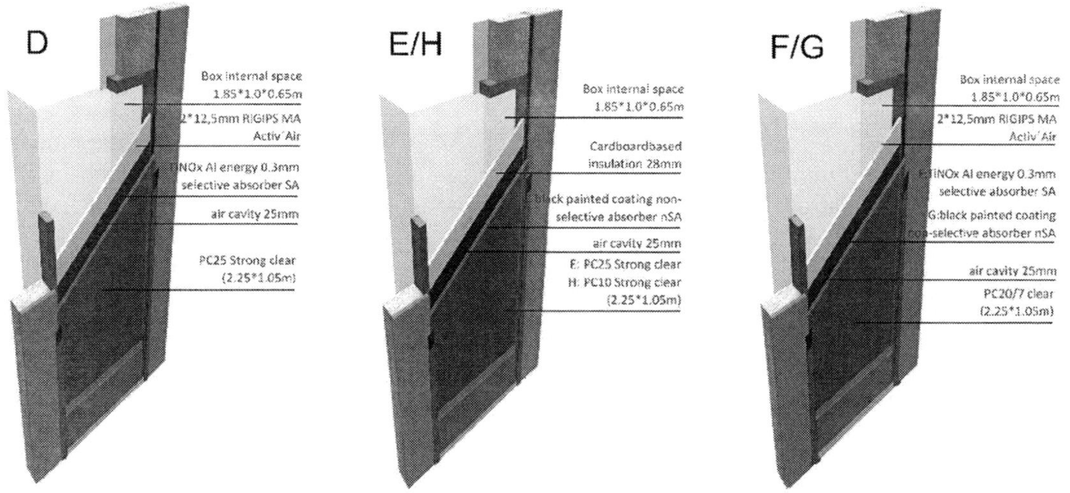

Figure 3. Test components D, E, F, G and H, key material layers, TIM and solar absorber types.

Components B and C integrate the most complex 32mm thick six-walls system with combined parallel and diagonal partitioning (PC32). Components D and E integrate 25mm thick three-wall panel with combined partitioning (PC25). Components F and G integrate 20 mm thick seven-wall panel (PC20). Basic parameters of the panels are described in Table 1. Second layer of the tested components is 25mm non-ventilated air gap. Third layer is made of a solar absorber. Two types of solar absorber are incorporated in the experimental components: selective absorber (SA) with low-e coating in components A, C, D and F and non-selective absorber (nSA) in components B, E, G and H. Reason for this is difference in the absorber's properties. The absorbers have similar solar absorbance, approx. 0.94 (SA) and 0.96 (nSA), however they significantly differ in thermal emissivity (0.06 and 0.94, respectively). Final layer of the components is the base layer for application of solar absorber. In real façade this layer would also fulfil the role of heat accumulator. For the purpose of the described experiment this layer is simplified to a 25mm plasterboard panel in components A, B, D, F and G. 28mm corrugated fiberboard panel is incorporated in the same role in components E and H for the purpose of performance comparison. Component C is used for evaluation of PCM application. Therefore the heat accumulating layer in this component is made of Rubitherm RT27 in plastic tanks enclosed between plasterboards.

3. Thermal performance of experimental PSFs

This section presents the results of standardized calculations employed to reproduce the theoretical thermal performance of presented PSF concepts. Calculated parameters consider the influence of different thermal and optical properties. In addition, the results of the PSF concepts with and without a selective absorber are also calculated. Firstly, measured thermal conductivities of Polycarbonate multi-wall systems are presented. Secondly, the thermal parameters of the whole concepts are calculated.
Thermal properties of the TIMs were evaluated with TLP 300-DTX-1P apparatus. It utilizes guarded hot plate (GHP) method to measure heat resistance and calculate equivalent thermal conductivity in compliance with ISO 8302 [17] and EN 12667 [18]. Results of the evaluation are summed in Table 1. Temperature dependency of the thermal conductivity is taken into account as well [19].

Table 1. Key material and measured thermal parameters of evaluated polycarbonate TIMs. [19]

	PC type	*Thickness* [mm]	*Weight* [kg.m^{-2}]	*Internal cells´ structure*	Equivalent thermal conductivity at mean temperature [W·m^{-1}·K^{-1}]		
					10°C	20°C	30°C
PC10	Clear 2walls	10.3	1.7	Parallel	0.0668	0.0712	0.0763
PC20	Clear 7walls	20.7	3.0	Parallel	0.0527	0.0551	0.0582
PC25	Clear 3walls	25.1	3.4	Parallel/Diagonal	0.0665	0.0702	0.0752
PC32	Clear 6walls	32.5	3.6	Parallel/Diagonal	0.0601	0.0634	0.0684

Thermal resistances and transmittances of individual parts as well as whole experimental components were calculated according to ISO 6946 [20] using thermal conductivity values described in Table 1. Results of these calculations are summed in Table 2. The table shows values representing individual "layers" (R_t is for TIM, R_g for air gap and solar absorber, R_i heat accumulation material) or parts (R_{te} for TIM and air gap coupled with solar absorber together and their calculated equivalent thermal conductivity λ_{te}) as well as whole components (R_c). For this study, dynamic effect of phase change integrated in test component C is not considered, thus only standard calculation is provided.

Table 2. Thermal resistance for individual layers and whole components.

[m^2·K·W^{-1}]	R_t	R_g	R_i	R_{te}	λ_{te}	R_c	U_c
A	0.145	0.57	0.125	0.795	0.044	1.050	0.95
B	0.513	0.18	0.125	0.733	0.078	0.988	1.01
C	0.513	0.57	0.275	1.123	0.051	1.528	0.65
D	0.357	0.57	0.125	0.968	0.052	1.223	0.82
E	0.357	0.18	0.444	0.578	0.086	1.152	0.87
F	0.376	0.57	0.125	0.986	0.046	1.241	0.81
G	0.376	0.18	0.125	0.596	0.075	0.851	1.17
H	0.145	0.18	0.444	0.365	0.096	0.939	1.06

The different performance of both solar absorber types is clearly visible in thermal resistance R_{te} and equivalent thermal conductivity values λ_{te}. The difference in those parameters of components A and H is 54% (with PC10), B and C is 35% (PC32), D and E represents 40% (PC25) and F with G corresponds to 40% (PC20). This difference means that for example the overall U-value of component A (with simplest PC10 TIM) bests the U-value of component B (with most complex PC32 TIM).

Component B with nSA has thermal resistance of 0.988 $m^2 \cdot K \cdot W^{-1}$. Compare this with the thermal resistance of 1.528 $m^2 \cdot K \cdot W^{-1}$ of component C with the same PC32 TIM combined with SA. The equivalent thermal conductivity of all components integrating SA from 0.044 (PC10) to 0.052 (PC25) approaches to recent conventional thermal insulations (e.g. expanded polystyrene reaches the value around 0.04).

4. Optical performance of polycarbonate TIM panels

Solar transmittance of polycarbonate TIMs presented in this paper is based on previous work [19]. Table 3 presents solar transmittance values based on laboratory spectrophotometer measurements using a Perkin Lambda 1050 UV/VIS/NIR spectrophotometer with a 150mm Spectralon integrating sphere. These spectral data are registered ranging from 200 nm to 3300 nm representing all spectral regions ($\tau_{\lambda\ SRS}UV$ for ultraviolet, $\tau_{\lambda\ SRS}\ VIS$ for visible and $\tau_{\lambda\ SRS}\ NIR$ for near infrared region respectively). Each sample was measured in the centre position of vertically oriented chambers. Spectral values and integrated Total Solar Transmittance (TST - τ_λ) parameters from 280 nm to 2 500 nm and declared values τ_{decl} are shown as well. The lowest value represents the system PC32 with solar transmittance of 43%, while the system PC10 has transmittance of 74%. It should be noted that solar transmittance obtained by laboratory spectral measurements does not correspond with producer-declared values. The difference is more than 10%.

Table 3. Solar transmittance of tested polycarbonate TIM panels [19].

	τ_λ ASTM G 173 [21]	$\tau_{\lambda\ SRS}UV$ 0.2 – 0.3 μm	$\tau_{\lambda\ SRS}\ VIS$ 0.3 – 0.8 μm	$\tau_{\lambda\ SRS}\ NIR$ 0.8 – 2.5 μm	τ_{decl}
PC10	0.74	0.08	0.79	0.77	0.84
PC20	0.50	0.04	0.54	0.52	0.62
PC25	0.48	0.07	0.50	0.50	0.63
PC32	0.43	0.03	0.46	0.45	0.53

5. Real outdoor performance and thermal response

As the proposed concept is based on various solar active materials, real outdoor testing may determine the maximum temperature that can be achieved in these test solar components and thus determine their potentials. There are many specific issues to take into account, such as inclined angular dependence, fluctuations of solar irradiation, overall solar distribution or orientation to the cardinal points. These issues could be solved all together with real-life monitoring of solar thermal performance. Results of full-size experiments in real climate condition demonstrate influence of different absorber efficiencies and effect of various geometrical and physical characteristics on the performance of individual PSF components. This part of the paper focuses on components A, B, C and D oriented to the South-East to analyze the performance of the components in direct solar exposure using contact digital thermometers Maxim DS18B20. Performance of the components is analyzed during representative clear sky day (24 hours) in winter and summer, with maximum sun height above the horizon of 31° and 59° respectively. The indoor climate was not controlled during the winter day, while during summer day the indoor air temperature in the test cell was maintained approx. at 21 °C. Average outdoor air temperature was T_e = -8.63 °C in winter (24.53 °C in summer), while interior temperature was T_{ai} = 5.16 °C in winter (21.11 °C in summer). Average indoor air temperature in individual TCBs in winter (summer) day reached: T_A = 6.99 °C (27.40 °C), T_B = 8.20 °C (26.77 °C), T_C = 6.98 °C (26.16 °C) and T_D = 10.20 °C (29.36 °C). Detailed temperature measurement results are shown in Figure 4. Overall, the best efficiency (represented by the highest indoor TCB temperature) was achieved with component D. It should be highlighted that test components A and B show similar summer performance (similarly to section 3) even though their composition is significantly different. This time, the reason is likely the application of rayWall90 prismatic glass in component A, which improves summer performance of this component. It should be also noted, that winter results were influenced by the fact that indoor conditions were not controlled. This means that PCM in component

C was not fully activated and the component could not perform at peak efficiency. Therefore, the effect of C in winter is represented by more stable temperature progression comparing to those with no PCM. When looking at minimum temperatures for summer test period, test component C shows the best results due to PCM integration. Its peak indoor temperature is more than 10 K lower compared to other components.

Figure 5 presents maximum temperatures achieved on solar absorber surfaces (T_{abs}) of particular components. Interestingly, component A exhibits almost the same maximum temperatures for both winter and summer days. Component B has maximum temperatures approx. 65 °C in winter and 80 °C in summer. Here the specific optical raster applied on the model A influence the internal temperature at the level comparable with the component B. The test component C has the highest surface temperature in both winter and summer. The winter temperature is beneficial, however in summer its overall performance could cause unnecessary overheating.

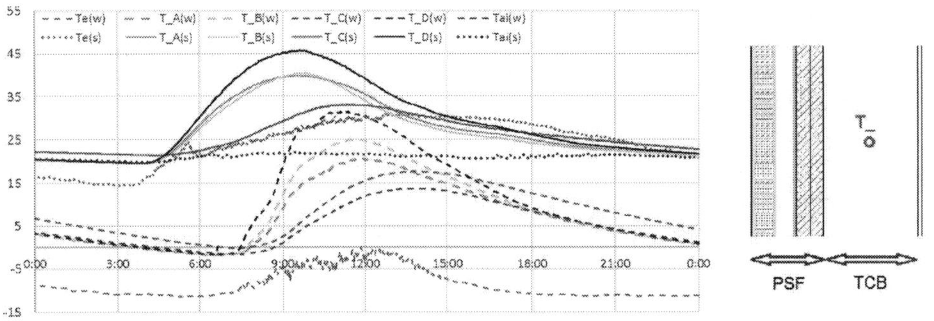

Figure 4. Internal test compensating box (TCB) temperatures; (w) winter day; (s) summer day.

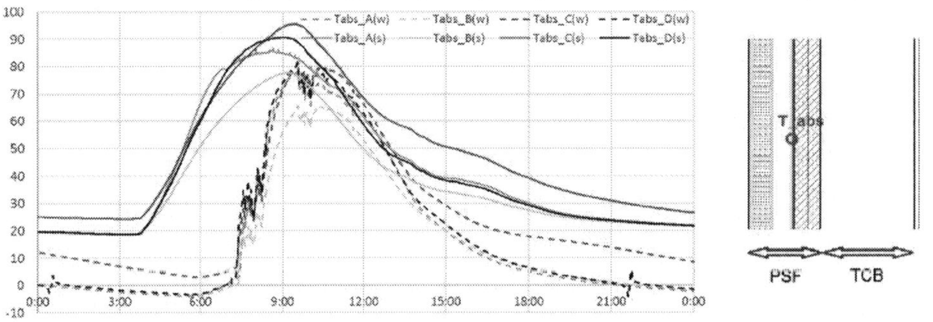

Figure 5. Absorber temperatures; (w) winter; (s) summer.

This is the reason for integration of PCM as a latent heat storage that would eliminate internal temperature fluctuations and limit the overheating. Last component D has the same absorber response in winter as C, however in summer it reaches the lower absorber temperature due to lower thermal capacity of plasterboard without PCM. Concerning only the polycarbonate TIMs, the temperature difference ΔTpc_ is a key element when temperature gradient needed to be considered for thermal calculations. Figure 6 presents results of maximum temperature difference obtained on polycarbonate systems in presented days. The maximum is achieved in winter case approaching 70K difference. When taking one particular type of PC32 into account, both test components B and C are in a strong contrast. The type of solar absorber significantly affects this aspect and the difference is almost 20K. In addition, measuring solar irradiance rates I_r and dynamic solar transmittance based on evaluating the amount of solar radiation penetrating through the multi-wall structure of the polycarbonate TIMs is provided as well (Figure 7). A specific methodical approach is applied in this relation to provide real data. During clear sky conditions, huge variation of penetrated part of solar irradiance can be observed. Depending on the particular TIM, the values significantly vary, which influences hourly solar transmittance values. Activation of the prismatic glass integrated in A is clearly demonstrated in Figure 7, however the internal structure of C indicates the same response.

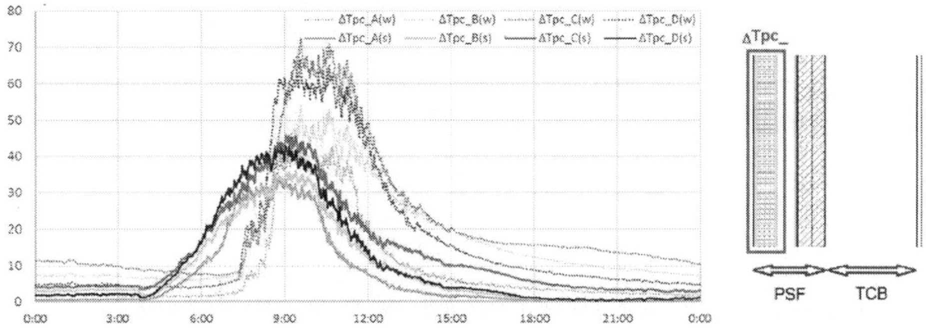

Figure 6. Temperature difference of polycarbonate systems; (w) winter; (s) summer

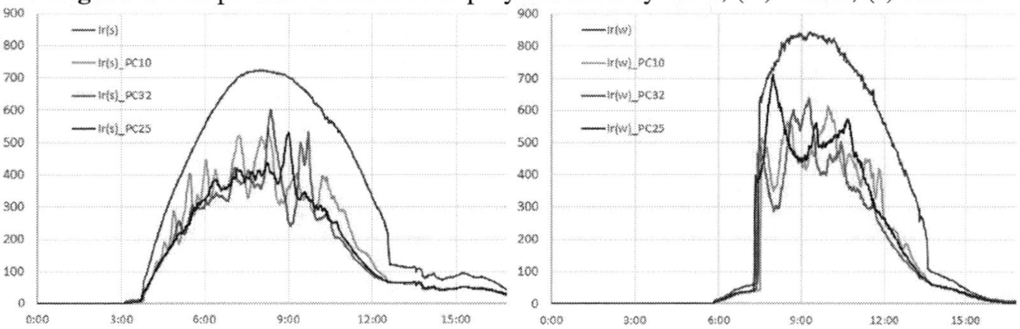

Figure 7. Solar irradiance rates penetrated through PC systems; (w) winter; (s) summer.

Conclusion

The paper presents the results of the solar and thermal characterization of proposed polycarbonate solar façade (PSF) concepts. The concepts evaluated impacts of different TIMs, selective and non-selective solar absorbers, as well as and latent thermal energy storage. Experimental monitoring of full-scale PSF components focused on the dynamic annual performance of TIMs coupled with solar absorbers as a fundamental step towards understanding overall interactions influencing the heat transfer of the presented concepts. Experimental results and calculations indicate that application of SA in proposed PSFs significantly improves their thermal parameters compared to application of nSA. Results show that when thermal resistance or equivalent thermal conductivity is considered, proposed concepts implementing the most complex PC32 TIM with SA (component C) provide comparable performance to conventional thermal insulation. When comparing the temperature response of the experimental PSF components invoked by solar radiation at its peak, a significant effect of particular solar absorber type and type of polycarbonate TIM is observed. Further research will be focused on detailed experimental analysis of long-term outdoor experiments and numerical modelling as well as the optimization of the described PSF concepts.

Acknowledgments

This research was supported by the project GJ 16-02430Y "Contemporary concepts of climatically active solar façades integrating advanced material solutions" supported by Czech Science Foundation.

References

[1] X. Liang, Z. Wang, M. Royapoor, Q. Wu, T. Roskilly (2017), Comparison of building performance between Conventional House and Passive House in the UK, *En. Proc.* **142** 1823
[2] S. Guillén-Lambea, B. Rodríguez-Soria, J. M. Marín (2016), Review of European ventilation

strategies to meet the cooling and heating demands of nearly zero energy buildings (nZEB)/Passivhaus. Comparison with the USA, *Renew Sust Energ Rev* **62** 561

[3] I. Zacà, D. D'Agostino, P. M. Congedo, C. Baglivo (2015), Assessment of cost-optimality and technical solutions in high performance multi-residential buildings in the Mediterranean area, *Energ Buildings* **102** 250

[4] T. Ekström, R. Bernardo, Å. Blomsterberg (2018), Cost-effective passive house renovation package for Swedish single-family houses from the 1960s and 1970s, *En. Buildings* **161** 89

[5] W. Ott, R. Bolliger (2015), Pitfalls in the Economic and Ecological Evaluation of Energy Related Building Renovation Strategies and Measures, *Energy Procedia* **78** 2340

[6] P. Ďurica, P. Juráš, R. Ponechal, D. Štaffenová (2016), Long-Time evaluation of thermal-Technical parameters of experimental wooden walls. In *CESB 2016 - Central Europe Towards Sustainable Building 2016*: *Innovations for Sustainable Future*, pp. 1063-1070.

[7] I. Visa, M. Moldovan, M. Comsit, M. Neagoe, A. Duta (2017), Facades Integrated Solar-thermal Collectors – Challenges and Solutions, *Energy Procedia* **112** 176

[8] R. Bolliger, W. Ott, S. von Grünigen (2015), Finding the Balance between Energy Efficiency Measures and Renewable Energy Measures in Building Renovation: An Assessment Based on Generic Calculations in 8 European Countries, *Energy Procedia* **78** 2372

[9] F. Favoino, F. Goia, M. Perino, V. Serra (2014), Experimental assessment of the energy performance of an advanced responsive multifunct. façade module, *Energ Buildings* **68** 647

[10] J. Cadafalch, R. Cònsul (2014), Detailed modelling of flat plate solar thermal collectors with honeycomb-like transparent insulation, *Solar Energy* **107** 202

[11] J. D. Osorio, A. Rivera-Alvarez, P. Girurugwiro, S. Yang, R. Hovsapian, J. C. Ordonez (2017), Integration of transparent insulation materials into solar collector devices, *Sol Energy* **147** 8

[12] N. D. Kaushika, K. Sumathy (2003) Solar transparent insulation materials: a review. *Renew Sust Energ Rev*. **7**(4) 317

[13] I. L. Wong, P. C. Eames, R. S. Perrera (2007), A review of transparent insulation systems and the evaluation of payback period for building applications, *Solar Energy* **81** 1058

[14] M. Casini (2016), Smart Buildings: Advanced Materials and Nanotechnology to Improve Energy-Efficiency and Environmental Performance, Elsevier: Cambridge, 2016, 363 p.

[15] E. Moretti, M. Zinzi, E. Belloni (2014) Polycarbonate panels for buildings: Experimental investigation of thermal and optical performance, *En Buildings* **70** 23

[16] M. Čekon, R. Slávik, K. Struhala, M. Formánek (2017) Experimental Full-Scale Test Cell Optimizing for Research of Novel Concepts towards Climatically Active Solar Façade Design, *Applied Mechanics and Materials* **861** 213

[17] ISO (1991), *Thermal insulation - Determination of steady-state thermal resistance and related properties - Guarded hot plate apparatus*, ISO 8302, International Organization for Standardization (ISO): Geneva, 1991, 47 p.

[18] ČSN EN (2001) *Thermal perforance of building materials and products – Deterination of thermal resistance by means of guarded hot plate and heat flow method*, ČSN EN 12667, Czech normalization institute (CNI): Prague, 2001, 60 p.

[19] M. Čekon, R. Slávik, J.Zach (2017) Experimental Analysis of Transparent Insulation Based on Polycarbonate Multi-Wall Systems: Thermal and Optical Performance, *Energy Procedia* **132** 502

[20] ISO (2017), *Building components and building elements — Thermal resistance and thermal transmittance — Calculation method*, ISO 6946, International Organization for Standardization (ISO): Geneva, 2017, 40 p.

[21] ASTM (2012), *Standard Tables for Reference Solar Spectral Irradiances: Direct Normal and Hemispherical on 37_ Tilted Surface;* ASTM G173-03*;* American Society for Testing and Materials (ASTM): West Conshohocken, PA, USA, 2012.

Institute of Physics
Dirac House, Temple Back
Bristol BS1 6BE UK

ISSN: 1757-8981
ISBN 978-1-5108-7379-7

CURRAN ASSOCIATES INC.
proceedings
.com